FUNDAMENTALS OF ACOUSTICS

THIRD EDITION

LAWRENCE E. KINSLER
Late Professor Emeritus
Naval Postgraduate School

AUSTIN R. FREY
Professor Emeritus
Naval Postgraduate School

ALAN B. COPPENS
Associate Professor of Physics
Naval Postgraduate School

JAMES V. SANDERS
Associate Professor of Physics
Naval Postgraduate School

John Wiley & Sons
New York   Chichester   Brisbane   Toronto   Singapore

Library of Congress Cataloging in Publication Data:
Fundamentals of acoustics.

   Includes index.
   1. Sound-waves. 2. Sound--Apparatus. 3. Archi-
tectural acoustics. 4. Underwater acoustics.
I. Kinsler, Lawrence E.
QC243.F86   1982   534   81-7463
ISBN 0-471-02933-5   AACR2

Printed in the United States of America

10 9 8 7 6 5 4 3

FOREWORD

It is with great regret that I report the untimely death of my good friend and co-author, Larry Kinsler. It is Dr. Kinsler who was primarily responsible for the revisions and alterations of the original text that appeared in the second edition of *Fundamentals of Acoustics*.

For the past thirty years my own interests have been in fields other than acoustics, and as a consequence I have not felt competent to undertake the task of writing a third edition. I have therefore asked two of my younger associates at the Naval Postgraduate School, Dr. A. B. Coppens and Dr. J. V. Sanders to do this for me. Both are highly competent acousticians, who have done important research in the field and have used both the original and the second editions of "Kinsler and Frey" in their teaching at the Postgraduate School.

I hope that the reader will agree with me that the changes and additions made by Drs. Coppens and Sanders have improved this text and have brought it up to date in this rapidly advancing field of physics and engineering.

Austin R. Frey

Monterey, California
January 1980

PREFACE

One purpose of this book is to present, in simple and concise form, the fundamental principles underlying the generation, transmission, and reception of acoustic waves. A second purpose is to apply these principles to a number of important fields of applied acoustics. The extensive developments of the past few decades have so broadened these fields that an exhaustive treatment of all their aspects could not be contained in any single volume. It has therefore been necessary to restrict the number of topics considered and to limit the extent to which each is developed.

In making this selection our primary aim has been to familiarize the student with the fundamental concepts and terminology of the subject and with the simpler analytical methods that are available for attacking acoustic problems. The first eight chapters of the book provide an analysis of the various types of vibration of solid bodies, and of the propagation of sound waves through fluid media. These chapters will suffice for a one-quarter course in the fundamentals of theoretical acoustics. The remaining chapters are concerned with a limited number of applications of acoustics. Those discussed have been selected either because of their outstanding importance, as concrete illustrations of the practical application of mathematical techniques developed in the earlier chapters, or because fundamental treatments are not readily available in other books. These chapters contain sufficient material to complete a year's course in theoretical and applied acoustics or, with proper selection of topics, a second quarter course. The topic selection is up to the instructor, but it may be necessary to include a few sections from omitted chapters.

The book may be studied with equal facility by advanced undergraduate or graduate students in science or engineering. The essential requirements are a knowledge of the fundamental principles of mechanics and electricity, and an understanding of the methods of calculus, including partial derivatives. Since the book is primarily intended as a textbook for classroom use, rather than as an encyclopedic reference work, no attempt has been made to include a complete bibliography, although numerous references are given, either where the treatment is necessarily incomplete or to provide an interested reader with a source containing more detailed information. We have attempted to derive most of the important equations from the fundamental laws of physics and to show in some detail not only the mathematical steps but also the logical processes involved in these derivations. The derivations of a few of the less important equations have been intentionally omitted, and are instead included as exercises for the students either implicitly or among the problems given at the end of each chapter.

Considerable attention has been paid to the selection of a comprehensive set of problems, since the ultimate check on the students' understanding of the subject is their ability to apply their knowledge to new situations. To assist those engaged in the self-study of this book, answers are provided for the odd-numbered problems. Tables of physical constants and functions and a glossary are given in the Appendix.

As far as has been convenient, the recommendations of the International Standardization Organization (ISO) have been used throughout this book. We have, however, deliberately retained a small number of superseded conventions in both the text and the problems. We feel that the student must be prepared over at least the next decade to extract information from the literature using these superseded conventions and be able to translate them into the ISO recommendations.

The writing of a third edition of this book has been largely the responsibility of the two new authors (A. B. Coppens and J. V. Sanders) and it has been carried out under the watchful eye of one of the original authors (A. R. Frey). It differs from the second edition in the following respects. Many chapters have been supplemented by the addition of somewhat more sophisticated material that reflects some of the modern concerns of acoustics. (These sections have asterisks next to them in the contents.) This material is almost always confined to the terminal sections of a chapter and, depending on the student's academic level, may be either omitted without interrupting the continuity of the presentation, or discussed later when the student's comprehension of acoustics has had time to mature.

The chapters on loudspeakers and microphones in the second edition have been combined into a single chapter on transduction, with emphasis placed on integrating the general theory with specific examples. The chapter on ultrasonic and sonar transducers has been regretfully omitted because we felt that it was not possible to do this subject justice in the space available. The material on absorption, hearing, architectural acoustics, and underwater sound has been considerably rewritten to bring it more up to date. Examples of the new material added include discussions of antiresonance, concert hall acoustics, detection theory, canonical equations, and normal mode propagation in the ocean. In response to the growing awareness of the important role that science and engineering should play in improving the quality of life, a new chapter has been added on environmental acoustics.

As an aid to those designing a new course or adapting an existing course to this third edition, sections within each chapter that we consider beyond the needs of many survey courses, have been set in smaller type. (We have been liberal in making such assignments, so it is suggested that such sections be reviewed before being omitted, since some are convenient introductions to material in later retained sections.)

We thank all those of our colleagues who have given criticisms and recommendations for this third edition. We would like to single out for special thanks Dr. Wayne Wright, whose thorough and comprehensive critique of the manuscript resulted in many beneficial improvements.

Austin R. Frey
Alan B. Coppens
James V. Sanders

Monterey, California
August 1980

CONTENTS

CHAPTER 3
VIBRATIONS OF BARS

CHAPTER 4
THE TWO-DIMENSIONAL WAVE EQUATION:
VIBRATIONS OF MEMBRANES AND PLATES

CHAPTER 5
THE ACOUSTIC WAVE EQUATION AND
SIMPLE SOLUTIONS

CHAPTER 6
TRANSMISSION PHENOMENA

CHAPTER 7
ABSORPTION AND ATTENUATION OF SOUND
WAVES IN FLUIDS

CHAPTER 8
RADIATION AND RECEPTION
OF ACOUSTIC WAVES

CHAPTER 9
PIPES, CAVITIES, AND WAVEGUIDES

CHAPTER 10
RESONATORS, DUCTS, AND FILTERS

CHAPTER 11
NOISE, SIGNAL DETECTION, HEARING, AND SPEECH

CHAPTER 12
ENVIRONMENTAL ACOUSTICS

CHAPTER 13
ARCHITECTURAL ACOUSTICS

CHAPTER 14
TRANSDUCTION

CHAPTER 15
UNDERWATER ACOUSTICS

APPENDIX

CHAPTER 1

FUNDAMENTALS
OF VIBRATION

1.1 INTRODUCTION. Acoustics may be defined as the generation, transmission, and reception of energy in the form of vibrational waves in matter. As the atoms or molecules of a fluid or solid are displaced from their normal configurations, an internal elastic restoring force arises. Examples include the tensile force produced when a spring is stretched, the increase in pressure produced when a fluid is compressed, and the transverse restoring force produced when a point on a stretched wire is displaced in a direction normal to its length. It is this elastic restoring force, coupled with the inertia of the system, that enables matter to participate in oscillatory vibrations and thereby generate and transmit acoustic waves.

Before beginning the discussion of acoustics, it is useful to choose a system of units. Acoustics encompasses such a wide range of scientific and engineering disciplines that this choice is not easy. A survey of the literature reveals a great lack of uniformity: writers use units common to their particular field of interest. Most of the early work has been reported in the CGS (centimeter-gram-second) system. Considerable engineering work has been reported in a mixture of metric and English units, and work in electro-acoustics and underwater acoustics is commonly reported in the MKS (meter-kilogram-second) system of units. Recently, a codification of the MKS system, the SI (Le Système International d'Unités), has been established as the accepted system. This is the system that we will favor in this book. A conversion table relating CGS and SI units is incorporated in Appendix A1. Throughout this text, "log" will represent logarithm to the base 10 and "ln" will represent logarithm to the base e ("ln" is frequently called the "natural logarithm").

The most familiar acoustic phenomenon is that associated with the sensation of sound. For the average young person, a vibrational disturbance is interpreted as sound if its frequency lies in the range of about 20 to 20,000 Hz (1 Hz = 1 hertz = 1 cycle/second). However, in a broader sense acoustics also includes the *ultrasonic* frequencies above 20,000 Hz and the *infrasonic* frequencies below 20 Hz. The natures of the vibrations associated with acoustics are many, including the simple sinusoidal vibrations produced by a tuning fork, the complex vibrations generated by a bowed violin string, and the nonperiodic motions associated with an explosion, to mention but a few. In studying vibrations it is advisable to begin with the simplest type: a one-dimensional sinusoidal vibration that has only a single frequency component (a pure tone).

1

1.2   THE SIMPLE OSCILLATOR.   If a mass m, fastened to a spring and constrained to move parallel to the spring, is displaced slightly from its rest position and released, the mass will vibrate. Measurement shows that the displacement of the mass from its rest position is a sinusoidal function of time. Sinusoidal vibrations of this type are called *simple harmonic vibrations*. A large number of vibrators used in acoustics can be modeled as simple oscillators. Loaded tuning forks and loudspeaker diaphragms, which are so constructed that at low frequencies their masses move as units, are but two examples. Even more complex vibrating systems have many of the characteristics of the simple systems and may often be modeled, to a first approximation, by simple harmonic oscillators.

The only physical restrictions placed on the equations for the motion of a simple harmonic oscillator are that the restoring force be directly proportional to the displacement (Hooke's law) and that there be no losses to attenuate the motion. When these restrictions apply the frequency of vibration is independent of amplitude and the motion is simple harmonic.

A similar restriction applies to more complex types of vibration, such as the transmission of an acoustic wave through a fluid. If the acoustic pressures are so large that they no longer are proportional to the displacements of the particles of fluid, it becomes necessary to replace the normal acoustic equations with more general equations which are much more complicated. With sounds of ordinary intensity this is not necessary, for even the noise generated by a large crowd at a football game rarely causes the amplitude of motion of the air molecules to exceed one-tenth of a millimeter, which is within the limit given above. The amplitude of the shock wave generated by a large explosion is, however, well above this limit, and hence the normal acoustic equations are not applicable.

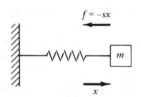

Fig. 1.1. Simple oscillator.

Returning to the simple oscillator shown in Fig. 1.1, let us assume that the restoring force f in newtons (N) can be expressed by the equation

$$f = -sx \qquad (1.1)$$

where x is the displacement in meters (m) of the mass m in kilograms (kg) from its rest position, s is the *stiffness* or *spring constant* in N/m, and the minus sign indicates that the force is opposed to the displacement. Substituting this expression for force into the general equation of linear motion

$$f = m \frac{d^2 x}{dt^2}$$

where d^2x/dt^2 is the acceleration of the mass, we obtain

$$\frac{d^2 x}{dt^2} + \frac{s}{m} x = 0$$

Both s and m are positive, so that we can define a constant

$$\omega_0 = \sqrt{s/m} \tag{1.2}$$

which casts our equation into the form

$$\boxed{\frac{d^2x}{dt^2} + \omega_0^2 x = 0} \tag{1.3}$$

This is an important linear differential equation whose solution is well-known and may be obtained by several methods.

One method is to assume a solution of the form

$$x = A_1 \cos \gamma t$$

Differentiation and substitution into (1.3) shows that this is a solution if $\gamma = \omega_0$. It may similarly be shown that

$$x = A_2 \sin \omega_0 t$$

is also a solution. The complete general solution is the sum of these two solutions,

$$x = A_1 \cos \omega_0 t + A_2 \sin \omega_0 t \tag{1.4}$$

where A_1 and A_2 are arbitrary constants and the parameter ω_0 is the *angular frequency* in radians per second (rad/s). Since there are 2π radians in one cycle, the frequency f_0 in hertz (Hz) is related to the angular frequency by

$$f_0 = \frac{\omega_0}{2\pi}$$

Thus, the frequency of the free vibration of the simple oscillator is $f_0 = \sqrt{s/m}/(2\pi)$. Note that either decreasing the stiffness or increasing the mass lowers the frequency. The *period T* of one complete vibration is given by

$$T = 1/f_0$$

1.3 INITIAL CONDITIONS. If at time $t = 0$ the mass has an initial displacement x_0 and an initial speed u_0, then the arbitrary constants A_1 and A_2 are fixed by these initial conditions, and the subsequent motion of the mass is completely determined. Direct substitution into (1.4) of $x = x_0$ at $t = 0$ will show that A_1 equals the initial displacement x_0. Differentiation of (1.4) and substitution of the initial speed at $t = 0$ gives $u_0 = \omega_0 A_2$, and (1.4) becomes

$$x = x_0 \cos \omega_0 t + \frac{u_0}{\omega_0} \sin \omega_0 t \tag{1.5}$$

Another form of (1.4) may be obtained by letting $A_1 = A \cos \phi$ and $A_2 = -A \sin \phi$, where A and ϕ are two new arbitrary constants. Substitution and simplification then gives

$$x = A \cos(\omega_0 t + \phi) \tag{1.6}$$

where A is the *amplitude* of the motion and ϕ is the *initial phase angle* of the motion. The values of A and ϕ are determined by the initial conditions and are

$$A = \sqrt{x_0^2 + \left(\frac{u_0}{\omega_0}\right)^2} \quad \text{and} \quad \phi = \tan^{-1}\left(-\frac{u_0}{\omega_0 x_0}\right)$$

Successive differentiation of (1.6) shows that the speed of the mass is

$$u = -U \sin(\omega_0 t + \phi) \tag{1.7}$$

where $U = \omega_0 A$ is the *speed amplitude*, and the acceleration of the mass is

$$a = -\omega_0 U \cos(\omega_0 t + \phi) \tag{1.8}$$

In these forms it is seen that the displacement lags $90°$ ($\pi/2$ rad) behind the speed and that the acceleration is $180°$ out of phase with the displacement (π rad), as shown in Fig. 1.2.

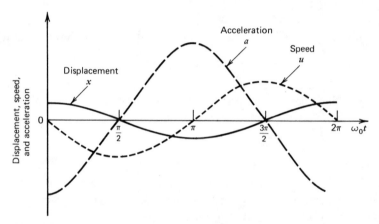

Fig. 1.2. *The speed u always leads the displacement x by $90°$. Acceleration a and displacement x are always $180°$ out of phase with each other. Plotted curves correspond to $\phi = 0$.*

1.4   ENERGY OF VIBRATION. The energy E of a system oscillating with simple harmonic motion of amplitude A and angular frequency ω_0 is the sum of the system's potential energy E_p and kinetic energy E_k. The potential energy is the work done in distorting the spring as the mass moves from its position of static equilibrium. Since the force exerted by the mass on the spring is in the direction of the displacement and equals $+sx$, the potential energy E_p stored in the spring is

$$E_p = \int_0^x sx \, dx = \tfrac{1}{2}sx^2$$

Expression of x by (1.6) gives

$$E_p = \tfrac{1}{2}sA^2 \cos^2(\omega_0 t + \phi) \tag{1.9}$$

The kinetic energy possessed by the mass is

$$E_k = \tfrac{1}{2}mu^2$$

Expression of u by (1.7) gives

$$E_k = \tfrac{1}{2}mU^2 \sin^2(\omega_0 t + \phi) \tag{1.10}$$

The total energy of the system is

$$E = E_p + E_k = \tfrac{1}{2}m\omega_0^2 A^2 \tag{1.11}$$

where use has been made of $s = m\omega_0^2$, $U = \omega_0 A$, and the trigonometric identity $\sin^2\sigma + \cos^2\sigma = 1$. The total energy can be rewritten in alternate forms,

$$E = \tfrac{1}{2}sA^2 = \tfrac{1}{2}mU^2$$

The total energy is a constant (independent of time) and is equal either to the maximum potential energy (when the mass is at its greatest displacement and is instantaneously at rest) or to the maximum kinetic energy (when the mass passes through its equilibrium position with maximum speed). Since the system was assumed to be free of external forces and not subject to any frictional forces, it is not surprising that the total energy does not change with time.

If all other quantities in the above equations are expressed in their MKS units, then E_p, E_k, and E will be in joules (J).

1.5   COMPLEX EXPONENTIAL METHOD OF SOLUTION.

Another approach to solving linear differential equations of the form (1.3) is to postulate

$$\mathbf{x} = \mathbf{A}e^{\gamma t}$$

Substitution gives $\gamma^2 = -\omega_0^2$, or $\gamma = \pm j\omega_0$ where $j = \sqrt{-1}$*. Thus, the general solution is

$$\mathbf{x} = \mathbf{A}_1 e^{j\omega_0 t} + \mathbf{A}_2 e^{-j\omega_0 t} \tag{1.12}$$

where $\mathbf{A}_1$ and $\mathbf{A}_2$ are to be determined by initial conditions, $\mathbf{x}(0) = x_0$ and $d\mathbf{x}(0)/dt = u_0$. This results in two equations

$$\mathbf{A}_1 + \mathbf{A}_2 = x_0 \qquad \text{and} \qquad \mathbf{A}_1 - \mathbf{A}_2 = \frac{u_0}{j\omega_0} = -j\frac{u_0}{\omega_0}$$

from which

$$\mathbf{A}_1 = \tfrac{1}{2}\left(x_0 - j\frac{u_0}{\omega_0}\right) \qquad \text{and} \qquad \mathbf{A}_2 = \tfrac{1}{2}\left(x_0 + j\frac{u_0}{\omega_0}\right)$$

Notice that $\mathbf{A}_1$ and $\mathbf{A}_2$ are complex conjugates, so there are really only two constants a and b where $\mathbf{A}_1 = a - jb$ and $\mathbf{A}_2 = a + jb$. This must be the case since the

---

* In this book **boldface type** indicates **complex** quantities and *italic type* represents *real* quantities, except for j. If readers are unacquainted with complex numbers, they should refer to Appendixes A2 and A3.

differential equation is of second order with two independent solutions and, therefore, with two arbitrary constants to be determined by two initial conditions. Substitution of A_1 and A_2 into (1.12) yields

$$x = x_0 \cos \omega_0 t + (u_0/\omega_0) \sin \omega_0 t$$

which is identical with (1.5). Satisfying the initial conditions which are both real caused the imaginary part of x to vanish as an automatic consequence.

In actual practice it is unnecessary to go through the mathematical steps required to make the imaginary part of the general solution vanish, for it is sufficient to note that the *real part of the complex solution is by itself a complete general solution* of the physical problem indicated by the original differential equation. Thus, for example, if we express $A_1 = a_1 + jb_1$ and $A_2 = a_2 + jb_2$ in (1.12) and, *before* applying initial conditions, take the real part, we have

$$\text{Re}\{x\} = (a_1 + a_2) \cos \omega_0 t - (b_1 - b_2) \sin \omega_0 t$$

Now, application of the initial conditions yields $a_1 + a_2 = x_0$ and $b_1 - b_2 = -u_0/\omega_0$ so that $\text{Re}\{x\}$ is identical with (1.5). Similarly, a complete solution is obtained if the displacement is written in the complex form

$$x = Ae^{j\omega_0 t} \tag{1.13}$$

where $A = a + jb$, and the real part only is considered

$$\text{Re}\{x\} = a \cos \omega_0 t - b \sin \omega_0 t$$

From the form (1.13), which will be used frequently throughout this book, it is particularly easy to obtain the complex speed $u = dx/dt$ and the complex acceleration $a = du/dt$ of the mass. The complex speed is

$$u = j\omega_0 Ae^{j\omega_0 t} = j\omega_0 x \tag{1.14}$$

and the complex acceleration is

$$a = -\omega_0^2 Ae^{j\omega_0 t} = -\omega_0^2 x \tag{1.15}$$

The expression $\exp(j\omega_0 t)$ may be thought of as a *phasor* of unit length rotating counterclockwise in the complex plane with an angular speed ω_0. Similarly, any complex quantity $A = a + jb$ may be represented by a phasor of length $A = (a^2 + b^2)^{1/2}$, making an angle $\phi = \tan^{-1}(b/a)$ with the positive real axis. Consequently, the expression $A \exp(j\omega_0 t)$ represents a phasor of length A and initial phase angle ϕ rotating in the complex plane with angular speed ω_0 (Fig. 1.3). The real part of this rotating phasor (its projection on the real axis) has magnitude

$$A \cos(\omega_0 t + \phi)$$

and varies harmonically with time.

From (1.14) we see that differentiation of x with respect to time gives $u = j\omega_0 x$, and hence the phasor representing speed leads that representing displacement by a phase angle of 90°. The projection of this phasor onto the real axis then represents the instantaneous speed, the speed amplitude being $\omega_0 A$. Equation (1.15) shows that

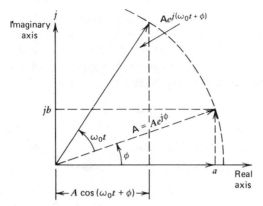

Fig. 1.3. *Physical representation of a phasor*
$\mathbf{A} \exp[j(\omega_0 t + \phi)]$.

the phasor **a** representing the acceleration is out of phase with the displacement phasor by π rad, or 180°. The projection of this phasor onto the real axis represents the instantaneous acceleration, the acceleration amplitude being $\omega_0^2 A$.

It will be the general practice in this textbook to analyze problems by the complex exponential method. The chief advantages of the procedure, as compared with the trigonometric method of solution, are its greater mathematical simplicity and the relative ease with which the phase relationships among the various mechanical and acoustic variables can be determined. However, care must be taken to obtain the *real* part of the complex solution in order to arrive at the correct physical equation.

1.6 DAMPED OSCILLATIONS. Whenever a real body is set into oscillation, dissipative (frictional) forces arise. These forces are of many types, depending on the particular oscillating system, but they will always result in a *damping* of the oscillations—a decrease in the amplitude of the free oscillations with time. Let us first consider the effect of a *viscous* frictional force f_r on a simple oscillator. Such a force is assumed to be proportional to the speed of the mass and to be directed so as to oppose the motion. It can be expressed as

$$f_r = -R_m \frac{dx}{dt} \qquad (1.16)$$

where R_m is a positive constant called the *mechanical resistance* of the system. It is evident that mechanical resistance has the units of newton-second per meter $(\mathrm{N \cdot s/m})$ or kilogram per second (kg/s).

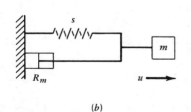

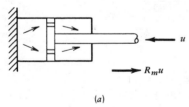

Fig. 1.4. (a) *Representative sketch of a dashpot.* (b) *A damped, free harmonic oscillator.*

A device that generates such a frictional force can be represented by a dashpot (shock absorber). This system is suggested in Fig. 1.4a. A simple harmonic oscillator subject to such a frictional force is usually diagrammed as in Fig. 1.4b.

If the effect of resistance is included, the equation of motion of an oscillator constrained by a stiffness force $-sx$ becomes

$$m\frac{d^2x}{dt^2} + R_m\frac{dx}{dt} + sx = 0$$

Dividing through by m and recalling that $\omega_0 = \sqrt{s/m}$, we have

$$\frac{d^2x}{dt^2} + \frac{R_m}{m}\frac{dx}{dt} + \omega_0^2 x = 0 \tag{1.17}$$

This equation may be solved by the complex exponential method.
Assume a solution of the form

$$x = Ae^{\gamma t}$$

and substitute into (1.17) to obtain

$$\left(\gamma^2 + \frac{R_m}{m}\gamma + \omega_0^2\right)Ae^{\gamma t} = 0$$

Since this must be true for all times,

$$\gamma^2 + \frac{R_m}{m}\gamma + \omega_0^2 = 0$$

or

$$\gamma = -\beta \pm \sqrt{\beta^2 - \omega_0^2} \tag{1.18}$$

where

$$\beta = \frac{R_m}{2m} \tag{1.19}$$

In most cases of importance in acoustics, the mechanical resistance R_m is small enough so that $\omega_0 > \beta$ and γ is complex. Also, notice that if $R_m = 0$ then

$$\gamma = \pm\sqrt{-\omega_0^2} = \pm j\omega_0$$

and the problem has been reduced to that of the undamped oscillator. This suggests defining a new constant ω_d by

$$\omega_d = \sqrt{\omega_0^2 - \beta^2} \tag{1.20}$$

Now, γ is given by

$$\gamma = -\beta \pm j\omega_d \tag{1.21}$$

and ω_d is seen to be the *natural angular frequency* of the damped oscillator. Notice that ω_d is always less than the natural angular frequency ω_0 of the same oscillator without damping.

The complete solution is the sum of the two solutions obtained above,

$$\mathbf{x} = e^{-\beta t}(\mathbf{A}_1 e^{j\omega_d t} + \mathbf{A}_2 e^{-j\omega_d t}) \tag{1.22}$$

As in the nondissipative case, the constants $\mathbf{A}_1$ and $\mathbf{A}_2$ are in general complex. As noted earlier, the real part of this complex solution is the complete general solution. One convenient form of this general solution is

$$x = Ae^{-\beta t}\cos(\omega_d t + \phi) \tag{1.23}$$

where A and ϕ are real constants determined by the initial conditions. Figure 1.5 displays the time history of the displacement of a damped harmonic oscillator for which $\phi = 0$.

The amplitude of the damped oscillator, now defined as $A\exp(-\beta t)$, is no longer constant, but decreases exponentially with time. As with the undamped oscillator, the frequency is independent of the amplitude of oscillation.

One measure of the rapidity with which the oscillations are damped by friction is the time required for the amplitude to decrease to $1/e$ of its initial value. This time τ is called the *relaxation time* (other names include *decay modulus, decay time, time constant,* and *characteristic time*) and is given by

$$\tau = \frac{1}{\beta} = \frac{2m}{R_m} \tag{1.24}$$

The quantity β is called the *temporal absorption coefficient*. (As with τ, there are a variety of names for β; we mention only one.) The smaller R_m, the larger τ is and the longer it takes for the oscillations to damp out.

If the mechanical resistance R_m is large enough, then $\omega_0 \leq \beta$ and the system is no longer oscillatory; when the mass is displaced it returns asymptotically to its rest position. If $\beta = \omega_0$, the system is known as *critically damped*.

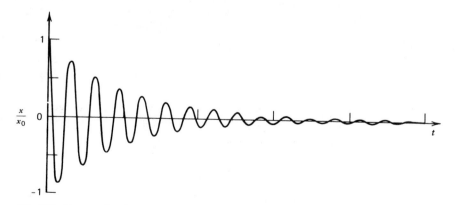

Fig. 1.5. *Decay of a damped, free harmonic oscillator :* $\beta/\omega_d = 0.05$, *initial conditions* $x_0 = 1$ *and* $u_0 = 0$.

The solution (1.23) is the real part of the complex solution

$$\mathbf{x} = \mathbf{A}e^{-\beta t}e^{j\omega_d t}$$

where $\mathbf{A} = A\exp(j\phi)$. If we rearrange the exponents,

$$\mathbf{x} = \mathbf{A}e^{j(\omega_d + j\beta)t}$$

we can define a *complex angular frequency*

$$\boldsymbol{\omega}_d = \omega_d + j\beta$$

whose real part is the angular frequency ω_d of the damped motion and whose imaginary part is the temporal absorption coefficient β. This convention of as-simulating the angular frequency and the absorption coefficient into a single complex quantity often proves useful in investigating damped vibrations, as we will see in subsequent chapters.

1.7 FORCED OSCILLATIONS.

A simple oscillator, or some equivalent system, is often driven by an *externally applied force* $f(t)$. The differential equation for the motion becomes

$$m\frac{d^2x}{dt^2} + R_m\frac{dx}{dt} + sx = f(t) \tag{1.25a}$$

Such a system is suggested in Fig. 1.6.

For the case of a sinusoidal driving force $f(t) = F\cos(\omega t)$ applied to the oscillator at some initial time, the solution of (1.25a) is the sum of two parts—a *transient* term containing two arbitrary constants and a *steady-state* term which depends on F and ω but does not contain any arbitrary constants. The transient term is obtained by setting F equal to zero. Since the resulting equation is identical with (1.17), the transient term is given by (1.23). Its angular frequency is ω_d. The arbitrary constants are determined by applying the initial conditions to the total solution. After a sufficient time interval, such that $\beta t \gg 1$, the damping term $\exp(-\beta t)$ makes this portion of the solution negligible, leaving only the steady-state term whose angular frequency ω is that of the driving force.

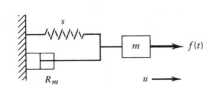

Fig. 1.6. *A damped, forced harmonic oscillator.*

To obtain the steady-state (particular) solution, it will be advantageous to replace the real driving force $F\cos(\omega t)$ by its equivalent complex driving force $\mathbf{f} = F\exp(j\omega t)$. The equation then becomes

$$m\frac{d^2\mathbf{x}}{dt^2} + R_m\frac{d\mathbf{x}}{dt} + s\mathbf{x} = Fe^{j\omega t} \tag{1.25b}$$

The solution of this equation gives the complex displacement $\mathbf{x}$. Since the real part of the complex driving force $\mathbf{f}$ represents the actual driving force $F\cos\omega t$, the real part of the complex displacement will represent the actual displacement.

Because $\mathbf{f} = F\exp(j\omega t)$ is periodic with angular frequency ω, it is plausible to assume that $\mathbf{x}$ must be also. Then, $\mathbf{x} = \mathbf{A}\exp(j\omega t)$ where $\mathbf{A}$ is in general complex. Equation (1.25b) becomes

$$(-\mathbf{A}\omega^2 m + j\mathbf{A}\omega R_m + \mathbf{A}s)e^{j\omega t} = Fe^{j\omega t}$$

Solving for $\mathbf{A}$ yields the complex displacement

$$\mathbf{x} = \frac{Fe^{j\omega t}}{j\omega[R_m + j(\omega m - s/\omega)]} \tag{1.26}$$

and differentiation gives the complex speed,

$$\mathbf{u} = \frac{Fe^{j\omega t}}{R_m + j(\omega m - s/\omega)} \tag{1.27}$$

These last two equations can be cast into somewhat simplier form if we define the *complex mechanical impedance* $\mathbf{Z}_m$ of the system

$$\mathbf{Z}_m = R_m + jX_m \tag{1.28}$$

where the *mechanical reactance* X_m is

$$X_m = \omega m - s/\omega \tag{1.29a}$$

The mechanical impedance $\mathbf{Z}_m = Z_m \exp(j\Theta)$ has magnitude

$$Z_m = \sqrt{R_m^2 + (\omega m - s/\omega)^2} \tag{1.29b}$$

and phase angle

$$\Theta = \tan^{-1} \frac{\omega m - s/\omega}{R_m} \tag{1.29c}$$

The dimensions of mechanical impedance are the same as those of mechanical resistance and are expressed in the same units, $N \cdot s/m$, often defined as *mechanical ohms*. It is to be emphasized that, although the mechanical ohm is analogous to the electrical ohm, these two quantities do not have the same units. The electrical ohm has the dimensions of voltage divided by current; the mechanical ohm has the dimensions of force divided by speed.

Using the definition of $\mathbf{Z}_m$ we may write (1.27) in the simplified form

$$\boxed{\mathbf{Z}_m = \frac{\mathbf{f}}{\mathbf{u}}} \tag{1.30}$$

which gives a most important physical meaning to the complex mechanical impedance: $\mathbf{Z}_m$ *is the ratio of the complex driving force* $\mathbf{f} = F\exp(j\omega t)$ *to the resultant complex speed* $\mathbf{u}$ *of the system at the point where the force is applied.*

If, for the driving frequency of interest, the complex impedance $\mathbf{Z}_m$ is known, then we can immediately obtain the complex speed

$$\mathbf{u} = \frac{\mathbf{f}}{\mathbf{Z}_m}$$

and make use of $\mathbf{u} = j\omega\mathbf{x}$ to obtain the complex displacement

$$\mathbf{x} = \frac{\mathbf{f}}{j\omega\mathbf{Z}_m}$$

Thus, knowledge of $\mathbf{Z}_m$ is equivalent to solving the differential equation.

The actual displacement is given by the real part of (1.26),

$$x = \frac{F \sin(\omega t - \Theta)}{\omega Z_m} \tag{1.31}$$

and the actual speed is given by the real part of (1.27) with the help of (1.29b) and (1.29c),

$$u = \frac{F \cos(\omega t - \Theta)}{Z_m} \tag{1.32}$$

The ratio F/Z_m gives the maximum speed of the driven oscillator and is the speed amplitude.

Equation 1.32 shows that Θ is the phase angle between the speed and the driving force. When this angle is positive, it indicates that the speed lags the driving force by the angle Θ. When this angle is negative, it indicates that the speed leads the driving force.

1.8 TRANSIENT RESPONSE OF AN OSCILLATOR.

Before continuing the discussion of the simple oscillator it will be well to consider the effect of superimposing the transient response on the steady-state condition. The complete general solution of (1.25b) is

$$x = Ae^{-\beta t} \cos(\omega_d t + \phi) + \frac{F}{\omega Z_m} \sin(\omega t - \Theta) \tag{1.33}$$

where A and ϕ are two arbitrary constants whose values are determined by the initial conditions.

As a special case let us assume that $x_0 = 0$ and $u_0 = 0$ at time $t = 0$ when the driving force is first applied, and that β is small compared to ω. Application of these conditions to (1.33) gives

$$A = \frac{F}{Z_m^2} \left(\frac{X_m^2}{\omega^2} + \frac{R_m^2}{\omega_d^2} \right)^{1/2}$$

and

$$\tan \phi = \frac{\omega R_m}{\omega_d X_m}$$

Representative curves showing the relative importance of the steady-state and transient motions in producing a combined motion are plotted in Fig. 1.7. The effect of the transient is apparent in the left-hand portion of these curves, but near the right-hand end the transient has been so damped that the final steady state is nearly reached. Curves for other initial conditions are analogous, in that the wave form is

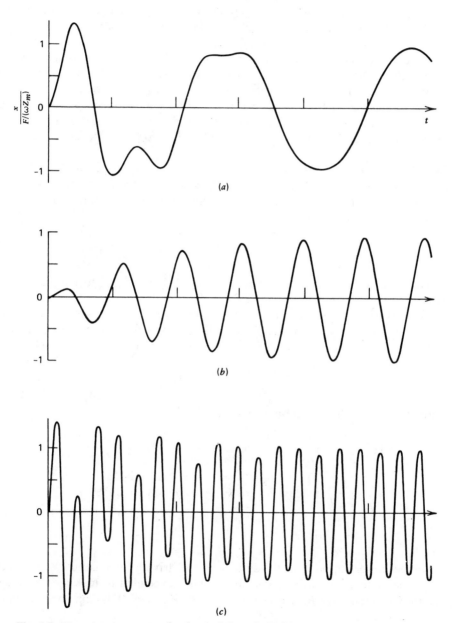

Fig. 1.7. *Transient response of a damped, forced oscillator with* $\beta/\omega_d = 0.1$ $x_0 = 0$, *and* $u_0 = 0$. *(a)* $\omega/\omega_d = 1/3$. *(b)* $\omega/\omega_d = 1.0$. *(c)* $\omega/\omega_d = 3$.

always somewhat irregular immediately after the application of the driving force, but it soon settles into the steady state.

Another important transient is the *decay transient* which results when the driving force is abruptly removed. The equation of this motion is that of the damped oscillator, (1.23), and its angular frequency of oscillation is ω_d, not ω. The constants giving the amplitude and phase angle of this motion depend on the part of its cycle in which the driving force is removed. It is impossible to remove the driving force without the appearance of a decay transient, although the effect will be negligible if the amplitude of the driving force is very slowly reduced to zero. The decay transient characteristics of mechanical vibrator elements are of particular importance when considering the fidelity of response of sound reproduction components such as loudspeakers and microphones. An example of an overly slow decay is a noticeable "hangover" of the fundamental frequency produced by some loudspeaker systems.

1.9 POWER RELATIONS The *instantaneous power* Π_i, in watts (W), supplied to the system in the *steady state* is equal to the product of the instantaneous driving force and the resulting instantaneous speed. Substituting the appropriate real expressions for force and speed,

$$\Pi_i = \frac{F^2}{Z_m} \cos \omega t \cos(\omega t - \Theta)$$

It should be noted that the instantaneous power Π_i is *not* equal to the real part of the *product* of the complex driving force $\mathbf{f}$ and the complex speed $\mathbf{u}$.

In most situations the *average power* Π being supplied to the system is of more significance than the instantaneous power. This average power is equal to the total work done per complete vibration, divided by the time of one vibration. Therefore,

$$\Pi = \frac{1}{T} \int_0^T \Pi_i \, dt = \langle \Pi_i \rangle_T$$

Substitution of Π_i in this equation gives

$$\Pi = \frac{F^2}{Z_m T} \int_0^T \cos \omega t \cos(\omega t - \Theta) \, dt$$

$$= \frac{F^2}{Z_m T} \int_0^T (\cos^2 \omega t \cos \Theta + \cos \omega t \sin \omega t \sin \Theta) \, dt$$

$$= \frac{F^2}{2Z_m} \cos \Theta \tag{1.34}$$

This average power supplied to the system by the driving force is not permanently stored in the system but is dissipated in the work expended in moving the system against the frictional force $R_m u$. Since $\cos \Theta = R_m/Z_m$, then (1.34) may be written as

$$\Pi = \frac{F^2 R_m}{2Z_m^{\;2}} \tag{1.35}$$

The average power delivered to the oscillator is a maximum when the mechanical reactance X_m vanishes, which from (1.29a) occurs when $\omega = \sqrt{s/m} = \omega_0$. At this frequency $\cos \Theta$ has its maximum value of unity ($\Theta = 0$) and Z_m its minimum value R_m.

1.10   MECHANICAL RESONANCE.

The (angular) frequency of *mechanical resonance* is defined as that at which the mechanical reactance X_m vanishes, $\omega_0 = (s/m)^{1/2}$. As has just been noted, this is the frequency at which a driving force will supply maximum power to the oscillator. In Sect. 1.2 it was also found to be the frequency of free oscillation of a similar undamped oscillator. At this frequency the mechanical impedance has its minimum value of $Z_m = R_m$, and is a pure real quantity. It is also the frequency of *maximum speed amplitude*. At this frequency (1.32) reduces to

$$u_{res} = \frac{F}{R_m} \cos \omega_0 t$$

and the displacement (1.31) reduces to

$$x_{res} = \frac{F}{\omega_0 R_m} \sin \omega_0 t$$

(Note that this frequency does not give the maximum displacement amplitude, which occurs at the angular frequency ω that makes $\omega[R_m^2 + (\omega m - s/\omega)^2]^{1/2}$ a minimum. It can be shown that this is $\omega = \sqrt{\omega_0^2 - 2\beta^2}$.)

If the average power supplied to the system as given by (1.35) is plotted as a function of the frequency of a driving force of constant amplitude, a curve similar to Fig. 1.8 is obtained. It has a maximum value of $F^2/2R_m$ at the resonance frequency ω_0 and falls at lower and higher frequencies. The sharpness of the peak of the power curve is primarily determined by R_m/m. If this ratio is small, the curve falls off very

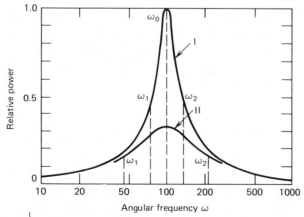

Fig. 1.8. *Resonance curves of a simple mechanical system. Curve* I *corresponds to* $Q = 1.8$. *Increasing* R_m *by a factor of 3 gives curve* II *with* $Q = 0.6$.

rapidly—a *sharp resonance*. If, on the other hand, R_m/m is large, the curve falls off more slowly and the system has a *broad resonance*. A more precise definition of the sharpness of resonance can be given in terms of the *quality factor Q* of the system, defined by

$$Q = \frac{\omega_0}{\omega_2 - \omega_1} \qquad (1.36)$$

where ω_2 and ω_1 are the two frequencies, respectively above and below resonance (ω_0) for which the average power has dropped to one-half its resonance value.

It is also possible to express Q in terms of the mechanical constants of the system. From (1.35) it is evident that the average power will be one-half of its resonance value whenever $Z_m^2 = 2R_m^2$. This corresponds to

$$R_m^2 + X_m^2 = 2R_m^2 \qquad \text{or} \qquad X_m = \pm R_m$$

Since $X_m = \omega m - s/\omega$, the two frequencies ω_2 and ω_1 that satisfy this requirement are given by

$$\omega_2 m - s/\omega_2 = R_m \qquad \text{and} \qquad \omega_1 m - s/\omega_1 = -R_m$$

The elimination of s between these equations gives

$$\omega_2 - \omega_1 = R_m/m$$

so that

$$Q = \omega_0 m/R_m \qquad (1.37)$$

Use of (1.24) for the relaxation time τ of this oscillator gives

$$Q = \tfrac{1}{2}\omega_0 \tau \qquad (1.38)$$

This shows that the sharpness of the resonance of the driven oscillator is directly related to the length of time it takes for the free oscillator to decay to $1/e$ of its initial amplitude. Furthermore, it can be shown that the number of oscillations taken for this decay is given by $(\omega_d/\omega_0)(Q/\pi)$, or about Q/π for weak damping. Thus, if an oscillator has a Q of 100 and a natural frequency 1000 Hz, it will take $(100/\pi)$ cycles or 32 ms to decay to $1/e$ of its initial amplitude. It should also be noted that $Q/(2\pi)$ is the ratio of the mechanical energy of the oscillator driven at its resonance frequency to the energy dissipated per cycle of vibration. Proof of this is left as an exercise.

Since the phase angle Θ is zero at resonance, when the oscillator is driven at resonance the speed is in phase with the driving force f. When the driving frequency ω is greater than ω_0 the phase angle is positive, and when ω approaches infinity the speed lags f by an angle that approaches $90°$. When ω is less than ω_0 the phase angle is negative, and as ω approaches zero the speed leads f by $90°$. Figure 1.9 shows the dependence of Θ on frequency for a typical oscillator. In systems having relatively small mechanical resistance, the phase angles of both speed and displacement vary rapidly in the vicinity of resonance.

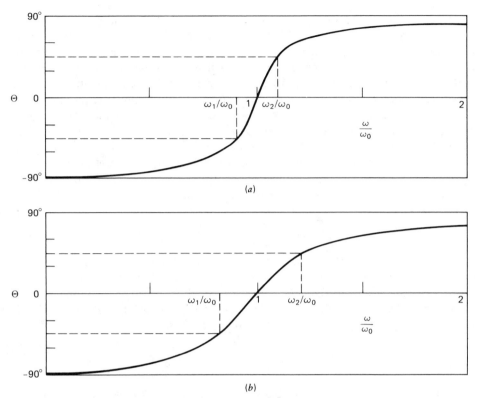

Fig. 1.9. *Representative curves of the phase angle* Θ *between the applied force f and the resulting speed u.* (*a*) $Q = 5$. (*b*) $Q = 2.5$.

1.11 MECHANICAL RESONANCE AND FREQUENCY.

Mechanical systems driven by periodic forces can be grouped into three different classes. (1) Sometimes it is desired that the system respond strongly to only *one* particular frequency. If the mechanical resistance of a simple oscillator is small, its impedance will be relatively large at all frequencies except those in the immediate vicinity of resonance, and such an oscillator will consequently respond strongly only in the vicinity of resonance. Some common examples are tuning forks, the resonators below the bars of a xylophone, and magnetostrictive sonar transducers. (2) In other applications it is desired that the system respond strongly to a series of discrete frequencies. The simple oscillator does not have this property, but mechanical systems that do behave in this manner can be designed. These will be considered in subsequent chapters of this book. (3) A third type of use requires that the system respond more or less uniformly to a wide range of frequencies. Examples include the vibrator elements of many electro-acoustic transducers: microphones, loudspeakers, hydrophones, many sonar transducers, and the sounding board of a piano.

In different applications, the quantity whose amplitude is supposed to be independent of frequency may be different. In some cases it is desired that the displacement amplitude be independent of frequency, in others it is the speed amplitude that is to be held constant, whereas in still others it is the amplitude of the acceleration that is to be invariant. By a suitable choice of the stiffness, mass, and mechanical resistance, a simple oscillator can be made to satisfy any of these requirements over a limited frequency range. These three special cases of frequency-independent driven oscillators are known as *stiffness-controlled*, *resistance-controlled*, and *mass-controlled* systems, respectively.

A *stiffness-controlled* system is characterized by a large value of the ratio s/ω for the frequency range over which the response is to be flat. Then in this range both ωm and R_m are negligible in comparison with s/ω, and $\mathbf{Z}_m$ is very nearly equal to $-js/\omega$, so that

$$x \approx (F/s)\cos \omega t \quad \text{and} \quad u \approx -(\omega F/s)\sin \omega t \quad (1.39)$$

It should be noted that, although the displacement amplitude is independent of frequency, the speed amplitude is not, nor is the acceleration amplitude.

A *resistance-controlled* system is one for which R_m is large in comparison with the reactance X_m. This will be true when an oscillator of relatively high mechanical resistance is operated in the vicinity of resonance. Then

$$x \approx \frac{F}{\omega R_m} \sin \omega t \quad \text{and} \quad u \approx \frac{F}{R_m} \cos \omega t \quad (1.40)$$

so that the speed amplitude is essentially independent of frequency, although both the displacement amplitude and the acceleration are not.

A *mass-controlled* system is characterized by a large value of ωm over the desired frequency range. Then s/ω and R_m are negligible and $\mathbf{Z}_m$ is approximately equal to $j\omega m$, so that

$$x \approx -\frac{F}{\omega^2 m} \cos \omega t \quad \text{and} \quad u \approx \frac{F}{\omega m} \sin \omega t$$

Neither the displacement amplitude nor the speed amplitude is independent of frequency, but

$$a = \frac{d^2 x}{dt^2} \approx \frac{F}{m} \cos \omega t \quad (1.41)$$

so that the acceleration amplitude is independent of frequency.

All driven mechanical vibrator elements are resistance-controlled for frequencies nearly equal to their resonant frequency, but for vibrators of low mechanical resistance the range of relatively flat response is extremely narrow. Similarly, all driven vibrators are stiffness-controlled for frequencies well below ω_0, and mass-controlled for frequencies well above ω_0. A suitable choice of mechanical constants will place any of these systems in the desired part of the frequency range, but the computed values are sometimes very difficult to attain in practice.

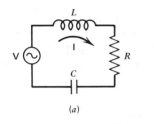

(a)

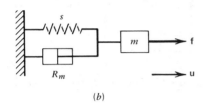

(b)

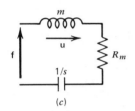

(c)

Fig. 1.10. Equivalent series systems.

1.12 EQUIVALENT ELECTRICAL CIRCUIT FOR A SIMPLE OSCILLATOR.

Many vibrating systems are mathematically equivalent to corresponding electrical systems. For example, consider a simple series electrical circuit containing inductance L, resistance R, and capacitance C, driven by an impressed sinusoidal voltage $V \cos \omega t$, as suggested in Fig. 1.10a. The differential equation for the current $\mathbf{I} = d\mathbf{q}/dt$ is

$$L\frac{d\mathbf{I}}{dt} + R\mathbf{I} + \frac{\mathbf{q}}{C} = \mathbf{V}$$

where $\mathbf{V} = V \exp(j\omega t)$ and the complex current $\mathbf{I}$ is the time rate of change of the complex charge $\mathbf{q}$. This equation may be written

$$L\frac{d^2\mathbf{q}}{dt} + R\frac{d\mathbf{q}}{dt} + \frac{\mathbf{q}}{C} = \mathbf{V}$$

which has the same form as (1.25b). Thus, the steady-state solution for $\mathbf{q}$ is

$$\mathbf{q} = \frac{\mathbf{V}}{j\omega[R + j(\omega L - 1/\omega C)]}$$

and the current has the form $\mathbf{I} = \mathbf{V}/\mathbf{Z}$ where $\mathbf{Z} = R + j(\omega L - 1/\omega C)$.

We see that the electrical circuit of Fig. 1.10a is the mathematical analog of the damped harmonic oscillator of Fig. 1.10b. The current $\mathbf{I}$ in the electrical system is equivalent to the speed $\mathbf{u}$ in the mechanical system, the charge $\mathbf{q}$ is equivalent to the displacement $\mathbf{x}$, and the applied voltage $\mathbf{V}$ is equivalent to the applied force $\mathbf{f}$. Furthermore, the impedance for these two systems have similar forms, with the mechanical resistance R_m analogous to the electrical resistance R, the mass m analogous to the electrical inductance L, and the mechanical stiffness s analogous to the reciprocal of the electrical capacitance C. By direct comparison with (1.25) et seq., it can be seen that the resonance frequency of the electrical circuit is

$$\omega_0 = 1/\sqrt{LC}$$

and the average power is

$$\Pi = \frac{V^2}{2Z} \cos \Theta$$

The elements in the electrical system (Fig. 1.10a) are said to be in *series* because they experience the same current. Similarly, the elements in the mechanical system (Fig. 1.10b) can be represented by the series circuit of Fig. 1.10c: they experience the same displacement and, therefore, the same speed.

If a simple mechanical oscillator is driven by a sinusoidal force applied to the normally fixed end of the spring as suggested by Fig. 1.11a, then the mass and the spring experience the same force and this combination represented by a *parallel* circuit, as shown in Fig. 1.11b. The speed of the driven end of the spring is equivalent to the current entering the parallel circuit, and the speed $\mathbf{u}_m$ of the mass is equivalent to the current flowing through the inductor.

Other equivalent systems are shown in Figs. 1.12 and 1.13.

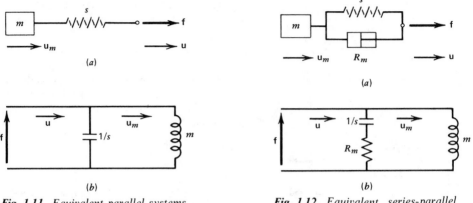

(a)

(b)

Fig. 1.11. *Equivalent parallel systems.*

(a)

(b)

Fig. 1.12. *Equivalent series-parallel systems.*

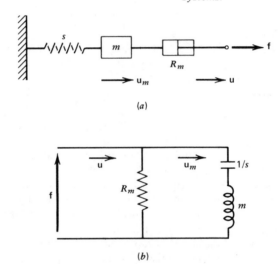

(a)

(b)

Fig. 1.13. *Equivalent series-parallel systems.*

1.13 LINEAR COMBINATIONS OF SIMPLE HARMONIC VIBRATIONS. In many important situations that arise in acoustics, the motion of a body is a linear combination of the vibrations induced separately by two or more simple harmonic excitations. It is easy to show that the displacement of the body is then the sum of the individual displacements resulting from each of the harmonic excitations. Combining the effects of individual vibrations by linear addition is valid for the majority of cases encountered in acoustics. In general, the presence of one vibration does not alter the medium to such an extent that the characteristics of other vibrations are disturbed. Consequently, the total vibration is obtained by a linear superposition of the individual vibrations.

One case is the combination of two excitations that have the same angular frequency ω. If the two individual displacements are given by

$$\mathbf{x}_1 = A_1 e^{j(\omega t + \phi_1)} \qquad \text{and} \qquad \mathbf{x}_2 = A_2 e^{j(\omega t + \phi_2)}$$

their linear combination $\mathbf{x}_1 + \mathbf{x}_2$ results in a motion $A \exp[j(\omega t + \phi)]$ where

$$Ae^{j(\omega t + \phi)} = (A_1 e^{j\phi_1} + A_2 e^{j\phi_2})e^{j\omega t}$$

Solution for A and ϕ can be accomplished easily if the addition of the phasors $A_1 \exp(j\phi_1)$ and $A_2 \exp(j\phi_2)$ is represented graphically, as in Fig. 1.14. From the

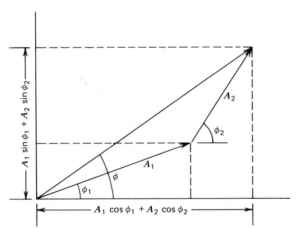

Fig. 1.14. *Phasor combination $A \exp(j\phi) = A_1\exp(j\phi_1) + A_2 \exp(j\phi_2)$ of two simple harmonic motions having identical frequencies.*

projections of each phasor on the real and imaginary axes, as indicated in the figure,

$$A = [(A_1 \cos \phi_1 + A_2 \cos \phi_2)^2 + (A_1 \sin \phi_1 + A_2 \sin \phi_2)^2]^{1/2} \qquad (1.42)$$

$$\tan \phi = \frac{A_1 \sin \phi_1 + A_2 \sin \phi_2}{A_1 \cos \phi_1 + A_2 \cos \phi_2} \qquad (1.43)$$

The real displacement is

$$x = A \cos(\omega t + \phi)$$

where A and ϕ are given by (1.42) and (1.43). The liner combination of two simple harmonic vibrations of identical frequency yields another simple harmonic vibration of this same frequency, having a different phase angle and an amplitude in the range $|A_1 - A_2| \le A \le A_1 + A_2$.

With the help of Fig. 1.14, it is clear that the addition of more than two phasors can be accomplished by drawing them in a chain, head to tail, and then taking their components on the real and imaginary axes. Thus, it may readily be shown that the vibration resulting from the addition of any number n of simple harmonic vibrations of identical frequency has amplitude given by

$$A = [(\sum A_n \cos \phi_n)^2 + (\sum A_n \sin \phi_n)^2]^{1/2} \qquad (1.44)$$

and phase angle ϕ obtained from

$$\tan \phi = \frac{\sum A_n \sin \phi_n}{\sum A_n \cos \phi_n} \tag{1.45}$$

Thus any linear combination of simple harmonic vibrations of identical frequency produces a new simple harmonic vibration of this same frequency. For example, when two or more sound waves overlap in a fluid medium, then at each point in the fluid the periodic sound pressures of the individual waves combine as described above.

The expression for the linear combination of two simple harmonic vibrations of *different* frequency is

$$x = A_1 e^{j(\omega_1 t + \phi_1)} + A_2 e^{j(\omega_2 t + \phi_2)} \tag{1.46}$$

where ω_1 is the angular frequency of one vibration and ω_2 that of the other. The resulting motion is not simple harmonic, so that it cannot be represented by a simple sine or cosine function. However, if the ratio of the larger to the smaller frequency is a rational number (commensurate), the motion is periodic with angular frequency given by the greatest common divisor of ω_1 and ω_2. Otherwise, the resulting motion is a nonperiodic oscillation which never repeats itself. The linear combination of three or more simple harmonic vibrations that have different frequencies has characteristics similar to those discussed for two.

The linear combination of two simple harmonic vibrations of nearly the same frequency is easy to interpret. If the angular frequency ω_2 is written as

$$\omega_2 = \omega_1 + \Delta\omega$$

then the combination of the two vibrations becomes

$$x = A_1 e^{j(\omega_1 t + \phi_1)} + A_2 e^{j(\omega_1 t + \Delta\omega t + \phi_2)}$$

This can be reexpressed as

$$x = [A_1 e^{j\phi_1} + A_2 e^{j(\phi_2 + \Delta\omega t)}] e^{j\omega_1 t}$$

and then cast into the form

$$x = A e^{j(\omega_1 t + \phi)} \tag{1.47}$$

where

$$A = [A_1^2 + A_2^2 + 2A_1 A_2 \cos(\phi_1 - \phi_2 - \Delta\omega t)]^{1/2} \tag{1.48a}$$

and

$$\tan \phi = \frac{A_1 \sin \phi_1 + A_2 \sin(\phi_2 + \Delta\omega t)}{A_1 \cos \phi_1 + A_2 \cos(\phi_2 + \Delta\omega t)} \tag{1.48b}$$

The resulting vibration may be regarded as *approximately* simple harmonic, with angular frequency ω_1, but with both amplitude A and phase ϕ *varying slowly* at a frequency of $\Delta\omega/(2\pi)$. It can be shown that the amplitude of the vibration waxes and wanes between the limits $(A_1 + A_2)$ and $|A_1 - A_2|$. The effect of the variation in phase angle is somewhat more complicated. It modifies the vibration rate in such a manner that its frequency is not strictly constant, but the average angular frequency may be shown to lie somewhere between ω_1 and ω_2, depending on the relative magnitudes of the amplitudes A_1 and A_2.

In the acoustic case of the simultaneous sounding of two pure tones of slightly different frequency, this variation in amplitude results in a rhythmic pulsing of the loudness of the sound known as *beating*. As an example let us consider the special case $A_1 = A_2$ and $\phi_1 = \phi_2 = 0$. The amplitude equation reduces to

$$A = A_1(2 + 2 \cos \Delta\omega t)^{1/2} \tag{1.49a}$$

Therefore, the amplitude ranges between $2A_1$ and zero, and the phenomenon of beating is very pronounced. The phase angle ϕ is given by

$$\tan \phi = \frac{\sin \Delta\omega t}{1 + \cos \Delta\omega t} \tag{1.49b}$$

Audible beats are heard whenever two sounds of nearly the same frequency strike the ear. This and other associated phenomena will be discussed in more detail in Chapter 11.

1.14 ANALYSIS OF COMPLEX VIBRATIONS BY FOURIER'S THEOREM.

In the preceding section we noted that the linear combination of two or more simple harmonic vibrations with commensurate frequencies leads to a complex vibration that has a frequency determined by the greatest common divisor of these frequencies. Conversely, by means of a powerful mathematical theorem originated by Fourier, it is possible to analyze any complex periodic vibration into a harmonic array of component frequencies.

Stated briefly, this theorem asserts that any single-valued periodic function may be expressed as a summation of simple harmonic terms whose frequencies are integral multiples of the repetition rate of the given function. Since the above restrictions are normally satisfied in the case of the vibrations of material bodies, the theorem is widely used in acoustics.

If a certain vibration of period T is represented by the function $f(t)$, then Fourier's theorem states that $f(t)$ may be represented by the harmonic series

$$f(t) = \tfrac{1}{2}A_0 + A_1 \cos \omega t + A_2 \cos 2\omega t + \cdots + A_n \cos n\omega t + \cdots$$
$$+ B_1 \sin \omega t + B_2 \sin 2\omega t + \cdots + B_n \sin n\omega t + \cdots \tag{1.50}$$

where $\omega = 2\pi/T$ and the A's and B's are constants to be determined.

The formulas for determining these constants (as derived in standard mathematical treatises on Fourier's theorem) are

$$A_n = \frac{2}{T} \int_0^T f(t) \cos n\omega t \, dt \tag{1.51a}$$

$$B_n = \frac{2}{T} \int_0^T f(t) \sin n\omega t \, dt \tag{1.51b}$$

Whether or not these integrations are feasible will depend on the nature and complexity of the function, $f(t)$. If this function exactly represents the combination of a finite number of pure sine and cosine vibrations, the series obtained by computing the above constants will contain only these terms. Analysis, for instance, of beats will yield only the two frequencies present. Similarly, the complex vibration constituting the sum of three pure musical tones will analyze into those frequencies alone. On the other hand, if the vibration is characterized by abrupt changes in slope, like saw-tooth waves or square waves, then the entire infinite series must be considered for a complete equivalence of motion. If $f(t)$ and df/dt are piecewise continuous over the interval $0 \le t \le T$, it is possible to show that the harmonic series is always convergent. However, jagged functions will require the inclusion of a large number of terms merely to achieve a reasonably good equivalence to the original function. Fortunately, the majority of vibrations encountered in acoustics are relatively smooth functions of time. In such cases, the convergence is rather rapid and only a few terms must be computed.

Depending on the nature of the function being expanded, some terms in the series may be absent. If the function $f(t)$ is symmetrical with respect to $f = 0$, the constant term A_0 will be absent. If the function $f(t)$ is *even*, $f(t) = f(-t)$, then all sine terms will be missing. An *odd* function, $f(t) = -f(-t)$, will cause all cosine terms to be absent.

In analyzing the perception of sound, a factor that enables us to reduce the number of higher frequency terms that must be computed is that the characteristics of the human ear are such that the subjective audible interpretation of a complex sound vibration is often only slightly altered if the higher frequencies are removed or ignored.

Let us apply the above equations in an analysis of the function represented by a saw-tooth vibration, as shown graphically in Fig. 1.15a. This function may be defined analytically as

$$f(t) = a[1 - (2t/T)]$$

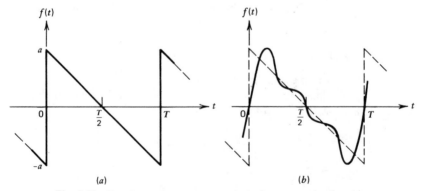

Fig. 1.15. *Fourier series representation of a sawtooth vibration.*

for the time interval $t = 0$ to $t = T$. A substitution of this function into (1.51) yields

$$A_n = 0 \qquad \text{and} \qquad B_n = \frac{a}{n\pi}$$

It should be noted that A_0 is zero because of symmetry of the motion about $f = 0$. All A_n's are zero since the function is odd. The complete harmonic series equivalent to the saw-tooth vibration is

$$f(t) = \frac{a}{\pi} \left(\sin \omega t + \frac{\sin 2\omega t}{2} + \cdots + \frac{\sin n\omega t}{n} + \cdots \right)$$

Plotted in Fig. 1.15b are results obtained by including only the first three terms of the series. Differences between the plots of Fig. 1.15 are quite apparent. It would be necessary to include at least 20 terms in plotting the series function if it were not to differ by more than 5 percent from a plot of the original analytical function.

1.15 TRANSIENTS AND FOURIER ANALYSIS. Two fundamental approaches are available for the analysis of transient motions: Laplace transforms and Fourier analysis. (Actually, these two approaches are closely related, the principle difference being the mathematical nomenclature.) While Laplace transforms are perhaps the more conventional approach, the underlying physics is somewhat hidden. We will, therefore, use the second approach, Fourier analysis.

It has been demonstrated in Sect. 1.14 that a repeating vibration of period T can be considered as a sum of sinusoidal vibrations whose frequencies are integral multiples of the fundamental frequency $f = 1/T$. If we now consider a nonrepeating transient as being one of a family whose sequential members are uniformly spaced a time T apart and then allow T to become infinitely large, the fundamental frequency of the motion must approach zero, and the summation over all harmonics must be replaced by an integration over all frequencies.

Thus, if $\mathbf{f}(t)$ is a transient disturbance, we can write the general expression

$$\mathbf{f}(t) = \int_{-\infty}^{\infty} \mathbf{g}(w)e^{jwt} \, dw \tag{1.52}$$

where w is the angular frequency. (We have chosen w rather than ω for notational reasons that will appear later, and because it is the "dummy" variable of integration.) The quantity $\mathbf{g}(w)$ is the *spectral density* of $\mathbf{f}(t)$. The integration region $-\infty < w < 0$ introduces the concept of "negative" frequency, but from

$$e^{j(\pm w)t} = \cos wt \pm j \sin wt$$

this is no more than a means of generating complex conjugates.

Given $\mathbf{f}(t)$, inversion of the integral to obtain the spectral density $\mathbf{g}(w)$ of the transient function yields

$$\mathbf{g}(w) = \frac{1}{2\pi} \int_{-\infty}^{\infty} \mathbf{f}(t)e^{-jwt} \, dt \tag{1.53}$$

(A proof, being rather mathematical and peripheral to our interests, will not be offered. Consult any standard text on Fourier transforms.) The pair (1.52) and (1.53) constitute one form of the *Fourier integral transforms*.

As an example, assume that $\mathbf{f}(t)$ represents a single extremely short but strong force such as striking an oscillator with a hammer or a drumhead with a drumstick. Such impulses can be approximated by the *Dirac delta function*, defined by

$$\delta(t) = 0 \qquad t \neq 0 \tag{1.54a}$$

$$\int_{-\infty}^{\infty} \delta(t)\, dt = 1 \tag{1.54b}$$

$\delta(t)$ could be represented by

$$\delta(t) = \begin{cases} 0 & |t| > \varepsilon/2 \\ 1/\varepsilon & |t| \leq \varepsilon/2 \end{cases} \tag{1.55}$$

in the limit $\varepsilon \to 0$.

Substitution of $\mathbf{f}(t) = \delta(t)$ into (1.53) yields

$$\mathbf{g}(w) = \frac{1}{2\pi} \int_{-\infty}^{\infty} \delta(t) e^{-jwt}\, dt$$

Now, use of (1.55) shows that, since $\delta(t)$ is nonzero only where $|t| \leq \varepsilon/2$, the limits can be replaced by $\pm\varepsilon/2$. As $\varepsilon \to 0$, $\exp(-jwt)$ can be replaced by its value at $w = 0$ which leaves the result

$$\mathbf{g}(w) = \frac{1}{2\pi} \int_{-\varepsilon/2}^{\varepsilon/2} \delta(t)\, dt = \frac{1}{2\pi}$$

Thus, all frequencies are equally present in $\delta(t)$.

Conversely, if we write $\mathbf{g}(w)$ as consisting only of a single frequency,
$$\mathbf{g}(w) = \delta(w - \omega)$$
then

$$\mathbf{f}(t) = \int_{-\infty}^{\infty} \delta(w - \omega) e^{jwt}\, dw = e^{j\omega t}$$

and the spectral density of a monofrequency signal is a delta function centered on that frequency.

The utility of this approach can be demonstrated by a simple exercise. Let $\mathbf{F}(t)$ be an impulsive force applied to an oscillator and express $\mathbf{F}(t)$ in terms of its Fourier components

$$\mathbf{F}(t) = \int_{-\infty}^{\infty} \mathbf{G}(w) e^{jwt}\, dw \tag{1.56a}$$

where the spectral density $\mathbf{G}(w)$ is found from (1.53). Each of these monofrequency force components

$$\mathbf{f}(w,t) = \mathbf{G}(w) e^{jwt}$$

will generate a monofrequency complex speed component $\mathbf{u}(w,t)$ given from (1.30) by

$$\mathbf{u}(w,t) = \frac{\mathbf{f}(w,t)}{\mathbf{Z}(w)} = \frac{\mathbf{G}(w)}{\mathbf{Z}(w)} e^{jwt}$$

where $\mathbf{Z}(w)$ is the input mechanical impedance of the oscillator at the angular frequency w. Now, $\mathbf{G}(w)/\mathbf{Z}(w)$ is the spectral density of the speed and the resultant transient speed $\mathbf{U}(t)$ of the oscillator is therefore

$$\mathbf{U}(t) = \int_{-\infty}^{\infty} \mathbf{u}(w,t)\, dw = \int_{-\infty}^{\infty} \frac{\mathbf{G}(w)}{\mathbf{Z}(w)} e^{jwt}\, dw \tag{1.56b}$$

It can be verified by direct substitution of (1.56b) into (1.25a) that $U(t)$ is the solution for the applied force $f(t) = F(t)$.

The physical interpretation of this approach is quite important and straightforward: If an arbitrary force is applied to a mechanical system, the resultant motion can be found by resolving the force into its individual frequency components, obtaining the motion resulting from each of these monofrequency components, and then assembling the resulting motion by combining the individual monofrequency motions. This is the very same case we encountered in periodic, nonharmonic forces, except that integrals must replace summations because the individual frequency components are not discrete, but are continuously distributed over a range of frequencies.

While evaluation of these integrals may be difficult and involve special techniques (such as calculus of residues) or approximations (such as the method of stationary phase), extensive tables of transformation pairs $f(t)$ and $g(w)$ have been compiled and are easily accessible.

PROBLEMS

1.1. Given two springs of stiffness s and two bodies of mass M, find the resonance frequencies of the systems sketched below.

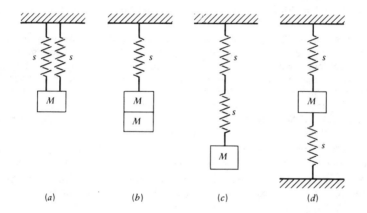

(a)   (b)   (c)   (d)

1.2. A simple undamped harmonic oscillator whose natural frequency is 5 rad/s is displaced a distance 0.03 m from its equilibrium position and released. Find (a) the initial acceleration, (b) the amplitude of the resulting motion, and (c) the maximum speed attained.

1.3. Given that the real part of $x = A \exp(j\omega t)$ is $x = A \cos(\omega t + \phi)$ show that the real part of x^2 does not equal x^2.

1.4. Find the real part, magnitude, and phase of (a) $\sqrt{x + jy}$, (b) $A \exp[j(\omega t + \phi)]$, and (c) $[1 + \exp(-2j\theta)] \exp(j\theta)$.

1.5. Given the two complex numbers $A = A \exp[j(\omega t + \theta)]$ and $B = B \exp[j(\omega t + \phi)]$, find (a) the real part of AB, (b) the real part of A/B, (c) the real part of A times the real part of B, (d) the phase of AB, and (e) the phase of A/B.

1.6. Given the complex numbers $A = x + jy$ and $B = X + jY$, find (a) the magnitude of A, (b) the magnitude of B, (c) the magnitude of AB, (d) the real part of AB, (e) the phase of AB, and (f) the real part of A/B.

1.7. A mass of 0.5 kg hangs on a spring. When an additional mass of 0.2 kg is attached to the spring, the spring is observed to stretch an additional 0.04 m. When the 0.2-kg mass is abruptly removed, the amplitude of the ensuing oscillations of the 0.5-kg mass is observed to decrease to $1/e$ of its initial value in 1 s. Compute values for R_m, ω_d, A, and ϕ.

1.8. The solution for a critically damped oscillator is $x = (A + Bt)\exp(-\beta t)$. Verify that this satisfies the equation of motion.

1.9. Show that if $\beta \ll \omega_0$ then $\omega_d \doteq \omega_0[1 - \frac{1}{2}(\beta/\omega_0)^2]$.

1.10. Show that $Z_m = \omega_0 m[(\omega/\omega_0 - \omega_0/\omega)^2 + 1/Q^2]^{1/2}$.

1.11. A mass m is fastened to one end of a horizontal spring of stiffness s, and a horizontal driving force $F \sin \omega t$ is applied to the other end of the spring. (a) Assuming no damping, determine the equation giving the motion of the driven end of the spring as a function of time t. (b) Show that the expression for the speed of this end of the spring is analogous to that giving the current into a parallel LC electrical circuit. (c) If the constants of the above system are $F = 3$ N, $s = 200$ N/m, and $m = 0.5$ kg, compute and plot curves showing how the displacement and speed amplitudes of the driven end of the spring vary with frequency in the range $0 < \omega < 100$ rad/s.

1.12. An undamped harmonic oscillator is driven with a force $F \sin \omega t$ where $\omega \neq \omega_0$. (a) Find the resultant speed of the mass if the mass is at rest at $t = 0$ when the force is applied. (b) Sketch the waveform of the speed if $\omega = 2\omega_0$. (c) If a small amount of damping is introduced and the driving frequency is far below resonance, show that the steady-state solution is approximated by $u \doteq (\omega F/s)\cos \omega t$.

1.13. Verify that in the steady state the power dissipated by the frictional force in the damped driven oscillator is equal to that being supplied by the driving force.

1.14. A mass of 0.5 kg hangs on a spring. The stiffness constant of the spring is 100 N/m, and the mechanical resistance of the system is 1.4 kg/s. The force driving the system is $f = 2 \cos 5t$. (a) What will be the steady-state values of the displacement amplitude, speed amplitude, and average power dissipation? (b) What is the phase angle between speed and force? (c) What is the resonance frequency and what would be the displacement amplitude, speed amplitude, and average power dissipation at this frequency and for the same force magnitude as in (a)? (d) What is the Q of the system, and over what range of frequencies will the power loss be at least 50 percent of its resonance value?

1.15. When a simple oscillator is driven at its resonance frequency, show that the ratio of the energy dissipated per cycle to the total mechanical energy present is $2\pi/Q$.

1.16. Derive equations that give the two frequencies corresponding to the half-power points on the power output curve of a driven simple oscillator. Show that they are given approximately by $\omega_0 \mp R_m/2m$.

1.17. (a) What is the general expression for the acceleration of a simple damped oscillator driven by a force $F \cos \omega t$? (b) Derive an expression that gives the frequency ω' which gives a maximum value to this acceleration.

1.18. Find the mechanical impedances, the mechanical resonance frequencies, and the equivalent electrical circuits for the following mechanical systems.

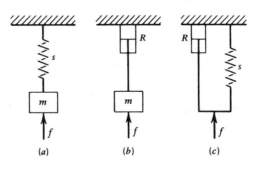

(a) (b) (c)

1.19. Show that the amplitude A_n of the displacement resulting from the linear addition of n harmonic vibrations all of the same amplitude A and frequency ω but having different initial phase angles of $\phi_1 = \varepsilon$, $\phi_2 = 2\varepsilon$, $\phi_3 = 3\varepsilon$, ..., $\phi_n = n\varepsilon$, is given by

$$A_n = \frac{A \sin(n\varepsilon/2)}{\sin(\varepsilon/2)}$$

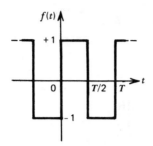

1.20. Show that the square waveform is represented by

$$f(t) = \frac{4}{\pi} \sum_{\text{odd } n} \frac{1}{n} \sin\left(2\pi n \frac{t}{T}\right)$$

1.21. (a) Find the spectral density $g(w)$ of the force $F \exp(j\omega t)$. (b) By substitution into (1.56b), solve for the resultant speed of a harmonic oscillator with mechanical impedance $\mathbf{Z}_m$.

1.22. One end of a dashpot of mechanical resistance R_m is attached to a wall and the other end is struck at $t = 0$ by a force $F(t) = \delta(t)$. (a) Obtain the resultant speed of the struck end and the resultant displacement by the Fourier transform technique. (b) Confirm the results of (a) by direct solution of (1.25a).

1.23. A mass M, connected to a rigid foundation by a spring and dashpot (spring constant s and mechanical resistance R_m), is constrained to move in one dimension. A second mass m rotates with angular frequency ω on an arm of length L about an axis perpendicular to the motion of the mass M. Find the resulting steady-state speed amplitude of the mass M.

CHAPTER 2

TRANSVERSE MOTION—
THE VIBRATING STRING

2.1 VIBRATIONS OF EXTENDED SYSTEMS. In the previous chapter it was assumed that the mass moves as a rigid body so that it could be considered concentrated at a single point. However, most vibrating bodies are not so simple. For example, the diaphragm of a loudspeaker has its mass distributed over its surface so that the cone does not move as a unit. The same occurs for a piano string and for the surface of a cymbal. Rather than beginning with the study of such complicated vibrations, we consider first the ideal vibrating string, the most readily visualized physical system involving the propagation of waves. Even this simple system is a hypothetical one; certain simplifying assumptions must be made that cannot be completely realized in practical cases. Nevertheless, the results obtained are extremely important because they yield a fundamental understanding of wave phenomena.

2.2 TRANSVERSE WAVES ON A STRING. Consider a string stretched to a tension T. If a portion of the string is suddenly displaced from its equilibrium position and released, it is observed that the displacement does .1ot remain fixed in its initial position, but instead breaks up into two separate disturbances that propagate along the string, one moving to the right and the other to the left with equal speed, as suggested by Fig. 2.1. Furthermore, it is observed that the speed of propagation of all *small displacements* is independent of the shape and amplitude of the initial displacement and depends only on the mass per unit length of the string and its tension. Experiment and theory show that this speed is given by $c = \sqrt{T/\rho_L}$ where c is in m/s, T is the tension in N, and ρ_L is the mass per unit length of the string in kg/m. A propagating transverse disturbance is referred to as a *transverse traveling wave*.

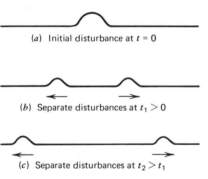

(a) Initial disturbance at $t = 0$

(b) Separate disturbances at $t_1 > 0$

(c) Separate disturbances at $t_2 > t_1$

Fig. 2.1. *Propagation of a transverse disturbance along a stretched string.*

2.3 THE ONE-DIMENSIONAL WAVE EQUATION. By considering the forces that tend to return the string to its equilibrium position, it is possible to derive a second-order, partial-differential equation known as a *wave equation*. Solutions of this wave equation satisfying the appropriate initial and boundary conditions will completely define the motion of the string.

Assume a string of uniform linear density ρ_L and negligible stiffness, stretched to a tension T great enough that the effects of gravity can be neglected. Also assume that there are no dissipative forces (such as those associated with friction or with the radiation of acoustic energy). Figure 2.2 isolates an infinitesimal element of the string with equilibrium position x and equilibrium length dx. If y (the transverse displacement of this element from its equilibrium position) is small, the tension T

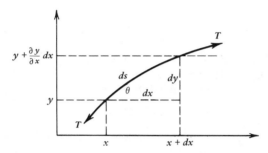

Fig. 2.2. *String segment.*

remains constant along the string and the difference between the y component of the tension at the two ends of the element is

$$df_y = (T \sin \theta)_{x+dx} - (T \sin \theta)_x \tag{2.1}$$

where θ is the angle between the tangent to the string and the x-axis, $(T \sin \theta)_{x+dx}$ is the value of $T \sin \theta$ at $x + dx$, and $(T \sin \theta)_x$ its value at x. Applying the Taylor's series expansion

$$f(x + dx) = f(x) + \left(\frac{\partial f}{\partial x}\right)_x dx + \frac{1}{2}\left(\frac{\partial^2 f}{\partial x^2}\right)_x dx^2 + \cdots \tag{2.2}$$

to (2.1) gives

$$df_y = \left[(T \sin \theta)_x + \frac{\partial(T \sin \theta)}{\partial x} dx + \cdots\right] - (T \sin \theta)_x = \frac{\partial(T \sin \theta)}{\partial x} dx$$

where we have retained only the lowest-order nonvanishing terms. If θ is small, $\sin \theta$ may be replaced by $\partial y/\partial x$ and the net transverse force on the element becomes

$$df_y = \frac{\partial\left(T \dfrac{\partial y}{\partial x}\right)}{\partial x} dx = T \frac{\partial^2 y}{\partial x^2} dx \tag{2.3}$$

Since the mass of the element is $\rho_L dx$ and its acceleration in the y direction is

$\partial^2 y/\partial t^2$, Newton's law gives

$$df_y = \rho_L \, dx \, \frac{\partial^2 y}{\partial t^2} \tag{2.4}$$

Combination of (2.3) and (2.4) then yields the *wave equation*

$$\frac{\partial^2 y}{\partial x^2} = \frac{1}{c^2} \frac{\partial^2 y}{\partial t^2} \tag{2.5}$$

where the constant c^2 is defined by

$$\boxed{c^2 = T/\rho_L} \tag{2.6}$$

2.4 GENERAL SOLUTION OF THE WAVE EQUATION.

Equation (2.5) is a second-order, partial differential equation. Its complete solution contains two arbitrary functions. The most general solution is

$$y(x, t) = y_1(ct - x) + y_2(ct + x) \tag{2.7}$$

where $y_1(ct - x)$ and $y_2(ct + x)$ are completely arbitrary functions of arguments $(ct - x)$ and $(ct + x)$, respectively. Possible examples of such arbitrary functions include $\log(ct \pm x)$, $(ct \pm x)^2$, $\sin[\omega(t \pm x/c)]$, $\exp[j\omega(t \pm x/c)]$, and $\sqrt{ct \pm x}$.

Consider the function $y_1(ct - x)$, and let $w = ct - x$. Then the first partial derivative with respect to time is

$$\frac{\partial y_1}{\partial t} = \frac{dy_1}{dw} \frac{\partial w}{\partial t} = c \frac{dy_1}{dw}$$

Repeating this partial differentiation with respect to time gives

$$\frac{\partial^2 y_1}{\partial t^2} = c^2 \frac{d^2 y_1}{dw^2} \tag{2.8a}$$

Furthermore,

$$\frac{\partial y_1}{\partial x} = \frac{dy_1}{dw} \frac{\partial w}{\partial x} = -\frac{dy_1}{dw}$$

and

$$\frac{\partial^2 y_1}{\partial x^2} = \frac{d^2 y_1}{dw^2} \tag{2.8b}$$

By combining (2.8a) and (2.8b) we obtain

$$\frac{\partial^2 y_1}{\partial x^2} = \frac{1}{c^2} \frac{\partial^2 y_1}{\partial t^2}$$

Thus, any function of argument $(ct - x)$ is a solution of the wave equation (2.5). Similarly, it can be shown that $y_2(ct + x)$ is also a solution. The sum of these two functions is the complete general solution of the wave equation.

2.5 WAVE NATURE OF THE GENERAL SOLUTION. Consider the solution $y_1(ct - x)$. At time t_1 the transverse displacement of the string is given by $y_1(ct_1 - x)$, as suggested by Fig. 2.3a. At a later time t_2 the shape of the string will be given by $y_1(ct_2 - x)$, as suggested by Fig. 2.3b. The particular transverse displacement $y_1(ct_1 - x_1)$ of the string that was found at x_1 when $t = t_1$ must be found at a position x_2 when $t = t_2$ where

$$ct_1 - x_1 = ct_2 - x_2$$

Thus, this particular displacement has moved a distance

$$x_2 - x_1 = c(t_2 - t_1)$$

to the right. Since the particular displacement chosen was arbitrary, any (and every) transverse displacement must move to the right with the same speed. This means that the shape of the disturbance remains unchanged and travels along the string to the right at a constant speed c. The function $y_1(ct - x)$ represents a *wave* traveling in the $+x$ direction. The speed with which the particular displacement $y_1(ct - x)$ travels along the string is called the *phase speed c*. It is important to note that, while

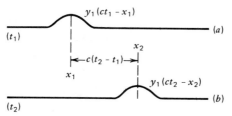

Fig. 2.3. *Speed of transverse wave propagation.*

the waveform moves with speed c, the material elements of the string move transversely about their equilibrium positions with speeds given by $\partial y_1/\partial t = c\, dy_1/dw$ where $w = ct - x$.

Similarly, it may be shown that $y_2(ct + x)$ represents a wave moving in the $-x$ direction with phase speed c. Note that the wave shape corresponding to each of the two arbitrary functions remains constant as the initial disturbance progresses along the string. This mathematical conclusion is never completely realized in practice, since the assumptions made in deriving the wave equation are never fulfilled for real strings which always have some bending stiffness and are acted on by dissipative forces; waves traveling along real strings become distorted. For the relatively flexible strings and low damping normally encountered in musical instruments, the rate of distortion is quite slight if the amplitude of the disturbance is small. For large amplitudes, on the other hand, the change of wave shape may be pronounced.

2.6 INITIAL VALUES AND BOUNDARY CONDITIONS. The functions $y_1(ct - x)$ and $y_2(ct + x)$ are determined by the initial values and boundary conditions. For the freely vibrating string, *initial values* at $t = 0$ are determined by the type and point of application of the exciting force applied to the string. For example, the initial wave shape set up by *striking* a string (when a piano is played) is quite different from that established by *plucking* a string (a harp or guitar) or in *bowing* a string (a violin); the functions representing the wave shape are consequently different. These functions are further determined by the *boundary conditions* existing at the ends of the string. Actual strings are always finite in length and must be fixed in some manner at their ends. For example, if the supports of the strings are rigid, the sum $y_1 + y_2$ is constrained to have a zero value at all times at the point of support. When a string is driven to *steady-state* conditions by a periodic external driving force, the functions y_1 and y_2 are periodic with the same frequency, but their other characteristics (such as amplitude of vibration) are determined by the point of application of the force and by the boundary conditions at the ends of the string.

2.7 REFLECTION AT A BOUNDARY. Assume that a string is rigidly supported at $x = 0$. Then $y_1(ct - x)$ and $y_2(ct + x)$ are no longer completely arbitrary, since their sum must be zero at all times at $x = 0$:

$$y(0, t) = y_1(ct - 0) + y_2(ct + 0) = 0 \tag{2.9}$$

so that

$$y_2(ct) = -y_1(ct)$$

The two functions are of the same form but opposite sign, and the displacement at any point on the string is

$$y(x, t) = y_1(ct - x) - y_1(ct + x)$$

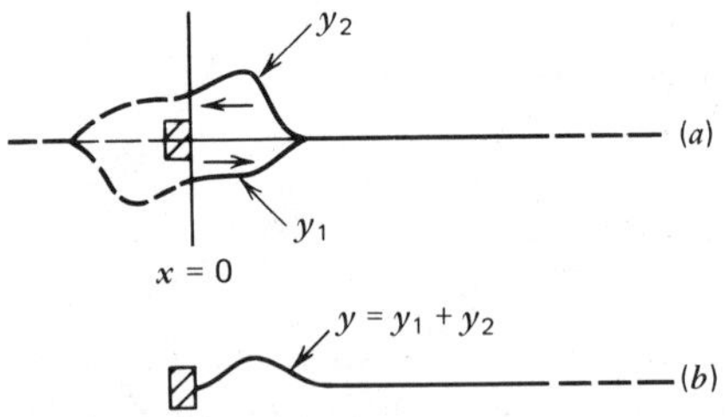

As may be seen in Fig. 2.4, this represents a wave $y_1(ct - x)$ traveling to the right, plus another wave $y_2 = -y_1(ct + x)$ of similar shape but opposite displacement traveling to the left. The process of reflection at a rigid boundary may be considered as one in which the wave moving to the left does not pass this boundary but is instead reflected into a similarly shaped wave of opposite displacement traveling to the right.

Fig. 2.4. *Reflection from a rigid end. In (a), the dashed-line segment of wave y_2 is shown as being reflected to become the solid-line segment of wave y_1. The resultant wave, shown in (b), has no displacement at $x = 0$.*

Another simple example of a boundary condition is an end supported so that there is *no transverse force* on the string. Such an end is termed a *free end*. The absence of a transverse force requires $T \sin \theta$ to vanish so that

$$\left(\frac{\partial y}{\partial x}\right)_{x=0} = 0 \tag{2.10}$$

The condition to be satisfied is thus

$$\left(\frac{\partial y_1}{\partial x}\right)_{x=0} + \left(\frac{\partial y_2}{\partial x}\right)_{x=0} = 0$$

Now,

$$\frac{\partial y_1}{\partial x} = \frac{dy_1}{d(ct - x)}(-1)$$

and

$$\frac{\partial y_2}{\partial x} = \frac{dy_2}{d(ct + x)}(+1)$$

so that at $x = 0$ we have

$$-\frac{dy_1}{d(ct)} + \frac{dy_2}{d(ct)} = 0$$

and the integration gives

$$y_1(ct) = y_2(ct)$$

The two functions are of the same form and sign, and the expression for the displacement at any point on the string is

$$y(x, t) = y_1(ct - x) + y_1(ct + x)$$

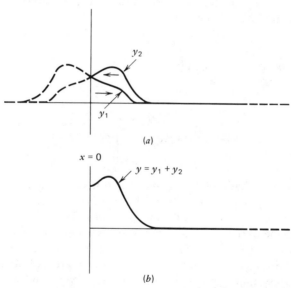

Fig. 2.5. *Reflection from a free end. In* (a), *the dashed-line segment of wave* y_2 *is shown as being reflected to become the solid-line segment of wave* y_1. *The resultant wave, shown in* (b), *has no slope at* $x = 0$.

The process of reflection at a free boundary may be considered as one in which the wave moving to the left does not pass the boundary but is instead reflected into a similarly shaped wave traveling to the right (Fig. 2.5).

2.8 FORCED VIBRATION OF AN INFINITE STRING.

The simplest type of vibration that can be set up in a string is that resulting from application of a *transverse sinusoidal driving force* to one end of an ideal string of infinite length. (Since all real strings are of finite length, this particular problem may seem to be of purely academic interest, but its analysis is justified because it not only furnishes a simple introduction to the study of vibrations of strings of finite length but also aids in the understanding of the transmission of acoustic waves.)

Consider an ideal string of infinite length extending to the right from $x = 0$, stretched to a tension T, with a transverse driving force $F \cos \omega t$ applied at the end $x = 0$. Assume that the end does not move in the x direction but is free to move in the y direction. As in the previous chapter, let us replace $F \cos \omega t$ with the complex force $\mathbf{f} = F \exp(j\omega t)$. Since the string extends infinitely far in the positive x direction and is excited into motion by the force at its left end, the solution must contain only waves moving to the right

$$\mathbf{y}(x, t) = \mathbf{y}_1(ct - x)$$

The boundary condition at $x = 0$ requires

$$\mathbf{y}(0, t) = \mathbf{A}e^{j\omega t}$$

where $\mathbf{A}$ is a complex constant (whose amplitude and phase eventually will be related to the driving force). Combination gives

$$y_1(ct) = \mathbf{A}e^{jk(ct)}$$

where the *wave number* k is defined by

$$k = \frac{\omega}{c} \tag{2.11a}$$

The solution for all x must be $\mathbf{A} \exp[jk(ct - x)]$ or

$$\mathbf{y}(x, t) = \mathbf{A}e^{j(\omega t - kx)} \tag{2.11b}$$

Figure 2.6*a* shows the shape of the string at two instants of time and Fig. 2.6*b* the time histories of two points on the string. The elements of the string execute simple harmonic motion about their equilibrium positions with frequency $f = \omega/(2\pi)$ and period $1/f$. The shape of the string at any instant is a sinusoid of amplitude $A = |\mathbf{A}|$. At fixed time, the shape is a function of x and, when x changes by an amount λ so that $k\lambda = 2\pi$, the displacement and slope of the string are as before. The distance λ between these corresponding points is called the *wavelength* and we see that

$$\lambda = \frac{2\pi}{k} \tag{2.12}$$

This waveform moves to the right with a phase speed $c = \sqrt{T/\rho_L}$, and is called a *harmonic traveling wave*. Because the waveform moves one wavelength in a time equal to one period, frequency and wavelength are related to the phase speed by

$$c = f\lambda \tag{2.13}$$

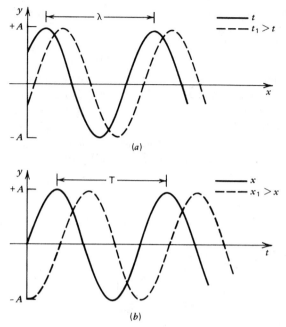

(a)

(b)

Fig. 2.6. *A harmonic wave traveling to the right.*
(a) Spatial behavior at two closely occurring times.
(b) Temporal behavior at two closely spaced positions.

an equation fundamental to all wave motion. Note that (2.13) can also be obtained from (2.11a) by expressing k as $2\pi/\lambda$ and ω as $2\pi f$.

To relate the amplitude of the wave to the driving force, consider the forces applied to the left end of the string, as shown in Fig. 2.7. Since there is no mass concentrated at the end of the string, the driver must provide the force necessary to exactly balance the tension: opposite to $T\cos\theta$ horizontally and opposing $T\sin\theta$ vertically, as suggested in the figure. Therefore, the total transverse force

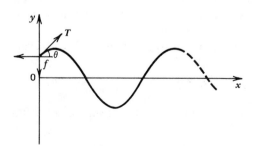

Fig. 2.7. *Forces acting on the end of a driven string.*

$(f + T \sin \theta)$ on the string at the left end must vanish, so that for small values of θ

$$\mathbf{f} = -T \left(\frac{\partial \mathbf{y}}{\partial x} \right)_{x=0} \tag{2.14}$$

where the minus sign denotes that the applied force must be directed downward when $(\partial \mathbf{y}/\partial x)_{x=0}$ is positive. Thus we see that the slope of the string at the forced end is determined by the applied force and the tension in the string, $(\partial \mathbf{y}/\partial x)_{x=0} = -\mathbf{f}/T$. (For example, if $\mathbf{f} = 0$ so that the end of the string is free to move transversely, then $(\partial \mathbf{y}/\partial x)_{x=0} = 0$. This is the boundary condition for a free end, as stated in (2.10).) Substitution of $\mathbf{f} = F \exp(j\omega t)$ and (2.11b) into (2.14) gives

$$F e^{j\omega t} = -T(-jk)A e^{j\omega t}$$

so that

$$\mathbf{y}(x, t) = \frac{F}{jkT} e^{j(\omega t - kx)}$$

and the particle speed $\mathbf{u} = \partial \mathbf{y}/\partial t$ becomes

$$\mathbf{u}(x, t) = \frac{F}{\rho_L c} e^{j(\omega t - kx)}$$

where $c = \sqrt{T/\rho_L}$

Now let us define the *input mechanical impedance* $\mathbf{Z}_{m0}$ of the string as the ratio of the driving force to the transverse speed of the string at $x = 0$.

$$\boxed{\mathbf{Z}_{m0} = \frac{\mathbf{f}}{\mathbf{u}(0, t)}} \tag{2.15a}$$

Then for the case of the infinite string

$$\mathbf{Z}_{m0} = \rho_L c \tag{2.15b}$$

The input impedance of an infinite string is a real quantity so that the mechanical load offered by the string is purely resistive. This is to be expected, since an infinite string can propagate energy only away from the driver. The input impedance of an infinite string is a function only of the tension of the string and its mass per unit length. Independent of the applied driving force, it is thus a characteristic property of the *string* and not of the *wave*. For this reason it is called the *characteristic mechanical impedance* of the string. It is analogous to the characteristic electrical impedance of an infinite transmission line.

The instantaneous power input to the string is $\Pi_i = fu$ evaluated at $x = 0$, or

$$\Pi_i = F \cos \omega t \left(\frac{F}{\rho_L c} \right) \cos \omega t$$

The time average over one cycle gives the average power input,

$$\Pi = \frac{F^2}{2\rho_L c} = \tfrac{1}{2}\rho_L c U_0^2 \tag{2.16}$$

where

$$U_0 = |\mathbf{u}(0, t)| = \frac{F}{\rho_L c}$$

is the speed amplitude of the string at $x = 0$.

2.9   FORCED VIBRATION OF A STRING OF FINITE LENGTH.

The behavior of a string of *finite* length forced at one end is considerably more complicated than that of the infinite string discussed in the previous section. The wave reflected from the support at the far end of the string interacts with the wave traveling toward the support and in turn is reflected from the driven end. However, when steady state is attained, the solution must be expressible in terms of two harmonic waves traveling in opposite directions

$$y(x, t) = \mathbf{A}e^{j(\omega t - kx)} + \mathbf{B}e^{j(\omega t + kx)} \tag{2.17}$$

where the complex amplitudes $\mathbf{A}$ and $\mathbf{B}$ are determined by the boundary conditions. Let us consider several classes of termination.

(a)   The Forced, Fixed String.   Assume that a string is driven at one end and rigidly fixed at the other. This is referred to as the forced, fixed string. At the left end, the boundary condition is (2.14)

$$Fe^{j\omega t} + T\left(\frac{\partial \mathbf{y}}{\partial x}\right)_{x=0} = 0$$

at all times. Substitution of (2.17) into this boundary condition gives

$$F + T(-jk\mathbf{A} + jk\mathbf{B}) = 0 \tag{2.18}$$

Since the string is rigidly supported at $x = L$ the displacement at this point is always zero, so that

$$\mathbf{A}e^{-jkL} + \mathbf{B}e^{jkL} = 0 \tag{2.19}$$

Solving (2.18) and (2.19) simultaneously for $\mathbf{A}$ and $\mathbf{B}$, we have

$$\mathbf{A} = \frac{F}{jkT}\frac{e^{jkL}}{e^{jkL} + e^{-jkL}} = \frac{Fe^{jkL}}{2jkT \cos kL}$$

and

$$\mathbf{B} = \frac{Fe^{-jkL}}{-2jkT \cos kL}$$

Substitution of these constants into (2.17) gives

$$y(x, t) = \frac{F}{2jkT \cos kL}\{e^{j[\omega t + k(L - x)]} - e^{j[\omega t - k(L - x)]}\} \tag{2.20}$$

or factoring the exp($j\omega t$) and simplifying,

$$\mathbf{y}(x, t) = \frac{F}{kT} \frac{\sin[k(L - x)]}{\cos kL} e^{j\omega t} \tag{2.21}$$

Thus we have two different but equivalent ways of looking at the solution: (2.20) can be interpreted as two waves of equal amplitude and wavelength traveling in opposite directions on the string. On the other hand, (2.21) describes a waveform that does not propagate along the string; instead, the string oscillates while the waveform remains stationary. Such a wave is called a *standing wave* and is mathematically characterized by an amplitude that depends on the position along the string. These two descriptions reveal that a combination of waves of equal amplitude traveling in opposite directions gives rise to a stationary vibration with a spatially dependent amplitude. This ability to view standing waves as combinations of traveling waves, and vice versa, will often be exploited in dealing with wave motion.

Consideration of the term $\sin[k(L - x)]$ in (2.21) shows that there are positions, called *nodes*, where the displacement is zero at all times. These positions are given by $k(L - x) = q\pi$ for $q = 0, 1, 2, \cdots \le kL/\pi$. The positions x_q of the nodes are then

$$x_q = L - \frac{q}{2}\lambda \qquad q = 0, 1, 2, \cdots \le 2L/\lambda$$

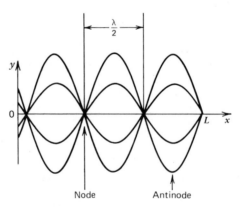

Fig. 2.8. *The shape of a string at several different times for a standing wave.*

A representative standing wave is shown in Fig. 2.8 where the instantaneous displacements of the string at various times have been sketched. The distance between nodes is $\lambda/2$. The moving portions of the string between the nodes are called *loops*, and the positions of maximum displacement are called *antinodes*.

Notice that the position of the driver with respect to the nodes varies as the frequency varies. If L is an integer times $\lambda/2$, a node will occur at the position of the driver. If the frequency is then increased, the wavelength decreases, causing the node to migrate away from the driver. An antinode will exist at the driver for driving frequencies such that L is an odd integer times $\lambda/4$.

The migration of the nodes with varying driving frequency is accompanied by some startling changes in the amplitude at the antinodes. The denominator of (2.21) becomes zero at driving frequencies such that $\cos kL = 0$,

$$kL = \frac{2n - 1}{2} \pi \qquad n = 1, 2, 3, \ldots$$

Since $\omega/k = c$, this gives

$$f_{rn} = \frac{2n - 1}{4} \frac{c}{L} \tag{2.22}$$

The string has its most violent vibration when the driving frequency has value f_{rn}. These are called *resonance frequencies*. The theoretically infinite amplitudes of vibration predicted by (2.21) at resonance do not occur in actual strings, because the assumption of small θ is violated. However, the amplitude will be a maximum at these frequencies. Notice that at resonance there is an antinode at the driven end, so that $u(0, t)$ is as large as possible.

Similarly, the frequencies for which the amplitude is a minimum are determined by the condition $\cos kL = \pm 1$,

$$kL = n\pi \qquad n = 1, 2, 3, \ldots$$

or

$$f_{an} = \frac{n}{2} \frac{c}{L} \tag{2.23}$$

(It will be observed that these minimum amplitudes decrease progressively with increasing frequency.) These frequencies are called *antiresonance frequencies*. At antiresonance, there is a node at the driven end, so that $u(0, t) = 0$. (In reality, the presence of dissipation causes $u(0, t)$ to be finite, but small; this will be studied in more detail later.)

The input mechanical impedance Z_{m0} is given by (2.15a),

$$Z_{m0} = \frac{Fe^{j\omega t}}{u(0, t)} \tag{2.24}$$

which for the case of the forced, fixed string yields

$$Z_{m0} = -j\rho_L c \cot kL \tag{2.25}$$

This impedance is a pure reactance so that no power is absorbed by the string. This is a consequence of the fact that with a rigid end there is no way for energy to leave the system.

A consideration of the variations in input impedance leads to the same conclusions as those arrived at from amplitude considerations: whenever $\cot kL = 0$, the input impedance is zero and the amplitude of vibration is consequently a maximum. *The resonance frequencies of any mechanical system are defined in general as those frequencies for which the input mechanical reactance goes to zero.* For the forced, fixed string this yields the resonance frequencies given by (2.22). At the antiresonance frequencies given by (2.23), $\mathbf{Z}_{m0}$ is infinite and the motion of the driven end of the string is infinitesimally small, although the remainder of the string is in motion. When $\mathbf{Z}_{m0}$ is not purely reactive, the specification of antiresonance becomes more complicated. This will be investigated in Chapter 3.

For very low frequencies the input impedance has the limiting value

$$\mathbf{Z}_{m0} \rightarrow -j\frac{\rho_L c}{kL} = -j\frac{T}{L}\frac{1}{\omega}$$

which is identical with the input impedance of a spring having the stiffness constant $s = T/L$.

Caution should be exercised in applying the concepts of resonance and antiresonance as developed above to a real driven string. In any physically realizable system, the driving force (usually originating from an electrical voltage) is transferred to the string through a transducer. This transducer has a mechanical impedance of its own which can significantly affect the behavior of the system. The full implications of this statement will be left until the discussion of the driven pipe in Chapter 9.

(b) The Forced, Mass-Loaded String.   If the string is terminated at $x = L$ not with a rigid support, but with one possessing inertance so that it behaves like a mass, as sketched in Fig. 2.9, then analysis of the motion becomes more complicated. As

Fig. 2.9. *The forced, mass-loaded string.*

before, the solution must still be of the form (2.17), and the boundary condition at $x = 0$ is still (2.14),

$$Fe^{j\omega t} + \rho_L c^2\left(\frac{\partial \mathbf{y}}{\partial x}\right)_{x=0} = 0 \tag{2.26}$$

where we have replaced T with $\rho_L c^2$.

The condition at $x = L$ is now different: the force applied to the mass must be $-T(\partial y/\partial x)_{x=L}$ since a negative slope at $x = L$ results in an upward force in the $+y$ direction. By Newton's second law this becomes

$$-\rho_L c^2\left(\frac{\partial \mathbf{y}}{\partial x}\right)_{x=L} = m\left(\frac{\partial^2 \mathbf{y}}{\partial t^2}\right)_{x=L} \tag{2.27}$$

Substitution of (2.17) into (2.26) yields

$$F = -\rho_L c^2(-jk\mathbf{A} + jk\mathbf{B})$$

as before, but substitution into (2.27) gives a new equation

$$-\rho_L c^2(-jk\mathbf{A}e^{-jkL} + jk\mathbf{B}e^{jkL}) = m(j\omega)^2(\mathbf{A}e^{-jkL} + \mathbf{B}e^{jkL})$$

Solution for $\mathbf{A}$ and $\mathbf{B}$ gives

$$\mathbf{A} = -\frac{F}{\omega\rho_L c}\frac{1 + (j\omega m/\rho_L c)}{(\omega m/\rho_L c)\cos kL + \sin kL}\frac{e^{jkL}}{2}$$

$$\mathbf{B} = -\frac{F}{\omega\rho_L c}\frac{1 - (j\omega m/\rho_L c)}{(\omega m/\rho_L c)\cos kL + \sin kL}\frac{e^{-jkL}}{2}$$

Notice that $\mathbf{A}$ and $\mathbf{B}$ are complex conjugates, so that the wave traveling to the left has the same amplitude as that traveling to the right. The complex speed of the string, $\mathbf{u} = \partial\mathbf{y}/\partial t = j\omega\mathbf{y}$, is

$$\mathbf{u} = -j\frac{F}{\rho_L c}\frac{\cos[k(L - x)] - (\omega m/\rho_L c)\sin[k(L - x)]}{(\omega m/\rho_L c)\cos kL + \sin kL}e^{j\omega t} \tag{2.28}$$

and the input mechanical impedance is

$$\mathbf{Z}_{m0} = j\rho_L c\frac{(\omega m/\rho_L c) + \tan kL}{1 - (\omega m/\rho_L c)\tan kL} \tag{2.29}$$

Again, $\mathbf{Z}_{m0}$ is purely reactive.

Resonance frequencies occur when the denominator of $\mathbf{u}$ vanishes, which is equivalent to

$$\tan kL = -(m/m_s)kL \tag{2.30}$$

where $m_s = \rho_L L$ is the mass of the string.

There is no explicit solution of this transcendental equation. For very small mass loading, however, $m \ll m_s$ so that $\tan kL \approx 0$ or $kL \approx n\pi$ which is the condition of resonance for a forced, free string. Such a result is obviously to be expected, since for very light loading the string is essentially free at the end $x = L$. Similarly, for heavy mass loading ($m \gg m_s$) the mass acts very much like a rigid support, and the resonance frequencies approach those of a forced, fixed string. The general case of intermediate mass loading can be solved readily with a hand calculator or by graphical means. If we plot the functions $\tan kL$ and $-(m/m_s)kL$ as functions of kL, the resonance frequencies will correspond to the values of kL for which the curves intersect.

For example, in the special case $m = m_s$, the values of kL satisfying (2.30) are

$$kL = 2.03, 4.91, 7.98, \cdots$$

as shown in Fig. 2.10. The lowest resonance frequency, given by $k_1 L = 2.03$, is $f_1 = (2.03/2\pi)(c/L)$. This is intermediate between the lowest resonance frequency $f_1 = \frac{1}{2}(c/L)$ of a forced, free string and the lowest resonance frequency $f_1 = \frac{1}{4}(c/L)$ of a forced, fixed string. The higher resonance frequencies are not integral multiples of the lowest resonance frequency. For example, the ratio of the frequency of the second resonance to that of the lowest resonance is $4.91/2.03 = 2.42$.

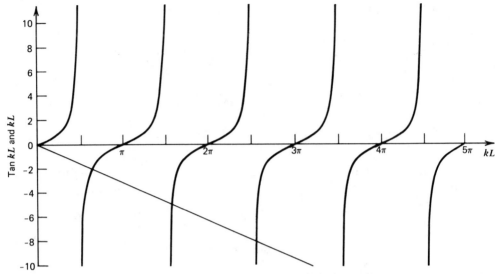

Fig. 2.10. *Graphical solution of the equation* tan $kL = -kL$. *The roots are* $kL = 2.03$, 4.91, 7.98, ...

The locations of the nodes on the string are altered by mass loading. The nodes fall where $\mathbf{u}(x,t) = 0$, found from the vanishing of the numerator of (2.28)

$$\tan[k(L - x_q)] = \frac{\rho_L c}{\omega m} \qquad q = 0, 1, 2, \ldots \leq 2L/\lambda$$

Since the righthand side of this equation gets larger as the frequency decreases, the node found at $x_0 = L$ for very high frequencies moves inward as the frequency is lowered, until at very low frequencies it is one-quarter wavelength from the end and there is an antinode at $x = L$. This means that the end at $x = L$ appears rigid at high frequencies and free at low frequencies.

(c) The Forced, Resistance-Loaded String. As a final example of the behavior of a forced string of finite length, assume that the end at $x = L$ is attached to a dashpot constrained to move transversely. The trial solution is (2.17) and the boundary condition at $x = 0$ is (2.14). Now, however, at $x = L$ we must have the force $R_m(\partial y/\partial t)_{x=L}$ balancing the force $-T(\partial y/\partial x)_{x=L}$, so

$$-\rho_L c^2 \left(\frac{\partial y}{\partial x}\right)_{x=L} = R_m \left(\frac{\partial y}{\partial t}\right)_{x=L} \qquad (2.31)$$

We could continue toward a solution as before, but a little subtlety will save us a lot of work. The solution must behave as $\exp(j\omega t)$, so that (2.31) can be rewritten as

$$-\rho_L c^2 \left(\frac{\partial y}{\partial x}\right)_{x=L} = \frac{R_m}{j\omega} \left(\frac{\partial^2 y}{\partial t^2}\right)_{x=L} \qquad (2.32)$$

Notice that if we replace m in (2.27) with $R_m/(j\omega)$ then (2.27) and (2.32) become the same.

Thus, we can use the formulas of the preceding example if we substitute R_m for $j\omega m$ everywhere. This yields new expressions for **A** and **B**,

$$\mathbf{A} = -\frac{F}{\omega\rho_L c} \frac{1 + (R_m/\rho_L c)}{(R_m/j\rho_L c)\cos kL + \sin kL} \frac{e^{jkL}}{2}$$

$$\mathbf{B} = -\frac{F}{\omega\rho_L c} \frac{1 - (R_m/\rho_L c)}{(R_m/j\rho_L c)\cos kL + \sin kL} \frac{e^{-jkL}}{2}$$

Thus **A** and **B** are no longer equal in amplitude. Indeed, we have $|\mathbf{B}|/|\mathbf{A}| = |\rho_L c - R_m|/(\rho_L c + R_m) \le 1$ so that the wave traveling to the left has smaller amplitude than that traveling to the right. This is physically pausible: since the dashpot dissipates energy, more must flow into it than out of it. This new result will have significant effects on the wave pattern of the string. The complex speed is found by substituting R_m for $j\omega m$ in (2.28)

$$\mathbf{u}(x,t) = \frac{F}{\rho_L c} \frac{\cos[k(L - x)] + j(R_m/\rho_L c)\sin[k(L - x)]}{(R_m/\rho_L c)\cos kL + j\sin kL} e^{j\omega t} \tag{2.33}$$

and the mechanical input impedance likewise is found from (2.29)

$$\mathbf{Z}_{m0} = \rho_L c \frac{(R_m/\rho_L c) + j\tan kL}{1 + (R_m/\rho_L c)j\tan kL} \tag{2.34}$$

Detailed analysis of this and similar forced, resonant systems will be undertaken in Chapter 3 where it will be seen that in general R_m or $j\omega m$ is replaced by $\mathbf{Z}_{mL}$, the mechanical impedance of the termination. Here we content ourselves with an observation concerning the wave pattern: the speed amplitude $U(x) = |\mathbf{u}(x,t)|$ is found from (2.33)

$$U(x) = \frac{F}{\rho_L c}\left\{\frac{\cos^2[k(L - x)] + (R_m/\rho_L c)^2 \sin^2[k(L - x)]}{(R_m/\rho_L c)^2 \cos^2 kL + \sin^2 kL}\right\}^{1/2} = \frac{F}{\rho_L c}\frac{\text{numerator}}{\text{denominator}}$$

The *numerator* varies between 1 and $R_m/(\rho_L c)$ as x decreases from L to 0, and the *denominator* has fixed finite value that depends on the driving frequency $\omega = kc$. Thus, $U(x)$ has relative maxima and minima, but *no* exact nulls.

2.10 NORMAL MODES OF THE FIXED, FIXED STRING.

Let us now turn our attention to a different class of solutions to the wave equation for finite strings. Rather than forcing the string into motion by driving one end, let the string be fixed at both ends and excited into motion by some initial displacement (or impact) along its length, much like a guitar when plucked or piano string when struck.

Since the string is fixed at both ends, the boundary conditions are $\mathbf{y} = 0$ at $x = 0$ and $x = L$. A trial solution which satisfies the wave equation is

$$\mathbf{y}(x, t) = \mathbf{A}e^{j(\omega t - kx)} + \mathbf{B}e^{j(\omega t + kx)}$$

and application of the boundary conditions gives

$$\mathbf{A} + \mathbf{B} = 0$$

and

$$\mathbf{A}e^{-jkL} + \mathbf{B}e^{jkL} = 0$$

The first of these requires $\mathbf{B} = -\mathbf{A}$ and this, substituted into the second, gives

$$\mathbf{A}e^{jkL} - \mathbf{A}e^{-jkL} = \mathbf{A}2j \sin kL = 0$$

This second boundary condition can be satisfied two ways. (1) Let $\mathbf{A} = 0$, but this gives $\mathbf{y} = 0$, the trivial solution of no motion. (2) Let $\sin kL = 0$. This choice requires that

$$kL = n\pi \qquad n = 1, 2, 3, \ldots$$

(The value $n = 0$ is not allowed, since for the string fixed at both ends this also corresponds to no motion.) This equation shows that only the discrete values $k = k_n = n\pi/L$ lead to nontrivial solutions. Furthermore, since $\omega/k = c$, only certain frequencies are allowed,

$$f_n = \frac{\omega_n}{2\pi} = \frac{n}{2}\frac{c}{L} \tag{2.35}$$

Thus, there is a family of solutions, each of the form

$$\mathbf{y}_n(x, t) = \mathbf{A}_n \sin k_n x \, e^{j\omega_n t} \tag{2.36a}$$

where $\mathbf{A}_n$ is the complex amplitude of the nth solution.

If we replace $\mathbf{A}_n$ with $A_n + jB_n$, then the real transverse displacement of the nth solution is

$$y_n(x, t) = (A_n \cos \omega_n t + B_n \sin \omega_n t)\sin k_n x \tag{2.36b}$$

The constants A_n and B_n must be determined from the initial conditions.

Application of the boundary conditions has limited the viable solutions of the wave equation to a series of discrete functions (2.36). These functions are called *eigenfunctions* or *normal modes*. Associated with each of these solutions is a unique frequency known as the *eigenfrequency*, *natural frequency*, or *normal mode frequency*. For the fixed, fixed string of our example, the eigenfrequencies are given by (2.35), and use of $\lambda_n f_n = c$ shows that $L = n\lambda_n/2$: an integral number of half-wavelengths encompasses the length of the fixed, fixed string.

The normal mode with the lowest eigenfrequency has $n = 1$ and is called the *fundamental mode*. Its eigenfrequency $f_1 = \frac{1}{2}(c/L)$ is called the *fundamental* or *first harmonic*. The eigenfrequencies with $n = 2, 3, \ldots,$ are called *overtones*. In the case of the fixed, fixed strings, $f_n = nf_1$, and the overtones are *harmonics*, integral multiples of the fundamental. Thus, the second harmonic is the *first* overtone, and so forth. (Another less confusing terminology, which should become more widely accepted, names the overtones as *partials*; by convention, the fundamental is the *first partial*, the second harmonic or first overtone is the *second partial*, and so forth.) As we will see in the next section, for more realistic boundary conditions the overtones of a freely vibrating string are not necessarily integral multiples of the fundamental. (Equivalently, the second and higher partials are not harmonic with the first.)

The complete solution for a rigidly supported, freely vibrating string is the sum of all the individual modes of vibration represented by (2.36)

$$y(x, t) = \sum_{n=1}^{\infty} (A_n \cos \omega_n t + B_n \sin \omega_n t)\sin k_n x \tag{2.37}$$

Assume that at time $t = 0$ the string is distorted from its normal linear configuration, the displacement at each point being

$$y(x, 0)$$

and the corresponding speed being

$$u(x, 0) = \left(\frac{\partial y}{\partial t}\right)_{t=0}$$

Now, if (2.37) is to represent the position of the string at all times, it must represent it at $t = 0$ so that

$$y(x, 0) = \sum_{n=1}^{\infty} A_n \sin k_n x \tag{2.38}$$

Its derivative with respect to time must also represent the speed at $t = 0$ and hence

$$u(x, 0) = \sum_{n=1}^{\infty} \omega_n B_n \sin k_n x \tag{2.39}$$

Applying the Fourier theorem to (2.38) gives

$$A_n = \frac{2}{L} \int_0^L y(x, 0) \sin k_n x \, dx \tag{2.40}$$

and, similarly, from (2.39) we have

$$B_n = \frac{2}{\omega_n L} \int_0^L u(x, 0) \sin k_n x \, dx \tag{2.41}$$

As an illustration, assume that the string is initially pulled aside a distance h at its center and is then released. Here $u(x, 0)$ is zero, and all coefficients B_n will be zero. The initial condition is written

$$y(x, 0) = 2 \frac{h}{L} x \qquad\qquad 0 \le x \le L/2$$

$$= 2 \frac{h}{L} (L - x) \qquad L/2 \le x \le L$$

and the coefficients A_n are given by

$$A_n = \frac{2}{L} \left[\int_0^{L/2} 2 \frac{h}{L} x \sin k_n x \, dx + \int_{L/2}^L 2 \frac{h}{L} (L - x) \sin k_n x \, dx \right]$$

Evaluation of the integrals gives

$$A_n = \frac{1}{n^2} \frac{8h}{\pi^2} \sin \frac{n\pi}{2} \tag{2.42}$$

Therefore $A_2 = A_4 = A_6 = \cdots = 0$, and $A_1 = 8h/\pi^2$, $A_3/A_1 = -1/9$, $A_5/A_1 = 1/25$, etc. The numerical values of the coefficients A_n determine the amplitude of the various harmonic modes of vibration of the string. Here, for example, all vibrations correspond to the even harmonics are absent. The absent harmonics correspond to standing waves, each having a node at the center where the string was initially pulled aside. In general, no harmonic is produced having a node at the point where the string is plucked.

If the string is struck a blow such that $u(x, 0)$ varies linearly along the string, but there is no initial displacement, then all the coefficients A_n are zero, and the coefficients B_n are given by (2.42). As with the plucked string, those harmonics will be absent whose modes of vibration have a node at the point initially struck.

2.11   EFFECTS OF MORE REALISTIC BOUNDARY CONDITIONS ON THE FREELY VIBRATING STRING.

Any yielding of the supports modifies the motion of the string, for the boundary conditions are no longer $y = 0$ at its ends, but rather that at these points the wave impedance of the string must be equal to the transverse mechanical impedance of the support.

Assume that the left end of the string is attached to a support whose mechanical impedance is Z_{m0}. For example, let the string be attached at $x = 0$ to an undamped harmonic oscillator. The mechanical impedance of the oscillator presented to the string at $x = 0$ is

$$Z_{m0} = j(\omega m - s/\omega)$$

This assumption, that the support can be replaced by the elements of a simple harmonic oscillator constrained to move transversely to the string, is representative of many real interactions wherein the support exhibits inertia and resilience.

The transverse force $\mathbf{f}_0$ exerted *by the string on the mass* is given by

$$\mathbf{f}_0 = T\left(\frac{\partial \mathbf{y}}{\partial x}\right)_{x=0}$$

Since the motion $\mathbf{u}_0$ of the mass is constrained to match that of the end the string,

$$\mathbf{u}_0 = \left(\frac{\partial \mathbf{y}}{\partial t}\right)_{x=0}$$

Newton's second law now yields

$$\mathbf{f}_0 - s\mathbf{y}(0, t) = m\left(\frac{\partial^2 \mathbf{y}}{\partial t^2}\right)_{x=0}$$

or

$$\mathbf{f}_0 = Z_{m0}\mathbf{u}_0$$

Thus, the boundary condition at $x = 0$ is $\mathbf{u}_0 = \mathbf{f}_0/Z_{m0}$ or

$$\mathbf{u}_0 = \frac{1}{Z_{m0}} T\left(\frac{\partial \mathbf{y}}{\partial x}\right)_{x=0}$$

Similarly, it may be shown that the condition at $x = L$ is

$$\mathbf{u}_L = -\frac{1}{Z_{mL}} T\left(\frac{\partial \mathbf{y}}{\partial x}\right)_{x=L}$$

where Z_{mL} is the mechanical impedance of the support at $x = L$.

If the mechanical impedance at the support is infinite, then the above requires $\mathbf{u} = 0$ and therefore $\mathbf{y} = 0$, the condition for a rigidly fixed end. If the support offers no restraint to the transverse motion of the string, its mechanical impedance is zero and the boundary condition must be $\partial \mathbf{y}/\partial x = 0$ at the support, the condition for a free end.

(a) The Fixed, Mass-Loaded String.   As an example, assume that the string is fixed at $x = 0$ and that the support at $x = L$ can be characterized by a mass m_L. The boundary conditions are

$$\mathbf{u}(0, t) = 0$$

and

$$\mathbf{u}(L, t) = -\frac{T}{j\omega m_L}\left(\frac{\partial \mathbf{y}}{\partial x}\right)_{x=L}$$

Application of the first boundary condition to the general harmonic solution (2.17) yields $\mathbf{A} = -\mathbf{B}$ so that the standing wave is

$$\mathbf{y}(x, t) = -2j\mathbf{A}\,\sin kx\,e^{j\omega t}$$

Substitution of this into the second boundary condition then gives

$$j\omega\,\sin kL = -\frac{Tk}{j\omega m_L}\cos kL$$

This can be rearranged,

$$\cot kL = \frac{m_L}{m_s}\,kL \tag{2.43}$$

where $m_s = \rho_L L$ is the mass of the string. Figure 2.11 illustrates the graphical solution of (2.43) for a few values of m_L/m_s. If m_L/m_s is large, the solutions approach $kL = n\pi$ and the string behaves as if it were fixed at both ends. As m_L/m_s is reduced, the allowed values of kL increase thereby *raising* the normal mode frequencies. Furthermore, these frequencies are no longer related by integers: *the overtones are not harmonics of the fundamental*. Since the frequencies are raised and there must be a node at $x = 0$, the last node at the other end of the string lies within $x = L$. As the mass gets smaller, this last node moves toward $L - \lambda/4$, and in the limit $m_L/m_s = 0$ there is an antinode at $x = L$.

(b) The Fixed, Resistance-Loaded String.   As a second (quite different) example, consider the effects on the standing wave of a support having finite resistance and no reactance. Assume that the string is fixed at $x = 0$ and attached at $x = L$ to a dashpot constrained to move transversely. The boundary conditions are

$$\mathbf{u}(0, t) = 0 \quad \text{and} \quad \mathbf{u}(L, t) = -\frac{T}{R_m}\left(\frac{\partial \mathbf{y}}{\partial x}\right)_{x=L}$$

Because of the damping provided by the dashpot, the standing wave will decay with time, and we introduce the complex angular frequency $\boldsymbol{\omega} = \omega + j\beta$ where ω is the angular frequency and β is the temporal absorption coefficient. Since there are no losses on the string except at the boundary, our solution must still satisfy the lossless wave equation (2.5) and, since $\partial^2 \mathbf{y}/\partial t^2 = -\boldsymbol{\omega}^2 \mathbf{y}$, we are led to the result $\partial^2 \mathbf{y}/\partial x^2 = -(\boldsymbol{\omega}/c)^2 \mathbf{y}$. This suggests solutions of the form

$$\mathbf{y}(x, t) = e^{j(\boldsymbol{\omega}t \pm \mathbf{k}x)}$$

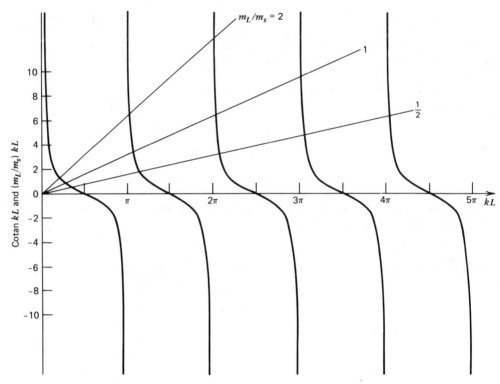

Fig. 2.11. *Graphical solution of the equation* $\cot kL = (m_L/m_s)kL$ *for* $m_L/m_s = \frac{1}{2}$, 1, *and* 2.

where the spatial factor has a *complex wave number* $\mathbf{k} = k + j\alpha$ given by

$$\boldsymbol{\omega}^2 = c^2\mathbf{k}^2$$

The quantities k and α must now be related to ω and β. Substitution of $\boldsymbol{\omega} = \omega + j\beta$ and $\mathbf{k} = k + j\alpha$ into this equation and collection of real and imginary parts gives

$$[\omega^2 - \beta^2 - c^2(k^2 - \alpha^2)] + j(2\omega\beta - 2c^2 k\alpha) = 0$$

The only way this equation can be satisfied is for both real and imaginary parts to vanish. This yields the pair of relations

$$\omega^2 - \beta^2 = c^2(k^2 - \alpha^2)$$

$$\omega\beta = c^2 k\alpha$$

For the cases usually encountered, the motion takes many cycles to decay so that $\beta \ll \omega$. For this restriction, the above relations have the approximate solutions $\omega/c \doteq k$ and $\beta/c \doteq \alpha$. The results allow use of the *approximation*

$$\boldsymbol{\omega} \doteq c\mathbf{k}$$

which simplifies the calculations.

Substitution of the trial solution

$$y(x, t) = Ae^{j(\omega t - kx)} + Be^{j(\omega t + kx)}$$

into the boundary conditions yields

$$y(x, t) = - 2jA \sin kx\, e^{j\omega t} \tag{2.44}$$

and

$$\sin kL = j\, \frac{\rho_L c}{R_m}\, \cos kL \tag{2.45}$$

This last equation must be satisfied by both its imaginary and real parts,

$$\cos kL \sinh \alpha L = (\rho_L c/R_m)\cos kL \cosh \alpha L$$

$$\sin kL \cosh \alpha L = (\rho_L c/R_m)\sin kL \sinh \alpha L$$

(Readers unfamiliar with trigonometric functions of complex angles should consult Appendix A3.) Solving simultanously, we get two possible solutions.

$$(1) \quad \sin kL = 0 \quad \text{and} \quad \tanh \alpha L = \rho_L c/R_m$$

or

$$(2) \quad \cos kL = 0 \quad \text{and} \quad \tanh \alpha L = R_m/(\rho_L c)$$

For weak damping, $R_m \ll \rho_L c$. This rules out the first possible solution because $\tanh x$ must always be less than unity. Then, since $\tanh x \approx x$ for $x \ll 1$, the second possibility yields (for weak damping)

$$\alpha L \approx \frac{R_m}{\rho_L c} \tag{2.46}$$

and $kL = (n - \frac{1}{2})\pi$ so that, since the end at $x = 0$ is fixed, the end at $x = L$ is an antinode.

In this limit of small damping we can use the approximations $\sinh \alpha x \approx \alpha x$ and $\cosh \alpha x \approx 1$ in (2.44) with the result that the particle displacement has magnitude

$$|y(x, t)| \approx 2Ae^{-\alpha ct}\sqrt{\sin^2 kx + (\alpha x)^2 \cos^2 kx} \tag{2.47}$$

The wave pattern resembles that of the fixed, fixed string, but the motion decays exponentially with a temporal absorption coefficient $\beta \approx \alpha c$ and the nodes (now defined as occurring where $\sin kx_q = 0$) are not exactly zero but have amplitudes $2\alpha x_q A \exp(-\alpha ct)$. Analogously, the antinodes have amplitude $2A \exp(-\alpha ct)$.

Up to this point we have neglected the effects of the surrounding medium on the motion of the string. One such effect is to provide a resistive force which opposes the motion. As with a simple oscillator, the effect of this frictional force is to damp out the free vibrations and reduce their frequency slightly. Part of the energy being dissipated by the string heats the surrounding medium and part goes into the radiation of sound energy.

Another effect of the medium is to add an effective additional mass per unit length to the string. This may not be negligible in a liquid medium, or at low frequencies in a gaseous medium. An analytical treatment of the effect of the medium on the vibrating string is not warranted; however, in Chapter 8 we consider the much more important case of its effect on the vibrations of a loudspeaker diaphragm.

It is just this similarity between the vibrations of a string and those of more complicated mechanical or acoustic systems that warrants such a thorough treatment of the vibrating string as given in this chapter. A knowledge of the various modes of vibration and other characteristics of a simple string will facilitate the future study of more complicated systems, where the mathematical difficulties might otherwise obscure the physical ideas. All the methods of attack employed in this chapter will be used over and over again in the complex situations to be encountered.

2.12   ENERGY OF VIBRATION OF A STRING.   A vibrating string contains kinetic energy because of the speeds with which its various portions are moving, and potential energy because the string must be stretched as it is deformed from its rest position. (Refer to Fig. 2.2.) The element of string between x and $x + dx$ has a mass $\rho_L\, dx$, and if it moves with speed $\partial y/\partial t$ its kinetic energy dE_k is

$$dE_k = \tfrac{1}{2}\rho_L \left(\frac{\partial y}{\partial t}\right)^2 dx$$

This element is also stretched to a length ds because of the deformation. Since the left end of the element has transverse displacement $y(x, t)$ and the right end is displaced $y(x, t) + (\partial y/\partial x)\, dx$, the increase in length is

$$ds - dx = \sqrt{dx^2 + \left(\frac{\partial y}{\partial x}\right)^2 dx^2} - dx = \left[\sqrt{1 + \left(\frac{\partial y}{\partial x}\right)^2} - 1\right] dx$$

For the small displacements assumed, this can be simplified by the approximation $\sqrt{1 + \varepsilon} \approx 1 + \tfrac{1}{2}\varepsilon$ to be

$$\frac{1}{2}\left(\frac{\partial y}{\partial x}\right)^2 dx$$

The product of the tension $T = \rho_L c^2$ of the string and the extension of the element gives the potential energy of deformation,

$$dE_p = \tfrac{1}{2}\rho_L c^2(\partial y/\partial x)^2\, dx$$

The energy per unit length dE/dx is the sum $dE_k/dx + dE_p/dx$,

$$\frac{dE}{dx} = \tfrac{1}{2}\rho_L c^2\left[\left(\frac{\partial y}{\partial x}\right)^2 + \left(\frac{1}{c}\frac{\partial y}{\partial t}\right)^2\right] \tag{2.48}$$

and the total energy of the string is the integration of the energy per unit length over the entire length of the string

$$E = \tfrac{1}{2}\rho_L c^2 \int_{\text{string}} \left[\left(\frac{\partial y}{\partial x}\right)^2 + \left(\frac{1}{c}\frac{\partial y}{\partial t}\right)^2\right] dx \tag{2.49}$$

As an example, assume that a string of length L is fixed at both ends and vibrating in its nth mode. Then, from (2.36) we have

$$\frac{\partial y}{\partial x} = k_n(A_n \cos \omega_n t + B_n \sin \omega_n t)\cos k_n x$$

and

$$\frac{1}{c}\frac{\partial y}{\partial t} = \frac{\omega_n}{c}(-A_n \sin \omega_n t + B_n \cos \omega_n t)\sin k_n x$$

so that the total mechanical energy of the entire string moving in this one mode of vibration is

$$E_n = \tfrac{1}{4}\rho_L L\omega_n^2(A_n^2 + B_n^2) = \tfrac{1}{4}m_s \omega_n^2(A_n^2 + B_n^2)$$

The quantity m_s is the total mass $\rho_L L$ of the string, and $(A_n^2 + B_n^2)^{\frac{1}{2}}$ is the maximum displacement amplitude of the nth harmonic. If we notice that

$$U_n = \omega_n\sqrt{A_n^2 + B_n^2}$$

is the maximum amplitude of the speed $\partial y/\partial t$ of the string, then the energy is seen to be

$$E_n = \tfrac{1}{4}m_s U_n^2$$

For this example, it can be seen that if the string is vibrating in two or more normal modes of vibration then the total energy of the string is the sum of the energies associated with each normal mode,

$$E = \sum E_n$$

If we apply this to the fixed, fixed string that is plucked in the middle, the displacement amplitudes are given by (2.42)

$$E_n = \tfrac{1}{4}m_s \omega_n^2\left(\frac{1}{n^2}\frac{8h}{\pi^2}\right)^2 \sin^2 \frac{n\pi}{2} \qquad n = 1, 3, 5, \ldots$$

Clearly, $E_n = 0$ for even n. (Recall that for the string plucked in the middle, no normal modes with nodes at $x = L/2$ can be excited.) Now, $\omega_n = k_n c = (n\pi/L)c$ so that

$$E = m_s c^2\left(\frac{4h}{\pi L}\right)^2 \sum_{\text{odd}} \frac{1}{n^2}$$

where the summation extends *only over the odd integers*. The energy of the fundamental mode of vibration is 9 times that of the third harmonic, 25 times that of the fifth harmonic, and so on. If all this energy were radiated in the form of acoustic waves, the sound intensities corresponding to each of the harmonics would be similarly related.

It is quite evident that variations in the position at which the string is plucked will alter the harmonic content and therefore the quality of the sound. It may be shown that the series representing the summation of all the energy in the various modes of vibration is equal to the work performed in initially deforming the string, as expected from conservation of energy.

2.13 OVERTONES AND HARMONICS.

As noted previously, the lowest natural frequency of a vibrating system is termed the *fundamental*, and the higher natural frequencies are termed *overtones*. We have also seen that, if the supports of a string are perfectly rigid, the overtones are harmonics; but, if otherwise, the overtones are in general not harmonic.

Nonharmonicity of the overtones is often encountered in musical instruments. For example, the vibration of a violin string is coupled to the sounding board by a bridge for more efficient radiation of energy. The bridge then acts as a reactive termination since it must flex somewhat to communicate motion to the sounding board. As a result, the natural frequencies resulting from the free vibration of the plucked string will be slightly nonharmonic.

Another effect that is sometimes important is that the free vibrations of many real strings differ from those of an idealized string: if the string has innate bending stiffness, as is true for a piano string, then the observed overtones will be higher than predicted on the basis of an ideal string. Since the effect of stiffness increases with increasing frequency, the higher overtones of a real piano string become increasingly sharp with respect to the fundamental. It is just this effect, plus the fact that the piano string is not exactly fixed at both ends, that gives the piano some of its distinctive tonal quality. Pianos are commonly designed so that the point of impact of the hammer is about one-seventh of the way from one end of the string. This would suggest that the seventh partial should be nearly absent. In reality, the situation is not so simple. It has been shown, for example, that the finite width of the hammer allows this partial to be excited. (On the other hand, when the more flexible guitar or violin string is plucked, partials are suppressed, as expected from the placement of the finger.) If the displacement of the piano string is studied after it is struck, it will be seen that the motion tends to decay (because of the loss of energy), the different overtones decaying at different rates. Equally significant, since the overtones are nonintegral multiples of the fundamental, the waveform is not stationary but shows considerable change in shape as the relative phases of the fundamental and overtones change with time. This is true for most percussive instruments, such as tympani, cymbals, *plucked* violin, piano, xylophone, woodblocks, and so forth. The sound they generate is a superposition of slowly decaying fundamental and more rapidly decaying nonharmonic overtones.

Instruments that are blown or bowed, on the other hand, like the oboe, *bowed* violin, organ, trumpet, and so forth, are *forced* vibrating systems. In this case, the forcing function is usually made up of harmonics. (The violin string is pulled to one side by the rosined hair of the bow until it snaps back and is reacquired by the bow; the motion is something like a sawtooth wave with a well-defined period and identical cycle-to-cycle behavior.) The motion of air within the clarinet and the associated periodic clapping of its reed are identically repeated cycle by cycle. Thus, the steady-state notes produced by these *driven* instruments consist of frequencies that are integral multiples of the lowest driving frequency. The relative amounts of these higher overtones control the tone color or *timbre* of the instrument. It must be remembered, however, that these forced vibrations are initiated at some definite time so that there is an initial time during which the transient vibrations are also strong enough to be heard. These affect the sound in much the same way as the transient solution affects the initial behavior of the forced, damped oscillator.

This transient is often of considerable importance in identifying a particular instrument: the ear seems to exhibit memory in that these initial effects aid in keeping track of each instrument even while the others are also sounding. Even though some individual instrument contributes negligible power to the output of a full orchestra, the mind is able to make use of these transient "fingerprints" as one aid in identifying a particular instrument in the general uproar.

PROBLEMS

2.1. What forms do the equations of motion for an idealized string take if: (a) the linear density varies with position, and (b) the string hangs vertically supported only at the upper end?

2.2. By direct substitution show that each of the following are solutions of the wave equation: (a) $f_1(x - ct)$, (b) $\ln[a(ct - x)]$, (c) $a(ct - x)^2$, and (d) $\cos[a(ct - x)]$. Similarly, show that each of the following are *not* solutions of the wave equation: (e) $a(ct - x^2)$ and (f) $at(ct - x)$.

2.3. Sketch $y = A \exp(-a|ct - x|)$ for $t = 0$, $t = 1$, and $t = 2$ s. Let $c = 5$ cm/s, $a = 3$ cm^{-1}, and $A = 1$ cm. What is the significance of the displacement of these curves?

2.4. Consider the waveform $y = 4 \cos(3t - 2x)$ propagating on a string of density 0.1 g/cm, where y and x are in centimeters and t is in seconds. (a) What is the amplitude, phase speed, frequency, wavelength, and wave number? (b) What is the particle speed of the element at $x = 0$ at $t = 0$?

2.5. Evaluate the mechanical impedance seen by the applied force driving a semi-infinite string at a distance L from a rigid end. Interpret the individual terms in the mechanical impedance.

2.6. Evaluate the mechanical impedance seen by the applied force driving a simple harmonic oscillator with an infinite string extending to the right from the mass.

2.7. A string is stretched between rigid supports a distance L apart. It is driven by a force $F \cos \omega t$ located at its midpoint. (a) What is the mechanical impedance at the midpoint? (b) Show that the amplitude of the midpoint is $(F/2kT) \tan(kL/2)$. (c) What is the amplitude of the displacement of the point $x = L/4$?

2.8. A string of density 0.01 kg/m is stretched with a tension of 5 N from a rigid support at one end to a device producing transverse periodic vibrations at the other end. The length of the string is 0.44 m, and it is observed that, when the driving frequency has a given value, the nodes are spaced 0.1 m apart and the maximum amplitude is 0.02 m. What are the (a) frequency and (b) amplitude of the driving force?

2.9. (a) Assume that a forced, fixed string is driven by a source that has constant speed amplitude $u(0, t) = U_0 \exp(j\omega t)$ where U_0 is independent of frequency. Find the frequencies of maximum amplitude of the standing wave. (b) Repeat for a source that has a constant displacement amplitude $y(0, t) = Y_0 \exp(j\omega t)$. (c) Contrast the results of (a) and (b) with the frequencies of mechanical resonance for the forced, fixed string. (d) Does mechanical resonance always coincide with maximum amplitude of the motion?

2.10. Given a string, fixed at both ends, with ρ_L, c, L, f, and T specified so that c and f are known numbers, obtain the phase speed c' in terms of c and the fundamental resonance f' in terms of f if another string of the same material is used but: (a) the length is doubled, (b) the density/length is quadrupled, (c) the cross-sectional area is doubled, (d) the tension is reduced to half, and (e) the diameter of the string is doubled.

2.11. A stretched string of length L is plucked at the position $L/3$ by producing an initial displacement h and then releasing the string. Determine the resulting amplitudes for the fundamental and the first three harmonic overtones. Sketch the wave shapes of these individual waves and the shape of the string resulting from the linear combination of these waves at $t = 0$. Repeat for $t = L/c$, where c is the transverse wave velocity of the string.

2.12. A mass of 0.2 kg is hung from a string of 0.05-kg mass and 1.0-m length. (*a*) What is the speed of transverse waves in the string? (Neglect weight of string in computing tension in string.) (*b*) What are the frequencies of the fundamental and first overtone modes of transverse vibration of the string? (*c*) When the string is vibrating at its first overtone, what is the relative amplitude of its displacement at the antinode to that of the mass?

2.13. Find the normal mode frequencies for a fixed, spring-loaded string when $T = sL$. Sketch the waveforms for the fundamental and the first overtone.

2.14. A string of linear density 0.01 kg/m and of 0.2-m length is stretched between rigid supports to a tension of 10 N. It is loaded at its center with a mass of 0.001 kg. (*a*) What is the fundamental frequency of the system? (*b*) What is the first overtone frequency of the system?

2.15. Show that the work done in displacing the center of a stretched string by an amount h equals the sum of the energies present in the various modes of vibration when the string is released.

2.16. A standing wave on a fixed, fixed string of length $L = 31.4$ cm and linear density 0.1 g/cm is given by $y = 2 \sin(x/5)\cos 3t$, where y and x are in centimeters and t is in seconds. (*a*) Find the phase speed, frequency, and wave number. (*b*) What is the amplitude of the particle displacement and speed at $x = L/2$ and $x = L/4$? (*c*) Find the energy density at these points. (*d*) How much energy is in the entire length of the string?

CHAPTER 3

VIBRATIONS OF BARS

3.1 LONGITUDINAL VIBRATIONS OF A BAR. Another important wave motion is the propagation of *longitudinal waves*, often encountered in solid bars (and, at low frequencies, in gas-filled tubes and ducts with rigid walls). As a longitudinal disturbance moves along a bar, the displacement of particles of the bar is essentially parallel to its axis. When the lateral dimensions of the bar are small compared with its length, each cross-sectional plane of the bar may be considered to move as a unit. (Actually the bar shrinks somewhat in a lateral direction as it expands longitudinally, but for thin bars this lateral motion may be neglected.)

A number of acoustic devices utilize longitudinal vibrations in bars. Frequency standards used for producing sounds of definite pitches may be constructed from circular rods of varying lengths. When longitudinal vibrations are excited in such rods, the frequency of vibration is observed to be inversely proportional to the length of the rod. Longitudinal vibrations in nickel tubes are often used to drive the vibrating diaphragm of a sonar transducer. Piezoelectric crystals may be cut so that the frequency of longitudinal vibration in the direction of the longest axis of the crystal is used either to control the frequency of an oscillating electric current or to drive an electroacoustic transducer.

A further reason for studying longitudinal vibrations of bars is that it aids in the understanding of acoustic waves. Not only are the mathematical expressions for the transmission of acoustic plane waves through fluid media very similar to those for the transmission of compressional waves along a bar, but if the fluid is confined to a rigid pipe, there is also a close correlation between the boundary conditions in the two cases.

3.2 LONGITUDINAL STRAIN. Consider a bar, of length L and uniform cross-sectional area S, subjected to longitudinal forces. The application of these forces will produce a longitudinal displacement ξ of each of the particles in the bar, and for long thin bars this displacement will be the same at all points in any particular cross section

$$\xi = \xi(x, t) \tag{3.1}$$

Let the coordinates of the left and right ends of the bar be $x = 0$ and $x = L$, and consider a short segment dx of the unstrained bar lying between x and $x + dx$. Assume that the application of forces to the bar causes the plane originally located at x to move a distance ξ to the right, and that similarly a plane originally located

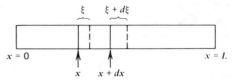

Fig. 3.1. *Longitudinal strain dξ/dx in a bar.*

at $x + dx$ to move a distance $\xi + d\xi$ to the right (Fig. 3.1). The convention adopted in this book is that a positive value of ξ signifies a displacement to the right, and a negative value a displacement to the left.

Since dx is assumed to be small, the displacement at $x + dx$ can be represented by the first two terms of a Taylor's series expansion of ξ about x,

$$\xi + d\xi = \xi + \left(\frac{\partial \xi}{\partial x}\right) dx$$

Since the left end of the segment has been displaced a distance ξ and the right end a distance $\xi + d\xi$, the increase in length of the segment is given by

$$(\xi + d\xi) - \xi = d\xi = \left(\frac{\partial \xi}{\partial x}\right) dx$$

The *strain* in the segment is defined as the ratio of its increase in length to its original length, or

$$\text{Strain} = \frac{\left(\frac{\partial \xi}{\partial x}\right) dx}{dx} = \frac{\partial \xi}{\partial x} \tag{3.2}$$

Note that since ξ is a function of both x and t we must use partial derivatives rather than total derivatives.

3.3 LONGITUDINAL WAVE EQUATION. Whenever a bar is strained, elastic forces are produced. These forces act across each cross-sectional plane in the bar and hold the bar together. Let $f = f(x, t)$ represent these longitudinal forces, where the convention is adopted of choosing a *positive* value of f to represent forces of *compression*, as indicated in Fig. 3.2, and a *negative* value to represent forces of *tension*. (This choice of sign makes the compression of a solid by a positive force analogous to the compression of a fluid by a positive increment in pressure.)

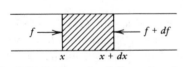

Fig. 3.2. *Compressional forces in a bar.*

The *stress* in the bar is defined as

$$\text{Stress} = \frac{f}{S} \tag{3.3}$$

For most materials, if the strain is small the stress is proportional to it. This relationship is known as Hooke's law,

$$\frac{f}{S} = -Y\frac{\partial \xi}{\partial x} \tag{3.4}$$

where Y, the *Young's modulus* or *modulus of elasticity*, is a characteristic property of the material. Since a positive stress results in a negative strain, the minus sign in (3.4) ensures a positive value for the constant Y. Values of Y for a number of common solids are given in Appendix A10. Rewriting (3.4), we obtain

$$\boxed{f = -SY\frac{\partial \xi}{\partial x}} \tag{3.5}$$

as an expression for the internal longitudinal forces in the bar.

If f represents the internal force at x, then $f + (\partial f/\partial x)\, dx$ represents the force at $x + dx$, and the net force to the right is

$$df = f - \left(f + \frac{\partial f}{\partial x}\, dx\right) = -\frac{\partial f}{\partial x}\, dx \tag{3.6a}$$

Substitution of (3.5) for f yields

$$df = SY\frac{\partial^2 \xi}{\partial x^2}\, dx \tag{3.6b}$$

The mass of the segment dx is $\rho S\, dx$ where ρ is the mass per unit volume of the bar. Therefore, the equation of motion of the segment is

$$(\rho S\, dx)\frac{\partial^2 \xi}{\partial t^2} = SY\frac{\partial^2 \xi}{\partial x^2}\, dx$$

Setting $c^2 = Y/\rho$, this equation becomes

$$\frac{\partial^2 \xi}{\partial x^2} = \frac{1}{c^2}\frac{\partial^2 \xi}{\partial t^2} \tag{3.7}$$

A comparison of this equation with the corresponding (2.5) for the transverse motion of a segment of a string shows that they are of identical form, with a longitudinal displacement ξ replacing the transverse displacement y. Therefore, (3.7) is a *one-dimensional longitudinal wave equation*.

The general solution of (3.7) is identical in form with that of the transverse wave equation,

$$\xi = \xi_1(ct - x) + \xi_2(ct + x) \tag{3.8a}$$

where the phase speed is

$$\boxed{c = \sqrt{Y/\rho}} \tag{3.8b}$$

The complex harmonic solution of (3.7) is

$$\xi = \mathbf{A}e^{j(\omega t - kx)} + \mathbf{B}e^{j(\omega t + kx)} \tag{3.9}$$

where $\mathbf{A}$ and $\mathbf{B}$ are complex amplitude constants and $k = \omega/c$ is the wave number.

Since Young's modulus Y is measured under conditions where the strained rod is allowed to alter its transverse dimensions, (3.8b) gives the phase speed only when the solid is in the form of a thin bar. In Chapter 6 it will be seen that, when the transverse dimensions of the solid are large compared to a wavelength, the bulk and shear moduli must be used in place of the Young's modulus to calculate the phase speed.

3.4 SIMPLE BOUNDARY CONDITIONS.

Let us now apply boundary conditions to a bar rigidly fixed at both ends, so that $\xi = 0$ at $x = 0$ and at $x = L$ for all times t. (The resulting expressions will be found to be identical with those obtained in Sect. 2.10 for a rigidly supported vibrating string.)

Application of $\xi = 0$ at $x = 0$ gives $\mathbf{A} + \mathbf{B} = 0$, so that $\mathbf{B} = -\mathbf{A}$ and (3.9) becomes

$$\xi(x, t) = \mathbf{A}e^{j\omega t}(e^{-jkx} - e^{jkx}) = -2j\mathbf{A}e^{j\omega t} \sin kx \tag{3.10}$$

The condition $\xi = 0$ at $x = L$ gives

$$\sin kL = 0 \tag{3.11}$$

or

$$k_n L = n\pi \qquad n = 1, 2, 3, \ldots \tag{3.12}$$

(the same as for a fixed, fixed string). The angular frequencies of the allowed modes of vibration are

$$\omega_n = \frac{n\pi c}{L} \qquad \text{or} \qquad f_n = \frac{n}{2}\frac{c}{L} \tag{3.13}$$

[identical with (2.35)]. The complex displacement ξ_n corresponding to the nth mode of vibration is

$$\xi_n(x, t) = -2j\mathbf{A}_n e^{j\omega_n t} \sin k_n x \tag{3.14a}$$

and the real part is

$$\xi_n(x, t) = (A_n \cos \omega_n t + B_n \sin \omega_n t)\sin k_n x \tag{3.14b}$$

where the real amplitude constants A_n and B_n are defined by $2\mathbf{A}_n = B_n + jA_n$. The complete solution is the sum of all separate harmonic solutions,

$$\xi(x, t) = \sum_{n=1}^{\infty} (A_n \cos \omega_n t + B_n \sin \omega_n t)\sin k_n x \tag{3.15}$$

If the initial conditions of displacement and speed of the bar are known, Fourier's theorem can be used, as in Sect. 2.10, to evaluate A_n and B_n.

Since a solid bar is very rigid, it is difficult to provide supports of greater rigidity, and hence the assumed boundary conditions are difficult to realize in practice. By contrast, a free-end condition may be achieved quite readily by supporting the bar on soft supports.

When a bar is free to move at an end, there can be no internal elastic force at the end, and hence $f = 0$ at this point. Since $f = -SY(\partial\xi/\partial x)$, this condition is equivalent to

$$\frac{\partial\xi}{\partial x} = 0 \tag{3.16}$$

at a free end.

Let us now consider the free, free bar. The condition $\partial\xi/\partial x = 0$ applied to (3.9) at $x = 0$ gives

$$-A + B = 0 \quad\text{or}\quad B = A$$

so that

$$\xi(x, t) = Ae^{j\omega t}(e^{-jkx} + e^{jkx}) = 2Ae^{j\omega t}\cos kx \tag{3.17}$$

Application of $\partial\xi/\partial x = 0$ at $x = L$ gives $\sin kL = 0$ or

$$\omega_n = n\pi(c/L) \qquad n = 1, 2, 3, \ldots \tag{3.18}$$

The frequencies of allowed vibration for a free, free bar are identical with those of (3.13) for a fixed, fixed bar. The complex displacement of the nth mode of vibration is

$$\xi_n(x, t) = 2A_n e^{j\omega_n t}\cos k_n x \tag{3.19}$$

and the real displacement is

$$\xi_n(x, t) = (A_n \cos \omega_n t + B_n \sin \omega_n t)\cos k_n x$$

where $2A_n = A_n - jB_n$. In contrast with the fixed, fixed bar which has nodes at either end, the free, free bar has antinodes at either end as indicated by the presence of the $\cos k_n x$ term in the above equation, instead of $\sin k_n x$ as in (3.14a). A comparison of the nodal patterns for these two types of support is given in Fig. 3.3. It should be observed that whenever an *antinode* occurs at the center of the bar the vibrations are *symmetrical* with respect to the center: when a segment of the bar to the left of center is displaced to the left, the corresponding segment to the right is also displaced the same distance to the left. Similarly, whenever there is a *node* at the center, the vibrations are *asymmetrical*.

A bar may be rigidly clamped at any of its nodal positions without interfering with the modes of vibration that have a node at this position. However, those modes of vibration not having a node at this position will be suppressed. It is impossible to find a position for clamping a free, free bar that will not eliminate at least some of the normal modes.

Next consider a free, fixed bar: free at $x = 0$ and rigidly fixed at $x = L$. Application of $\partial\xi/\partial x = 0$ to (3.9) gives (3.17), and application of $\xi = 0$ at $x = L$ yields

$$\cos kL = 0 \tag{3.20}$$

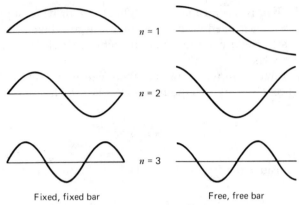

Fig. 3.3. *Standing wave patterns in a bar.*

The allowed frequencies are those satisfying

$$k_n L = (2n - 1)\pi/2 \qquad n = 1, 2, 3, \ldots$$

or

$$f_n = \frac{2n - 1}{4} \frac{c}{L} \tag{3.21}$$

The frequency of the fundamental is half that of a similar free, free bar, and only the odd-numbered harmonics are present; the frequency of the first overtone of a free, fixed bar is three times that of its fundamental. Because of the absence of even harmonics, the quality of the sound produced by a vibrating free, fixed bar differs from that produced by a free, free bar.

3.5   THE FREE, MASS-LOADED BAR.   In many practical applications, a vibrating bar is neither rigidly fixed nor completely free to move at its ends. Instead, it may be loaded with some kind of mechanical impedance, most commonly of the mass-controlled type.

To analyze this type of constraint, consider a bar that is free at $x = 0$ and is loaded with a concentrated mass m at $x = L$. (Ideally, this mass should be a point mass, for otherwise it will not move as a unit but will instead have waves propagated through it.) The boundary condition $\partial\xi/\partial x = 0$, applied to (3.9) at $x = 0$, leads again to

$$\xi = 2Ae^{j\omega t} \cos kx$$

The boundary condition at $x = L$ is obtained by the following argument: Since a *positive* value for f was chosen to indicate *compression* of the bar, the reaction to such a force will accelerate the mass attached to the right end of the bar toward the right. Since the mass is attached to the bar, the end of the bar and the mass must experience the same acceleration. Thus, the boundary condition must be

$$(\mathbf{f})_{x=L} = m\left(\frac{\partial^2\xi}{\partial t^2}\right)_{x=L} \tag{3.22a}$$

or, with the help of (3.5),

$$-SY\left(\frac{\partial \xi}{\partial x}\right)_{x=L} = m\left(\frac{\partial^2 \xi}{\partial t^2}\right)_{x=L} \tag{3.22b}$$

Applying the above boundary conditions to ξ gives

$$kSY \sin kL = -m\omega^2 \cos kL$$

or

$$\tan kL = -\frac{m\omega c}{SY} \tag{3.23}$$

There is no explicit solution of this transcendental equation. For very small mass loading, however, $m \approx 0$ so that $\tan kL \approx 0$ or $kL \approx n\pi$, which is the condition for the allowed frequencies of a free, free bar. Such a result is obviously to be expected, since for very light loadings the bar is essentially free at both ends. Similarly, for heavy mass loadings the mass acts very much like a rigid support, and the allowed frequencies approach those of a free, fixed bar.

It should be noted that in practice the process of "fixing" the end of a bar actually consists in loading it with a large mass, the mass of the support. For light bars a heavy support will act essentially as an infinite mass, and hence like a rigid clamp, but for heavy bars it may be very difficult, if not impossible, to approximate the rigidly clamped condition.

The general case of mass loading can be solved most readily by graphical means. It will be convenient to replace Y by its equivalent expression ρc^2, as given by (3.8b), and to let $m_b = \rho SL$ represent the mass of the bar. Then (3.23) becomes

$$\tan kL = -\frac{m}{m_b} kL \tag{3.24}$$

This transcendental equation is identical with (2.30) which was developed in Chapter 2 for the forced, mass-loaded string (except that m_b, the mass of the bar, replaces m_s, the mass of the string). Analysis proceeds exactly as before; if we choose the special case $m_b = m$, then the allowed values of kL solving (3.24) are $kL = 2.03$, 4.91, 7.98, ... The nodes of the vibrations must occur where

$$\cos kx = 0$$

and the fundamental mode, for which $kL = 2.03$, yields a node at

$$2.03x/L = \pi/2 \qquad \text{or} \qquad x = 0.77L \tag{3.25}$$

In contrast with the free, free bar, the node is no longer at the center, but has shifted toward the loading mass, as suggested by Fig. 3.4. Such a bar could be supported at this nodal position without interfering with the fundamental mode of vibration.

Clearly, as the value of m_b changes from $m_b \ll m$ to $m_b \gg m$, the position of the node of the fundamental shifts from $x \approx L/2$ to $x \approx L$. Thus, the larger the mass attached to a free, mass-loaded bar, the more the nodes of each normal mode of vibration are shifted toward the mass-loaded end.

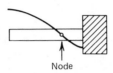

Node

Fig. 3.4. *Fundamental mode of vibration of a free, mass-loaded bar.*

Notice that the overtones of the free, mass-loaded bar are not harmonics. The presence of nonharmonic overtones is sometimes advantageous in practical applications of longitudinally vibrating bars. As an illustration, consider a mass-loaded nickel tube which is driven magnetostrictively by alternating currents in a coil mounted on the tube and is intended to generate a pure tone. Unless the current produced by the oscillator-amplifier unit driving the tube is well filtered, harmonic frequency components other than the desired fundamental will be present in the output. However, since the overtone frequencies of the mass-loaded tube are not harmonics of the fundamental, they will not be resonant with the harmonics of the driving current, and hence will be weakly excited, if at all.

3.6 THE FREELY VIBRATING BAR: GENERAL BOUNDARY CONDITIONS. For a freely vibrating bar with arbitrary loading on each end, the normal modes of vibration can be determined in terms of the mechanical impedance at each end of the bar. If the mechanical impedance of the support at $x = 0$ is $\mathbf{Z}_{m0}$, the force acting on this support due to the bar is

$$\mathbf{f}_0 = -\mathbf{Z}_{m0}\,\mathbf{u}(0,\,t) \qquad (3.26a)$$

where the minus sign arises because a positive compressive force in the bar leads to an acceleration of the support to the left. On the other hand, a positive compressive force at the end $x = L$ leads to an acceleration of the adjacent support to the right so that the force acting on this support is

$$\mathbf{f}_L = +\mathbf{Z}_{mL}\,\mathbf{u}(L,\,t) \qquad (3.26b)$$

where $\mathbf{Z}_{mL}$ is the mechanical impedance of the support at the left end of the bar.

These equations can be expressed in terms of the particle displacement by using (3.5) to replace the compressive force and writing $\mathbf{u} = \partial\xi/\partial t$,

$$\left(\frac{\partial\xi}{\partial x}\right)_{x=0} = \frac{\mathbf{Z}_{m0}}{\rho_L c^2}\left(\frac{\partial\xi}{\partial t}\right)_{x=0} \qquad (3.27a)$$

and

$$\left(\frac{\partial\xi}{\partial x}\right)_{x=L} = -\frac{\mathbf{Z}_{mL}}{\rho_L c^2}\left(\frac{\partial\xi}{\partial t}\right)_{x=L} \qquad (3.27b)$$

where $\rho_L = \rho S$ is the density per unit length of the bar.

The choice of a trial solution satisfying the lossless wave equation for the bar and the boundary conditions (3.27) depends on the natures of the impedances $\mathbf{Z}_{m0}$ and $\mathbf{Z}_{mL}$. If these loads are purely reactive, there can be no loss of acoustical energy so that there is no temporal or spatial damping. An appropriate trial solution would then be (3.9). We may go one step further and notice that, since there are no losses, the wave traveling to the right must possess the same energy as that going to the left. The amplitudes must therefore be equal, $|\mathbf{A}| = |\mathbf{B}|$. Application of the boundary conditions (3.27) then amounts to determining the phase angles of these complex amplitudes.

On the other hand, if either or both of Z_{m0} and Z_{mL} have resistive components, a more general trial solution must be assumed. As was noted in the earlier discussion, in Sect. 2.11(b) on the freely vibrating string terminated by a resistive support, the presence of resistance requires that there be temporal damping. This means that the temporal behavior of the vibrating bar must be described by a complex angular frequency $\omega = \omega + j\beta$ whose real part is the angular frequency of vibration ω and whose imaginary part is the temporal absorption coefficient β. Since there are no internal losses in the bar, the wave equation is still (3.7). Thus, we postulate

$$\xi(x, t) = (Ae^{-jkx} + Be^{jkx})e^{j\omega t} \tag{3.28}$$

where k is determined by $\omega^2 = c^2 k^2$. As before, if the losses are small, the approximation

$$\omega \approx ck \tag{3.29}$$

can be used to simplify the manipulations. Application of the boundary conditions (3.27) to the generalized trial solution (3.28) and use of (3.29) yields the pair of equations

$$A - B = -\frac{Z_{m0}}{\rho_L c}(A + B)$$

$$Ae^{-jkL} - Be^{jkL} = \frac{Z_{mL}}{\rho_L c}(Ae^{-jkL} + Be^{jkL})$$

The first equation is solved for B in terms of A and this substituted into the second equation. The results are

$$B = \frac{1 + (Z_{m0}/\rho_L c)}{1 - (Z_{m0}/\rho_L c)} A \tag{3.30a}$$

$$\tan kL = j\frac{(Z_{m0}/\rho_L c) + (Z_{mL}/\rho_L c)}{1 + (Z_{m0}/\rho_L c)(Z_{mL}/\rho_L c)} \tag{3.30b}$$

Given the impedances Z_{m0} and Z_{mL}, the properties of the vibration have been obtained, although explicit solution is not in general easy. Any resistive component in Z_{m0} or Z_{mL} causes the argument of the tangent to be complex, introducing calculational difficulties in solving this transcendental equation.

3.7 FORCED VIBRATIONS OF A BAR: RESONANCE AND ANTIRESON-ANCE REVISITED.

In discussing the behavior of a forced, loaded string (Sect. 2.9), we defined resonance to occur when the speed amplitude was as large as possible and antiresonance to occur when the speed amplitude was as small as possible. It was there seen that resonance corresponded to the vanishing of the input mechanical reactance and that antiresonance corresponded (for purely reactive loads) to the reactance becoming infinite. We will now investigate these concepts in more detail and show that they must be modified for loads with a nonzero resistive component. We will find that, for the case of small resistance and small reactance loading, resonance and antiresonance occur at frequencies close to those previously derived; at resonance the speed amplitude is maximized and at antiresonance it is minimized, but both resonance and antiresonance correspond to the vanishing of the input reactance.

Assume that a bar of length L is driven at $x = 0$ with a force $\mathbf{f} = F_0 \exp(j\omega t)$ and is terminated at $x = L$ by a support possessing a mechanical impedance $\mathbf{Z}_{mL}$. We assume the trial solution (3.9). The boundary condition at the forced end is (3.5)

$$F_0 e^{j\omega t} = -\rho_L c^2 \left(\frac{\partial \xi}{\partial x} \right)_{x=0} \tag{3.31}$$

where $\rho_L = \rho S$ and $Y = \rho c^2$. At the loaded end, the boundary condition is $\mathbf{f}_L = \mathbf{Z}_{mL}\mathbf{u}(L, t)$

$$\left(\frac{\partial \xi}{\partial x} \right)_{x=L} = -\frac{\mathbf{Z}_{mL}}{\rho_L c^2} \left(\frac{\partial \xi}{\partial t} \right)_{x=L} \tag{3.32}$$

Either direct application of these boundary conditions to (3.28) or argument by analogy will determine A and B and the input mechanical impedance.

Let us argue by analogy. Direct comparison of (3.31) and (3.32) with (2.26) and (2.27) reveals that the boundary conditions are identical if we substitute $\mathbf{Z}_{mL}$ for $j\omega m$. Since the trial solution remains unchanged, the same substitution into (2.29) gives a generalized form of the input impedance

$$\mathbf{Z}_{m0} = \rho_L c \frac{(\mathbf{Z}_{mL}/\rho_L c) + j \tan kL}{1 + (\mathbf{Z}_{mL}/\rho_L c)j \tan kL} \tag{3.33}$$

Decomposing into real and imaginary parts gives

$$\frac{\mathbf{Z}_{m0}}{\rho_L c} = \frac{[r_L(\tan^2 kL + 1)] - j[x_L \tan^2 kL + (r_L^2 + x_L^2 - 1)\tan kL - x_L]}{(r_L + x_L)^2 \tan^2 kL - 2x_L \tan kL + 1} \tag{3.34}$$

where the normalized load impedance is defined by

$$\frac{\mathbf{Z}_{mL}}{\rho_L c} = \frac{R_L}{\rho_L c} + j \frac{X_L}{\rho_L c} = r_L + jx_L \tag{3.35}$$

It is left as an exercise to show that for $r_L = 0$ the input impedance vanishes for frequencies such that $\tan kL = -x_L$ and becomes infinite when $\tan kL = 1/x_L$. Since the driving force is constant, the vanishing of the input impedance $\mathbf{Z}_{m0} = \mathbf{f}/\mathbf{u}_0$ means that the speed amplitude at the point of application of the force is infinite: the condition for mechanical resonance. On the other hand, if the input impedance becomes infinite, the speed amplitude at the driver goes to zero: the condition for antiresonance.

If the load resistance r_L is not zero, these results must be modified and these modifications will lead to more general definitions of resonance and antiresonance. Assume that the mechanical support at the right end of the bar has small losses so that $r_L \ll 1$ and also that it is very yielding so that $|x_L| \ll 1$. We know that this should be the longitudinal-wave analog of the forced, free string. Because both r_L and x_L are small, the input reactance vanishes when

$$x_L \tan^2 kL - \tan kL - x_L \approx 0$$

This expression is quadratic in x_L and yields

$$\tan kL \approx \frac{1}{2x_L}\left(1 \pm \sqrt{1 + 4x_L^2}\right) \tag{3.36}$$

The square root in (3.36) can be approximated, $(1 + \varepsilon)^{1/2} \approx 1 + \varepsilon/2$, and we have the approximate results

$$\tan kL \approx \frac{1 \pm (1 + 2x_L^2)}{2x_L} \approx \begin{cases} -x_L \\ 1/x_L \end{cases}$$

From the discussion following (3.35), the first choice $\tan kL \approx -x_L$ gives

$$\mathbf{Z}_{mL}/(\rho_L c) \sim 0$$

and yields frequencies similar to the resonance frequencies $f_{rn} = (n/2)(L/c)$ for a forced, nearly free bar. The second choice, $\tan kL \approx 1/x_L$ gives $\mathbf{Z}_{mL}/(\rho_L c) \gg 1$ and yields frequencies similar to the antiresonance frequencies $f_{an} = [(2n - 1)/2](L/c)$ for a forced, nearly free bar.

Thus, both resonance and antiresonance frequencies occur when the input mechanical reactance vanishes. To resolve this apparent paradox, let us evaluate the input resistance at resonance and antiresonance. Substitution of $\tan kL \approx -x_L$ into (3.34) yields

$$\frac{\mathbf{Z}_{m0}}{\rho_L c} \approx \frac{r_L(x_L^2 + 1)}{(r_L + x_L)^2 x_L^2 + 2x_L^2 + 1} \approx r_L$$

at resonance, and substitution of $\tan kL \approx 1/x_L$ yields

$$\frac{\mathbf{Z}_{m0}}{\rho_L c} \approx \frac{r_L(1/x_L^2 + 1)}{(r_L + x_L)^2/x_L^2 - 1} \approx \frac{1}{r_L + 2x_L}$$

at antiresonance. The input resistance is small at resonance and large at antiresonance.

Thus, if there are losses in the system (as is always the case in reality), then at *resonance* the input reactance vanishes and the input resistance is *small*, whereas at *antiresonance* the input reactance also vanishes but the input resistance is *large*.

These definitions are consistent with the standing wave being of large amplitude at resonance and small amplitude at antiresonance; for the forced, nearly free bar, there must be an antinode close to the end $x = L$ so that the maximum speed amplitude of the bar is nearly $U_L = |\mathbf{u}(L, t)|$. The power transmitted from the bar into the load at $x = L$ is approximated by

$$\Pi \approx \tfrac{1}{2}U_L^2 R_L$$

while the power sent into the bar at resonance or antiresonance is

$$\Pi = \frac{1}{2}\frac{F_0^2}{R_0}$$

where R_0 is the input mechanical resistance found from (3.34). Since the bar itself is assumed to be lossless, these two powers must be equal and we can solve for the approximate antinodal speed amplitude,

$$U_L \approx \frac{F_0}{\sqrt{R_0 R_L}}$$

Now, substitution of the appropriate values for R_0 yields

$$U_L \approx \frac{F_0}{R_L} \qquad \text{at resonance}$$

and

$$U_L \approx \frac{F_0}{\rho_L c} \sqrt{\frac{R_L + 2X_L}{R_L}} \qquad \text{at antiresonance}$$

Since we have assumed $R_L \ll \rho_L c$ and $|X_L| \ll \rho_L c$, it is clear that the standing wave at resonance has much greater amplitude than at antiresonance.

3.8 TRANSVERSE VIBRATIONS OF A BAR. A bar is capable of vibrating transversely as well as longitudinally, and the internal coupling between strains makes it difficult to produce one motion without the other. For example, if a long thin bar is supported at its center and set into vibration by a hammer blow directed as nearly as possible along the axis of the bar, it is usually found that the unavoidable slight eccentricity of the blow results in the establishment of predominantly transverse vibrations, rather than the desired longitudinal vibrations.

Consider a straight bar of length L, having a uniform cross section S with bilateral symmetry. Let the x coordinate measure positions along the bar, and the y coordinate the transverse displacements of the bar from its normal configuration. When the bar is bent as indicated in Fig. 3.5, the lower part is compressed and the upper part is stretched. Somewhere between the top and the bottom of the bar there will be a *neutral axis* whose length remains unchanged. (If the cross section of the bar is symmetrical about a horizontal plane, this neutral axis will coincide with the central axis of the bar.)

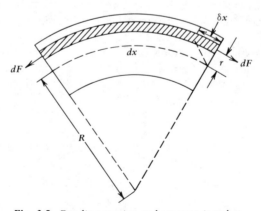

Fig. 3.5. *Bending strains and stresses in a bar.*

Now consider a segment of the bar of length dx, and assume that the bending of the bar is measured by the radius of curvature R of the neutral axis. Let $\delta x = (\partial \xi / \partial x)\, dx$ be the increment of length, due to bending, of a filament of the bar located at a distance r from the neutral axis. Then the longitudinal force df is given by

$$df = -Y\, dS\, \frac{\delta x}{dx} = -Y\, dS\, \frac{\partial \xi}{\partial x} \tag{3.37}$$

where dS is the cross-sectional area of the filament. The value of δx for the particular filament considered in Fig. 3.5 is positive, so that df is a tension, and consequently negative. For filaments below the neutral axis δx is negative, giving a positive force of compression.

Now from the geometry $(dx + \delta x)/(R + r) = dx/R$ and hence $\delta x/dx = r/R$. Substitution into (3.37) yields

$$df = -\frac{Y}{R}\, r\, dS \tag{3.38}$$

The total longitudinal force $f = \int df$ is zero, negative forces above the neutral axis being canceled by positive forces below the neutral axis. However, a bending moment M is present in the bar,

$$M = \int r\, df = -\frac{Y}{R} \int r^2\, dS$$

If we define a constant κ by

$$\kappa^2 = \frac{\int r^2\, dS}{S}$$

then

$$M = -\frac{Y S \kappa^2}{R} \tag{3.39}$$

The constant κ can be thought of as the radius of gyration of the cross-sectional area S, by analogy with the definition of the radius of gyration of a solid. (The value of κ for a bar of rectangular cross section is $t/\sqrt{12}$, where t is the thickness of the bar measured in the y direction. For a circular rod of radius a, $\kappa = a/2$.)

The radius of curvature R is not in general a constant but is rather a function of position along the neutral axis. If the displacements y of the bar are limited to small values, $\partial y/\partial x \ll 1$, then we may use the approximate relation

$$R = \frac{[1 + (\partial y/\partial x)^2]^{3/2}}{\partial^2 y/\partial x^2} \approx \frac{1}{\partial^2 y/\partial x^2} \tag{3.40}$$

Substitution of (3.40) into (3.39) yields

$$M = -Y S \kappa^2\, \frac{\partial^2 y}{\partial x^2} \tag{3.41}$$

In the situation illustrated in Fig. 3.5, the curvature is such as to make $\partial^2 y/\partial x^2$ negative, and the bending moment M is consequently positive. It is apparent that to obtain the curvature illustrated, the torque applied to the *left* end of the segment dx must act in a counterclockwise or *positive* angular direction, so that (3.41) gives the torque acting on the left end of the segment both as to magnitude and as to direction. Similarly, the torque acting on the *right* end of the segment must be clockwise, with the result that it is *negative* and is therefore represented both in direction and in magnitude by $-M$.

3.9 TRANSVERSE WAVE EQUATION.   The effect of distorting the bar is to produce not only bending moments but also shear forces. Consider an upward shear force F_y acting on the *left* end of the segment dx as positive (Fig. 3.6). Then the associated shear force acting on the right end of the segment must be downward, and is consequently negative. When a bent bar is in a condition of *static* equilibrium, the torques and shear forces acting on any

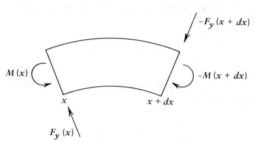

Fig. 3.6. *Bending moments and shear forces in a bar.*

segment must be so related as to produce no net turning moment. Taking moments about the left end of the segment of Fig. 3.6, we have

$$M(x) - M(x + dx) = F_y(x + dx)\, dx \tag{3.42}$$

For segments of small length dx, $M(x + dx)$ and $F_y(x + dx)$ can be expanded in Taylor's expansions about x, and this yields

$$F_y = -\frac{\partial M}{\partial x} = YS\kappa^2 \frac{\partial^3 y}{\partial x^3} \tag{3.43}$$

where second-order terms in dx have been dropped.

This relation between the shear force F_y and the bending moment M has been derived for a condition of static equilibrium. For transverse vibrations of a bar the equilibrium is dynamic, rather than static, and the right-hand side of (3.42) must equal the rate of increase of angular momentum of the segment. However, if the displacement and the slope of the bar are limited to small values, the variations in angular momentum may be neglected, and (3.43) serves as an adequate approximation for the relation between F_y and y.

The net upward force dF_y acting on the segment dx is then given by

$$dF_y = F_y(x) - F_y(x + dx) = -\frac{\partial F_y}{\partial x}\, dx = -YS\kappa^2 \frac{\partial^4 y}{\partial x^4}\, dx \tag{3.44}$$

This force will give the segment an upward acceleration, and, since the mass of the segment is $\rho S\, dx$, the equation of motion is

$$(\rho S\, dx)\frac{\partial^2 y}{\partial t^2} = -YS\kappa^2 \frac{\partial^4 y}{\partial x^4}\, dx$$

or

$$\frac{\partial^2 y}{\partial t^2} = -\kappa^2 c^2 \frac{\partial^4 y}{\partial x^4}$$

(3.45)

where, as for longitudinal waves, $c = \sqrt{Y/\rho}$ from (3.8a). One significant difference between this differential equation and the simpler equation for the transverse waves on a string is the presence of a fourth partial derivative with respect to x, rather than a second partial. As a result of this difference, functions of the form $f(ct - x)$ are *not* solutions of (3.45), a fact that can be shown by direct substitution of $f(ct - x)$ as an assumed solution. Transverse waves *do not* travel along the bar with a constant speed c and unchanging shape.

Assume that (3.45) may be solved by separation of variables, and write the complex transverse displacement as

$$y = \Psi(x)e^{j\omega t}$$

(3.46)

Upon substitution into (3.45), the exponential function of time cancels out, leaving a new *total* differential equation involving Ψ as a function of x only,

$$\frac{d^4\Psi}{dx^4} = \frac{\omega^2}{\kappa^2 c^2}\Psi$$

or letting

$$v = \sqrt{\omega c \kappa}$$

(3.47)

$$\frac{d^4\Psi}{dx^4} = \frac{\omega^4}{v^4}\Psi$$

(3.48)

Now assume that Ψ can be expressed as an exponential of the form $\Psi = A \exp(\gamma x)$, and substitute into (3.48),

$$\gamma^4 = (\omega/v)^4$$

This can be satisfied by four values $\gamma = \pm\omega/v,\ \pm j\omega/v$.

The complete solution is then given by the sum of these four solutions,

$$\Psi = Ae^{\omega x/v} + Be^{-(\omega x/v)} + Ce^{j\omega x/v} + De^{-(j\omega x/v)}$$

where A, B, C, and D are complex amplitude constants. The solution for the displacements y is therefore

$$y = e^{j\omega t}(Ae^{\omega x/v} + Be^{-(\omega x/v)} + Ce^{j\omega x/v} + De^{-(j\omega x/v)})$$

(3.49)

None of the individual terms in (3.49) represents waves moving with a speed c. For example, the last term represents a wave disturbance moving to the right with a phase speed v, but from (3.47) it can be seen that v is itself a function of frequency, so that waves of different frequencies travel with different speeds. In a complex wave containing various frequency components, the high-frequency components travel with greater speeds and hence outrun the low-frequency components, thereby altering the shape of the wave. A precise definition of what is meant by the speed of such a wave is consequently difficult. However, each frequency component of the complex wave progresses at its own speed v, the so-called *phase speed* of the component. This situation is analogous to the transmission of light through glass, where the different component frequencies of a complex light beam travel with different speeds, and *dispersion* results. A vibrating bar is a *dispersive medium* for transverse waves.

The actual solution of (3.45) is the real part of (3.49). It may be conveniently obtained by employing hyperbolic and trigonometric identities (see Appendix A3)

$$y = \cos(\omega t + \phi)\left[A \cosh \frac{\omega x}{v} + B \sinh \frac{\omega x}{v} + C \cos \frac{\omega x}{v} + D \sin \frac{\omega x}{v} \right] \tag{3.50}$$

where A, B, C, and D are real constants. Although these constants are related to the complex constants $\mathbf{A}$, $\mathbf{B}$, $\mathbf{C}$, and $\mathbf{D}$, the relationships are unimportant, since in practice A, B, C, and D are always directly evaluated through the application of initial and boundary conditions.

3.10 BOUNDARY CONDITIONS. Since (3.50) contains twice as many arbitrary constants as the corresponding equation for the transverse vibrations of a string, the determination of these constants requires twice as many boundary conditions. This need is fulfilled by the existence of *pairs* of boundary conditions at the ends of the bar. The particular forms of these conditions depend on the nature of the support and include the following.

(a) Clamped End. If the end of the bar is rigidly clamped, both the displacement and the slope must be zero at the end at all times t. The boundary conditions are therefore

$$y = 0 \qquad \text{and} \qquad \frac{\partial y}{\partial x} = 0 \tag{3.51}$$

(b) Free End. At a free end there can be neither an externally applied torque nor a shearing force, and hence both M and F_y are zero in a plane located an infinitesimal distance from the end. However, the displacement and slope are not constrained, except by the general restriction that their values must be small. Then from (3.41) and (3.43), the boundary conditions are

$$\frac{\partial^2 y}{\partial x^2} = 0 \qquad \text{and} \qquad \frac{\partial^3 y}{\partial x^3} = 0 \tag{3.52}$$

3.11 BAR CLAMPED AT ONE END. Assume that a bar of length L is rigidly clamped at $x = 0$ and is free at $x = L$. Then applying the two conditions of (3.51) at $x = 0$ to the general solution of (3.50) we obtain $0 = A + C$ and $0 = B + D$ so that the general solution reduces to

$$y = \cos(\omega t + \phi)\left[A\left(\cosh \frac{\omega x}{v} - \cos \frac{\omega x}{v} \right) + B\left(\sinh \frac{\omega x}{v} - \sin \frac{\omega x}{v} \right) \right] \tag{3.53}$$

A further application of the two conditions of (3.52) at $x = L$ gives

$$A\left(\cosh \frac{\omega L}{v} + \cos \frac{\omega L}{v} \right) = -B\left(\sinh \frac{\omega L}{v} + \sin \frac{\omega L}{v} \right)$$

and

$$A\left(\sinh \frac{\omega L}{v} - \sin \frac{\omega L}{v} \right) = -B\left(\cosh \frac{\omega L}{v} + \cos \frac{\omega L}{v} \right)$$

It is impossible for both of these equations to be true for all frequencies, although at certain frequencies they become equivalent. To determine these allowed frequencies, divide one equation by the other, thus cancelling out the constants A and B. Then cross-multiply and simplify by using the identities $\cos^2 \theta + \sin^2 \theta = 1$ and $\cosh^2 \theta - \sinh^2 \theta = 1$. This gives

$$\cosh \frac{\omega L}{v} \cos \frac{\omega L}{v} = -1 \tag{3.54}$$

It would be possible to obtain the roots of this transcendental equation by plotting curves of $\cosh(\omega L/v)$ and $-\sec(\omega L/v)$ as functions of $\omega L/v$, and then determining their intersections. However, such a procedure is impractical except for small values of $\omega L/v$, since the hyperbolic cosine increases in an approximately exponential manner with increasing $\omega L/v$. A more convenient form of (3.54) can be obtained by application of the identities

$$\tan \frac{\theta}{2} = \sqrt{\frac{1 - \cos \theta}{1 + \cos \theta}} \quad \text{and} \quad \tanh \frac{\theta}{2} = \sqrt{\frac{\cosh \theta - 1}{\cosh \theta + 1}}$$

Then (3.54) becomes

$$\cot \frac{\omega L}{2v} = \pm \tanh \frac{\omega L}{2v} \tag{3.55}$$

Figure 3.7 is a graph of the functions $\cot(\omega L/2v)$ and $\pm\tanh(\omega L/2v)$, plotted against $\omega L/2v$. From the points of intersection of these curves it is apparent that the frequencies corresponding to the allowed modes of vibration are given by

$$\frac{\omega L}{2v} = \frac{\pi}{4}(1.194, 2.988, 5, 7, \ldots) \tag{3.56}$$

The numerical value of the hyperbolic tangent approaches unity for all angles greater than π. Therefore, for such angles the roots of equation (3.55) are given to a close approximation by the angles $\omega L/2v = (2n - 1)\pi/4$, where $n = 3, 4, 5, \ldots$, and the cotangent of $\omega L/2v$ is consequently equal to ± 1. The accurate values $1.194\pi/4$ and $2.988\pi/4$ must be used for the two lowest allowed frequencies.

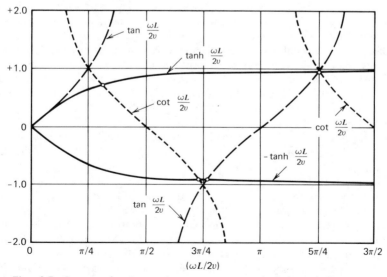

Fig. 3.7. *Curves showing tangent, cotangent, and hyperbolic tangent functions.*

Substituting $v = \sqrt{\omega c \kappa}$ into (3.56) and squaring both sides, we have

$$f = \frac{\pi c \kappa}{8L^2} (1.194^2, 2.988^2, 5^2, 7^2, \ldots) \tag{3.57}$$

as the allowed frequencies of transverse vibration of a clamped, free bar. Therefore, the application of boundary conditions limits the allowed modes of free vibration of a finite bar to a discrete set of frequencies, just as it does for a vibrating string. However, in contrast with the string, the overtone frequencies of a bar are not harmonics of its fundamental; (3.57) yields the first column of Table 3.1.

Table 3.1. Transverse vibration characteristics of a clamped, free bar (100 cm)

Frequency	Phase Speed	Wavelength (cm)	Nodal Positions (cm from clamped end)
f_1	v_1	335.0	0
$6.267f_1$	$2.50v_1$	133.4	0, 78.3
$17.55f_1$	$4.18v_1$	80.0	0, 50.4, 86.8
$34.39f_1$	$5.87v_1$	57.2	0, 35.8, 64.4, 90.6

The first overtone has a frequency higher than the sixth harmonic of a string of the same fundamental frequency. If a bar is struck in such a manner that the amplitudes of vibration of some of the overtones are appreciable, the sound produced has a metallic quality. However, these high-frequency overtones are rapidly damped out, so that the initial sound is soon mellowed into a nearly pure tone, whose frequency is that of the fundamental. A struck tuning fork exhibits the above characteristics of an initial metallic sound, which rapidly dies out, leaving a nearly pure tone.

The vibrating reeds used as frequency standards in frequency meters, and as components in low-frequency electrical filters, are other applications of clamped, free bars. It is possible to adjust the fundamental resonance frequency of such reeds either by varying their thickness, and thereby κ, or by varying their length. It is to be noted that as the length is doubled, the frequency is divided by four.

The distribution of nodal points along a transversely vibrating clamped, free bar is much more complex than in the examples previously considered, for the nodes are not evenly spaced at intervals of $\lambda/2$ but have instead an irregular spacing. Furthermore there are *three* distinct types of nodal points, that is, positions where $y = 0$ at all times. The point at which the bar is clamped is one node, characterized by the conditions $y = 0$ and $\partial y/\partial x = 0$. The next group of nodal points is characterized by $y = 0$ and $\partial^2 y/\partial x^2 \approx 0$. These so-called *true* nodes are located near points of inflection of the bar. Also, the spacing between *true* nodes is very nearly $\lambda/2$. A third type of nodal point is that occurring at the node adjacent to the free end of the bar, where $y = 0$. A point of inflection, $\partial^2 y/\partial x^2 = 0$, does not occur near this nodal position but instead is shifted out to the free end. It is also to be noted that the vibrational amplitude at the various antinodal positions is not the same for each antinode, because that at the free end is always the greatest.

Table 3.1 gives the nodal positions for transverse vibrations of a bar 100 cm in length, which is clamped at $x = 0$ and is free at $x = 100$. The ratios of the frequency f and the phase speed v to their fundamental values are also tabulated, together with the wavelength $\lambda = v/f$ in centimeters for each frequency component. The increase in speed with frequency of the overtone is quite apparent. It should also be noted that the wavelengths are not in general

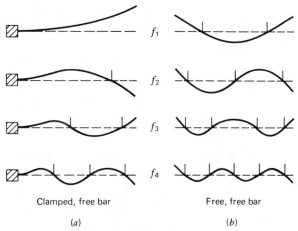

Clamped, free bar   Free, free bar

(a)   (b)

Fig. 3.8. *Transverse modes of vibration of a bar.*

equal to twice the distance between nodes. However, the nodal spacing between *true* nodes for the third overtone mode of vibration is 28.6 cm (64.4 − 35.8) which, within the accuracy of the data supplied, equals $\lambda/2$.

3.12   BAR FREE AT BOTH ENDS.   Another important type of transverse vibration is that of a *free, free* bar. The boundary conditions are satisfied at $x = 0$ if $A - C = 0$ and $B - D = 0$. Application of the same conditions at $x = L$ restricts the allowed frequencies to those satisfying

$$\tan\left(\frac{\omega L}{2v}\right) = \pm\tanh\left(\frac{\omega L}{2v}\right) \tag{3.58}$$

A consideration of Fig. 3.7 shows that the allowed frequencies are given by

$$f = \frac{\pi c\kappa}{8L^2}(3.0112^2, 5^2, 7^2, 9^2, \ldots) \tag{3.59}$$

and again the overtones are not harmonics of the fundamental.

Table 3.2, which is similar to Table 3.1, gives information concerning the frequencies, phase speeds, and nodal positions of a free, free bar 100 cm long. An inspection of Fig. 3.8b shows that the modes of vibration corresponding to the fundamental f_1 and all additional odd-numbered frequencies, f_3, f_5, etc., are symmetrical about the center. The slope $\partial x/\partial y$ at the center is always zero, giving a true antinode. In contrast, the even-numbered frequencies f_2, f_4, f_6, etc., correspond to asymmetrical modes of vibration with respect to the center. In all modes the nodal positions are symmetrically distributed about the center. The bar may be supported on a knife edge, or clamped by knife-edge clamps, at any nodal point without interfering with the mode of vibration having a node at this point. A knife-edge type of support or clamp is required, since it must merely restrict the displacement to zero and must not restrict the changes in slope that occur at a node.

The bars of a xylophone are supported at points corresponding to the nodal positions of the fundamental. Since the nodes of the accompanying overtones will not in general be located at these same two points, the overtones will be rapidly damped out, leaving the pure

Table 3.2. Transverse vibration characteristics of a free, free bar (100 cm)

Frequency	Phase Speed	Wavelength (cm)	Nodal Positions (cm from end)
f_1	v_1	133.0	22.4, 77.6
$2.756f_1$	$1.66v_1$	80.0	13.2, 50.0, 86.8
$5.404f_1$	$2.32v_1$	57.2	9.4, 35.6, 64.4, 90.6
$8.933f_1$	$2.99v_1$	44.5	7.3, 27.7, 50.0, 72.3, 92.7

tone of the fundamental. This is one of a number of factors that contribute to the mellow pure tonal quality of a xylophone.

The theory of a free, free bar may be used qualitatively to explain the behavior of tuning forks. Such a fork is essentially a bar bent into the shape of a letter U. This bending as well as the mass-loading effect of the stem attached to the center of the bar brings about a closer spacing of the two nodes present, when vibrating at its fundamental frequency. Compare Fig. 3.9 with Fig. 3.8b. As has been previously mentioned, when a tuning fork is struck, the frequencies corresponding to the higher modes of vibration rapidly damp out, leaving a pure sinusoidal vibration at the fundamental frequency. Since the stem partakes of the antinodal motion at the center of the original free, free bar, the radiation efficiency of a tuning fork is greatly enhanced by either touching its stem to a surface of large area, such as a table top, or by attaching it to a resonator box tuned to its fundamental frequency.

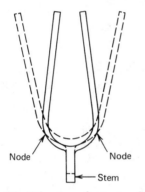

Fig. 3.9. Vibration of a tuning fork.

If a bar is rigidly clamped at both ends, the boundary conditions that $y = 0$ and $\partial y/\partial x = 0$ at $x = 0$ and at $x = L$ lead to the same set of allowed frequencies as for a free, free bar. However, as is to be expected, the arrangement of the nodal positions is different.

PROBLEMS

3.1. A bar of length L is rigidly fixed at $x = 0$ and free to move at $x = L$. (a) Show that only odd integral harmonic overtones are allowed. (b) Determine the fundamental frequency of the bar, if it is composed of steel and has a length of 0.5 m. (c) If a static force F is applied to the free end of the bar so as to displace this end h, show that, when the bar vibrates longitudinally subsequent to the release of this force, the amplitudes of the various harmonic vibrations are given by $A_n = [8h/(n\pi)^2]\sin(n\pi/2)$. (d) Determine these amplitudes for the above steel bar, if the force is 5000 N and the cross-sectional area of the bar is 0.00005 m^2.

3.2. A 2-kg mass is hanging on a steel wire of 0.00001-m^2 cross-sectional area and 1.0-m length. (a) Compute the fundamental frequency of vertical oscillation of the mass by considering it to be a simple oscillator. (b) Compute the fundamental frequency of vertical oscillation of the mass by considering the system to be that of a longitudinally vibrating bar fixed at one end and mass-loaded at the other. (c) Show that for $kL < 0.2$, the equation derived in (b) reduces to (1.2).

3.3. A steel bar of 0.0001-m^2 cross-sectional area and 0.25-m length is free to move at $x = 0$ and is loaded with 0.15 kg at $x = 0.25$ m. (a) Compute the fundamental frequency of longitudinal vibrations of the above mass-loaded bar. (b) Determine the position at which the bar may be clamped so as to cause the least interference with its fundamental mode of vibration. (c) When this bar is vibrating in its fundamental mode, what is the ratio of the displacement amplitude of the free end to that of the mass-loaded end? (d) What is the frequency of the first overtone of this bar?

3.4. A steel bar of 0.2-m length and 0.04-kg mass is loaded at one end with 0.027 kg and at the other end with 0.054 kg. (a) Calculate the fundamental frequency of longitudinal vibration of this system. (b) Calculate the position of the node in the bar. (c) Calculate the ratio of the displacement amplitudes at the two ends of the bar.

3.5. A thin bar of length L and mass M is rigidly fixed at one end and free at the other. What mass m must be attached to the free end in order to decrease the fundamental frequency of longitudinal vibration by 25 percent from its fixed, free value?

3.6. Determine an expression giving the fundamental frequency of longitudinal vibrations of a fixed, free bar of length L and mass m, if the reaction of the fixture corresponds to a mechanical reactance of $-js/\omega$ (stiffness).

3.7. A long thin bar of length L is driven by a longitudinal force $F \cos \omega t$ at $x = 0$ and is free at $x = L$. (a) Derive the equation that gives the amplitude of the standing waves set up in the bar. (b) What is the expression giving the input mechanical impedance of such a bar of length L. (c) What is the input mechanical impedance of a similar bar of infinite length? (d) If the material of the bar is aluminum, the length is 1.0 m, the cross-sectional area is $0.0001\,\text{m}^2$, and the amplitude of the driving force is 10 N, plot the amplitude of the driven end of the bar of part (a) as a function of frequency over the range from 200 to 2000 Hz.

3.8. Show that $v = \sqrt{\omega c \kappa}$ has the dimensions of a speed. For what frequency will the transverse vibrations of an aluminum rod of 0.01-m diameter have the same phase speed as that of longitudinal vibrations in the rod?

3.9. A steel rod of 0.005-m radius has a length of 0.5 m. (a) What is its fundamental frequency of free, free transverse vibrations? (b) If the displacement amplitude at the center of the rod is 2 cm when vibrating in its fundamental mode, what is the displacement amplitude at the ends?

CHAPTER 4

THE TWO-DIMENSIONAL WAVE EQUATION: VIBRATIONS OF MEMBRANES AND PLATES

4.1 VIBRATION OF A PLANE SURFACE. We now consider vibrations of systems that extend in two dimensions, such as a drumhead or the diaphragm of a microphone. While analysis of such systems may seem more complicated because two spatial coordinates are needed to locate a point on the surface and a third is needed to specify its displacement, the equation of motion (subject to the same simplifying assumptions as invoked in the previous two chapters) will be the two-dimensional generalization of the one-dimensional wave equation for a string or bar. In addition, solutions to this new equation will display all the now-familiar properties of waves. In the previous chapters we noted that the application of boundary conditions limits the allowed frequencies to a discrete set. Similar restrictions will be found to apply to the free vibrations of membranes, but boundary conditions must now include not only the type of support but also the shape of the perimeter of the membrane.

Finally, it should be mentioned that the generalization to two dimensions necessitates the choice of the system of coordinates to be used. Although different two-dimensional coordinate systems result in different *appearing* equations, these equations can be shown to be equivalent. However, choice of the coordinate system to match the boundary conditions, such as Cartesian coordinates for a rectangular boundary and polar coordinates for a circular boundary, will greatly simplify the labor involved in obtaining and interpreting a solution. Unfortunately, the available number of coordinate systems is strictly limited and, consequently, the number of solvable membrane problems is similarly limited.

4.2 THE WAVE EQUATION FOR A STRETCHED MEMBRANE. To obtain the simplest equation of motion, assumptions about the physical properties of the membrane will be made that are analogous to the assumptions made in deriving the equation of motion for a string and a bar. The membrane is assumed to be thin and uniform with negligible stiffness, to be perfectly elastic with no damping, and to vibrate with small displacement amplitudes. Let ρ_S be the *surface density* of the membrane (mass per unit area) expressed in kilograms per square meter, and let $\mathscr{T}$ be the membrane tension per unit length in newtons per meter of length so that the material on opposite sides of a line segment of length dl will tend to be pulled apart with a force $\mathscr{T}\, dl$. This tension is assumed to be distributed uniformly throughout the membrane.

The equation of motion will be developed in Cartesian coordinates with the transverse displacement of a point expressed as $y = y(x, z, t)$. The force acting on a displaced rectangular surface element of area $dS = dx\, dz$ is then the sum of the net transverse forces acting on the edges parallel to the x and z axes. From the incremental area of the membrane shown in Fig. 4.1 and by argument analogous to

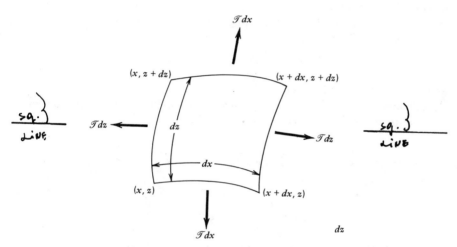

Fig. 4.1. *Element of a vibrating membrane.*

that used in Sect. 2.3 for the transverse force acting on a segment of string, the net force on the element $dx\, dz$ due to the pair of tensions $\mathscr{T}\, dz$ is

$$\mathscr{T}\, dz\left[\left(\frac{\partial y}{\partial x}\right)_{x+dx} - \left(\frac{\partial y}{\partial x}\right)_{x}\right] = \mathscr{T}\,\frac{\partial^2 y}{\partial x^2}\, dx\, dz$$

and that due to the pair of tensions $\mathscr{T}\, dx$ is $\mathscr{T}(\partial^2 y/\partial z^2)\, dx\, dz$. Equating the sum of these two terms to the product of the element's mass $\rho_S\, dx\, dz$ by its acceleration $\partial^2 y/\partial t^2$ gives

$$\mathscr{T}\left(\frac{\partial^2 y}{\partial x^2} + \frac{\partial^2 y}{\partial z^2}\right) dx\, dz = \rho_S\, dx\, dz\,\frac{\partial^2 y}{\partial t^2} \tag{4.1}$$

or

$$\frac{\partial^2 y}{\partial x^2} + \frac{\partial^2 y}{\partial z^2} = \frac{1}{c^2}\frac{\partial^2 y}{\partial t^2} \tag{4.2}$$

where

$$c = \sqrt{\mathcal{T}/\rho_S}$$

(4.3)

Equation (4.2) may be expressed in a general form, appropriate to any coordinate system, as

$$\nabla^2 y = \frac{1}{c^2} \frac{\partial^2 y}{\partial t^2}$$

(4.4)

where ∇^2 is the *Laplacian operator*, two-dimensional in this case. Equation (4.4) is the *two-dimensional wave equation*. The *form* of the Laplacian depends on the choice of the coordinate system.

In the Cartesian coordinates used above, the Laplacian is expressed as

$$\nabla^2 = \frac{\partial^2}{\partial x^2} + \frac{\partial^2}{\partial z^2}$$

This form is appropriate for rectangular membranes.

For a circular membrane it is preferable to express the Laplacian in polar coordinates (r, θ) where $x = r \cos \theta$ and $z = r \sin \theta$,

$$\nabla^2 = \frac{\partial^2}{\partial r^2} + \frac{1}{r} \frac{1}{\partial r} + \frac{1}{r^2} \frac{\partial^2}{\partial \theta^2}$$

Substitution into (4.4) leads to

$$\frac{\partial^2 y}{\partial r^2} + \frac{1}{r} \frac{\partial y}{\partial r} + \frac{1}{r^2} \frac{\partial^2 y}{\partial \theta^2} = \frac{1}{c^2} \frac{\partial^2 y}{\partial t^2}$$

(4.5)

as the wave equation appropriate for transverse vibrations of a circular membrane.

Solutions to (4.4) will be shown to have all the properties we have come to associate with waves, but generalized to two dimensions.

For calculating normal modes it is conventional to recast (4.4) by assuming the solution

$$y = \Psi e^{j\omega t}$$

(4.6a)

where Ψ is a function only of position. Substitution into (4.4) and definition of $k = \omega/c$ yields

$$\nabla^2 \Psi + k^2 \Psi = 0$$

(4.6b)

the *time-independent wave equation* or *Helmholtz equation*. The solutions (4.6a) to (4.6b) for a membrane of specified shape and boundary conditions are the normal modes of vibration of that membrane.

4.3 FREE VIBRATIONS OF A RECTANGULAR MEMBRANE WITH FIXED RIM.   Assume a stretched membrane fixed at its edges at $x = 0$, $x = L_x$, $z = 0$, and $z = L_z$. The appropriate boundary conditions are

$$y(0, z, t) = y(L_x, z, t) = y(x, 0, t) = y(x, L_z, t) = 0$$

(4.7)

Let us assume a solution of the form

$$\mathbf{y}(x, z, t) = \mathbf{\Psi}(x, z)e^{j\omega t} \tag{4.8}$$

so that (4.2) reduces to the Helmholtz equation (4.6b) in Cartesian coordinates,

$$\frac{\partial^2 \mathbf{\Psi}}{\partial x^2} + \frac{\partial^2 \mathbf{\Psi}}{\partial z^2} + k^2 \mathbf{\Psi} = 0 \tag{4.9}$$

Furthermore, let us assume that $\mathbf{\Psi}$ is the product of two functions, each dependent on only one of the dimensions,

$$\mathbf{\Psi}(x, z) = \mathbf{X}(x)\mathbf{Z}(z) \tag{4.10}$$

then (4.9) becomes

$$\frac{1}{\mathbf{X}}\frac{d^2\mathbf{X}}{dx^2} + \frac{1}{\mathbf{Z}}\frac{d^2\mathbf{Z}}{dz^2} + k^2 = 0 \tag{4.11}$$

The first term in this expression, being a function of x alone, is independent of z. Therefore, the second term must also be independent of z: otherwise, the three terms cannot sum to zero for all z. The first two terms must thus each be constant for all x and z. This results in the pair of equations

$$\frac{d^2\mathbf{X}}{dx^2} + k_x^2 \mathbf{X} = 0 \tag{4.12a}$$

$$\frac{d^2\mathbf{Z}}{dz^2} + k_z^2 \mathbf{Z} = 0 \tag{4.12b}$$

where k_x^2 and k_z^2 are the respective constants related by

$$k_x^2 + k_z^2 = k^2 \tag{4.13}$$

Solutions of (4.12) are seen to be sinusoids, so that

$$\mathbf{y}(x, z, t) = \mathbf{A}\,\sin(k_x x + \phi_x)\sin(k_z z + \phi_z)e^{j\omega t} \tag{4.14}$$

where ϕ_x and ϕ_z must be determined by the boundary conditions and $|\mathbf{A}|$ is the maximum amplitude of transverse displacement. For the case in question, the conditions $y(0, z, t)$ and $y(x, 0, t)$ yield $\phi_x = \phi_y = 0$, and the conditions $y(L_x, z, t)$ and $y(x, L_z, t)$ require $\sin k_x L_x = 0$ and $\sin k_z L_z = 0$. Thus, the normal modes of vibration are given by

$$\mathbf{y}(x, z, t) = \mathbf{A}\sin k_x x \sin k_z z\, e^{j\omega t} \tag{4.15}$$

where

$$k_x = n\pi/L_x \qquad n = 1, 2, 3, \ldots \tag{4.16a}$$

$$k_z = m\pi/L_z \qquad m = 1, 2, 3, \ldots \tag{4.16b}$$

These two equations are seen to limit each of the constants k_x and k_z to a discrete set of values. These limitations, in turn, restrict the characteristic frequencies for the allowed modes of free vibration to

$$f_{nm} = \frac{\omega_{nm}}{2\pi} = \frac{c}{2}\sqrt{\left(\frac{n}{L_x}\right)^2 + \left(\frac{m}{L_z}\right)^2} \qquad (4.17)$$

This is no more than the two-dimensional extension of the results of the free vibrations of the lossless finite string fixed at both ends.

The fundamental frequency is obtained by substitution of $n = 1$ and $m = 1$ into (4.17). Those overtones corresponding to $n = m$ will be harmonics of the fundamental, while those for which $n \neq m$ may not be. Figure 4.2 illustrates a number of the possible modes for a rectangular membrane. The light areas vibrate $180°$ out of time phase with the shaded areas. The normal modes are denoted by the ordered pair

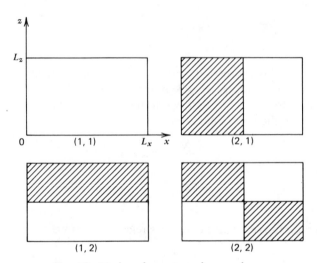

Fig. 4.2. *Modes of a rectangular membrane.*

(n, m). Since the nodal lines are lines of zero displacement, it is possible to insert rigid supports along any of them without affecting the nodal pattern for the particular frequency involved.

4.4 FREE VIBRATIONS OF A CIRCULAR MEMBRANE FIXED AT THE RIM.

For a circular membrane it is best to use the wave equation in the polar-coordinate form of (4.5). If the membrane is fixed at its circular boundary then the transverse displacement must vanish at the rim. If the radius of the membrane is at $r = a$, the required boundary condition is

$$y(a, \theta, t) = 0$$

In order to solve (4.5) let us assume that its harmonic solution may be expressed as the product of three terms, each a function of only one variable

$$y(r, \theta, t) = R(r)\Theta(\theta)e^{j\omega t} \qquad (4.18)$$

The boundary condition is now

$$R(a) = 0$$

Upon making this substitution, (4.5) becomes the Helmholtz equation in polar coordinates,

$$\Theta \frac{d^2R}{dr^2} + \frac{\Theta}{r}\frac{dR}{dr} + \frac{R}{r^2}\frac{d^2\Theta}{d\theta^2} + k^2R\Theta = 0 \tag{4.19a}$$

where $k = \omega/c$. Next, multiply each term in this equation by $r^2/(\Theta R)$, and move terms containing r to one side of the equality sign and those containing θ to the other. The result is

$$\frac{r^2}{R}\left(\frac{d^2R}{dr^2} + \frac{1}{r}\frac{dR}{dr}\right) + k^2r^2 = -\frac{1}{\Theta}\frac{d^2\Theta}{d\theta^2} \tag{4.19b}$$

The left-hand side of this equation, a function of r alone, cannot be equal to the right-hand side, a function of θ alone, unless both functions equal some constant. If we let this constant be m^2, then the right-hand side becomes

$$\frac{d^2\Theta}{d\theta^2} = -m^2\Theta$$

which has a harmonic solution

$$\Theta(\theta) = \cos(m\theta + \gamma) \tag{4.20}$$

where γ is the initial phase angle. The azimuthal coordinate θ is periodic, repeating itself after 2π, 4π, etc. Consequently, if the displacement y is to be a single-valued function of position, then $y(r, \theta, t)$ must equal $y(r, \theta + 2\pi, t)$, which restricts the constant m to integral values, $m = 0, 1, 2, 3, \ldots$. Equation (4.19b) now becomes *Bessel's equation*

$$\frac{d^2R}{dr^2} + \frac{1}{r}\frac{dR}{dr} + \left(k^2 - \frac{m^2}{r^2}\right)R = 0 \tag{4.21}$$

Solutions to this equation are the transcendental functions called *Bessel functions* of the *first kind* $J_m(kr)$ and *second kind* $Y_m(kr)$ of order m,

$$R(r) = AJ_m(kr) + BY_m(kr) \tag{4.22a}$$

Some properties of Bessel functions are summarized in Appendix A4. They are oscillating functions whose amplitudes diminish as kr increases, and the $Y_m(kr)$ become unbounded in the limit $kr \to 0$. Thus, while (4.22a) is the general solution to (4.21), since the circular membrane extends across the origin ($r = 0$), the solution y must be finite at $r = 0$. This requires $B = 0$ so that

$$R(r) = AJ_m(kr) \tag{4.22b}$$

(If, on the other hand, the membrane is stretched between an inner radius and an outer rim, so that the membrane *excludes* the origin, then both terms in (4.22a) would have to be used to supply the two arbitrary constants needed for the two boundary conditions.) Application of the boundary condition $\mathbf{R}(a) = 0$ requires $J_m(ka) = 0$. If the values of the argument of the mth order Bessel function J_m which cause that function to equal zero are designated by j_{mn}, then we have $J_m(j_{mn}) = 0$, so that the allowed values of k again assume discrete values given by $k_{mn} = j_{mn}/a$. Values of j_{mn} for some of the zeros of some of the Bessel functions are given in Appendix A5.

The normal modes of vibration are therefore

$$y_{mn}(r,\ \theta,\ t) = \mathbf{A}_{mn} J_m(k_{mn}r)\cos(m\theta + \gamma_{mn})e^{j\omega_{mn}t} \tag{4.23a}$$

where

$$k_{mn} a = j_{mn} \tag{4.23b}$$

and the natural frequencies are

$$f_{mn} = \frac{1}{2\pi}\frac{j_{mn}\,c}{a} \tag{4.23c}$$

The physical motion of the normal mode designated by the integers (m, n) is the real part of (4.23a),

$$y_{mn}(r,\ \theta,\ t) = A_{mn} J_m(k_{mn}r)\cos(m\theta + \gamma_{mn})\cos(\omega_{mn} t + \phi_{mn}) \tag{4.23d}$$

where $\mathbf{A}_{mn} = A_{mn}\exp(j\phi_{mn})$. The azimuthal phase angle γ_{mn} is one of the arbitrary constants of the solution; for each normal mode it determines those directions along which radial nodal lines of zero displacement will appear and in its turn depends on the azimuthal angle at which the membrane is initially excited.

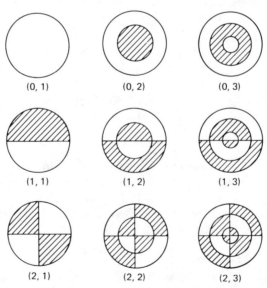

Fig. 4.3. *Modes of a circular membrane.*

Figure 4.3 illustrates a number of the simpler modes of vibration with $\gamma_{mn} = 0$. These modes are designated by the ordered pair (m, n) where the integer m determines the number of *radial nodal lines* and the integer n determines the number of *azimuthal nodal circles*. It should be noted that $n = 1$ is the minimum allowed value of n and corresponds to a mode of vibration in which the azimuthal nodal circle occurs only at the boundary of the membrane where $r = a$.

For each value of m there exists a whole sequence of allowed radial modes of vibration of increasing frequency. When $m = 0$, the allowed frequencies correspond to the condition that $J_0(k_{0n} a) = 0$. For $m = 1$ the sequence of allowed frequencies is determined by $J_1(k_{1n} a) = 0$, for $m = 2$ by $J_2(k_{2n} a) = 0$, etc. Table 4.1 contains a few of these frequencies f_{mn} expressed relative to the fundamental frequency f_{01} as given by (4.23c). Note that none of the overtones are harmonics of the fundamental.

Table 4.1. Relative frequencies of a circular membrane

$f_{01} = 1.0 f_{01}$	$f_{11} = 1.593 f_{01}$	$f_{21} = 2.135 f_{01}$
$f_{02} = 2.295 f_{01}$	$f_{12} = 2.917 f_{01}$	$f_{22} = 3.500 f_{01}$
$f_{03} = 3.598 f_{01}$	$f_{13} = 4.230 f_{01}$	$f_{23} = 4.832 f_{01}$

4.5 SYMMETRIC VIBRATIONS OF A CIRCULAR MEMBRANE FIXED AT THE RIM.

In many situations that can be described by the motion of a circular membrane fixed at its perimeter, it turns out that the vibrations having circular symmetry are of greatest importance. Let us, therefore, confine our attention to those solutions

$$y_{0n} = A_{0n} J_0(k_{0n} r) e^{j\omega_{0n} t} \tag{4.24}$$

which are independent of θ. Because only $m = 0$ is involved, we drop this subscript and retain only the index n.

The boundary condition at the edge of a fixed circular membrane is $y = 0$ at $r = a$, so that $J_0(ka) = 0$, and hence

$$k_n a = j_{0n} = 2.405, \ 5.520, \ 8.654, \ 11.792, \ldots \tag{4.25}$$

The fundamental frequency is therefore

$$f_1 = \frac{\omega_1}{2\pi} = \frac{k_1 c}{2\pi} = \frac{2.405}{2\pi a} c = \frac{2.405}{2\pi a} \sqrt{\frac{\mathcal{T}}{\rho_S}} \tag{4.26}$$

The ratios of the overtone to fundamental frequencies are obtained from (4.25)

$$f_2 = (5.520/2.405) f_1 = 2.295 f_1$$

$$f_3 = 3.598 f_1 \tag{4.27}$$

$$f_4 = 4.90 f_1$$

. . . .

As noted previously, the overtones are not harmonics of the fundamental.

Taking the real part of $y_1 = A_1 J_0(k_1 r) \exp(j\omega_1 t)$, we obtain the general expression for the displacement of the membrane when it is vibrating in its fundamental mode:

$$y_1 = A_1 \cos(\omega_1 t + \phi_1) J_0(2.405 r/a) \tag{4.28}$$

where $A_1 = |A_1|$ is the displacement amplitude at the center of the membrane and k_1 has been replaced by $2.405/a$. The complete solution for *symmetric* vibrations is

$$y = \sum A_n \cos(\omega_n t + \phi_n) J_0(k_n r) \tag{4.29}$$

For all symmetric modes of vibration other than the fundamental, inner nodal circles will occur at those radial distances for which $J_0(k_n r)$ vanishes. Considering, for example, the first overtone, the J_0 function is zero for

$$k_2 r = 5.520\, r/a = 2.405$$

or

$$r = 0.436a$$

An inspection of Fig. 4.3 shows that when the central part of the membrane is displaced up, the adjacent ring is displaced down, and vice versa. Consequently, a membrane vibrating at frequencies other than its fundamental produces little net displacement of the surrounding air. For this reason, the vibrating drumhead of a kettledrum has a low efficiency of sound production for its overtone frequencies.

One parameter for judging the efficiency of each normal mode in producing sound is the average displacement amplitude of the surface of the membrane when it is vibrating in a single normal mode. From (4.29) the average displacement amplitude $\langle \Psi_n \rangle_S$ of the nth symmetric normal mode is

$$\langle \Psi_n \rangle_S = \frac{1}{\pi a^2} \int_S A_n J_0(k_n r)\, dS$$

$$= \frac{1}{\pi a^2} \int_0^a A_n J_0(k_n r) 2\pi r\, dr \tag{4.30}$$

which when integrated (see Appendix A4) becomes

$$\langle \Psi_n \rangle_S = \frac{2 A_n}{k_n a} J_1(k_n a) \tag{4.31}$$

[Notice that for all modes other than the symmetric ones, the angular dependance $\cos(m\theta + \gamma)$ guarantees that the average displacement is zero. Thus, reverting to our previous double subscript notation, we have $\langle \Psi_{mn} \rangle_S = 0$ for all $m \neq 0$.]

Let us determine the value of $\langle \Psi_1 \rangle_S$ for the fundamental mode of vibration

$$\langle \Psi_1 \rangle_S = \frac{2 A_1}{k_1 a} J_1(k_1 a) = \frac{2 A_1}{2.405} J_1(2.405) = 0.432 A_1$$

Consequently, a rigid flat piston of radius a and displacement amplitude $0.432 A_1$ will displace the same volume of air as will the membrane when vibrating at its fundamental frequency. Similarly, $\langle \Psi_2 \rangle_S = -0.123 A_2$, where the negative sign indicates that the average displacement amplitude is oppositely directed to the displacement at the center. These examples show that when the displacement amplitudes at the center are equal, $A_1 = A_2$, the fundamental mode of vibration is more than three times as effective for displacing air as is the first overtone.

In many problems encountered in the study of sources of sound waves, the characteristics of the sound wave are found to depend on the amount of air displaced, the *volume displacement amplitude*, and not on the exact shape of the moving surface. The radiating source may then be replaced by an *equivalent simple piston*, such that the product of its area, S_{eq}, and displacement amplitude, ξ_{eq}, equals the volume displacement amplitude of the true source. The volume displacement amplitude of a simple piston which is equivalent to the above membrane vibrating in its fundamental mode is

$$S_{eq} \xi_{eq} = 0.432 A_1 \pi a^2$$

Real membranes do not vibrate with each mode having constant amplitude. The effect of damping forces, such as those arising within the membrane from internal frictional forces as well as external forces associated with the radiation of energy in the form of sound waves, causes the amplitude of each mode to decrease exponentially as $\exp(-\beta_n t)$. This decrease may be derived in a manner similar to that used in Sect. 1.6 for a simple oscillator. In general, the damping constant β_n increases with frequency so that the higher frequencies damp out more quickly than does the fundamental.

4.6 THE KETTLEDRUM. The resistive damping force just mentioned is but one of a number of forces that may act on the surface of a membrane and influence its vibration. Another is that caused by changes in pressure occurring within a closed space behind the drumhead of a drum or the diaphragm of a condenser microphone as the volume of the entrapped gas is altered by vibration of the membrane.

For example, consider a kettledrum, which consists of a membrane stretched tightly over the open end of a hemispherical vessel. As the drumhead vibrates, the air in the vessel is alternately compressed and expanded. If the radial velocity of transverse waves along the membrane is considerably less than the speed of sound in air, the pressure resulting from the compression and expansion of the air in the vessel is nearly uniform over the entire extent of the membrane and thus depends only on the average displacement $\langle y \rangle_S$. The increment in volume of the enclosed air is $dV = \pi a^2 \langle y \rangle_S$, where a is the radius of the drumhead. If the equilibrium volume inside the vessel is V_0 and the equilibrium pressure is $\mathscr{P}_0$, then for *adiabatic* changes in volume the new pressures $\mathscr{P}$ and volumes V are related by

$$\mathscr{P} V^\gamma = \mathscr{P}_0 V_0^\gamma = \text{constant} \qquad (4.32)$$

where γ is the ratio of the specific heat of the entrapped air at constant pressure to its specific heat at constant volume (see Appendix A9). Differentiation shows that the excess pressure $d\mathscr{P}$ inside the vessel will be

$$d\mathscr{P} \doteq -\frac{\gamma \mathscr{P}_0}{V_0} dV = -\frac{\gamma \mathscr{P}_0}{V_0} \pi a^2 \langle y \rangle_S \qquad (4.33)$$

This generates an additional force $d\mathscr{P} \, dx \, dz$ on each incremental area $dx \, dz$ of the membrane.

Inclusion of this force in (4.1) modifies (4.4), which now becomes

$$\nabla^2 y - \frac{1}{c^2} \frac{\gamma \mathscr{P}_0}{\rho_s V_0} \pi a^2 \langle y \rangle_s = \frac{1}{c^2} \frac{\partial^2 y}{\partial t^2} \tag{4.34}$$

In this equation $\langle y \rangle_s$ is an integral function of all the allowed modes of vibration, including their relative amplitudes and phases. A general solution of (4.34) is consequently too involved to be considered. However, if only one mode of vibration is present, solution is greatly simplified.

Recall that for all normal modes that depend on θ, the average displacement amplitude is zero; none of these modes will be affected by the pressure fluctuation in the cavity. Thus only the symmetric modes involving J_0 need to be examined. If only one frequency ω is present, we may assume a solution of the form

$$y = \Psi e^{j\omega t}$$

where Ψ is a function of r only. Then (4.34) becomes

$$\frac{d^2 \Psi}{dr^2} + \frac{1}{r} \frac{d\Psi}{dr} + k^2 \Psi = \frac{\gamma \mathscr{P}_0}{\mathscr{T} V_0} \int_0^a \Psi 2\pi r \, dr \tag{4.35}$$

To solve this differential equation we assume that it is satisfied by a function of the form

$$\Psi = A[J_0(kr) - J_0(ka)] \tag{4.36}$$

A solution of this type is suggested by the fact that, if the right-hand integral term of (4.35) were not present, the resulting solution would involve $J_0(kr)$, whereas the presence of this integral term involving the radius a may be expected to introduce some function of a, such as $J_0(ka)$, as an additional term in the solution. The assumed solution has the further desirable property of satisfying the boundary condition that $\Psi = 0$ when $r = a$, regardless of the particular value assigned to k.

With this assumption, the right-hand term of (4.35) may be integrated and becomes

$$\frac{2\pi \gamma \mathscr{P}_0}{\mathscr{T} V_0} A \left[\frac{r J_1(kr)}{k} - \frac{r^2}{2} J_0(ka) \right]_0^a$$

$$= \frac{\pi a^2 \gamma \mathscr{P}_0}{\mathscr{T} V_0} A \left[\frac{2 J_1(ka)}{ka} - J_0(ka) \right] = \frac{\pi a^2 \gamma \mathscr{P}_0}{\mathscr{T} V_0} A J_2(ka) \tag{4.37}$$

Direct substitution of (4.36) into (4.35) then shows that the assumed solution is allowed, provided that

$$-k^2 J_0(ka) = \frac{\pi a^2 \gamma \mathscr{P}_0}{\mathscr{T} V_0} J_2(ka)$$

or

$$J_0(ka) = -B \frac{J_2(ka)}{(ka)^2} \tag{4.38}$$

where

$$B = \frac{\pi a^4 \gamma \mathscr{P}_0}{\mathscr{T} V_0}$$

is a nondimensional constant measuring the relative importance of the restoring forces, owing to the compression of the air in the vessel and to the tension applied to the membrane. This constant B is small if either the volume of the vessel or the tension in the membrane is large. In the limit, where this constant approaches zero, the allowed frequencies are those corresponding to $J_0(ka) = 0$, as previously determined for free vibrations of the membrane.

Table 4.2 lists the values of ka satisfying (4.38) for selected values of B ranging from 0 to 10. It will be seen that the presence of the vessel has raised the numerical magnitudes of the allowed values of ka, and consequently those of the allowed frequencies. This is to be expected, since the additional term in (4.34) is proportional to the displacement and is therefore one of stiffness. It is also to be noted that the effect on the fundamental frequency is much more pronounced than it is on the other modes of vibration. This is also to be expected, since the average displacement amplitude becomes smaller and smaller as the membrane vibrates in higher modes and consequently in a greater number of oppositely phased segments.

Since only the frequencies of the modes y_{0n} are affected by the pressure fluctuations of the vessel, the area πa^2 of the drumhead and the volume V_0 of the vessel are parameters that can be varied to alter the tonal qualities of the kettledrum. Variation of B affects the *relative* values of the f_{0n} frequencies, and altering a and V_0 such that a^4/V_0 remains constant will vary the nonsymmetric overtones $f_{mn}(m \neq 0)$ with respect to the symmetric ones.

Table 4.2. Frequencies of a kettledrum

B	$k_1 a$	$k_2 a$	$k_3 a$
0	2.405	5.520	8.654
1	2.545	5.54	8.657
2	2.68	5.55	8.660
5	3.02	5.59	8.67
10	3.485	5.67	8.69

4.7 FORCED VIBRATIONS OF A MEMBRANE. Let us next consider a circular membrane that is acted on by a sinusoidal driving pressure, uniformly distributed over one side only. Then this pressure is $p = P \cos \omega t$, or in complex notation,

$$\mathbf{p} = Pe^{j\omega t} \tag{4.39}$$

and the equation of motion becomes

$$\frac{\partial^2 \mathbf{y}}{\partial t^2} = c^2 \nabla^2 \mathbf{y} + \frac{P}{\rho_S} e^{j\omega t} \tag{4.40}$$

If we assume a steady-state solution of the form

$$\mathbf{y} = \mathbf{\Psi} e^{j\omega t} \tag{4.41}$$

then

$$\nabla^2 \mathbf{\Psi} + k^2 \mathbf{\Psi} = -\frac{P}{\rho_S c^2} = -\frac{P}{\mathcal{T}} \tag{4.42}$$

where

$$k = \omega/c$$

It should be noted that for the driven membrane, the angular frequency ω may have any value, so that k is not restricted to the discrete sets of values discussed in preceding sections for free vibrations.

The complete solution of (4.42) is the sum of two terms, one being the general solution of the equation $\nabla^2 \Psi_h + k^2 \Psi_h = 0$ and the other being the particular solution

$$\Psi_p = -P/(k^2 \mathcal{T})$$

Then

$$\Psi = A J_0(kr) - P/(k^2 \mathcal{T}) \tag{4.43}$$

Application of the boundary condition $\Psi = 0$ at $r = a$ gives

$$A = \frac{P}{k^2 \mathcal{T}} \cdot \frac{1}{J_0(ka)} \tag{4.44}$$

and the assumed equation for the displacement of the membrane becomes

$$y = \frac{P}{k^2 \mathcal{T}} \left[\frac{J_0(kr)}{J_0(ka)} - 1 \right] e^{j\omega t} \tag{4.45}$$

Correspondingly, the amplitude of displacement at any coordinate position on the membrane is given by

$$\Psi = \frac{P}{\mathcal{T}} \left[\frac{J_0(kr) - J_0(ka)}{k^2 J_0(ka)} \right] \tag{4.46}$$

An inspection of this equation shows that the amplitude of the displacement is directly proportional to that of the driving force and inversely proportional to the tension $\mathcal{T}$. The dependence of amplitude of vibration, at any coordinate position, on frequency is given by the relatively complicated expression within the square bracket of (4.46). Whenever the driving frequency ω corresponds to any of the free oscillation frequencies of (4.26) or (4.27), the function $J_0(ka) = 0$, so that an infinite amplitude is indicated. However, in all practical cases there are damping forces, which may be represented in (4.40) by a term of the type $-(R/\rho_S)(\partial y/\partial t)$, and which limit the amplitudes at these frequencies to finite maximum values.

The most important practical application of a driven membrane is that of the circular diaphragm of a condenser microphone. Here the incident sound wave, acting on a tightly stretched metallic membrane placed above a metal plate, produces an approximately uniform driving force. As the membrane is displaced, the electrical capacitance between the membrane and the adjacent plate is varied in such a manner as to generate an output voltage which is a linear function of the average displacement of the membrane. This microphone will be considered in Chapter 14.

The average displacement of the driven membrane is

$$\langle y \rangle_s = \frac{e^{j\omega t} \int_0^a \frac{P}{k^2 \mathcal{T}} \left[\frac{J_0(kr)}{J_0(ka)} - 1 \right] 2\pi r \, dr}{\pi a^2} = \frac{P}{k^2 \mathcal{T}} \frac{J_2(ka)}{J_0(ka)} e^{j\omega t} \tag{4.47}$$

At low frequencies, such that ka is smaller than unity,

$$J_0(ka) \approx \frac{1 - (ka)^2}{4}$$

and

$$J_2(ka) \approx \frac{(ka)^2}{8}\left[1 - \frac{(ka)^2}{12}\right]$$

giving

$$\frac{J_2(ka)}{J_0(ka)} \approx \frac{(ka)^2}{8}\left[1 + \frac{(ka)^2}{6}\right]$$

Substitution of this expression into (4.47) gives for the average displacement at low frequencies

$$\langle y \rangle_s \approx \frac{Pa^2}{8\mathscr{T}}\left[1 + \frac{(ka)^2}{6}\right]e^{j\omega t} \tag{4.48}$$

Therefore, as long as the driving frequency is low enough so that ka is smaller than unity, the output of a condenser microphone is nearly independent of frequency. In this frequency range there are also no resonance difficulties, which first arise when $ka = 2.405$. Replacing k by

$$k = 2\pi f/c = 2\pi f/\sqrt{\mathscr{T}/\rho_s}$$

and assuming that the limiting frequency of uniform response of an ideal condenser microphone is given by $ka < 1$, then

$$f < \frac{1}{2\pi a}\sqrt{\frac{\mathscr{T}}{\rho_s}} \tag{4.49}$$

This upper frequency limit may be increased either by increasing the tension $\mathscr{T}$ or by decreasing the radius a. Excellent small condenser microphones having uniform output over a wide range of frequencies have been produced. However, it should be noted that either an increase in $\mathscr{T}$ or a decrease in a reduces the amplitude of the average displacement $\langle y \rangle_s$ and, consequently, the voltage output of the microphone.

If a damping force $-(R/\rho_s)(\partial y/\partial t)$ is introduced into (4.40), the resulting solution for the displacement y is identical in form with (4.45), except that k must be replaced with $\mathbf{k}$ where

$$\mathbf{k}^2 = (\omega/c)^2 - j\omega R/\mathscr{T} \tag{4.50}$$

It can be shown that the presence of the term $-j\omega R/\mathscr{T}$ in this expression reduces the average displacement at resonance to a finite value. By a proper choice of damping resistance it is possible to take advantage of this reduction in the amplitude near resonance so as to extend the range of fairly uniform output up to, and somewhat beyond, the first resonance frequency. Commercial microphones are available in which this method of extending the frequency range is actually employed.

A response curve showing the normalized average displacement amplitude of a driven membrane without dissipation as computed by means of (4.47) is given in Fig. 4.4. It will be noted that the amplitude becomes infinite for $ka = 2.405$. The graph also gives a typical curve illustrating the reduction in amplitude resulting from the presence of a damping force. Both these curves indicate zero response at the frequency for which $ka = 5.136$, the value making $J_2(ka) = 0$. At a frequency about 60 percent above the first resonance frequency, a circular nodal line appears at the outer edge of the membrane, and as the frequency is further increased this line shrinks toward the center. The displacement of the central part of the membrane is out of phase with the driving force, while that of the outer part is in phase, so that as the driving frequency increases and the nodal circle shrinks there is an increasing

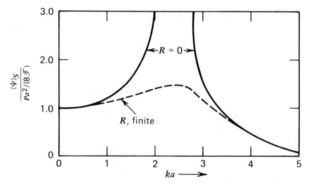

Fig. 4.4. *Average displacement response* $\langle \Psi \rangle_S$ *of a driven membrane as a function of frequency.*

tendency for the average displacement of the two zones to cancel each other. When $ka = 5.136$ there is complete cancellation, so that the average displacement, and hence the output, is zero.

4.8   VIBRATION OF THIN PLATES.   The essential difference between the vibration of a membrane and a thin plate is that in a membrane the restoring force is due entirely to the tension applied to the membrane, whereas in a thin plate the restoring force is due entirely to the stiffness of the diaphragm, no tension being applied. This same difference exists between the transverse restoring forces in strings and bars.

Our analysis of the plate will be limited to the *symmetrical vibrations* of a *uniform circular diaphragm.* Since a rigorous development of the equation of motion applicable to this case is too involved to be given here, the resulting equation will merely be written down and will then be justified by analogy with the equations for bars and membranes. This equation is

$$\frac{\partial^2 y}{\partial t^2} = -\frac{\kappa^2 Y}{\rho(1 - \sigma^2)} \nabla^2(\nabla^2 y) \tag{4.51}$$

where ρ is the volume density of the material, σ is its *Poisson's ratio,* Y is Young's modulus, and κ is a surface *radius of gyration.* For a circular plate of uniform thickness d

$$\kappa = d/\sqrt{12}.$$

Since the restoring force acting on a circular plate depends on its elastic resistance to bending, we would expect the coefficient of the right-hand term in the above differential equation to be similar to that for the transverse vibration of a bar (3.45), $-\kappa^2 Y/\rho$. This assumption is not strictly correct, for there is a tendency for the sheet to curl up sidewise when it is bent down lengthwise. This curling results from the lateral expansion that accompanies a longitudinal compression and produces an increase in the effective stiffness in the sheet. The negative ratio of the lateral strain $\partial\zeta/\partial z$ which accompanies a longitudinal strain $\partial\xi/\partial x$ is known as *Poisson's ratio* σ,

$$\sigma = -\frac{\partial\zeta/\partial z}{\partial\xi/\partial x}$$

Since a positive longitudinal strain of tension is always accompanied by a negative lateral strain of compression, σ is a positive number. Values of Poisson's ratio for a number of materials are given in Appendix A10. Note that σ is about 0.3 for most materials. The effective increase in stiffness of the sheet resulting from the curling yields the factor $(1 - \sigma^2)^{-1}$ in (4.51).

Let us assume periodic vibration,

$$y = \Psi e^{j\omega t} \tag{4.52}$$

where Ψ is a function only of r. Substitution into (4.51) yields

$$\nabla^2(\nabla^2\Psi) - K^4\Psi = 0 \tag{4.53a}$$

where

$$K^4 = \frac{\omega^2\rho(1 - \sigma^2)}{\kappa^2 Y} \tag{4.53b}$$

Substitution of the relationship

$$\nabla^2\Psi = -K^2\Psi \tag{4.54a}$$

into (4.53a) reveals that if Ψ satisfies (4.54a), it also is a solution to (4.53a). This is also true if Ψ satisfies

$$\nabla^2\Psi = K^2\Psi \tag{4.54b}$$

Thus, the complete solution of (4.53a) is the sum of the four independent solutions to

$$\nabla^2\Psi \pm K^2\Psi = 0$$

Equation (4.54a) is the Helmholtz equation with circular symmetry. Its solutions are $J_0(Kr)$ and $Y_0(Kr)$, but the latter must be discarded since we require finite displacement at the center of the plate ($r = 0$). Equation (4.54b) must be solved by $J_0(jKr)$ and $Y_0(jKr)$, and again the latter must be discarded. Now, $J_0(jx)$ is customarily written as $I_0(x)$, a modified Bessel function of the first kind, and the solution of (4.53a) is

$$\Psi = AJ_0(Kr) + BI_0(Kr) \tag{4.55}$$

Some of the properities of the modified Bessel functions $I_m(x)$ are given in Appendix A4. These functions are not solutions of Bessel's equation, but the related equation

$$\frac{d^2y}{dx^2} + \frac{1}{x}\frac{dy}{dx} - \left(1 + \frac{m^2}{x^2}\right)y = 0 \tag{4.56}$$

It should be noted that, whereas $J_0(x)$ is an alternating function, $I_0(x)$ increases continuously with x, so that in contrast to $J_0(x)$, which resembles a damped cosine function, $I_0(x)$ resembles a hyperbolic cosine. Values of the functions I_0, I_1, and I_2 are given in Appendix A5.

To evaluate the constants **A** and **B** we must know the manner in which the diaphragm is supported. The most common type of support is rigid clamping of the diaphragm all around its circumference at $r = a$. This is equivalent to

$$\Psi = 0 \quad \text{and} \quad \frac{\partial \Psi}{\partial r} = 0 \qquad \text{at } r = a \tag{4.57}$$

These conditions give

$$AJ_0(Ka) = BI_0(Ka) \tag{4.58a}$$

and

$$-AJ_1(Ka) = BI_1(Ka) \tag{4.58b}$$

and dividing one by the other gives

$$\frac{J_0(Ka)}{J_1(Ka)} = -\frac{I_0(Ka)}{I_1(Ka)} \tag{4.59}$$

which determines the allowed values of K.

Since both I_0 and I_1 are positive for all values of Ka, solutions occur only when J_0 and J_1 are of opposite sign. From the tables of Bessel functions it can be seen that this equation is satisfied by

$$Ka = 3.20,\ 6.30,\ 9.44,\ 12.57, \ldots \tag{4.60}$$

or approximately by

$$Ka = n\pi \qquad n = 1, 2, 3, \ldots$$

with the approximation becoming increasingly better with increasing n.

From (4.53b) it is evident that

$$\omega = \kappa K^2 \sqrt{\frac{Y}{\rho(1 - \sigma^2)}}$$

and substituting $3.20/a$ for K gives

$$f_1 = \frac{\omega_1}{2\pi} = \frac{3.2^2}{2\pi a^2} \frac{d}{\sqrt{12}} \sqrt{\frac{Y}{\rho(1 - \sigma^2)}} = 0.47 \frac{d}{a^2} \sqrt{\frac{Y}{\rho(1 - \sigma^2)}} \tag{4.61}$$

where d is the thickness of the diaphragm. The frequencies of the other modes of vibration are not harmonics of the fundamental and are given by

$$f_2 = \left(\frac{6.3}{3.2}\right)^2 f_1 = 3.88 f_1$$

$$f_3 = 8.70 f_1$$

etc.

The natural frequencies are spread much farther apart than are those of the circular membrane.

The actual displacement of a thin circular plate vibrating in its fundamental mode is

$$y_1 = \cos(\omega_1 t + \phi_1)\left[A_1 J_0\left(\frac{3.2}{a}r\right) + B_1 I_0\left(\frac{3.2}{a}r\right)\right]$$

From (4.58a)

$$B_1 = -A_1 \frac{J_0(K_1 a)}{I_0(K_1 a)} = -A_1 \frac{J_0(3.2)}{I_0(3.2)} = +0.0555 A_1$$

and therefore

$$y_1 = A_1 \cos(\omega_1 t + \phi_1)\left[J_0\left(\frac{3.2}{a}r\right) + 0.0555 I_0\left(\frac{3.2}{a}r\right)\right] \qquad (4.62)$$

Note that the amplitude at the center is not A_1, but is instead $1.0555 A_1$.

A comparison of the shape function

$$J_0\left(\frac{3.2}{a}r\right) + 0.0555 I_0\left(\frac{3.2}{a}r\right)$$

for a thin circular plate vibrating in its fundamental mode, with the corresponding shape function

$$J_0\left(\frac{2.405}{a}r\right)$$

for a membrane, shows that the relative displacement of the plate near its edge is much smaller than that of a similar membrane. Consequently, we should expect the ratio of its average amplitude to the amplitude at the center to be less than for the membrane. The average displacement amplitude is

$$\langle \Psi_1 \rangle_S = 0.326 A_1 = 0.309 y_0$$

where $y_0 = 1.0555 A_1$ is the amplitude at the center of the plate. The circular plate may therefore be replaced by an equivalent flat piston such that

$$S_{eq} \xi_{eq} = 0.309 y_0 \pi a^2 \qquad (4.63)$$

The treatment of loaded and driven plates is analogous to that of membranes, and the response curves for a uniform driving force are similar to those shown in Fig. 4.4, with large amplitudes occurring at the fundamental resonance frequency unless there is considerable damping. Condenser microphones may be constructed with a thin circular plate instead of a stretched membrane. However, it is difficult to design a vibrating-plate type of microphone having adequate sensitivity and at the same time a sufficiently high resonant frequency. For this reason membranes are used in most condenser microphones, although plate diaphragms have been successful in miniature condenser microphones.

The most important utilization of the vibrating thin plate is in the diaphragms of ordinary telephone microphones and receivers. Although the response of these devices is not uniform over a wide range of frequencies, they give adequate intelligibility and are simple and rugged in their construction. Another application is in sonar transducers used for producing sounds in water at frequencies below 1 kHz; sound is generated by the motion of relatively thin circular steel plates driven by alternations in the magnetic field of an adjacent electromagnet.

PROBLEMS

Except when otherwise noted, all membranes should be considered to be fixed at their rims.

4.1 A square membrane of width a vibrates at its fundamental frequency with an amplitude A at its center. (a) Derive a general expression that gives its average displacement amplitude. (b) Derive a general expression for locating points on the membrane having an amplitude of 0.5 A. (c) Compute and plot a few points as given by the equation derived in part (b). Do they form a circle?

4.2. A rectangular membrane has a width a and a length b. If $b = 2a$, compute the ratio of each of the first four overtone frequencies relative to the fundamental frequency.

4.3. A square membrane with sides of length L, uniform surface density ρ_S, and uniform tension $\mathcal{T}$ is fixed on three sides and free on the other. (a) Find the frequency of the fundamental mode. (b) Write a general expression for the natural frequencies and one for the normal modes. (c) Sketch the nodal patterns for the three normal modes with the lowest natural frequencies.

4.4. Although it may be hard to do physically, it is not hard to imagine a circular membrane with a free rim. (a) Write the general expression for the normal modes. (b) Sketch the nodal patterns for the three normal modes with the lowest natural frequencies. (c) Find the frequencies of these three normal modes in terms of the tension and surface density.

4.5. A steel membrane of 0.02-m radius and 0.0001-m thickness is stretched to a tension of 20,000 N/m. (a) For circularly symmetrical vibration, what is the frequency of the second overtone mode? (b) What are the radii of the two nodal circles when the membrane is vibrating at the above frequency? (c) When the membrane is vibrating at the above frequency, the displacement amplitude at the center is observed to be 0.0001 m. What is the average displacement amplitude?

4.6. A circular membrane of 0.25-m radius has an area density of 1.0 kg/m^2 and is stretched to a tension of 25,000 N/m. (a) Compute the four lowest frequencies of free vibration. (b) For each of these frequencies locate any nodal circles.

4.7. A circular membrane of 1-cm radius and 0.2-kg/m^2 area density is stretched to a linear tension of 4000 N/m. When vibrating in its fundamental mode, the amplitude at the center is observed to be 0.01 cm. (a) What is its fundamental frequency? (b) What is the maximum volume of air displaced by the membrane?

4.8. Show that the total energy of a circular membrane when vibrating in its fundamental mode is given by $0.135\pi a^2 \rho_S \omega^2 A_1^2$, where a is its radius, ρ_S the area density, ω the angular frequency of vibration, and A_1 the amplitude at its center.

4.9. The maximum tensile stress that may be applied to aluminum is 2×10^8 Pa and to steel is 10^9 Pa. (a) What is the maximum fundamental frequency of a stretched aluminum membrane of 0.01-m radius? (b) of a steel membrane of equal radius? (Note that for thin membranes these maximum frequencies are independent of the thickness.)

4.10. A circular membrane is acted on uniformly over its surface by a damping force per unit area of $-(R\partial y/\partial t)$. Introduce this term into (4.5) in a manner consistent with the dimensions of the terms in the latter, and solve the resulting equation so as to show that the amplitude of the resulting free vibrations are damped exponentially as $\exp(-\tfrac{1}{2}Rt/\rho_S)$.

4.11. The circular membrane of a kettledrum has a radius of 0.25 m, its area density is 1.0 kg/m^2, and it is stretched to a tension of 10,000 N/m. (a) What is its fundamental frequency without a backing vessel? (b) What is its fundamental frequency if the backing vessel is a hemispherical bowl of 0.25-m radius? Assume the backing vessel to be filled with air at a pressure of 10^5 Pa, having a ratio of specific heats of 1.4.

4.12. (a) Compute and plot the shape of the driven circular membrane when being driven at a frequency one-half of its fundamental frequency. (b) Similarly, compute and plot the shape of the membrane when being driven at twice its fundamental frequency.

4.13. An undamped circular membrane of 0.02-m radius, having an area density of 1.5 kg/m^2 and stretched to a tension of 950 N/m, is driven by a pressure 6000 cos ωt Pa acting uniformly over its surface. (*a*) Compute and plot the amplitude of the displacement at the center as a function of frequency in the interval from 0 to 1000 Hz. (*b*) Compute and plot the shape of the membrane when driven at a frequency of 400 Hz. (*c*) Repeat part (*b*) for a frequency of 1000 Hz.

4.14. The diaphragm of a condenser microphone consists of a circular sheet of aluminum of 0.03-m diameter and 0.00002-m thickness. It may be stretched to a maximum tensile stress of 2×10^8 Pa. (*a*) What is the maximum tension in newtons per meter to which this aluminum sheet may be stretched? (*b*) What will be its fundamental frequency when stretched to this tension? (*c*) What will be the displacement amplitude at its center when acted on by a sound wave of 500 Hz having a pressure amplitude of 2.0 Pa? (*d*) What will be the average displacement amplitude under these conditions?

4.15. If the volume of air trapped behind the diaphragm of the condenser microphone of Problem 4.14 is 3×10^{-7} m^3, by what percentage will its fundamental frequency be raised? Assume $P_0 = 10^5$ Pa and $\gamma = 1.4$.

4.16. By integration over its surface show that the average displacement amplitude of a thin plate when vibrating in its fundamental mode of vibration equals $0.309A$, where A is the displacement amplitude at the center of the circular plate. Assume the plate to be rigidly clamped at its rim.

4.17. The diaphragm of a telephone receiver consists of a circular sheet of steel, 4 cm in diameter and 0.02 cm thick. (*a*) If it is rigidly clamped at its rim, what is its fundamental frequency of vibration? (*b*) What will be the effect on this frequency of doubling the thickness of the diaphragm? (*c*) Of doubling the diameter?

4.18. To what tension would the diaphragm of Problem 4.17 need to be stretched if its fundamental frequency, considered as resulting from the restoring forces of tension alone, were to equal that resulting from stiffness forces alone?

4.19. (*a*) Determine the ratio of the constants B_2/A_2 for a thin circular plate clamped at its rim and vibrating in its first overtone mode of vibration. (*b*) Express the resulting motion in the form of an equation analogous to equation (4.62). (*c*) Plot the shape function of the diaphragm when vibrating in this mode. (*d*) What is the ratio of the radius of the nodal circle to the radius of the plate when vibrating in this mode?

4.20. The vibrating circular steel plate of an electromagnetic-drive sonar transducer is rigidly clamped at its rim and has a radius of 0.1 m and a thickness of 0.005 m. What is its fundamental frequency of vibration?

CHAPTER 5

THE ACOUSTIC WAVE EQUATION AND SIMPLE SOLUTIONS

5.1 INTRODUCTION. Acoustic waves that produce the sensation of sound are one of a variety of pressure disturbances that can propagate through a compressible fluid. There are also ultrasonic and infrasonic waves whose frequencies are beyond the audible limits, high-intensity waves (such as those present near jet engines and missiles) which may produce a sensation of pain rather than of sound, and shock waves generated by explosions and supersonic aircraft.

Acoustic waves in inviscid fluids are longitudinal waves: the molecules move back and forth in the direction of propagation of the wave, producing adjacent regions of compression and rarefaction similar to those produced by longitudinal waves in a bar. Fluids exhibit fewer constraints to deformations than do solids. As a result, the restoring force responsible for propagating a wave is simply the pressure change that occurs when a fluid is compressed or expanded.

Previously, we have been able to limit the difficulty of the mathematics by restricting our waves to one or two dimensions. It is now appropriate to discuss the behavior of waves in three dimensions. Once the three-dimensional wave equation has been developed, it will be convenient to look at some simple examples, starting with the simplest case of plane waves.

The characteristic property of plane waves is that each acoustic variable (particle displacement, density, pressure, etc.) has constant amplitude on any given plane perpendicular to the direction of wave propagation. Plane waves may be produced readily in a fluid that is confined in a rigid pipe through the action of a low-frequency vibrating piston located at one end of the pipe. Since wave fronts of any type of divergent wave in a homogeneous medium become nearly planar far from their source, we may expect that the properties of divergent waves will, at large distances, also become very similar to those of plane waves.

The following symbols* will be used :

$\vec{r}$ = equilibrium position of a particle of the fluid at (x, y, z)

$$\vec{r} = x\hat{x} + y\hat{y} + z\hat{z}$$

---

* The symbols $\hat{x}$, $\hat{y}$, and $\hat{z}$ represent the *unit vectors* in the x, y, and z directions, respectively.

$\vec{\xi}$ = particle displacement from the equilibrium position

$$\vec{\xi} = \xi_x \, \hat{x} + \xi_y \, \hat{y} + \xi_z \, \hat{z}$$

$\vec{u}$ = *particle velocity*

$$\vec{u} = \frac{\partial \vec{\xi}}{\partial t} = u_x \, \hat{x} + u_y \, \hat{y} + u_z \, \hat{z}$$

ρ = *instantaneous density* at any point
ρ_0 = constant equilibrium density of the fluid
s = *condensation* at any point

$$s = (\rho - \rho_0)/\rho_0$$

$\mathscr{P}$ = instantaneous pressure at any point
$\mathscr{P}_0$ = constant equilibrium pressure in the fluid
p = excess pressure or *acoustic pressure* at any point

$$p = \mathscr{P} - \mathscr{P}_0$$

c = phase speed of the wave
Φ = *velocity potential*

$$\vec{u} = \nabla \Phi$$

T_K = temperature in kelvins (K)
T = temperature in degrees Celsius (or centigrade) (°C)

$$T + 273.15 = T_K$$

The term *particle of the fluid* means a volume element large enough to contain millions of molecules so that the fluid may be thought of as a continuous medium, yet small enough that all acoustic variables may be considered nearly constant throughout the volume element. The molecules of a fluid do not have fixed mean positions in the medium; even without the presence of a wave, they are in constant motion, with average velocities far in excess of any particle velocity associated with the wave motion. However, a small volume may be treated as an unchanging unit since those molecules leaving its confines are replaced by an equal number possessing (on the average) identical properties, so that the macroscopic properties of the element remain unchanged. As a consequence, it is possible to speak of particle displacements and velocities when discussing acoustic waves in fluids, as was done for elastic waves in solids. In the following analysis the effects of gravitational forces will be neglected, so that ρ_0 and $\mathscr{P}_0$ have uniform values throughout the fluid. The fluid is also assumed to be homogeneous, isotropic, and perfectly elastic; no dissipative effects such as those arising from viscosity or heat conduction are present. Finally, the analysis will be limited to waves of relatively small amplitude so that changes in density of the medium will be small compared with its equilibrium value, $|s| \ll 1$. These assumptions are necessary, to arrive at the simplest theory for sound in fluids. We are indeed fortunate that experiments have shown that this simplest theory adequately describes most common acoustical phenomena. However, one must remember that there are interesting situations where these assumptions are violated and the theory must be modified.

5.2 THE EQUATION OF STATE. The equation of state for a fluid relates the internal restoring forces to the corresponding deformations, as was done for oscillators, strings, and bars. As before, we will search for a linear relationship which, while simplifying the development, restricts the amount of allowed deformation. For fluid media, the equation of state must relate three physical quantities describing the thermodynamic behavior of the fluid. For example, the equation of state for a *perfect gas*

$$\mathscr{P} = \rho r T_K \tag{5.1}$$

gives the relationship between the total pressure $\mathscr{P}$ in pascals (Pa), the density in kilograms per cubic meter (kg/m^3), and the absolute temperature T_K in kelvins. The quantity r is a constant whose value depends on the particular gas involved. This equation is general and describes any thermodynamic process for a perfect gas (see Appendix A9).

Greater simplification can be achieved if the thermodynamic process is *restricted*. For example, if the fluid is contained within a vessel whose walls are highly thermally conductive, then slow variations in the volume of the vessel will result in thermal energy being transferred between the walls and the fluid. If the walls have sufficient thermal capacity, they and the fluid will remain at a constant temperature. In this case, the perfect gas is described by the *isothermal* equation of state

$$\frac{\mathscr{P}}{\mathscr{P}_0} = \frac{\rho}{\rho_0} \qquad \text{(perfect gas)}$$

On the other hand, it is found experimentally that *acoustic* processes are nearly *adiabatic*: there is insignificant exchange of thermal energy from one particle of fluid to another. Under these conditions, the entropy (and not the temperature) of the fluid remains nearly constant. The behavior of the perfect gas under these conditions is described by the *adiabatic* equation of state

$$\frac{\mathscr{P}}{\mathscr{P}_0} = \left(\frac{\rho}{\rho_0}\right)^{\gamma} \qquad \text{(perfect gas)} \tag{5.2}$$

where γ is the ratio of the specific heats (see Appendix A9, Eq. (A9.12)). For the acoustic disturbance of the fluid to be adiabatic, neighboring elements of the fluid must not exchange thermal energy. This means that the thermal conductivity of the fluid and the temperature gradients of the disturbance must be small enough that no significant thermal flux occurs during the time of the disturbance. For the frequencies and amplitudes usually of interest in acoustics, this is the case. The major effect of finite thermal conductivity is to dissipate very small fractions of the acoustic energy so that the disturbance attenuates slowly with time or distance. These effects will be considered in Chapter 7.

For fluids other than a perfect gas, the adiabatic equation of state is more complicated. In these cases it is preferable to determine experimentally the isentropic relationship between pressure and density fluctuations. Given this relationship, we write a Taylor's expansion

$$\mathscr{P} = \mathscr{P}_0 + \left(\frac{\partial \mathscr{P}}{\partial \rho}\right)_{\rho_0} (\rho - \rho_0) + \frac{1}{2}\left(\frac{\partial^2 \mathscr{P}}{\partial \rho^2}\right)_{\rho_0} (\rho - \rho_0)^2 + \cdots \qquad (5.3)$$

where the partial derivatives are constants determined for the adiabatic compression and expansion of the fluid about its equilibrium density. If the fluctuations are small, only the lowest order term in $(\rho - \rho_0)$ need be retained. This gives a linear relationship between the pressure fluctuation and the change in density

$$\mathscr{P} - \mathscr{P}_0 \doteq \mathscr{B}\left(\frac{\rho - \rho_0}{\rho_0}\right) \qquad (5.4)$$

where $\mathscr{B} = \rho_0(\partial \mathscr{P}/\partial \rho)_{\rho_0}$ is the *adiabatic bulk modulus*. In terms of the acoustic pressure p and the condensation s, (5.4) can be reexpressed as

$$\boxed{p \doteq \mathscr{B}s} \qquad (5.5)$$

The essential restriction is that the condensation must be small, $|s| \ll 1$.

5.3   THE EQUATION OF CONTINUITY.   To relate the motion of the fluid to its compression or dilatation, we need a functional relationship between the particle velocity $\mathring{u}$ and the instantaneous density ρ. Consider a small rectangular-parallelepiped volume element $dV = dx\, dy\, dz$ which is *fixed* in space and through which elements of the fluid travel. The net rate with which mass flows into the volume through its surface must equal the rate with which the mass within the volume increases. Referring to Fig. 5.1, we see that the net influx of mass into this *spatially fixed* volume, resulting from flow in the x direction, is

$$\left\{\rho u_x - \left[\rho u_x + \frac{\partial(\rho u_x)}{\partial x}\, dx\right]\right\} dy\, dz = -\frac{\partial(\rho u_x)}{\partial x}\, dV$$

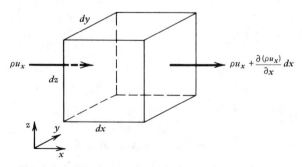

Fig. 5.1. *Mass flow in the x direction through a fixed volume dV.*

Similar expressions give the net influx for the y and z directions, so that the total influx must be

$$-\left[\frac{\partial(\rho u_x)}{\partial x} + \frac{\partial(\rho u_y)}{\partial y} + \frac{\partial(\rho u_z)}{\partial z}\right] dV \equiv -[\nabla \cdot (\rho \vec{u})] \, dV$$

where $\nabla \cdot$ is the *divergence* operator.* The rate with which the mass increases in the volume is $(\partial \rho / \partial t) \, dV$. Since the net influx must equal the rate of increase, we obtain

$$\frac{\partial \rho}{\partial t} + \nabla \cdot (\rho \vec{u}) = 0 \tag{5.6}$$

the *equation of continuity*. Notice that it is nonlinear; the second term involves the product of particle velocity and instantaneous density, both of which are acoustic variables. However, if we write $\rho = \rho_0(1 + s)$, use the fact that ρ_0 is a constant in both space and time, and assume that s is very small, (5.6) becomes

$$\boxed{\frac{\partial s}{\partial t} + \nabla \cdot \vec{u} \doteq 0} \tag{5.7}$$

the *linearized continuity equation*.

Before obtaining a force equation, let us combine the equations of state and continuity. If (5.7) is integrated with respect to time, we have

$$\int \left(\frac{\partial s}{\partial t} + \nabla \cdot \vec{u}\right) dt = \text{constant}$$

The integration constant must be zero since acoustic quantities vanish if there is no acoustic disturbance. Furthermore, since $\int \nabla \cdot \vec{u} \, dt = \nabla \cdot \int \vec{u} \, dt = \nabla \cdot \int (\partial \vec{\xi}/\partial t) \, dt = \nabla \cdot \vec{\xi}$, this becomes

$$s = -\nabla \cdot \vec{\xi} \tag{5.8}$$

and combination with the equation of state (5.5) gives

$$p \doteq -\mathscr{B}\nabla \cdot \vec{\xi}$$

Notice that if we consider one-dimensional waves (as was done in Chapter 3), the compressive force is $f = pS$, $\nabla \cdot \vec{\xi}$ reduces to $\partial \xi_x / \partial x$, and the above equation becomes $f = -S\mathscr{B} \, \partial \xi_x / \partial x$, which is equivalent to (3.5) with $Y = \mathscr{B}$. Thus, $p = -\mathscr{B}\nabla \cdot \vec{\xi}$ is the three-dimensional analog of $f = -SY(\partial \xi / \partial x)$, written in terms of pressure rather than compressive force.

5.4   THE SIMPLE FORCE EQUATION: EULER'S EQUATION.

In real fluids, the existence of viscosity and the failure of acoustic processes to be perfectly adiabatic introduce dissipative terms. However, since we have already neglected the

---

* If readers are unfamiliar with the operator ∇, they should refer to Appendix A7.

effects of thermal conductivity in the equation of state, we also ignore the effects of viscosity and consider the fluid to be *inviscid*.

Consider a fluid element $dV = dx\, dy\, dz$ which *moves with the fluid*, containing a specified mass dm of fluid. The net force $d\vec{f}$ on the element will accelerate it according to Newton's second law $d\vec{f} = \vec{a}\, dm$. In the absence of viscosity, the net force experienced by the element in the x direction is

$$df_x = \left[\mathscr{P} - \left(\mathscr{P} + \frac{\partial \mathscr{P}}{\partial x} dx \right) \right] dy\, dz = -\frac{\partial \mathscr{P}}{\partial x} dV$$

Analogous expressions for df_y and df_z allow us to write the complete vector force $d\vec{f} = df_x\, \hat{x} + df_y\, \hat{y} + df_z\, \hat{z}$ as

$$d\vec{f} = -\nabla \mathscr{P}\, dV$$

The acceleration of the fluid is a little more complicated. The particle velocity $\vec{u}$ is a function of both time and space. When the fluid element with velocity $\vec{u}(x, y, z, t)$ at (x, y, z) and time t moves to a new location $(x + dx, y + dy, z + dz)$ at a later time $t + dt$, its new velocity is $\vec{u}(x + dx, y + dy, z + dz, t + dt)$. Thus the acceleration is

$$\vec{a} = \lim_{dt \to 0} \frac{\vec{u}(x + u_x\, dt, y + u_y\, dt, z + u_z\, dt, t + dt) - \vec{u}(x, y, z, t)}{dt}$$

The move from the former position to the new one allows us to relate the incrementals through the velocity components of the element,

$$dx = u_x\, dt \qquad dy = u_y\, dy \qquad dz = u_z\, dz$$

Since we are assuming all incrementals very small, the new velocity can be expressed by the first terms of its Taylor's expansion

$$\vec{u}(x + u_x\, dt, y + u_y\, dt, z + u_z\, dt, t + dt) =$$

$$\vec{u}(x, y, z, t) + \frac{\partial \vec{u}}{\partial x} u_x\, dt + \frac{\partial \vec{u}}{\partial y} u_y\, dt + \frac{\partial \vec{u}}{\partial z} u_z\, dt + \frac{\partial \vec{u}}{\partial t} dt$$

and the acceleration of the chosen fluid element is

$$\vec{a} = \frac{\partial \vec{u}}{\partial t} + u_x \frac{\partial \vec{u}}{\partial x} + u_y \frac{\partial \vec{u}}{\partial y} + u_z \frac{\partial \vec{u}}{\partial z}$$

If we define the vector operator $(\vec{u} \cdot \nabla)$ as

$$(\vec{u} \cdot \nabla) = u_x \frac{\partial}{\partial x} + u_y \frac{\partial}{\partial y} + u_z \frac{\partial}{\partial z}$$

then $\vec{a}$ can be written more succinctly

$$\vec{a} = \frac{\partial \vec{u}}{\partial t} + (\vec{u} \cdot \nabla)\vec{u} \qquad (5.9)$$

Now, since the mass dm of the element is $\rho\,dV$, substitution into $d\vec{f} = \vec{a}\,dm$ gives

$$-\nabla\mathscr{P} = \rho\left[\frac{\partial\vec{u}}{\partial t} + (\vec{u}\cdot\nabla)\vec{u}\right] \tag{5.10}$$

This nonlinear, inviscid force equation is *Euler's equation*. It can be simplified if we require $|s| \ll 1$ and $|(\vec{u}\cdot\nabla)\vec{u}| \ll |\partial\vec{u}/\partial t|$. Then ρ can be replaced with ρ_0, and the term $(\vec{u}\cdot\nabla)\vec{u}$ can be dropped in (5.10). Finally, we can substitute $\nabla\mathscr{P} = \nabla p$, since $\mathscr{P}_0$ is a constant, and obtain

$$\rho_0\frac{\partial\vec{u}}{\partial t} \doteq -\nabla p \tag{5.11}$$

This is the *linear inviscid force equation*, valid for acoustic processes of small amplitude.

5.5 THE LINEARIZED WAVE EQUATION. The three equations (5.5), (5.7), and (5.11) must be combined to yield a single differential equation with one dependent variable. The particle velocity can be eliminated between (5.7) and (5.11). Take the divergence of (5.11)

$$\rho_0\nabla\cdot\frac{\partial\vec{u}}{\partial t} = -\nabla\cdot(\nabla p) = -\nabla^2 p$$

where ∇^2 is the three-dimensional Laplacian operator. Next, take the time derivative of (5.7) and use $\partial(\nabla\cdot\vec{u})/\partial t = \nabla\cdot(\partial\vec{u}/\partial t)$,

$$\frac{\partial^2 s}{\partial t^2} + \nabla\cdot\frac{\partial\vec{u}}{\partial t} = 0$$

Combination gives

$$\nabla^2 p = \rho_0\frac{\partial^2 s}{\partial t^2}$$

Use of the equation of state (5.5) to eliminate s yields

$$\nabla^2 p = \frac{1}{c^2}\frac{\partial^2 p}{\partial t^2} \tag{5.12}$$

where c is defined by

$$c = \sqrt{\mathscr{B}/\rho_0} \tag{5.13}$$

Equation (5.12) is the linearized, lossless wave equation for the propagation of sound in fluids. As should be obvious from the discussion of longitudinal waves in bars, c is the phase speed for acoustic waves in fluids.

Use of (5.13) allows the equation of state to be written in a more convenient form

$$p = \rho_0 c^2 s$$ (5.14)

Because p and s are proportional, the condensation satisfies the wave equation. Since the density ρ and the condensation are linearly related, the instantaneous density also satisfies the wave equation.

Since the curl of the gradient of a function must vanish, $\nabla \times \nabla f = 0$, from (5.11) the particle velocity must be irrotational, $\nabla \times \hat{u} = 0$. This means that it can be expressed as the gradient of a scalar function Φ,

$$\hat{u} = \nabla \Phi$$ (5.15)

where Φ is defined as the *velocity potential*. The physical meaning of this important result is that the acoustical excitation of an inviscid fluid involves no rotational flow; there are no effects such as boundary layers, shear waves, or turbulence. In real fluids, for which there is finite viscosity, the particle velocity is not curl-free everywhere, but for most acoustic processes the presence of small rotational effects is confined to the vicinity of boundaries and exerts little influence on the propagation of sound.

If we substitute (5.15) into (5.11) to obtain

$$\rho_0 \frac{\partial}{\partial t} \nabla \Phi = -\nabla p$$

or

$$\nabla \left(\rho_0 \frac{\partial \Phi}{\partial t} + p \right) = 0$$

and notice that the quantity in parentheses can be chosen to vanish identically if there is no acoustic excitation, then

$$p = -\rho_0 \frac{\partial \Phi}{\partial t}$$ (5.16)

Substitution of this equation into (5.12) and integrating with respect to time will show that Φ also satisfies the wave equation.

5.6 SPEED OF SOUND IN FLUIDS. By combining (5.4) and (5.13), we get a thermodynamic expression for the speed of sound

$$c = \sqrt{\left(\frac{\partial \mathscr{P}}{\partial \rho} \right)_{adiabatic}}$$ (5.17)

where the partial derivative is evaluated at equilibrium conditions of pressure and density. It is a characteristic property of the fluid dependent on the thermodynamic variables of temperature, pressure, and density. For ordinary acoustic waves corresponding to those normally audible to the human ear it is independent of frequency.

When a sound wave propagates through a *perfect gas*, the adiabatic gas law relating pressure and density may be utilized to derive an important special form of (5.17). Direct differentiation of (5.2) leads to

$$\left(\frac{\partial \mathscr{P}}{\partial \rho}\right)_{adiabatic} = \gamma \frac{\mathscr{P}}{\rho}$$

If this expression is now evaluated at ρ_0 and substituted into (5.17),

$$c = \sqrt{\gamma \mathscr{P}_0 / \rho_0} \tag{5.18}$$

is obtained.

Included in the appendix are values of γ and ρ_0 for various gases at $0°C$ and standard pressure $\mathscr{P}_0 = 1$ atm $= 1.013 \times 10^5$ Pa. Substitution of the appropriate values for air gives

$$c_0 = \sqrt{\frac{1.402 \times 1.013 \times 10^5}{1.293}} = 331.6 \text{ m/s}$$

as the theoretical value for the speed of sound in air at $0°C$. This is in excellent agreement with measured values and thereby supports our earlier assumption that acoustic processes in a fluid are adiabatic.

For most gases at constant temperature the ratio of $\mathscr{P}_0 / \rho_0$ is nearly independent of pressure: a doubling of pressure is accompanied by a doubling of the density of the gas, so that the speed of sound in a gas does not change with variations in the barometric pressure.

An alternate expression for the speed of sound in a perfect gas is found from (5.1) and (5.18) to be

$$c = \sqrt{\gamma r T_K} \tag{5.19}$$

The speed is proportional to the square root of the absolute temperature. In terms of the speed c_0 at $0°C$, this becomes

$$c = c_0 \sqrt{T_K / 273} = c_0 \sqrt{1 + T/273} \tag{5.20}$$

Theoretical prediction of the speed of sound for liquids is considerably more difficult than for gases. However, it is possible to show theoretically that $\mathscr{B} = \gamma \mathscr{B}_T$ where $\mathscr{B}_T$ is the isothermal bulk modulus. Since $\mathscr{B}_T$ is much easier to measure experimentally than $\mathscr{B}$, a convenient expression for the speed of sound in liquids is then obtained from (5.13) and the above,

$$c = \sqrt{\gamma \mathscr{B}_T / \rho_0} \tag{5.21}$$

where γ, $\mathscr{B}_T$, and ρ_0 all vary with the temperature and pressure of the liquid. Since no simple theory is available for predicting these variations, they must be measured experimentally and the resulting speed of sound expressed as a numerical formula.

For example, in distilled water a simplified formula is

$$c(\mathscr{P}, t) = 1402.7 + 488t - 482t^2 + 135t^3 + (15.9 + 2.8t + 2.4t^2)(\mathscr{P}_G/100) \qquad (5.22)$$

where $\mathscr{P}_G$ is the gauge pressure in bars and $t = T/100$, with T in degrees Celsius. This equation is accurate to within 0.05 percent for $0 \le T \le 100°C$ and $0 \le \mathscr{P}_G \le 200$ bar (1 bar $= 10^5$ Pa).

5.7 HARMONIC PLANE WAVES. If all the acoustic variables are functions of only one spatial coordinate, the phase of any variable is a constant on any plane perpendicular to this coordinate. Such a wave is called a *plane wave*. If the coordinate system is chosen so that this plane wave propagates along the x axis, the wave equation reduces to

$$\frac{\partial^2 p}{\partial x^2} = \frac{1}{c^2}\frac{\partial^2 p}{\partial t^2} \qquad (5.23)$$

where $p = p(x, t)$. By direct comparison with (2.5), we see that all the mathematical discussion of the solutions for transverse waves in Sects. 2.4 and 2.5 can be applied here and need not be repeated. Let us therefore proceed directly to harmonic plane waves and the interrelationships among the acoustic variables.

The complex form of the harmonic solution for the acoustic pressure of a plane wave is

$$\mathbf{p} = \mathbf{A}e^{j(\omega t - kx)} + \mathbf{B}e^{j(\omega t + kx)} \qquad (5.24)$$

and the associated particle velocity, from (5.11),

$$\mathbf{\dot{u}} = \left[\frac{\mathbf{A}}{\rho_0 c}e^{j(\omega t - kx)} - \frac{\mathbf{B}}{\rho_0 c}e^{j(\omega t + kx)}\right]\hat{x} \qquad (5.25)$$

is entirely in the direction of propagation.

If we write

$$\mathbf{p}_+ = \mathbf{A}e^{j(\omega t - kx)}$$
$$\mathbf{p}_- = \mathbf{B}e^{j(\omega t + kx)} \qquad (5.26)$$

then (5.24) and (5.25) yield the particle speeds

$$\mathbf{u}_+ = +\frac{\mathbf{p}_+}{\rho_0 c} \qquad \text{and} \qquad \mathbf{u}_- = -\frac{\mathbf{p}_-}{\rho_0 c} \qquad (5.27)$$

Use of (5.14) and (5.16) shows that

$$\mathbf{s}_+ = +\frac{\mathbf{p}_+}{\rho_0 c^2} \qquad \text{and} \qquad \mathbf{s}_- = -\frac{\mathbf{p}_-}{\rho_0 c^2} \qquad (5.28)$$

and

$$\mathbf{\Phi}_+ = -\frac{1}{j\omega\rho_0}\mathbf{p}_+ \qquad \text{and} \qquad \mathbf{\Phi}_- = +\frac{1}{j\omega\rho_0}\mathbf{p}_- \qquad (5.29)$$

For a plane wave travelling in some arbitrary direction, it is plausible to try a solution of the form

$$\mathbf{p} = Ae^{j(\omega t - k_x x - k_y y - k_z z)} \tag{5.30}$$

Substitution into (5.12) shows that this is acceptable if

$$\sqrt{k_x^2 + k_y^2 + k_z^2} = \omega/c \tag{5.31a}$$

If we define the *propagation vector*

$$\vec{k} = k_x \hat{x} + k_y \hat{y} + k_z \hat{z} \tag{5.31b}$$

which has magnitude ω/c and a position vector

$$\vec{r} = x\hat{x} + y\hat{y} + z\hat{z}$$

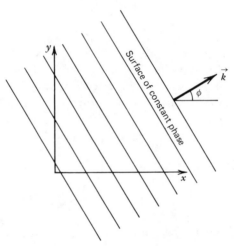

Fig. 5.2. *An oblique plane wave.*

which gives the location of the point (x, y, z) with respect to the origin $(0, 0, 0)$ of the coordinate system, then our trial solution (5.30) can be expressed by

$$\mathbf{p} = Ae^{j(\omega t - \vec{k} \cdot \vec{r})} \tag{5.32}$$

and the surfaces of constant phase are given by $\vec{k} \cdot \vec{r} = constant$. Since $\vec{k} = \nabla(\vec{k} \cdot \vec{r})$ is a vector perpendicular to the surfaces of constant phase, $\vec{k}$ points in the direction of propagation. The magnitude of $\vec{k}$ is the *wave number* k and k_x/k, k_y/k, and k_z/k are the *direction cosines* of $\vec{k}$ with respect to the x, y, and z axes. As a special case, for a plane wave with the propagation vector $\vec{k}$ parallel to the $z = 0$ plane, we have $k_z = 0$, and the direction cosines become $k_x/k = \cos \phi$ and $k_y/k = \sin \phi$ where ϕ is the angle of elevation above the x axis, as suggested by Fig. 5.2.

5.8 ENERGY DENSITY. The energy transported by acoustic waves through a fluid medium is of two forms; the *kinetic energy* of the moving particles and the *potential energy* of the compressed fluid. Consider a small fluid element that moves with the fluid and occupies volume V_0 of the undisturbed fluid. The kinetic energy of this element is

$$E_k = \tfrac{1}{2}\rho_0 V_0 u^2 \tag{5.33}$$

where $\rho_0 V_0$, the mass of the element, is calculated using the density and volume of the undisturbed fluid. The change in potential energy associated with a volume change from V_0 to V is

$$E_p = -\int_{V_0}^{V} p \, dV \tag{5.34a}$$

where the negative sign indicates that the potential energy will increase as work is done on the element when its volume is decreased by action of a positive acoustic pressure p. To carry out this integration it is necessary to express all variables under the integral sign in terms of one variable, p for example. From conservation of mass we have $\rho V = \rho_0 V_0$ so that

$$dV = -\frac{V}{\rho} \, d\rho \doteq -\frac{V_0}{\rho_0} \, d\rho$$

Now, use of $p = \rho_0 c^2 s$ and $s = (\rho - \rho_0)/\rho_0$ gives

$$dV = -\frac{V_0}{\rho_0 c^2} \, dp$$

Substitution of this into (5.34a) and integration of the acoustic pressure from 0 to p gives

$$E_p = \frac{1}{2} \frac{p^2}{\rho_0 c^2} V_0 \tag{5.34b}$$

The total acoustic energy of the volume element is

$$E = E_k + E_p = \tfrac{1}{2}\rho_0\left(u^2 + \frac{p^2}{\rho_0^2 c^2}\right) V_0$$

and the *instantaneous energy density* $\mathscr{E}_i = E/V_0$ in joules per cubic meter (J/m^3) is

$$\mathscr{E}_i = \tfrac{1}{2}\rho_0\left(u^2 + \frac{p^2}{\rho_0^2 c^2}\right) \tag{5.35}$$

The instantaneous particle speed and acoustic pressure are functions of both position and time, and consequently the instantaneous energy density $\mathscr{E}_i$ is not constant throughout the fluid. The time average of $\mathscr{E}_i$ gives the *energy density* $\mathscr{E}$ at any point in the fluid

$$\mathscr{E} = \langle \mathscr{E}_i \rangle_t = \frac{1}{t} \int_0^t \mathscr{E}_i \, dt \tag{5.36}$$

where the time interval is one period T of a harmonic wave.

These expressions apply to any acoustic wave. However, to proceed further, it is necessary to know the relationship between p and u. For a plane harmonic wave traveling in the $\pm x$ direction, reference to (5.27) shows that $p = \pm \rho_0 cu$ so that (5.35) gives

$$\mathscr{E}_i = \rho_0 u^2 = pu/c$$

and if P and U are the amplitudes of the acoustic pressure and particle speed,

$$\mathscr{E} = \tfrac{1}{2}PU/c = \tfrac{1}{2}P^2/(\rho_0 c^2) = \tfrac{1}{2}\rho_0 cU^2 \tag{5.37}$$

It must be noticed that in any other case (for example, spherical or cylindrical waves, or standing waves in a room), the pressure and particle speed in (5.35) must be the *real* quantities obtained from the *superposition* of all waves present. In these more complicated cases, there is no guarantee that $p = \pm \rho_0 cu$ nor that the energy density is given by $\mathscr{E} = \tfrac{1}{2}PU/c$. It is true, however, that $\mathscr{E} = \tfrac{1}{2}PU/c$ is approximately correct for progressive waves when the surfaces of constant phase become so close to planar that the radius of curvature is much greater than a wavelength. This occurs, for example, for spherical and cylindrical waves at great distances (many wavelengths) from their sources.

5.9 ACOUSTIC INTENSITY. The *acoustic intensity* I of a sound wave is defined as the *average* rate of flow of energy through a unit area normal to the direction of propagation. Its fundamental units are watts per square meter (W/m^2). The instantaneous rate at which work is done per unit area by one element of fluid on an adjacent element is pu. The intensity is the time average of this rate,

$$I = \langle pu \rangle_t = \frac{1}{t} \int_0^t pu \, dt \tag{5.38}$$

where the integration is taken over a time corresponding to the period of one complete cycle. To evaluate this integral for any particular wave, it is necessary to know the relationship between p and u.

For a plane harmonic wave traveling in the positive x direction, $p = \rho_0 cu$ so that

$$I = \tfrac{1}{2}P_+ U_+ = \tfrac{1}{2}P_+^2/(\rho_0 c) \tag{5.39a}$$

On the other hand, for a plane harmonic wave traveling in the negative x direction, we have $p = -\rho_0 cu$ and

$$I = -\tfrac{1}{2}P_- U_- = -\tfrac{1}{2}P_-^2/(\rho_0 c) \tag{5.39b}$$

To emphasize the similarity of (5.39) with corresponding equations for electromagnetic waves and voltage waves on transmission lines, as well as to write them in a more practical form, let us express them in terms of effective (root-mean-square) amplitudes. If we define F_e as the *effective amplitude* of a *periodic* quantity $f(t)$, then

$$F_e = \sqrt{\frac{1}{T} \int_0^T f^2(t) \, dt}$$

where T is the period of the motion. For harmonic waves this yields

$$P_e = P/\sqrt{2} \qquad \text{and} \qquad U_e = U/\sqrt{2} \tag{5.40}$$

so that

$$I_\pm = \pm P_e U_e = \pm P_e^2/(\rho_0 c) \tag{5.41}$$

for a plane wave traveling in either the $+x$ or $-x$ direction.

Once more it must be emphasized that, while (5.38) is completely general, $I_\pm = \pm P_e U_e$ are exact only for plane harmonic waves and approximate for diverging waves at great distances from their sources.

5.10 SPECIFIC ACOUSTIC IMPEDANCE. The ratio of acoustic pressure in a medium to the associated particle speed is the *specific acoustic impedance*

$$\boxed{\mathbf{z} = \frac{\mathbf{p}}{\mathbf{u}}} \tag{5.42}$$

For plane waves this ratio is

$$\mathbf{z} = \pm \rho_0 c \tag{5.43}$$

the plus or minus sign depending on whether propagation is in the plus or minus direction. The MKS unit of specific acoustic impedance is Pa $\cdot$ s/m.* In later sections it will become apparent that the product $\rho_0 c$ often has greater significance as a characteristic property of the medium than does either ρ_0 or c individually. For this reason $\rho_0 c$ is called the *characteristic impedance* (*resistance*) of the medium.

Although the specific acoustic impedance of the medium is a real quantity for progressive plane waves, this is not true for standing plane waves or for diverging waves. In general, $\mathbf{z}$ will be found to be complex

$$\mathbf{z} = r + jx \tag{5.44}$$

where r is called the *specific acoustic resistance* and x the *specific acoustic reactance* of the medium for the particular wave being considered.

The characteristic impedance of a medium for acoustic waves is analogous to the index of refraction n of a transparent medium for light waves, to the wave impedance $\sqrt{\mu/\varepsilon}$ of a dielectric medium for electromagnetic waves, and to the characteristic impedance Z_0 of an electric transmission line.

Numerical values of $\rho_0 c$ for various fluids, and some solids, are given in Appendix A10.

At a temperature of 20°C and atmospheric pressure the density of air is 1.21 kg/m^3 and the speed of sound is 343 m/s, giving the standard characteristic impedance of *air*

$$(\rho_0 c)_{20} = 415 \text{ Pa} \cdot \text{s/m} \tag{5.45a}$$

---

* The unit of specific acoustic impedance is often given as *rayl*, where 1 MKS rayl = 1 Pa $\cdot$ s/m, established in honor of John William Strutt, Baron Rayleigh.

At 20°C and one atmosphere, the speed of sound in *distilled water*, as computed from (5.22), is 1482.3 m/s and its density is 998.2 kg/m^3, resulting in a characteristic impedance of

$$(\rho_0 c)_{20} = 1.48 \times 10^6 \text{ Pa} \cdot \text{s/m} \tag{5.45b}$$

5.11   SPHERICAL WAVES. Expressed in spherical coordinates the Laplacian operator is

$$\nabla^2 = \frac{\partial^2}{\partial r^2} + \frac{2}{r}\frac{\partial}{\partial r} + \frac{1}{r^2 \sin\theta}\frac{\partial}{\partial\theta}\left(\sin\theta\frac{\partial}{\partial\theta}\right) + \frac{1}{r^2 \sin^2\theta}\frac{\partial^2}{\partial\phi^2}$$

where $x = r\sin\theta\cos\phi$, $y = r\sin\theta\sin\phi$, and $z = r\cos\theta$, as shown in Appendix A7. If the waves have spherical symmetry, the acoustic pressure p is a function of radial distance and time but not of the angular coordinates θ and ϕ. Then this equation simplifies to

$$\nabla^2 = \frac{\partial^2}{\partial r^2} + \frac{2}{r}\frac{\partial}{\partial r}$$

and the wave equation for spherically symmetric pressure fields is

$$\frac{\partial^2 p}{\partial r^2} + \frac{2}{r}\frac{\partial p}{\partial r} = \frac{1}{c^2}\frac{\partial^2 p}{\partial t^2} \tag{5.46}$$

(Conservation of energy and the relationship $I = P^2/(2\rho_0 c)$ lead us to expect that the pressure amplitude should fall off as $1/r$, so that the quantity rp might have amplitude independent of r.) Rewriting (5.46) with rp treated as the dependent variable results in

$$\frac{\partial^2 (rp)}{\partial r^2} = \frac{1}{c^2}\frac{\partial^2 (rp)}{\partial t^2} \tag{5.47}$$

If the product rp in this equation is considered as a single variable, the equation is of the same form as the plane wave equation with the general solution

$$rp = f_1(ct - r) + f_2(ct + r)$$

or

$$p = \frac{1}{r}f_1(ct - r) + \frac{1}{r}f_2(ct + r)$$

The first term represents a spherical wave diverging from a point source at the origin with speed c; the second term represents a wave converging on the origin.

The most important diverging spherical waves are harmonic. Such waves are represented in complex form by

$$\mathbf{p} = \frac{\mathbf{A}}{r}e^{j(\omega t - kr)} \tag{5.48}$$

Use of the relationships developed in Sect. 5.5 for a general wave allows the other acoustic variables to be expressed in terms of the pressure

$$\mathbf{\Phi} = -\frac{1}{j\omega\rho_0}\,\mathbf{p} \tag{5.49}$$

$$\hat{\mathbf{u}} = \nabla\mathbf{\Phi} = \left(1 - \frac{j}{kr}\right)\frac{\mathbf{p}}{\rho_0 c}\,\hat{r} \tag{5.50}$$

The observed acoustic variables are obtained by taking the real parts of (5.48) through (5.50).

It is apparent from (5.50) that, in contrast with plane waves, the particle speed is *not* in phase with the pressure. The specific acoustic impedance is not $\rho_0 c$, but rather

$$\mathbf{z} = \rho_0 c\,\frac{kr}{\sqrt{1 + (kr)^2}}\,e^{j\theta} \tag{5.51}$$

or

$$\mathbf{z} = \rho_0 c\,\cos\theta\,e^{j\theta} \tag{5.52}$$

where

$$\cot\theta = kr \tag{5.53}$$

A geometrical representation of θ is given in Fig. 5.3. As is true with many other acoustic phenomena, the product of k and r is the determining factor, rather than the magnitude of either. Since $kr = 2\pi r/\lambda$, the phase angle θ is a function of the ratio of the source distance to the wavelength. When the distance from the source is only a small fraction of a wavelength, the phase difference between the complex pressure and particle speed is large. On the other hand, at distances corresponding to a considerable number of wavelengths, $\mathbf{p}$ and $\mathbf{u}$ are very nearly in phase, and the spherical wave then assumes the characteristics of a plane wave. This behavior is to be expected, since the wave fronts of all spherical waves become essentially plane at great distances from their source.

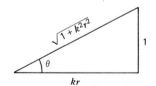

Fig. 5.3. *The relationship between θ and kr.*

Separating (5.51) into real and imaginary parts, we have

$$\mathbf{z} = \rho_0 c\,\frac{(kr)^2}{1 + (kr)^2} + j\rho_0 c\,\frac{kr}{1 + (kr)^2} \tag{5.54}$$

The first term is the *specific acoustic resistance*, and the second term contains the *specific acoustic reactance*. Both terms approach zero for very small values of kr, but for very large values of kr the resistive term approaches $\rho_0 c$, while the reactive term approaches zero. When $kr = 1$, both the specific acoustic resistance and reactance are equal to $\rho_0 c/2$ and the specific acoustic reactance has its maximum value.

The absolute magnitude z of the specific acoustic impedance is equal to the ratio of the pressure amplitude P of the wave to its speed amplitude U,

$$z = \frac{P}{U} = \rho_0 c \, \cos \theta \tag{5.55}$$

and the relationship between pressure and speed amplitude may be written as

$$P = \rho_0 c \, U \cos \theta \tag{5.56}$$

For large values of kr, $\cos \theta$ approaches unity, and the relationship between pressure and speed is then the same as that given for a plane wave. As the distance from the source of a spherical acoustic wave to the point of observation is decreased, both kr and $\cos \theta$ decrease, so that larger and larger particle speeds are associated with a given pressure amplitude. For very small distances from a point source of sound, the particle speed corresponding to even very low acoustic pressures becomes impossibly large: a source small compared to a wavelength is inherently incapable of generating waves of large intensity.

Let us rewrite (5.48) as

$$\mathbf{p} = \frac{A}{r} \, e^{j(\omega t - kr)} \tag{5.57}$$

where without any loss in generality we have chosen a new origin of time such that the complex amplitude $\mathbf{A}$ becomes a real constant A. Then A/r is the *pressure amplitude* of the wave. It should be noted that the pressure amplitude in an undamped spherical wave is not constant, as it is for an undamped plane wave, but decreases inversely with the distance from the source. The actual pressure is the real part of (5.57),

$$p = \frac{A}{r} \, \cos(\omega t - kr) \tag{5.58}$$

Since $\mathbf{u} = \mathbf{p}/\mathbf{z}$, the corresponding complex expression for the particle speed is

$$\mathbf{u} = \frac{A}{r\mathbf{z}} \, e^{j(\omega t - kr)} \tag{5.59}$$

In replacing $\mathbf{z}$ by (5.52) and then taking the real part of the resulting expression, we have for the actual particle speed

$$u = \frac{1}{\rho_0 c} \frac{A}{r} \frac{1}{\cos \theta} \cos(\omega t - kr - \theta) \tag{5.60}$$

It is apparent that, since θ is a function of kr, the speed amplitude

$$U = \frac{1}{\rho_0 c} \frac{A}{r} \frac{1}{\cos \theta} \tag{5.61}$$

is not inversely proportional to the distance from the source. As a result it is usually advantageous to treat problems involving spherical waves in terms of pressure amplitude rather than speed amplitude.

For a harmonic spherical wave (5.38) yields

$$I = \frac{1}{T}\int_0^T P \cos(\omega t - kr)U \cos(\omega t - kr - \theta)\, dt = \frac{PU \cos \theta}{2} = \frac{P^2}{2\rho_0 c} \qquad (5.62)$$

where the factor $\cos \theta$ is analogous to the power factor of an alternating-current circuit. Notice that the formula $I = P^2/(2\rho_0 c)$ has been found to be *exactly* true for both plane and spherical waves. In the case of the spherical wave, this is consistent with the argument following (5.46), and in fact verifies that P falls off *exactly* as $1/r$ for a lossless medium.

The average rate at which energy flows through a closed spherical surface of radius r surrounding a source of symmetrical spherical waves is

$$\Pi = 4\pi r^2 I = 4\pi r^2 P^2/(2\rho_0 c) \qquad (5.63)$$

or since $P^2 = A^2/r^2$

$$\Pi = 2\pi A^2/(\rho_0 c) \qquad (5.64)$$

The average rate of energy flow through any spherical surface surrounding the origin is independent of the radius of the surface, a conclusion that is consistent with conservation of energy in a lossless medium.

5.12.   DECIBEL SCALES.   It is customary to describe sound pressures and intensities through the use of logarithmic scales known as *sound levels*. One reason for doing this is the very wide range of sound pressures and intensities encountered in our acoustic environment; audible intensities range from approximately 10^{-12} to 10 W/m^2. The use of a logarithmic scale compresses the range of numbers required to describe this wide range of intensities. A second reason is that humans judge the relative loudnesses of two sounds by the ratio of their intensities, a logarithmic behavior.

The most generally used logarithmic scale for describing sound levels is the decibel scale. The *intensity level IL* of a sound of intensity I is defined by

$$IL = 10 \log(I/I_{ref}) \qquad (5.65)$$

where I_{ref} is a reference intensity, IL is expressed in *decibels referenced to I_{ref}* (*dB re I_{ref}*), and "log" represents logarithm to the base 10.

We have shown in Sects. 5.9 and 5.11 that intensity and effective pressure of progressive plane and spherical waves are related by $I = P_e^2/(\rho_0 c)$. Consequently, the intensities in (5.65) may be replaced by expressions for pressure, leading to the *sound pressure level*

$$SPL = 20 \log (P_e/P_{ref}) \qquad (5.66)$$

where SPL is expressed in dB *re P_{ref}*; P_e is the measured effective pressure of the sound wave, and P_{ref} is the reference effective pressure. If we choose $I_{ref} = P_{ref}^2/(\rho_0 c)$, then IL re $I_{ref} = SPL$ re P_{ref}.

There is a multiplicity of units used to specify pressures throughout the various scientific disciplines and many of these are found in acoustics. In addition, a number of reference levels, of various degrees of antiquity, are encountered. Let us first catalog a few units:

CGS units
1 dyne/cm^2, also called the microbar (μbar) (The microbar was originally 10^{-6} atmospheres but is now defined to be *identically* 1 dyne/cm^2.)

MKS units
1 pascal (Pa), identical to 1 N/m^2

Others
1 atmosphere = 1.013×10^5 Pa = 1.013×10^6 μbar
1 kilogram/cm^2 (kgf/cm^2) = 0.968 atm = 0.981×10^5 Pa

Equivalents
1 μbar = 0.1 N/m^2 = 10^5 μPa

The reference standard for airborne sounds is 10^{-12} W/m^2, which is approximately the intensity of a 1000 Hz pure tone that is just barely audible to a person with unimpaired hearing. Substitution of this intensity into (5.39) shows that it corresponds to a peak pressure amplitude of

$$P = \sqrt{2\rho_0 cI} = 2.89 \times 10^{-5} \text{ Pa}$$

or a corresponding effective (root-mean-square) pressure of

$$P_e = P/\sqrt{2} = 20.4 \ \mu\text{Pa} \tag{5.67}$$

The latter pressure, rounded off to 20 μPa, is often used as a reference for sound pressure levels in air. Since this pressure is almost exactly equivalent to the effective pressure corresponding to a reference intensity of 10^{-12} W/m^2 used in (5.65), essentially identical numerical results are obtained by use of either (5.65) or (5.66) when plane or spherical progressive waves are being measured in air. However, in certain more complex sound fields, such as standing wave patterns, intensity and pressure are no longer simply related by (5.41) and (5.62); as a consequence, (5.65) and (5.66) will not yield identical results. Since the voltage outputs of microphones and hydrophones commonly used in acoustic measurements are proportional to pressure, acoustic pressure is the most readily measured variable in a sound field. For this reason, sound pressure levels are more widely used in specifying sound levels than are intensity levels.

Three different pressures are encountered as reference pressures for specifying sound pressure levels in underwater acoustics. One is an effective pressure of 20 μPa (the same as the reference pressure in air). The second reference pressure is 1 μbar and the third is 1 μPa. Of these three, the last is now preferred.

This plurality of reference pressures can lead to confusion unless care is taken to always specify the reference pressure being used: *SPL re* 20 μPa, *re* 1 μPa, or *re* 1 μbar. Table 5.1 summarizes the various methods of expressing the level of a sound in decibels.

From the above discussion, note that a given acoustic pressure in air corresponds to a much higher intensity than does the same acoustic pressure in water.

Table 5.1. References and conversions

Medium	Reference	Nearly equivalent to
Air	10^{-12} W/m^2	20 μPa
	20 μPa = 0.0002 μbar	10^{-12} W/m^2
Water	1 μbar = 10^5 μPa	6.76×10^{-9} W/m^2
	0.0002 μbar = 20 μPa	2.70×10^{-16} W/m^2
	1 μPa	6.76×10^{-19} W/m^2

SPL re 1 μbar + 100 = SPL re 1 μPa
SPL re 0.0002 μbar − 74 = SPL re 1 μbar
SPL re 0.0002 μbar + 25 = SPL re 1 μPa

Since (5.41) or (5.62) shows that for a given pressure amplitude, intensity is inversely proportional to the characteristic impedance of the medium, the ratio of the intensities in air to that in water for the *same acoustic pressure* is $1.48 \times 10^6/415 = 3570$. On the other hand, if we compare two acoustic waves of the *same frequency and particle displacement*, the intensity of the one in water is 3570 times that in air.

5.13 RAYS AND WAVES. Plane waves of infinite spatial extent do not exist in nature. The nearest thing to a plane wave that can be produced in the laboratory is the wave that travels down a rigid-walled tube when the frequency is low enough that the wavelength is much greater than the cross-sectional dimensions of the tube. Even here the influence of viscosity at the walls will introduce small but measureable deviations. In the real world, instead of plane waves, we find beams of sound whose cross-sectional areas and directions of propagation may change as the beams traverse the medium. In such circumstances, we frequently find it useful to think of *rays* rather than *waves*. A *ray* can be defined as a line everywhere perpendicular to the surfaces of constant phase. Its usefulness lies in the intuitive feeling, mathematically justified under certain conditions, that energy is carried along a ray. In many cases, especially where c is a function of space or where the wave is restricted to a limited solid angle (such as the beam of sound from a highly directional source), description in terms of rays is much easier than in terms of wave fronts. However, rays are not exact replacements for waves but only approximations that are valid under certain rather restrictive conditions.

Let us begin by first considering a plausible solution to the wave equation

$$\nabla^2 p = \frac{1}{c^2}\frac{\partial^2 p}{\partial t^2} \tag{5.68}$$

where $c = c(x, y, z)$. For a beam of finite aperture traversing either a homogeneous fluid ($c = constant$) or an inhomogeneous fluid (c = a function of position), we must expect that the amplitude of this wave will vary with position and that the surfaces of constant phase can be complicated functions of space. As a trial solution try

$$\mathbf{p}(x,\ y,\ z,\ t) = A(x,\ y,\ z)e^{j\omega[t-\Gamma(x,\ y,\ z)/c_0]} \tag{5.69}$$

where A has units of pressure, Γ has units of length, and c_0 is a constant value of the phase speed to be defined later.

The values of (x, y, z) satisfying $\Gamma = constant$ define the surfaces of constant phase and, as noted in the discussion of Sect. 5.7, $\nabla\Gamma$ is everywhere perpendicular to each surface of constant phase. For example, notice that if $A = constant$ and $\Gamma = x$, (5.69) becomes $\mathbf{p} = A \; \exp[j\omega(t - x/c_0)]$, a plane wave solution of (5.68) if $c = constant = c_0$. Furthermore, notice that $\nabla\Gamma = \hat{x}$ has unit magnitude and points in the direction of propagation (always in the x direction in this simple example).

Substitution of the trial solution into (5.68) gives

$$\frac{\nabla^2 A}{A} - \left(\frac{\omega}{c_0}\right)^2 \nabla\Gamma \cdot \nabla\Gamma + \left(\frac{\omega}{c}\right)^2 - j\frac{\omega}{c_0}\left(2\frac{\nabla A}{A} \cdot \nabla\Gamma + \nabla^2\Gamma\right) = 0 \qquad (5.70a)$$

If A and $\nabla\Gamma$ vary slowly enough that $|A^{-1}\nabla^2 A| \ll (\omega/c)^2$, $|\nabla^2\Gamma| \ll \omega/c$, and $|A^{-1}\nabla A \cdot \nabla\Gamma| \ll \omega/c$ all terms in (5.70a) except the second and third can be considered small. Then (5.70a) simplifies to

$$\boxed{\nabla\Gamma \cdot \nabla\Gamma \doteq n^2} \qquad (5.70b)$$

where

$$n(x, y, z) = \frac{c_0}{c(x, y, z)} \qquad (5.70c)$$

is the *refractive index* and the constant c_0 is an arbitrary reference speed. Equation (5.70b) is called the *Eikonal equation*.

Sufficient conditions for the inequalities required to reduce (5.70a) to (5.70b) are that (1) the amplitude of the wave must not change appreciably in distances comparable to a wavelength, and (2) the speed of sound must not change appreciably in distances comparable to a wavelength. If we consider a beam of sound traveling through a fluid, the first condition states that the Eikonal equation can be applied to the central portion of the beam where A is not rapidly varying. At the edges of the beam, however, A may rapidly decay to zero over distances of a wavelength and then (5.70b) is no longer valid. The failure of the Eikonal equation reveals itself in the *diffraction* of sound at the edges of the beam. This is the acoustic analog of the familiar diffraction of light through a slit or pinhole. The second condition requires that the speed of sound be so slowly varying with space that the local direction of propagation of the wavefronts does not change significantly over distances of a wavelength; the *refraction* of the sound beam must not be too rapid.*

This means that (5.70b) is in general an accurate description of acoustic propagation only in the limit of high frequencies (short wavelength); how high the frequency must be depends on the rapidity of spatial variations of c and A.

If A and Γ are slowly varying functions of space, then the waves described by (5.69) are similar to plane waves over small regions of space, but over larger distances these waves may be observably different from plane waves. Furthermore, from (5.70b) $\nabla\Gamma$ defines, point by point, the direction of travel of each ray so that solution of the Eikonal equation gives the trajectories of the ray paths traversed by the acoustic energy.

For example, assume that $\nabla\Gamma$ can be written as

$$\nabla\Gamma = n(\cos\phi \; \hat{x} + \sin\phi \; \hat{y}) \qquad (5.71)$$

---

* For simplicity, we have given *sufficient* conditions for the validity of (5.70b). More rigorous, *necessary* conditions can be stated, but their physical meanings are less direct. Indeed, there are examples of propagating waves that do not satisfy the sufficient conditions, but for which (5.70b) is valid.

so that we are dealing with propagation in the x, y plane. The angle of elevation ϕ above the x axis will be a function of position; therefore, the local direction of propagation of the wave will be different at different locations. At some instant of time, $\nabla\Gamma$ evaluated at a point on a surface of constant phase indicates the direction in which that portion of the surface will advance (see Fig. 5.4). As time advances, this surface element advances and the magnitude and direction of $\nabla\Gamma$ for this portion of the surface will change. In this way, each portion of the wave can be seen to trace out a trajectory in space. Each of these trajectories is a ray path, and the portion of the wave traveling along that path defines a ray. By labeling the *distance*

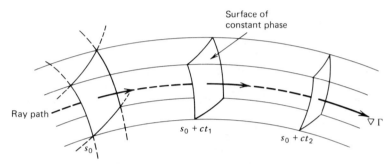

Fig. 5.4. *A time history of a portion of a surface of constant phase.*

along the ray path by a coordinate s, we see that the way $\nabla\Gamma$ changes along a ray path can be determined from the vector differential $d(\nabla\Gamma)/ds$. Examining this, component by component, in the x direction we have

$$\frac{d}{ds}\left(\frac{\partial\Gamma}{\partial x}\right) = \frac{\partial}{\partial x}\left(\frac{\partial\Gamma}{\partial x}\right)\frac{dx}{ds} + \frac{\partial}{\partial y}\left(\frac{\partial\Gamma}{\partial x}\right)\frac{dy}{ds}$$

$$= \frac{\partial}{\partial x}\left(\frac{\partial\Gamma}{\partial x}\right)\frac{dx}{ds} + \frac{\partial}{\partial x}\left(\frac{\partial\Gamma}{\partial y}\right)\frac{dy}{ds}$$

$$= \cos\phi\,\frac{\partial}{\partial x}(n\cos\phi) + \sin\phi\,\frac{\partial}{\partial x}(n\sin\phi) = \frac{\partial n}{\partial x}$$

Similarly, in the y direction,

$$\frac{d}{ds}\left(\frac{\partial\Gamma}{\partial y}\right) = \frac{\partial n}{\partial y}$$

so that the behavior of $\nabla\Gamma$ is found from

$$\frac{d}{ds}(\nabla\Gamma) = \nabla n \tag{5.72}$$

Although this derivation was carried out in two dimensions, the same argument can be applied in three dimensions, except that the cosine and sine must be replaced by direction cosines. This complicates the development, but leads to the same vector equation (5.72).

A very powerful consequence of (5.72) is *Snell's law*. A simple statement of this law can be obtained if we let the speed of sound be a function only of x. Then $n = n(x)$ and the components of (5.72) become

$$\frac{d}{ds}\left(\frac{c_0}{c}\sin\phi\right) = 0 \tag{5.73a}$$

and

$$\frac{d}{ds}\left(\frac{c_0}{c}\cos\phi\right) = -\frac{c_0}{c^2}\frac{dc}{dx} \tag{5.73b}$$

It is important to notice that, since ∇c points in the x direction, our choice of ϕ is such that when $\phi = 0$, $\vec{k}$ is *parallel* to ∇c. Integration of the first equation gives

$$\boxed{\frac{\sin\phi}{c(x)} = constant} \tag{5.74}$$

which is a form of Snell's law. It states that the direction of propagation given by ϕ for a ray is uniquely determined. If the ray has some angle ϕ_0 at a value $x = x_0$ where the speed of sound is c_0, then the angle of this same ray when it reaches any other value x where the speed of sound is $c(x)$ is determined by

$$\frac{\sin\phi}{c(x)} = \frac{\sin\phi_0}{c_0}$$

[The importance of (5.74) will be revealed in Chapter 15, which is devoted to underwater acoustics.]

5.14 THE INHOMOGENEOUS WAVE EQUATION.

In previous sections we developed a wave equation that applied to regions of space not containing any sources of acoustic energy. However, a source must be present to generate any acoustic disturbance. If the source is *external* to the region of interest, it can be taken into account by introducing time-dependent boundary conditions, as we did in Chapters 2, 3, and 4. Alternately, the hydrodynamic equations can be modified to include *source terms*. There are many possible types of sources, but only two will be considered here.

(1) If mass is being injected into the space at a rate per unit volume $G(\vec{r}, t)$, the linearized equation of continuity becomes

$$\rho_0 \frac{\partial s}{\partial t} + \rho_0 \nabla \cdot \vec{u} = G(\vec{r}, t) \tag{5.75}$$

This $G(\vec{r}, t)$ is generated by any closed surface that changes volume, such as a loudspeaker in an enclosed cabinet.

(2) If there are *body forces* present in the fluid, then a body force per unit volume $\vec{F}(\vec{r}, t)$ must be included in Euler's equation. The linearized equation of motion then becomes

$$\rho_0 \frac{\partial \vec{u}}{\partial t} + \nabla p = \vec{F}(\vec{r}, t) \tag{5.76}$$

An example of this kind of force is that produced by a body that oscillates back and forth without any change in volume, such as the cone of an unbaffled loudspeaker.

If these two modifications are combined with the linearized equation of state, an *inhomogeneous* wave equation is obtained,

$$\nabla^2 p - \frac{1}{c^2}\frac{\partial^2 p}{\partial t^2} = -\frac{\partial G}{\partial t} + \nabla \cdot \vec{F} \tag{5.77}$$

For all regions of space without sources, the right-hand side of (5.77) vanishes, leaving the homogeneous wave equation.

5.15 THE POINT SOURCE. The monofrequency spherical wave given by (5.57) is a solution to the homogeneous wave equation (5.12) everywhere *except* at $r = 0$. (This is consistent with the fact that there must be a source at $r = 0$ to generate the wave.) However, (5.57) does satisfy the inhomogeneous wave equation

$$\nabla^2 \mathbf{p} - \frac{1}{c^2} \frac{\partial^2 \mathbf{p}}{\partial t^2} = -4\pi A \, \delta(\vec{r}) e^{j\omega t} \tag{5.78}$$

for *all* r; the three-dimensional delta function $\delta(\vec{r})$ is defined by

$$\int_V \delta(\vec{r}) \, dV = \begin{cases} 1 & \text{if } \vec{r} = 0 \text{ is inside } V \\ 0 & \text{if } \vec{r} = 0 \text{ is outside } V \end{cases} \tag{5.79}$$

To prove this, multiply both sides of (5.78) by dV, integrate over a volume V that includes $\vec{r} = 0$, and then use Gauss' law (see Appendix A8) to reduce the volume integral to a surface integral and (5.79) to evaluate the delta function integral:

$$\int_S \nabla \mathbf{p} \cdot \hat{n} \, dS + \frac{1}{c^2} \int_V \frac{\partial^2 \mathbf{p}}{\partial t^2} \, dV = -4\pi A e^{j\omega t}$$

where $\hat{n}$ is the unit outward normal to the surface S of V. By substituting (5.57) for $\mathbf{p}$ and carrying out the surface integration over a sphere centered on $\vec{r} = 0$ with vanishingly small radius (so that the volume integration vanishes), we can complete the proof.

It is trivial to generalize to a point source located at $\vec{r} = \vec{r}_0$: Making the appropriate change of variable in (5.48) results in

$$\mathbf{p} = \frac{A}{|\vec{r} - \vec{r}_0|} \exp[\, j(\omega t - k|\vec{r} - \vec{r}_0|)] \tag{5.80a}$$

which is a solution to

$$\nabla^2 \mathbf{p} - \frac{1}{c^2} \frac{\partial^2 \mathbf{p}}{\partial t^2} = -4\pi A \, \delta(\vec{r} - \vec{r}_0) e^{j\omega t} \tag{5.80b}$$

This may appear to be a difficult way to account for sources, but it will be seen that the incorporation of a point source directly into the wave equation is, in the proper circumstances, a considerable mathematical simplification. (See Sects. 15.13 to 15.15 as one example.) We will, however, use this formalism only when necessary, utilizing in most cases methods that are more closely related to elementary physical intuition.

PROBLEMS

5.1. (a) Linearize (5.2) by assuming $s \ll 1$. Then, by comparing this result with (5.4), obtain the adiabatic bulk modulus of a perfect gas in terms of $\mathscr{P}_0$ and γ. (b) With the help of (5.1) applied to equilibrium conditions, obtain the temperature dependence of $\mathscr{B}$ at constant volume.

5.2 Another form of the perfect gas law is $\mathscr{P}V = n\mathscr{R}T_K$ where n is the number of moles and $\mathscr{R}$ is the universal gas constant. Obtain a relationship between r and $\mathscr{R}$.

5.3. An incompressible fluid is defined by the equation of state $(\partial \rho / \partial t) + \vec{u} \cdot \nabla \rho = 0$. (a) Show that for an incompressible fluid the equation of continuity reduces to $\nabla \cdot \vec{u} = 0$. (b) Write Euler's equation for the flow of an incompressible fluid. (c) What is c for an incompressible fluid?

5.4. (a) By means of (5.22), determine the speed of sound in distilled water at atmospheric pressure and a temperature of $30°C$. (b) What is the rate of change of the speed of sound in water with respect to temperature at this temperature?

5.5. (a) For a perfect gas, does c vary with the equilibrium pressure? With the instantaneous pressure in an acoustic process? (b) Find c for a perfect gas that obeys the *isothermal* equation of state. (c) Compare the above value of c to that for air at $20°C$.

5.6. If $\vec{u} = \hat{x} U \exp[j(\omega t - kx)]$ show that the ratio of $|(\vec{u} \cdot \nabla)\vec{u}|$ to $|\partial \vec{u}/\partial t|$ is U/c, the *Mach number*.

5.7. For a plane wave $u = U \exp[j(\omega t - kx)]$, find expressions for the Mach number U/c in terms of (a) P, ρ_0, and c. (b) In terms of s.

5.8. (a) Derive an equation expressing the adiabatic temperature rise ΔT produced in a gas by an acoustic pressure p. (b) What is the amplitude of the temperature fluctuations produced by a sound of 10 W/m^2 intensity in air at $20°C$ and standard atmospheric pressure?

5.9. If $\mathbf{p} = P \exp[j(\omega t - kx)]$ find (a) the acoustic density, (b) the particle speed, (c) the velocity potential, (d) the energy density, and (e) the intensity.

5.10. Given a small source of spherical waves in air, for a radial distance of 10 cm, compute the difference in phase angle between pressure and particle velocity at frequencies of 10, 100, and 1000 Hz. Similarly compute the magnitude of the specific acoustic impedance for these conditions.

5.11. For a spherical wave $\mathbf{\Phi} = r^{-1} \cos(kr) \exp(j\omega t)$, find (a) the acoustic pressure, (b) the particle speed, (c) the specific acoustic impedance, (d) the instantaneous intensity, and (e) the intensity.

5.12. For a plane traveling wave in air with a frequency of 171 Hz and a sound pressure level of 40 dB *re* 20 μPa, find (a) the acoustic pressure amplitude, (b) the intensity, (c) the acoustic particle speed amplitude, and (d) the acoustic density amplitude.

5.13. A plane sound wave in air of 100 Hz has a peak acoustic pressure amplitude of 2 Pa. (a) What is its intensity? Its intensity level? (b) What is its peak particle displacement amplitude? (c) What is its peak particle speed amplitude? (d) What is its effective or rms pressure? (e) What is its sound pressure level *re* 20 μPa?

5.14. An acoustic wave has a sound pressure level of 80 dB *re* 1 μbar. Find (a) the sound pressure level *re* 1 μPa, and (b) the sound pressure level *re* 20 μPa.

5.15. (a) Show that a plane wave having an effective acoustic pressure of 1 μbar in air has an intensity level of 74 dB. (b) Find the intensity (W/m^2) produced by an acoustic plane wave in water of 120 dB sound pressure level relative to 1 μbar. (c) What is the ratio of the sound pressure in water for a plane wave to that of a similar wave in air of equal intensity?

5.16. (a) Determine the energy density and effective pressure of a plane wave in air of 70-dB intensity level. (b) Determine the energy density and effective pressure of a plane wave in water if its sound pressure level is 70 dB *re* 1 μbar.

5.17. (a) Show that the characteristic impedance $\rho_0 c$ of a gas is inversely proportional to the square root of its absolute temperature T_K. (b) What is the characteristic impedance of air at $0°C$? At $80°C$? (c) If the pressure amplitude of a sound wave remains constant, what is its percent change in intensity as the temperature increases from $0°C$ to $80°C$? (d) What would be the corresponding change in measured intensity level? In pressure level?

5.18. Cavitation may take place at the face of a sonar transducer when the sound pressure amplitude being produced exceeds the hydrostatic pressure in the water. (a) For a hydrostatic pressure of 200,000 Pa, what is the highest intensity that may be radiated without producing cavitation? (b) What is the sound pressure level of this sound re 1 μbar?

5.19. (a) Show that, if c is a function only of x, then $d\phi/ds = [(\sin \phi_0)/c_0] \, dc/dx$, where ϕ_0 is the angle of elevation of the ray where $c = c_0$. (b) If $dc/dx = g$, a constant, find the radius of curvature R of the ray. Is R a constant? (c) If the temperature of air decreases linearly with height z, verify that $c(z) \doteq c_0 - gz$ where $g > 0$. If the temperature decreases 5 C$^\circ$/km, find the radius of curvature of a ray that is horizontal at $z = 0$ (assume $c_0 =$ 340 m/s). At what horizontal range will this ray have risen to a height of 10 m?

5.20. Show that in spherical coordinates $\delta(\vec{r}) = (4\pi r^2)^{-1} \, \delta(r)$.

5.21. (a) Show that $p = (A/r)f(t - r/c)$ is a solution of the inhomogeneous wave equation $\nabla^2 p - c^{-2} \, \partial^2 p/\partial t^2 = 4\pi A \, \delta(\vec{r})f(t)$. (b) Show that $p = r^{-1} \, \delta(t - r/c)$ is a solution of the inhomogeneous wave equation $\nabla^2 p - c^{-2} \, \partial^2 p/\partial t^2 = 4\pi \, \delta(\vec{r}) \, \delta(t)$.

CHAPTER 6

TRANSMISSION PHENOMENA

6.1 CHANGES IN MEDIA. When an acoustic wave traveling in one medium encounters the boundary of a second medium, reflected and transmitted waves are generated. Discussion of this phenomenon is greatly simplified if it is assumed that both the incident wave and the boundary between the media are planar and that all media are fluids. The complications that arise when one of the media is a solid will be left to the latter part of this chapter. However, it is worthwhile to note that for normal incidence a large class of solids obeys the same equations developed for fluids. The only modification needed is that the speed of sound in the solid must be based on the *bulk* modulus. (The bulk modulus is used rather than Young's modulus because, unlike a bar, an extended solid is not free to change its transverse dimensions.) Values of the bulk speeds of sound in some solids are listed in Appendix A10.

The ratios of the pressure amplitudes and intensities of the reflected and transmitted waves to those of the incident wave depend both on the characteristic acoustic impedances and speeds of sound in the two media and on the angle the incident wave makes with the normal to the interface. Let the incident and reflected waves travel in a fluid of characteristic acoustic impedance $r_1 = \rho_1 c_1$ where ρ_1 is the equilibrium density of the fluid and c_1 the phase speed in the fluid. Let the transmitted wave travel in a fluid of characteristic acoustic impedance $r_2 = \rho_2 c_2$. If the complex pressure amplitude of the incident wave is $\mathbf{P}_i$, that of the reflected wave $\mathbf{P}_r$, and that of the transmitted wave $\mathbf{P}_t$, then we can define the *pressure transmission and reflection coefficients*

$$\mathbf{T} = \mathbf{P}_t/\mathbf{P}_i \tag{6.1a}$$

$$\mathbf{R} = \mathbf{P}_r/\mathbf{P}_i \tag{6.1b}$$

Since the intensity of a harmonic plane progressive wave is $P^2/(2r)$, the *intensity transmission and reflection coefficients* are real and are defined by

$$T_I = \frac{I_t}{I_i} = \frac{r_1}{r_2}\,|\mathbf{T}|^2 \tag{6.2a}$$

$$R_I = \frac{I_r}{I_i} = |\mathbf{R}|^2 \tag{6.2b}$$

Most real situations are characterized by beams of sound with finite cross-sectional area. As we have seen in the previous chapter, if a beam is many wavelengths wide it can be described by parallel rays and thus can be approximated by a plane wave of *finite* extent. While one can expect anomalies resulting from diffraction at the edges of the beam, if the cross-sectional area is sufficiently great, these effects can be ignored and the equations developed in this chapter safely utilized.

The power transmitted by a *beam* of sound is calculated by multiplying the acoustic intensity by the cross-sectional area of the beam. If an incident beam of cross-sectional area A_i is obliquely incident on a boundary, the cross-sectional area A_t of the transmitted beam is not generally the same as that of the incident beam. (It will be shown later in this chapter that the cross-sectional area of the reflected and incident beams is equal under all circumstances.) The *power transmission and reflection coefficients* are defined by

$$T_\pi = \frac{A_t}{A_i} \, T_I = \frac{A_t}{A_i} \frac{r_1}{r_2} \, |\mathbf{T}|^2 \tag{6.3a}$$

$$R_\pi = R_I = |\mathbf{R}|^2 \tag{6.3b}$$

As a consequence of the conservation of energy, the power in the incident beam must be shared between the reflected and the transmitted beams so that

$$R_\pi + T_\pi = 1 \tag{6.3c}$$

6.2 TRANSMISSION FROM ONE FLUID TO ANOTHER: NORMAL INCIDENCE.

As suggested in Fig. 6.1, let the plane $x = 0$ be the boundary between fluid I of characteristic acoustic impedance r_1 and fluid II of characteristic acoustic impedance r_2. Let there be an *incident wave* traveling in the positive x direction,

$$\mathbf{p}_i = \mathbf{P}_i \, e^{j(\omega t - k_1 x)} \tag{6.4a}$$

which, when striking the boundary, generates a *reflected wave*

$$\mathbf{p}_r = \mathbf{P}_r \, e^{j(\omega t + k_1 x)} \tag{6.4b}$$

and a *transmitted wave*

$$\mathbf{p}_t = \mathbf{P}_t \, e^{j(\omega t - k_2 x)} \tag{6.5}$$

The transmitted wave has the same frequency as the incident wave but, because of the different phase speeds c_1 and c_2, the wave numbers $k_1 = \omega/c_1$ in fluid I and $k_2 = \omega/c_2$ in fluid II are different.

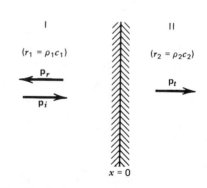

Fig. 6.1. *Reflection and transmission of plane waves normally incident on a boundary.*

There are two boundary conditions that must be satisfied for all times at all points on the boundary: (1) the acoustic pressures on both sides of the boundary are equal and (2) the particle velocities normal to the boundary are equal. The first condition, *continuity of pressure*, means that there can be no *net* force on the (massless) plane separating the fluids. The second condition, *continuity of normal velocity*, requires that the fluids remain in contact. The pressure and normal particle velocity in fluid I are $\mathbf{p}_i + \mathbf{p}_r$ and $(\mathbf{u}_i + \mathbf{u}_r)\hat{x}$ so that the two boundary conditions are

$$\mathbf{p}_i + \mathbf{p}_r = \mathbf{p}_t \qquad \text{at } x = 0 \qquad (6.6a)$$

and

$$\mathbf{u}_i + \mathbf{u}_r = \mathbf{u}_t \qquad \text{at } x = 0 \qquad (6.6b)$$

Division of (6.6a) by (6.6b) yields

$$\frac{\mathbf{p}_i + \mathbf{p}_r}{\mathbf{u}_i + \mathbf{u}_r} = \frac{\mathbf{p}_t}{\mathbf{u}_t} \qquad \text{at } x = 0 \qquad (6.6c)$$

which is a statement of the continuity of *normal specific acoustic impedance* across the boundary.

Since a plane wave has $\mathbf{p}/\mathbf{u} = \pm r$, depending on the direction of propagation, (6.6c) becomes

$$r_1 \frac{\mathbf{p}_i + \mathbf{p}_r}{\mathbf{p}_i - \mathbf{p}_r} = r_2$$

which leads directly to the reflection coefficient

$$\mathbf{R} = \frac{r_2 - r_1}{r_2 + r_1} = \frac{1 - r_1/r_2}{1 + r_1/r_2} \qquad (6.7a)$$

Then, since (6.6a) is equivalent to $1 + \mathbf{R} = \mathbf{T}$, we have

$$\mathbf{T} = \frac{2r_2}{r_2 + r_1} = \frac{2}{1 + r_1/r_2} \qquad (6.7b)$$

The intensity reflection and intensity transmission coefficients follow directly from (6.2)

$$R_I = \left(\frac{r_2 - r_1}{r_2 + r_1}\right)^2 = \left(\frac{1 - r_1/r_2}{1 + r_1/r_2}\right)^2 \qquad (6.8a)$$

and

$$T_I = \frac{4r_1 r_2}{(r_1 + r_2)^2} = 4\frac{r_1/r_2}{(1 + r_1/r_2)^2} \qquad (6.8b)$$

Since the cross-sectional areas of all the beams are equal, the power coefficients in (6.3) are equal to the intensity coefficients.

From (6.7a), $\mathbf{R}$ is always real: positive when $r_1 < r_2$ and negative when $r_1 > r_2$. Consequently, at the boundary the acoustic pressure of the reflected wave is either in phase or 180° out of phase with that of the incident wave. When the characteristic acoustic impedance of fluid II is greater than that of fluid I (a wave in air incident on the air-water interface), a positive pressure in the incident wave is reflected as a positive pressure. On the other hand, if $r_1 > r_2$ (a wave in water incident on the water-air interface), a positive pressure is reflected as a negative pressure. Note that when $r_1 = r_2$ then $\mathbf{R} = 0$, and there is complete transmission.

From (6.7b), it is seen that **T** is real and positive irrespective of the relative magnitudes of r_1 and r_2. Consequently, at the boundary the acoustic pressure of the transmitted wave is *always* in phase with that of the incident wave. Study of (6.8b) reveals that whenever r_1 and r_2 have widely separated values, the intensity transmission coefficient is small. In addition, from the symmetries of (6.8) it is apparent that the intensity reflection and transmission coefficients are *independent* of the direction of the wave. For example, they are the same from water into air as from air into water. This is a special case of the *principle of acoustic reciprocity*.

For $r_1/r_2 \to 0$, the wave is reflected with only a slight reduction in amplitude and no change in phase. The resulting wave in fluid II ($\mathbf{p}_t$) has a maximum pressure amplitude almost twice $\mathbf{P}_i$ and a particle speed at the boundary that is almost zero. Because of this latter fact, such a boundary is termed *rigid*.

For $r_1/r_2 \to \infty$, the amplitude of the reflected wave is again almost equal to that of incident wave. The resulting wave in fluid II will have a pressure amplitude approaching zero. Since the acoustic pressure at the boundary is nearly zero, such a boundary is termed *pressure release*.

6.3 TRANSMISSION THROUGH A LAYER: NORMAL INCIDENCE.
Assume that a layer of uniform thickness L lies between two dissimilar fluids and that a plane wave is normally incident on its boundary, as suggested in Fig. 6.2. Let the characteristic impedances of the fluids be r_1, r_2, and r_3, respectively.

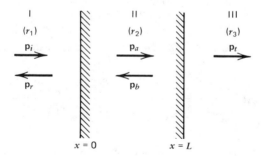

Fig. 6.2. *Reflection and transmission of plane waves normally incident on a layer.*

When an incident signal in fluid I arrives at the boundary between fluids I and II, some of the energy is reflected and some is transmitted into the second fluid. The portion of the wave transmitted will proceed through fluid II to interact with the boundary between fluids II and III, where again some of the energy is reflected and some transmitted. The reflected wave proceeds back to the boundary between fluids I and II, and the whole process is repeated.

If the duration of the incident signal is less than $2L/c_2$, an observer in either fluid I or III will detect a series of echoes separated in time by $2L/c_2$ whose amplitudes can be calculated by applying the results of the previous section the appropriate number of times.

On the other hand, if the incident wave train is long compared to $2L$, and monofrequency, it can be assumed to be

$$\mathbf{p}_i = \mathbf{P}_i \, e^{j(\omega t - k_1 x)} \tag{6.9a}$$

The various transmitted and reflected waves now combine so that in the steady-state condition the wave reflected into fluid I may be represented by

$$\mathbf{p}_r = \mathbf{P}_r \, e^{j(\omega t + k_1 x)} \tag{6.9b}$$

the transmitted and reflected waves in fluid II by

$$\mathbf{p}_a = \mathbf{A} e^{j(\omega t - k_2 x)} \tag{6.10a}$$

and

$$\mathbf{p}_b = \mathbf{B} e^{j(\omega t + k_2 x)} \tag{6.10b}$$

respectively, and the wave transmitted into fluid III by

$$\mathbf{p}_t = \mathbf{P}_t \, e^{j(\omega t - k_3 x)} \tag{6.11}$$

Continuity of the normal specific acoustic impedance at $x = 0$ gives

$$\frac{\mathbf{P}_i + \mathbf{P}_r}{\mathbf{P}_i - \mathbf{P}_r} = \frac{r_2}{r_1} \frac{\mathbf{A} + \mathbf{B}}{\mathbf{A} - \mathbf{B}}$$

Similarly, at $x = L$

$$\frac{\mathbf{A} e^{-jk_2 L} + \mathbf{B} e^{jk_2 L}}{\mathbf{A} e^{-jk_2 L} - \mathbf{B} e^{jk_2 L}} = \frac{r_3}{r_2}$$

Algebraic manipulation then reveals the form of the pressure reflection coefficient

$$\mathbf{R} = \frac{\left(1 - \dfrac{r_1}{r_3}\right)\cos k_2 L + j\left(\dfrac{r_2}{r_3} - \dfrac{r_1}{r_2}\right)\sin k_2 L}{\left(1 + \dfrac{r_1}{r_3}\right)\cos k_2 L + j\left(\dfrac{r_2}{r_3} + \dfrac{r_1}{r_2}\right)\sin k_2 L} \tag{6.12}$$

The intensity transmission coefficient is found by using (6.3) and noting that $A_t = A_i$. Thus, for normal incidence $T_I = 1 - |\mathbf{R}|^2$ which yields

$$T_I = \frac{4}{2 + \left(\dfrac{r_3}{r_1} + \dfrac{r_1}{r_3}\right)\cos^2 k_2 L + \left(\dfrac{r_2^2}{r_1 r_3} + \dfrac{r_1 r_3}{r_2^2}\right)\sin^2 k_2 L} \tag{6.13}$$

There exist a number of special forms of (6.13) which are of particular interest.

(1) If the final fluid is the same as the initial fluid, $r_1 = r_3$ and

$$T_I = \frac{1}{1 + \dfrac{1}{4}\left(\dfrac{r_2}{r_1} - \dfrac{r_1}{r_2}\right)^2 \sin^2 k_2 L} \tag{6.14a}$$

If, in addition, $r_2 \gg r_1$, (6.14a) further simplifies to

$$T_I \approx \frac{1}{1 + \dfrac{1}{4}\left(\dfrac{r_2}{r_1}\right)^2 \sin^2 k_2 L} \tag{6.14b}$$

The latter situation applies, for example, to the transmission of sound from air in one room through a solid wall into air in an adjacent room. (For normal incidence, a solid can be treated as a fluid, since there is no opportunity for a transverse, or shear, wave to occur.) It also applies to sound waves in water passing through a steel plate into water on the opposite side of the plate. The solid materials forming the walls of rooms have such large characteristic impedances relative to air that $(r_2/r_1)\sin k_2 L \gg 2 \cos k_2 L$ for all reasonable frequencies and thicknesses of walls. Therefore, when the fluid medium is air, (6.14b) further simplifies to

$$T_I \approx \left(\frac{2r_1}{r_2 \sin k_2 L}\right)^2 \tag{6.14c}$$

Finally, for all situations, excepting those of high frequencies and very thick walls, $k_2 L \ll 1$ and, consequently, $\sin k_2 L$ may be replaced by $k_2 L$ so that (6.14c) becomes

$$T_I \approx \left(\frac{2}{k_2 L}\frac{r_1}{r_2}\right)^2 \tag{6.15}$$

(To check the validity of the above assumption, note that at 1000 Hz the value of $k_2 L$ for a 0.1-m thick concrete wall is $k_2 L = 2\pi \times 1000 \times 0.1/3100 = 0.2$.) Notice that for a solid wall of given composition the transmitted pressure is inversely proportional to the thickness L and, therefore, also inversely proportional to the mass per unit area of the wall. This behavior is observed to be approximately true for certain kinds of commonly encountered walls. In the case of solid panels in water, both terms occurring in the denominator of (6.14b) usually are significant, so that the complete equation must be used. However, for either thin panels or low frequencies such that $(r_2/r_1)\sin k_2 L \ll 1$, (6.14b) simplifies to

$$T_I \approx 1 \tag{6.16}$$

This behavior is used in the design of free-flooding streamlined domes for housing sonar transducers. These domes are constructed of reinforced panels of either stainless steel or rubber and of such small thickness that little loss takes place as sound is transmitted through the dome into exterior seawater.

(2) Another special form of (6.13) is obtained by assuming that the intermediate fluid has a larger characteristic impedance than either fluid I or fluid III but has such small thickness and moderate characteristic impedance that $r_2 \sin k_2 L \ll 1$ and $\cos k_2 L \approx 1$. Then, (6.13) reduces to

$$T_I \approx \frac{4r_1 r_3}{(r_1 + r_3)^2} \tag{6.17}$$

This is equivalent to (6.8b) which gives the intensity transmission coefficient for a wave moving directly from fluid I into fluid III. It is therefore apparent that a *thin* membrane of solid material of appropriate characteristic impedance may be used in preventing two gases or two liquids from mixing and yet not interfere with sound transmission between them. In particular, notice that if $r_1 = r_3$ then there is total transmission from fluid I to fluid III; it is as if fluid II did not exist.

(3) Returning to the general form of T_I in (6.13), we see that, if $k_2 L = n\pi$, (6.13) reduces to (6.17) at the specific frequencies

$$f = n \frac{c_2}{2L} \tag{6.18}$$

(and narrow bands of surrounding frequencies). Notice that for these frequencies $L = n\lambda_2/2$; the intermediate layer is an integral number of half-wavelengths thick. This is in contrast with the transmission through a *thin* layer, where (6.17) is independent of the intervening fluid for *all* frequencies below some upper limit determined by $r_2 \sin k_2 L \ll 1$.

(4) Finally, if $k_2 L \approx (2n - 1)\pi/2$ where n is any integer, then $\cos k_2 L = 0$ and $\sin k_2 L \approx 1$, so that (6.13) becomes

$$T_I \approx \frac{4r_1 r_3}{\left(r_2 + \dfrac{r_1 r_3}{r_2} \right)^2} \tag{6.19}$$

for frequencies very close to $f = (n - \frac{1}{2})c_2/(2L)$. As an interesting special case, notice that (6.19) yields $T_I = 1$ when $r_2 = \sqrt{r_1 r_3}$. It is therefore possible to obtain total transmission of acoustic power from one medium to another through the use of an intermediate medium whose characteristic impedance is the geometric mean of the other two. However, this action is selective, since it occurs only for bands of frequencies centered about the particular frequencies for which $f = (n - \frac{1}{2})c_2/(2L)$, which means that the thickness of the intervening layer is $L = (n - \frac{1}{2})\lambda_2/2$. This technique of obtaining complete transmission of acoustic power through the use of a quarter-wavelength thick intermediate layer is similar to the method of making nonreflective glass lenses by coating them with a quarter-wavelength thick layer of some suitable material. Another example is the use of quarter-wavelength sections to match an antenna to an electrical transmission line.

The procedure for deriving the transmission and reflection coefficients for a plane wave normally incident on a system of two or more layers separated by parallel plane boundaries is an extension of the techniques described in this section.

The impedance $\mathbf{z}_2$ presented to fluid I by any number of layers can be expressed in terms of the pressure reflection coefficient. The boundary between fluid I and fluid II corresponds to an impedance given by

$$\mathbf{z}_2 = \left. \frac{\mathbf{p}_i + \mathbf{p}_r}{\mathbf{u}_i + \mathbf{u}_r} \right|_{x=0} \tag{6.20}$$

Division of numerator and denominator by $\mathbf{p}_i$ and use of the relation $\mathbf{p}_\pm = \pm r\mathbf{u}_\pm$ results in

$$z_2 = r_1 \frac{1 + R}{1 - R} \tag{6.21}$$

In this way a multilayered fluid boundary to the right of fluid I can be replaced by a single boundary at $x = 0$ whose impedance, given by (6.21), has both real and imaginary components.

6.4 TRANSMISSION FROM ONE FLUID TO ANOTHER: OBLIQUE INCIDENCE.

Assume that the boundary separating two fluids is the plane $x = 0$ and that the incident, reflected, and transmitted waves make the respective angles θ_i, θ_r, and θ_t with the x axis, as shown in Fig. 6.3. Then the incident, reflected, and transmitted waves are

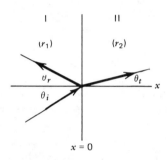

$$\mathbf{p}_i = \mathbf{P}_i e^{j(\omega t - k_1 x \cos \theta_i - k_1 y \sin \theta_i)} \tag{6.22}$$

$$\mathbf{p}_r = \mathbf{P}_r e^{j(\omega t + k_1 x \cos \theta_r - k_1 y \sin \theta_r)} \tag{6.23}$$

and

$$\mathbf{p}_t = \mathbf{P}_t e^{j(\omega t - k_2 x \cos \theta_t - k_2 y \sin \theta_t)} \tag{6.24}$$

The reason for writing θ_t as a complex quantity will emerge as the discussion proceeds.

Fig. 6.3. *Oblique incidence of plane waves on a fluid-fluid boundary.*

Applying the condition of continuity of pressure at the boundary $x = 0$ yields

$$\mathbf{P}_i e^{-jk_1 y \sin \theta_i} + \mathbf{P}_r e^{-jk_1 y \sin \theta_r} = \mathbf{P}_t e^{-jk_2 y \sin \theta_t} \tag{6.25}$$

Since this expression must be true for all y, the exponents must all be equal. This means that

$$\boxed{\sin \theta_i = \sin \theta_r} \tag{6.26}$$

so that the angle of incidence is equal to the angle of reflection, and

$$\boxed{\frac{\sin \theta_i}{c_1} = \frac{\sin \theta_t}{c_2}} \tag{6.27}$$

a statement of *Snell's law*. Since the exponents in (6.25) are all equal, the condition for continuity of pressure on the boundary reduces to

$$1 + R = T \tag{6.28}$$

The condition for continuity of the normal component of particle velocity at the boundary is

$$\mathbf{u}_i \cos \theta_i + \mathbf{u}_r \cos \theta_r = \mathbf{u}_t \cos \theta_t$$

Replacing each velocity by the appropriate value of $\mathbf{p}/r$, and recalling that $\theta_i = \theta_r$, we have

$$1 - \mathbf{R} = \frac{r_1 \cos \theta_t}{r_2 \cos \theta_i} \mathbf{T} \tag{6.29}$$

Equations (6.28) and (6.29) can be combined to eliminate $\mathbf{T}$, giving

$$\mathbf{R} = \frac{(r_2/r_1) - (\cos \theta_t/\cos \theta_i)}{(r_2/r_1) + (\cos \theta_t/\cos \theta_i)} \tag{6.30}$$

where Snell's law reveals

$$\cos \theta_t = \sqrt{1 - \sin^2 \theta_t}$$

$$= \sqrt{1 - (c_2/c_1)^2 \sin^2 \theta_i} \tag{6.31}$$

Equation (6.30) is often referred to as the *Rayleigh reflection coefficient*.

It is important to note three implications of (6.31):

(1) If $c_1 > c_2$, the angle of transmission θ_t is *real* and *less* than the angle of incidence. A transmitted beam exists in the second fluid and this beam is bent *toward* the normal to the boundary for all angles of incidence.

(2a) If $c_1 < c_2$ and $\theta_i < \theta_c$ where θ_c is the *critical angle* defined by

$$\boxed{\sin \theta_c = \frac{c_1}{c_2}} \tag{6.32}$$

the angle of transmission is again real but *greater* than the angle of incidence; a transmitted beam exists, but the beam is bent *away* from the normal for all angles of incidence less than the critical angle.

(2b) If $c_1 < c_2$ and $\theta_i > \theta_c$, the transmitted wave assumes a very peculiar form. From (6.31) we see that $\sin \theta_t$ is real and greater than unity, so that $\cos \theta_t$ is now pure imaginary. Inspection of (6.24) then reveals that the transmitted pressure has the form

$$\mathbf{p}_t = \mathbf{P}_t e^{-\gamma x} e^{j(\omega t - k_1 y \sin \theta_i)}$$

where

$$\gamma = k_2 \sqrt{(c_2/c_1)^2 \sin^2 \theta_i - 1}$$

The transmitted wave propagates in the y direction (*parallel* to the boundary) and has an amplitude that decays *perpendicular* to the direction of propagation. It is left as an exercise to show that for this case the incident wave is totally reflected.

If we return to the most general case, (6.26) shows that the reflected and incident beams have the same cross-sectional areas so that the power reflection coefficient is

$$R_\pi = |\mathbf{R}|^2$$

The power transmission coefficient can be most readily computed from (6.3c), giving

$$T_\pi = \begin{cases} \dfrac{4(r_2/r_1)\cos\theta_t/\cos\theta_i}{(r_2/r_1 + \cos\theta_t/\cos\theta_i)^2} & \theta_t \text{ real} & (6.33a) \\ \\ 0 & \theta_t \text{ imaginary} & (6.33b) \end{cases}$$

Notice that (6.33a) applies when either $\theta_i < \theta_c$ or $c_1 > c_2$.

For the case (6.33a), when $r_2/r_1 = \cos\theta_t/\cos\theta_i$, the power reflection coefficient (6.30) is zero—all the incident power is transmitted. If this condition is combined with (6.27) to eliminate θ_t, then

$$\sin\theta_I = \sqrt{\frac{(r_2/r_1)^2 - 1}{(r_2/r_1)^2 - (c_2/c_1)^2}} = \sqrt{\frac{1 - (r_1/r_2)^2}{1 - (\rho_1/\rho_2)^2}} \qquad (6.34)$$

gives the angle of incidence for which there is no reflection and therefore complete transmission. This angle θ_I known as the *angle of intromission*, will exist only if $r_1 < r_2$ and $\rho_1 < \rho_2$ or $r_1 > r_2$ and $\rho_1 > \rho_2$. If there is a critical angle, it can be seen that the angle of intromission is less than the critical angle.

If the angle of incidence approaches 90°, *grazing incidence*, $\cos\theta_i \to 0$ and (6.30) is reduced to $R \approx -1$. Consequently, at grazing incidence there is complete reflection of the incident acoustic energy irrespective of the relative characteristic acoustic impedances of the two fluids.

Figures 6.4 to 6.7 show typical behaviors for the reflection coefficient for all possible conditions.

The reflection that takes place in seawater from a sand or silt bottom is a good example of reflection associated with two fluids in contact. Such behavior is to be expected since the saturated sand or silt is more like a fluid than a solid in its inability to transmit shear waves. As a first approximation, (6.30) may be used for computing a reflection coefficient. The sand or silt is assumed to have a density ρ_2 and to transmit unattenuated plane waves with a speed c_2. Measured values of ρ_2 for sand and silt range from $1.5\rho_1$ to $2.0\rho_1$, where ρ_1 is the density of seawater; values of c_2 range from $0.9\,c_1$ to $1.1\,c_1$, where c_1 is the speed of sound in seawater.

Extension of the techniques of this section to one or more parallel fluid layers involves no new concepts, but rapidly becomes mathematically complex.

6.5 REFLECTION AT THE SURFACE OF A SOLID: NORMAL INCIDENCE.

The reflection of plane waves in a fluid from the surface of a solid is somewhat more involved than that between two fluids. In order to generalize our analysis of reflection from a solid surface, let us ignore the characteristics of the plane wave that penetrates into the solid. Instead, we will describe the behavior of this wave in terms of $\mathbf{z}_n$, the *normal specific acoustic impedance* of the solid

$$\mathbf{z}_n = \frac{\mathbf{p}}{\mathbf{\dot{u}} \cdot \hat{n}} \qquad (6.35)$$

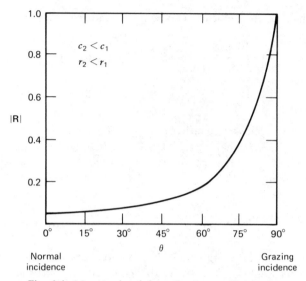

Fig. 6.4. *Magnitude of the reflection coefficient for reflection from a slow bottom ($c_2/c_1 = 0.9$) with $r_2/r_1 = 0.9$.*

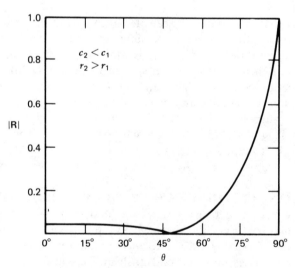

Fig. 6.5. *Magnitude of the reflection coefficient for reflection from a slow bottom ($c_2/c_1 = 0.9$) with $r_2/r_1 = 1.1$. Note angle of intromission at $46.4°$.*

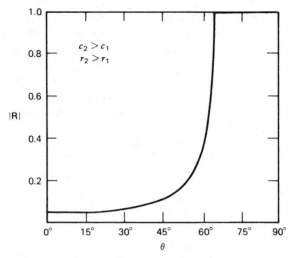

Fig. 6.6. *Magnitude of the reflection coefficient for reflection from a fast bottom* $(c_2/c_1 = 1.1)$ *with* $r_2/r_1 = 1.1$. *Note critical angle at* 65.6°.

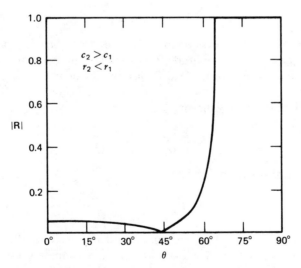

Fig. 6.7. *Magnitude of the reflection coefficient for reflection from a fast bottom* $(c_2/c_1 = 1.1)$ *with* $r_2/r_1 = 0.9$. *Note angle of intromission at* 43.2° *and critical angle at* 65.6°.

where $\hat{n}$ is the unit vector perpendicular to the interface. Since acoustic pressure is not always in phase with particle velocity at the surface of a solid, the normal specific acoustic impedance may be complex,

$$z_n = r_n + jx_n \tag{6.36}$$

with r_n the resistive component and x_n the reactive component.

For normal incidence, the normal component $\hat{u} \cdot \hat{n}$ is simply the particle speed. The continuity equations

$$\mathbf{p}_i + \mathbf{p}_r = \mathbf{p}_t \qquad \text{at } x = 0 \tag{6.37a}$$

and

$$\mathbf{u}_i + \mathbf{u}_r = \mathbf{u}_t \qquad \text{at } x = 0 \tag{6.37b}$$

then can be combined and z_n written as

$$z_n = \frac{\mathbf{p}_i + \mathbf{p}_r}{\mathbf{u}_i + \mathbf{u}_r} \qquad \text{at } x = 0 \tag{6.37c}$$

Substitution of the expressions for pressure and particle speed gives

$$z_n = r_1 \frac{1 + \mathbf{R}}{1 - \mathbf{R}} \tag{6.37d}$$

which when rearranged gives the pressure reflection coefficient

$$\mathbf{R} = \frac{(r_n - r_1) + jx_n}{(r_n + r_1) + jx_n} \tag{6.38}$$

Note that when $x_n = 0$, (6.38) becomes equivalent to (6.7a). Except in those cases where z_n is real, the pressure amplitude $\mathbf{P}_r$ is complex and consequently the reflected wave at the boundary may either lead or lag the incident wave by angles ranging from $0°$ to $180°$.

The intensity reflection coefficient is

$$R_I = \frac{(r_n - r_1)^2 + x_n^2}{(r_n + r_1)^2 + x_n^2} \tag{6.39}$$

and the intensity transmission coefficient is

$$T_I = \frac{4r_1 r_n}{(r_n + r_1)^2 + x_n^2} \tag{6.40}$$

Now consider reflection from the surfaces of relatively rigid nonporous solids such as steel, glass, sealed concrete, and so forth. In an infinite solid material, and in finite solids of transverse dimensions much larger than the wavelength of the acoustic wave, two types of elastic waves may be propagated, plane longitudinal and plane shear waves. In such an isotropic solid having large transverse cross section, the speed of sound is no longer $\sqrt{Y/\rho_0}$ as in Chapter 3; here the so-called bulk or plate speed of longitudinal waves must be used,

$$c = \sqrt{\frac{\mathcal{B} + \frac{4}{3}\mathcal{G}}{\rho_0}} \tag{6.41}$$

where $\mathcal{B}$ and $\mathcal{G}$ are the bulk and shear moduli of the solid and ρ_0 is its density. Numerical values of the bulk speed for various solids are given in the appendix, along with corresponding characteristic impedances $\rho_0 c$. (It should be noted that the bulk speed given for each material is always higher than that given for longitudinal waves in thin bars.)

6.6 REFLECTION AT THE SURFACE OF A SOLID: OBLIQUE INCIDENCE. No single simple method is available for analyzing the reflection of plane waves obliquely incident on the surface of a solid. Because of the differences in the porosity and internal elastic structure of various solids, the nature of the transmitted wave varies widely and thereby influences the reflected wave. For instance, the wave transmitted into the solid may be refracted (1) so that it is propagated effectively only perpendicular to the surface, (2) in a manner similar to plane waves entering a second fluid, or (3) into two waves, longitudinal bulk waves traveling in one direction and transverse shear waves traveling at a lower speed and in a different direction.

(1) The first type of refraction occurs for the so-called *normally reacting* or *locally acting* surfaces. One example of this type of refraction is that occurring in anisotropic solids, where waves propagated parallel to the surface travel with a much lower speed than those propagated perpendicular to the surface. This is typical of solids having a honeycomb structure in which the speed of compressional waves through the fluid contained in capillary pores oriented at right angles to the surface is much higher than that from pore to pore through the solid material of the structure. This type of refraction also will occur in an isotropic solid when the speed of longitudinal wave propagation in the solid is small compared with that in the fluid. When $c_2 \ll c_1$, Snell's law requires that $\theta_t \ll \theta_i$, resulting in a marked bending of the transmitted wave toward the normal direction. Many of the higher sound-absorbing materials

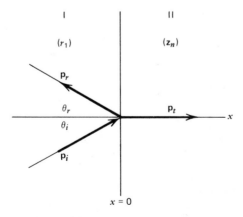

Fig. 6.8. *Oblique incidence of plane waves on a normally reacting solid.*

used in building construction, such as acoustic tile, perforated panels, and so on, appear to behave as normally reacting surfaces.

Since for such surfaces the refracted ray travels nearly normal to the surface, irrespective of the particular angle of incidence, it is reasonable to assume that the ratio of acoustic pressure acting on the surface to the component of fluid velocity normal to the surface will be independent of the direction of the incident wave. Consequently, reflection from such a surface is most readily discussed in terms of the normal specific acoustic impedance z_n of its surface as defined by (6.35). The condition of continuity of normal specific impedance applied at the surface of Fig. 6.8 therefore can be expressed as

$$\frac{\mathbf{p}_i + \mathbf{p}_r}{\mathbf{u}_i \cos \theta_i + \mathbf{u}_r \cos \theta_r} \tag{6.42}$$

or

$$r_1 \frac{\mathbf{P}_i + \mathbf{P}_r}{\mathbf{P}_i - \mathbf{P}_r} = \mathbf{z}_n \cos \theta_i \tag{6.43}$$

Solving this equation for **R** gives

$$\mathbf{R} = \frac{\mathbf{z}_n \cos \theta_i - r_1}{\mathbf{z}_n \cos \theta_i + r_1} \tag{6.44}$$

This equation is similar in form to (6.38) and differs only in that $\mathbf{z}_n \cos \theta_i$ replaces $\mathbf{z}_n$. As a consequence, equations for the reflection and transmission coefficients applicable to this more general case may be obtained from those developed in Sect. 6.5 by replacing $\mathbf{z}_n$ with $\mathbf{z}_n \cos \theta_i$.

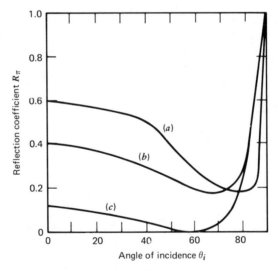

Fig. 6.9. *Dependence of the reflection coefficient on the angle of incidence for typical normally reacting solids. Curve (a) corresponds to $r_n/r_1 = x_n/r_1 = 4$. Curve (b) corresponds to $r_n/r_1 = x_n/r_1 = 2$. Curve (c) corresponds to $r_n/r_1 = 2, x_n/r_1 = 0$.*

For instance, the power reflection coefficient is

$$R_\pi = \frac{(r_n \cos \theta_i - r_1)^2 + (x_n \cos \theta_i)^2}{(r_n \cos \theta_i + r_1)^2 + (x_n \cos \theta_i)^2} \qquad (6.45)$$

and the power transmission coefficient is

$$T_\pi = \frac{4 r_1 r_n \cos \theta_i}{(r_n \cos \theta_i + r_1)^2 + (x_n \cos \theta_i)^2} \qquad (6.46)$$

For most solid materials $r_n > r_1$, so that as θ_1 increases an angle will be reached where $r_n \cos \theta_i = r_1$. When this occurs the power reflection coefficient will be near its minimum value. In particular, if x_n were zero, R_π would be zero and T_π would be unity. For $\theta_i \approx 90°$, R_π approaches unity. Plotted in Fig. 6.9 are curves for the reflection coefficient R_π as a function of the angle of incidence θ_i for a few assumed values of the nondimensional parameters r_n/r_1 and x_n/r_1.

(2) The second type of refraction is similar to the reflection and refraction occurring between two fluids as discussed in Sect. 6.4

(3) The third type of refraction occurs for rigid elastic solids. A detailed discussion requires consideration of the propagation of both elastic shear and compressional waves in the solid and is not warranted in a textbook on fundamentals. Information on this type of refraction is available in the references.[1-3]

PROBLEMS

6.1. A plane wave of 50 Pa effective (rms) pressure and 1 kHz is incident normally on the water-air boundary at the surface of the water. (a) What is the effective pressure of the plane wave transmitted into the air? (b) What is the intensity of the incident wave in the water and of the wave transmitted into the air? (c) Express, as a decibel reduction, the ratio of the intensity of the transmitted wave in air to that of the incident wave in water. (d) Answer the same three questions for the above sound wave incident on a thick layer of ice. (e) What is the sound power reflection coefficient from the layer of ice?

6.2. If a plane wave is reflected from the ocean floor at normal incidence with a level 20 dB below that of the incident wave, what are the possible values of the specific acoustic impedance of the fluid bottom material?

6.3. (a) A plane wave in seawater is normally incident on the water-air interface. Find the pressure and intensity transmission coefficients. (b) Repeat (a) for a wave in air normally incident on the air-water interface. (c) For (a) and (b) find the change in pressure and intensity levels if P_{ref} and I_{ref} are the same in both media.

6.4. (a) What must be the thickness of, and the speed of sound in, a plastic layer having a density of 1500 kg/m³, if it is to transmit plane waves at a frequency of 20 kHz from water into steel with no reflection? (b) What would be the intensity reflection coefficient back into water for normally incident waves impinging on an infinitely thick layer of this plastic?

6.5. Given a 2 kHz plane wave in water to impinge normally on a wide steel plate of 1.5-cm thickness, (a) what is the transmission loss, expressed in decibels, through the steel plate into water on the opposite side. (b) What is the sound power reflection coefficient of this

[1] Officer, *Introduction to the Theory of Sound Transmission*, McGraw-Hill (1958).

[2] Ewing, Jardetzky, and Press, *Elastic Waves in Layered Media*, McGraw-Hill (1957).

[3] Brekhovskikh, *Waves in Layered Media*, Academic Press (1960).

plate? (c) Repeat (a) and (b) for a 1.5-cm thick slab of sponge rubber having a density of 500 kg/m^3 and which propagates longitudinal waves with a speed of 1000 m/s.

6.6. Given the task of maximizing the transmission of sound waves from water into steel, (a) what is the optimum characteristic impedance of the material to be placed between the water and the steel? (b) What must be the density of, and sound speed in, a layer of 1-cm thickness that will produce 100 percent transmission at a frequency of 20 kHz?

6.7. Sketch the amplitude and phase of the pressure reflection coefficient for the case $c_2 = c_1$ and $\rho_2 > \rho_1$. Identify any significant features.

6.8. For plane wave reflection from a fluid-fluid interface it is observed that at normal incidence the pressure amplitude of the reflected wave is one-half that of the incident wave (no phase information is recorded). As the angle of incidence is increased, the amplitude of the reflected wave first decreases to zero and then increases until at 30° the reflected wave is as strong as the incident wave. Find the density and sound speed in the second medium if the first medium is water.

6.9. A plane wave traveling from air into hydrogen gas, across a thin membrane separating the two gases, is refracted by 40° from its original direction. (a) What is the angle of incidence in the air? (b) What is the sound power transmission coefficient?

6.10. A plane wave in water of 100 Pa peak pressure amplitude is incident at 45° on a mud bottom having a density of 2000 kg/m^3 and a sound speed of 1000 m/s. Compute (a) the angle at which the refracted ray is transmitted into the mud, (b) the peak pressure amplitude of the transmitted ray, (c) the peak pressure amplitude of the reflected ray, and (d) the sound power reflection coefficient.

6.11. Plane waves in water of 100 Pa effective (rms) pressure are incident normally on a sand bottom. The sand has a density of 2000 kg/m^3 and a sound speed of 2000 m/s. (a) What is the effective pressure of the wave reflected back into the water? (b) What is the effective pressure of the wave transmitted into the sand? (c) What is the sound power coefficient of reflection from the sand? (d) What is the smallest angle of incidence at which all of the sound energy will be reflected?

6.12. A sand bottom in seawater is characterized by a density of 1700 kg/m^3 and a sound speed of 1600 m/s. (a) What is the critical angle of incidence corresponding to total reflection. (b) For what angle of incidence is the sound power reflection coefficient equal to 0.25? (c) What is the power reflection coefficient for normal incidence?

6.13. A certain acoustic tile panel is characterized by a normal specific acoustic impedance of $900 - j1200$ Pa·s/m. (a) For what angle of incidence in air will the sound power reflection coefficient be a minimum? (b) What is the power reflection coefficient for an angle of incidence of 80°? (c) What is the power reflection coefficient for normal incidence?

6.14. Consider a wall that reflects plane waves in a manner similar to a normally reacting surface of normal specific acoustic impedance $z_n = r_1 + j\omega\rho_S$ where r_1 is the characteristic impedance of the air and ρ_S is the area density of the wall in kilograms per square meter. Derive a general equation for its sound power reflection coefficient as a function of the incident angle θ. For a wall with an area density of 2 kg/m^2, compute and plot the power reflection coefficient at 100 Hz as a function of θ.

CHAPTER 7

ABSORPTION AND ATTENUATION OF SOUND WAVES IN FLUIDS

7.1 INTRODUCTION.   In previous chapters no consideration has been given to the *dissipation* of acoustic energy. In many situations dissipation takes place so slowly that it can be ignored for small distances or short times. Ultimately, however, all acoustic energy is degraded into thermal energy. The sources of this dissipation may be divided into two general categories: those due to losses *in the medium* and those associated with losses *at the boundaries* of the medium. The first is important when the volume of the fluid is large, as in the transmission of sound in the earth's atmosphere and oceans, through large ducts, and within large auditoriums. The second is important in the opposite extreme—with porous materials, thin ducts, and small rooms. Losses in the medium may be divided into three basic types: *viscous* losses, *heat conduction* losses, and losses associated with *molecular exchanges* of energy. *Viscous* losses result whenever there is relative motion between adjacent portions of the medium, such as during the compressions and expansions that accompany the transmission of a sound wave. These may be thought of as frictional losses. *Heat conduction* losses result from the conduction of thermal energy (heat) between higher temperature condensations and lower temperature rarefactions. *Molecular exchanges* of energy that can lead to absorption include the conversion of kinetic energy of the molecules into (a) stored *potential energy* (as in a structural rearrangement of adjacent molecules in some cluster), (b) *internal rotational and vibrational energies* (for polyatomic molecules), or (c) *energies of association and dissociation* between different ionic species and complexes in ionized solutions (magnesium sulphate and boric acid complexes in seawater).

Each of these absorption processes is characterized by a *relaxation time*, which measures the amount of time for the particular process to be nearly completed. For example, when the period of the acoustic cycle is greater than, or comparable with, the time required for a portion of the compression energy of the fluid to be converted into internal energy of molecular vibration, then during the expansion cycle some of this energy will be delayed in its restoration, resulting in a tendency toward pressure equalization and an attendant attenuation of the wave.

So far in this book, we have assumed that the fluid is a *continuum* having directly observable properties such as pressure, density, compressibility, specific heat, and temperature, without being concerned with its molecular structure. In this same spirit, by use of *viscosity*, Stokes[1] developed the first successful theory of sound absorption. Subsequently, Kirchhoff[2] utilized the property of *thermal conductivity* to develop an additional contribution to generate what is called the *classical sound absorption* in fluids. In more recent times, as more accurate sound absorption measurements were made, it became evident that explanations of sound absorption from this viewpoint were inadequate in some fluids. Consequently, it became necessary to adopt a *microscopic* view and consider such phenomena as the binding energies within and between molecules to develop additional absorption mechanisms. These latter mechanisms are commonly referred to as *molecular* or *relaxation* types of sound absorption. (In point of fact, all loss mechanisms are relaxational in nature, but often certain effects of the relaxations are not observed in the range of frequencies and temperatures usually encountered.) For a discussion more complete than that presented in this chapter, the reader is referred to the literature.[3-5]

7.2 A PHENOMENOLOGICAL APPROACH TO ABSORPTION.

A consequence of ignoring any loss mechanisms is that the acoustic pressure p and the condensation s are *in phase*, related by the linear equation of state (5.14),

$$p = \rho_0 c^2 s \qquad (7.1)$$

We now show that one way to introduce losses is to modify this equation of state to allow for a *delay* between the application of a sudden pressure change and the attainment of the resulting equilibrium condensation given by (7.1).

One simple relation yielding such an effect is a modified equation of state first considered by Stokes,

$$p = \rho_0 c^2 \left(1 + \tau \frac{\partial}{\partial t}\right) s \qquad (7.2)$$

where τ is the *relaxation time*. To visualize the effects resulting from this relationship, assume that the fluid is at rest until time $t = 0$, when a sudden increase of pressure P_0 is applied (Fig. 7.1a). The solution to (7.2) that satisfies this condition is

$$s = \frac{P_0}{\rho_0 c^2} (1 - e^{-t/\tau}) \qquad t \geq 0 \quad (7.3)$$

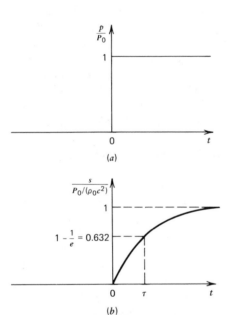

Fig. 7.1. *Response of a relaxing fluid to a sudden increase in pressure.*

[1] Stokes, *Trans. Cambridge Phil. Soc.*, **8**, 287 (1845).

[2] Kirchhoff, *Pogg. Ann. Phys.*, **134**, 177 (1868).

[3] Markham, Beyer, and Lindsay, *Reviews of Modern Physics*, **23**, 533, (1951).

[4] Herzfeld and Litovitz, *Absorption and Dispersion of Ultrasonic Waves*, Academic Press (1959).

[5] *Physical Acoustics*, Vol. IIA, ed. Mason, Academic Press (1965).

As shown in Fig. 7.1*b*, the condensation rises from zero to its equilibrium value $P_0/(\rho_0 c^2)$, reaching 0.632 of its final value at a time $t = \tau$.

If the modified equation of state (7.2) is combined with the linearized, lossless force equation

$$\rho_0 \frac{\partial \check{u}}{\partial t} = -\nabla p \tag{7.4}$$

and the linearized continuity equation

$$\nabla \cdot \check{u} = -\frac{1}{\rho_0} \frac{\partial \rho}{\partial t} = -\frac{\partial s}{\partial t} \tag{7.5}$$

a modified wave equation is obtained

$$\left(1 + \tau \frac{\partial}{\partial t}\right) \nabla^2 p = \frac{1}{c^2} \frac{\partial^2 p}{\partial t^2} \tag{7.6}$$

The quantity c^2 in (7.2) and (7.6) is the thermodynamic speed of sound, determined from $c^2 = (\partial p/\partial \rho)_{ad}$, as seen in (5.17). It is not necessarily the phase speed c_p because of the additional term $\tau \partial(\nabla^2 p)/\partial t$ in (7.6).

If discussion is now restricted to cases for which the acoustic quantities behave as $\exp(j\omega t)$, (7.6) reduces to the *lossy Helmholtz equation*

$$(\nabla^2 + \mathbf{k}^2)\mathbf{p} = 0 \tag{7.7}$$

where

$$\mathbf{k} = \frac{\omega}{c} \frac{1}{\sqrt{1 + j\omega\tau}} \tag{7.8}$$

By substituting the definition

$$\mathbf{k} = k - j\alpha \tag{7.9}$$

into (7.8) and by collecting real and imaginary parts, we obtain two equations

$$(k^2 - \alpha^2) + 2\omega\tau\alpha k = \left(\frac{\omega}{c}\right)^2 \tag{7.10}$$

$$2\alpha k = \omega\tau(k^2 - \alpha^2) \tag{7.11}$$

Elimination of k between these two equations yields

$$\alpha = \frac{\omega}{c} \frac{1}{\sqrt{2}} \left[\frac{\sqrt{1 + (\omega\tau)^2} - 1}{1 + (\omega\tau)^2}\right]^{1/2} \tag{7.12}$$

Elimination of α between these same equations yields

$$k = \frac{\omega}{c} \frac{1}{\sqrt{2}} \left[\frac{\sqrt{1 + (\omega\tau)^2} + 1}{1 + (\omega\tau)^2}\right]^{1/2} \tag{7.13}$$

Solution of (7.7) for a plane wave traveling in the $+x$ direction gives

$$\mathbf{p} = P_0 e^{j(\omega t - \mathbf{k}x)}$$

where P_0 is the pressure amplitude at $x = 0$. With the use of (7.9) this becomes

$$\mathbf{p} = P_0\, e^{-\alpha x} e^{j(\omega t - kx)} \tag{7.14}$$

Thus, the traveling wave has a *phase speed*

$$\boxed{c_p = \frac{\omega}{k}} \tag{7.15}$$

For this case, use of (7.13) reveals

$$c_p = c\sqrt{2}\left[\frac{1 + (\omega\tau)^2}{\sqrt{1 + (\omega\tau)^2} + 1}\right]^{1/2} \tag{7.16}$$

It is obvious from (7.16) that the phase speed c_p is not equal to c, but is a function of frequency. When the phase speed depends on frequency, the propagation is termed *dispersive*. Furthermore, there is an amplitude $P(x)$ that decays as $P_0\exp(-\alpha x)$. In complete analogy with traveling waves on a damped string, α is the *absorption coefficient* and k is the familiar *propagation constant*.

Substitution of (7.14) into (7.4) and (7.5) allows calculation of the associated particle speed

$$\mathbf{u} = \frac{k}{\omega\rho_0}\mathbf{p} = \frac{1}{\rho_0 c_p}\left(1 - j\,\frac{\alpha}{k}\right)\mathbf{p} \tag{7.17}$$

The specific acoustic impedance for this traveling wave is

$$\mathbf{z} = \frac{\mathbf{p}}{\mathbf{u}} = \rho_0 c_p\,\frac{1}{1 - j\alpha/k} \tag{7.18}$$

Calculation of the acoustic intensity for a monofrequency wave from the fundamental definition

$$I = \frac{1}{T}\int_{t}^{t+T} pu\, dt \tag{7.19}$$

results in

$$I = \frac{(P_0\, e^{-\alpha x})^2}{2\rho_0 c_p} = I(0)e^{-2\alpha x} \tag{7.20}$$

Attenuation coefficients for waves decaying with distance are frequently expressed in *nepers per meter*. Note that the neper (Np) is a dimensionless unit. When $x = 1/\alpha$ the pressure amplitude has dropped to $1/e$ of its initial value P_0.

The change in *intensity level* of the attenuated wave, expressed in decibels, is given by

$$IL(0) - IL(x) = 10\log\frac{I(0)}{I(x)} = 10\log e^{2\alpha x}$$

$$= 8.7\alpha x \equiv ax$$

Consequently, $a = 8.7\alpha$ is a measure of the spatial rate of decrease in intensity level expressed in decibels per meter (dB/m).

If the angular frequency ω is low enough and the relaxation time short enough so that $\omega\tau \ll 1$, then (7.12) becomes

$$\alpha \doteq \tfrac{1}{2}\omega^2\tau/c \tag{7.21}$$

and the absorption coefficient is proportional to the square of the frequency. As a consequence, it has become the practice to plot data in the form α/f^2 as a function of frequency. Any departure from a horizontal line is a measure of the deviation of experimental results from the predictions of (7.21). Furthermore, the phase speed simplifies to

$$c_p \doteq c\left[1 + \tfrac{3}{8}(\omega\tau)^2\right] \tag{7.22}$$

In this case, the dispersion is of second order, $O^2(\omega\tau)$, in $\omega\tau$ so that it is only slight and the phase speed is virtually identical with c. Combination of (7.21) and (7.22) in this limit reveals

$$\alpha/k \doteq \tfrac{1}{2}\omega\tau \tag{7.23}$$

so that α/k and $\omega\tau/2$ are equivalent. Note also that when $\alpha/k \ll 1$ the specific acoustic impedance (7.18) is well approximated by

$$z \doteq e^{j\alpha/k}\rho_0 c \tag{7.24}$$

This is almost identical with the lossless result $\rho_0 c$ and reduces to it in the limit $\alpha/k \to 0$.

7.3 COMPLEX SOUND SPEED.

Before proceeding to a discussion of specific absorption mechanisms, it is appropriate to develop the widely accepted convention of a *complex phase speed* for *monofrequency* acoustic processes. If the *dynamic* equation of state (7.2) is written

$$\mathbf{p} = \rho_0 \mathbf{c}^2\mathbf{s} \tag{7.25}$$

then the usual manipulations lead to a modified wave equation of the form

$$\nabla^2\mathbf{p} = \frac{1}{\mathbf{c}^2}\frac{\partial^2\mathbf{p}}{\partial t^2} \tag{7.26}$$

and a lossy Helmholtz equation exactly of the form (7.7) if we define

$$\mathbf{k} = \frac{\omega}{\mathbf{c}} \tag{7.27}$$

The usefulness of this concept is that, if it is possible to obtain a *monofrequency* equation of state in the form (7.25), then $\mathbf{c}$ can be used in (7.27) to calculate $\mathbf{k}$, which in turn yields α and k from (7.9) and c_p from (7.15).

As an example, consider the Stokes equation of state (7.2). If we assume that all motion is monofrequency, so that all acoustic quantities vary as $\exp(j\omega t)$, then (7.2) becomes

$$\mathbf{p} = \rho_0 c^2 (1 + j\omega\tau)\mathbf{s}$$

and the complex phase speed is

$$\mathbf{c} = c\sqrt{1 + j\omega\tau}$$

Then

$$\mathbf{k} = \frac{\omega}{\mathbf{c}} = \frac{\omega}{c}\frac{1}{\sqrt{1 + j\omega\tau}}$$

and the discussion proceeds as that following (7.8).

We will find this a useful convention in dealing with molecular exchange processes where it is easier to find an equation of state than to work from the equation of motion.

7.4 THE CLASSICAL ABSORPTION COEFFICIENT.

If the properties of a viscous fluid are considered, it is necessary to redevelop the force equation in tensor formalism, which is beyond the scope of this book. The result of this more general derivation is the *Navier-Stokes* equation[6]

$$\rho\left[\frac{\partial \vec{u}}{\partial t} + (\vec{u} \cdot \nabla)\vec{u}\right] = -\nabla\rho + \left(\tfrac{4}{3}\eta + \eta_B\right)\nabla(\nabla \cdot \vec{u}) - \eta\nabla \times \nabla \times \vec{u} \qquad (7.28)$$

where η and η_B are parameters depending on the fluid and its temperature. They both have units of pascal seconds (Pa · s).

The *shear viscosity coefficient* η can be measured directly by subjecting the fluid to shearing forces. Experimentally, it is observed to be independent of frequency, depending only on the temperature, for almost all fluids over the range of physical parameters of practical interest. (This result has been theoretically verified for the ideal gas from application of statistical mechanics. While η is manifest most directly in shear flow, it is actually a measure of the diffusion of momentum by molecules from regions of fluid possessing higher velocities to adjoining regions possessing lower velocities; it is therefore active in producing absorption even in pure longitudinal motion.)

The *bulk* (or *volume*) *viscosity* η_B is a measure of the mechanical energy lost by a fluid subjected to pure compression or dilatation. The bulk viscosity appears to be zero for monatomic gases and nonzero only in those fluids in which energy can be transferred by processes other than those associated with the translational kinetic energy of the molecules.

The term $\eta\nabla \times \nabla \times \vec{u}$ represents the dissipation of acoustic energy involving turbulence, streaming, and vorticity. While these effects can be important in hydrodynamics and nonlinear acoustics, they are usually confined to small regions of a sound field and are found to be negligible for linear acoustics. Ignoring this term is

---

[6] Morse and Ingard, *Theoretical Acoustics*, McGraw-Hill (1968).

equivalent to assuming *irrotational flow*, $\nabla \times \vec{u} = 0$. If we also ignore nonlinear effects and make the *Stokes assumption*

$$\eta_B = 0$$

then (7.28) becomes

$$\rho_0 \frac{\partial \vec{u}}{\partial t} = -\nabla p + \tfrac{4}{3}\eta\nabla(\nabla \cdot \vec{u}) \tag{7.29}$$

Now, use of (7.1) and (7.5) yields

$$\left(1 + \frac{\tfrac{4}{3}\eta}{\rho_0 c^2}\frac{\partial}{\partial t}\right)\nabla^2 p = \frac{1}{c^2}\frac{\partial^2 p}{\partial t^2} \tag{7.30}$$

Direct comparison with (7.6) shows that we can identify a relaxation time

$$\tau_S = \frac{4}{3}\frac{\eta}{\rho_0 c^2} \tag{7.31}$$

and that the absorption coefficient and phase speed are given by (7.12) and (7.16). In particular, (7.21) reveals the low frequency approximation

$$\alpha_S \doteq \frac{2}{3}\frac{\omega^2}{\rho_0 c^3}\eta \tag{7.32}$$

which is useful for all except extremely high ultrasonic frequencies or exceptionally viscous fluids.

Since $\vec{u}$ is assumed irrotational, it can be derived from a scalar potential Φ, as in the lossless case, (5.15). Manipulation of (7.29) and the linearized equations of state and continuity relates p and Φ by

$$\left(1 + \tau_S \frac{\partial}{\partial t}\right)p = -\rho_0 \frac{\partial \Phi}{\partial t} \tag{7.33a}$$

Since p and $\partial p/\partial t$ both satisfy (7.30), which is linear and homogeneous, then so also must $\partial\Phi/\partial t$ and Φ. Thus, both p and Φ satisfy the lossy wave equation (7.30) even though they are related through the somewhat involved (7.33a). For monofrequency motion (7.33a) becomes

$$\mathbf{p} = -\frac{j\omega\rho_0}{1 + j\omega\tau_S}\Phi \tag{7.33b}$$

which can be approximated by $\mathbf{p} \doteq -j\omega\rho_0\Phi$ to first order in $\omega\tau_S$.

Another classical mechanism producing absorption is *thermal conduction*. Derivation of the relaxation time and absorption coefficient associated with thermal conduction losses requires rather extensive use of thermodynamics, so we will simply quote the results. The relaxation time is

$$\tau_\kappa = \frac{1}{\rho_0 c^2}\frac{\kappa}{C_{\mathscr{P}}} \tag{7.34}$$

where κ is the thermal conductivity of the fluid and $C_{\mathscr{P}}$ its heat capacity at constant pressure. The absorption coefficient associated with thermal conductivity alone is

$$\alpha_\kappa \doteq \frac{1}{2} \frac{\omega^2}{\rho_0 c^3} (\gamma - 1) \frac{\kappa}{C_{\mathscr{P}}} \qquad (\omega\tau_\kappa \ll 1) \tag{7.35}$$

for frequencies of practical importance. The quantity γ is the ratio of heat capacities $C_{\mathscr{P}}/C_V$. For gases the absorption associated with heat conduction is somewhat less than that for viscous absorption but of the same magnitude. On the other hand, for most *nonmetallic* liquids the absorption produced by thermal conductivity is negligible compared with that from viscosity.

When the absorption is small, it is possible to assume that viscous and thermal conduction losses act independently in producing attenuation of sound. This means that the combined *classical absorption coefficient* is the sum

$$\alpha(\text{classical}) \doteq \frac{\omega^2}{2\rho_0 c^3} \left[\frac{4}{3} \eta + (\gamma - 1) \frac{\kappa}{C_{\mathscr{P}}} \right] \tag{7.36}$$

Based on this discussion, we will state without proof that the effects of separate mechanisms for absorption of sound combine so that the total absorption coefficient is nearly the sum of the absorption coefficients of the individual loss mechanisms calculated as if they were operating alone,

$$\boxed{\alpha = \sum_i \alpha_i} \tag{7.37}$$

This is true as long as $\alpha/k \ll 1$.

Table 7.1 contains comparative data on calculated and observed values of the

Table 7.1. Acoustic absorption in fluids

All data for T = 20°C and $\mathscr{P}_0$ = 1 atm	α/f^2 (Np · s²/m)			
	Shear viscosity	Thermal conductivity	Classical	Observed
Gases	Multiply all values by 10^{-11}			
Argon	1.08	0.77	1.85	1.87
Helium	0.31	0.22	0.53	0.54
Oxygen	1.14	0.47	1.61	1.92
Nitrogen	0.96	0.39	1.35	1.64
Air (dry)	0.99	0.38	1.37	α/f peaks at 40 Hz
Carbon dioxide	1.09	0.31	1.40	α/f peaks at 30 kHz
Liquids	Multiply all values by 10^{-15}			
Glycerin	3000.0	—	3000.0	3000.0
Mercury	—	6.0	6.0	5.0
Acetone	6.5	0.5	7.0	30.0
Water	8.1	—	8.1	25.0
Seawater	8.1	—	8.1	α/f peaks at 1.2 kHz and 136 kHz

absorption coefficient α in nepers per meter (Np/m) for representative gases and liquids. As to be expected, the absorption observed for the monatomic gases such as argon and helium is in good agreement with the classical absorption coefficient (shear viscosity and thermal conduction). The classical absorption is also in good agreement for highly viscous liquids such as glycerin and highly conducting liquid metals such as mercury. However, the classical absorption falls short of the observed results in polyatomic gases and in most common liquids.

Calculation of the relaxation times τ_S and τ_κ for representative fluids for which the classical absorption coefficient accurately describes the observed attenuation shows that typical values are about 10^{-10} s in gases and less than about 10^{-12} s for all liquids except highly viscous ones like glycerin. Thus the frequencies for which the assumption $\omega\tau \ll 1$ fails lie in the very high ultrasonic range.

7.5 MOLECULAR THERMAL RELAXATION.

As shown in the previous section, continuum mechanics predicts the existence of the *classical* absorption mechanisms of viscosity and thermal conductivity. However, it does not provide a means of predicting either the *values* of the viscosity and heat conduction coefficients or their *temperature dependences*. These quantities can be calculated for simple fluids from statistical mechanics. Further mechanisms for acoustic absorption can be predicted by taking into account the internal structure of the molecules and the interactions between them that lead to internal vibrations, rotations, ionizations, and short-range ordering.

The oldest and most successful of the many theoretical approaches to these problems is that treating *molecular thermal relaxation* in fluids composed of polyatomic molecules. In this theory it is acknowledged that, in addition to the three degrees of *translational freedom* each molecule possesses, there are also internal degrees of freedom associated with the *rotation* and *vibration* of these molecules. The time necessary for energy to be transferred from translational motion of the molecule into internal states compared to the period of the acoustic process determines how much acoustic energy will be converted to thermal energy during the transitions. If the relaxation time τ of an internal energy state is long compared to the time for changes in the acoustic variables to take place ($\omega\tau \gg 1$), this internal energy state will not be populated and there will be no absorption. On the other hand, if the relaxation time is short ($\omega\tau \ll 1$), the internal state will always be in equilibrium with the translational states and, again, there will be no absorption. However, absorption will take place for intermediate frequencies ($\omega\tau \sim 1$).

If all the internal states are always in equilibrium, the heat capacity C_V will be

$$C_V = \mathscr{R}\left[\frac{3}{2} + \frac{1}{2}\sum_{i=1}^{N} H_i(T)\right] \tag{7.38}$$

where each $H_i(T)$ is the fraction of the ith state populated at the temperature T. The 3/2 gives the contribution for the three translational degrees of freedom of each molecule (these states are fully populated at all nonzero temperatures). The summation covers the N internal degrees that are available for internal energy storage, and the $H_i(T)$ accounts for the fact that these internal energy states cannot be significantly populated unless the temperature is above some specific temperature T_i for each internal state. Each $H_i(T)$ asymptotically approaches zero at low temperatures and unity at sufficiently high temperatures and changes rapidly only for temperatures close to the specific value T_i for that state. [The $H_i(T)$ are not step functions because at any given temperature the molecules possess a *distribution* of kinetic energies, so that the energy of a given molecule may be greater or less than the average energy.] For example, at room temperature $H_i(T)$ will be near unity for most rotational states, near unity or appreciably less for the lower vibrational states, and extremely small for higher vibrational levels and electronic excitations which require very high temperatures (and therefore very large energy exchanges) to be appreciably excited.

Acoustic absorption due to a given internal state will be appreciable only when the period of the sound is close to the relaxation time for the state. Then, a phase delay exists between the acoustic pressure and the temperature. As we will see, this leads to an equation of state for monofrequency motion wherein the speed of sound is a complex quantity. For most fluids commonly encountered at normal temperatures, the relaxation times for the rotational states are very short (usually only a few collisions are necessary to bring the rotational states into equilibrium with the translational ones). Relaxation times for vibrational states are much longer (many collisions are required). However, as noted above, these states require higher temperatures to be significantly excited and, therefore, may not be important at normal temperatures.

As mentioned in Sect. 7.3, if we obtain a dynamic equation of state of the form (7.25) for monofrequency motion, then analysis of **c** will yield the absorption and dispersion associated with the loss mechanism. Since these internal energy states affect the heat capacity C_V, if we can relate the complex speed of sound to a complex heat capacity we have accomplished our goal. Recall from the discussion in Chapter 5 that the adiabatic equation of state for acoustical processes in an ideal gas is given by $\mathscr{P}/\mathscr{P}_0 = (\rho/\rho_0)^\gamma$. This can be written in the form $p = \rho_0 c^2 s$, where the speed of sound c is given by (5.19), $c = \sqrt{\gamma r T_K}$.* It is shown in Appendix A9 that for an ideal gas $C_\mathscr{P}$ and C_V are simply related by $C_\mathscr{P} = C_V + \mathscr{R}$. Thus, for a complex heat capacity $\mathbf{C}_V$ the speed of sound is also complex and can be written as

$$\mathbf{c} = c\sqrt{\boldsymbol{\gamma}/\gamma} \tag{7.39}$$

where complex $\boldsymbol{\gamma}$ is

$$\boldsymbol{\gamma} = 1 + \mathscr{R}/\mathbf{C}_V \tag{7.40}$$

With these definitions, we have achieved the necessary dynamic equation of state (7.25). Calculation of α and k proceeds as discussed in Sec. 7.3. Thus, the problem has been reduced to obtaining $\mathbf{C}_V$.

To begin, it is necessary to develop a little nonequilibrium thermodynamics. We follow an approach developed more completely in the excellent text by Morse and Ingard.[6] Restrict attention to the ith internal state. The rate of change of the energy

---

* Here $r = \mathscr{R}/M$, where $\mathscr{R}$ is the *universal gas constant* in J/(mol · K) and M is the mass in kg.

stored in this internal state is proportional to the difference between the energy that would be stored under equilibrium conditions $E_i(eq)$ and the amount E_i that is stored at some instant of time. This is expressed mathematically by the differential equation

$$\frac{dE_i}{dt} = \frac{1}{\tau} [E_i(eq) - E_i] \tag{7.41}$$

where τ is a proportionality factor. For example, if the system is changed instantaneously from a thermodynamic configuration for which the equilibrium energy stored in the internal state is E_0 to one in which the amount stored will be $E_0 + \Delta E_i$, then solution for E_i gives

$$E_i = E_0 + (1 - e^{-t/\tau})\Delta E_i \qquad t > 0$$

$$= E_0 \qquad\qquad\qquad t < 0$$

and we can now identify τ as the relaxation time. The *equilibrium* thermodynamic heat capacity associated with this state is given by

$$\Delta E_i = C_i \Delta T$$

where $C_i = \frac{1}{2}\mathcal{R}H_i(T)$ and ΔT and ΔE_i are the equilibrium values of the change in temperature and internal energy, attained in the limit $t/\tau \to \infty$. Now, assume that the process is not allowed to attain equilibrium so that $E_i(eq)$ is a fluctuating quantity to which E_i is always trying to adjust. For a monofrequency acoustic process where the energy input is such that the equilibrium temperature will vary as $\Delta T = T \exp(j\omega t)$, we can write $\Delta E_i = C_i T \exp(j\omega t)$, so that

$$\mathbf{E}_i(eq) = E_0 + C_i T e^{j\omega t} \tag{7.42a}$$

The oscillatory (particular) solution of (7.41) with $\mathbf{E}_i(eq)$ given by (7.42a) can be verified to be

$$\mathbf{E}_i = E_0 + \frac{C_i}{1 + j\omega\tau} T e^{j\omega t} \tag{7.42b}$$

If we define the *complex heat capacity* $\mathbf{C}_i$ for the internal state in this monofrequency process by

$$\mathbf{E}_i = E_0 + \mathbf{C}_i T e^{j\omega t} \tag{7.43a}$$

then combination of (7.42b) and (7.43a) yields

$$\mathbf{C}_i = \frac{C_i}{1 + j\omega\tau} \tag{7.43b}$$

For N internal degrees of freedom, the combined dynamic heat capacity (including the contribution from translational motion) becomes

$$\mathbf{C}_V = \mathcal{R}\left[\frac{3}{2} + \frac{1}{2}\sum_{i=1}^{N} \frac{H_i(T)}{1 + j\omega\tau_i}\right] \tag{7.44}$$

which reduces to (7.38) if $\omega\tau_i \ll 1$.

In the simplest case of only one internal degree of freedom contributing significantly to the heat capacity, we have

$$\mathbf{C}_V = C_e + \frac{C_i}{1 + j\omega\tau_M}$$

where τ_M is the relaxation time for molecular processes, and for economy of notation we have

$$C_e = \tfrac{3}{2}\mathcal{R} \qquad \text{and} \qquad C_i = \tfrac{1}{2}\mathcal{R}H_i(T) \qquad (7.45a)$$

The equilibrium thermodynamic heat capacity C_V must be the limit of $\mathbf{C}_V$ for $\omega\tau \to 0$,

$$C_V = C_e + C_i \qquad (7.45b)$$

Having obtained $\mathbf{C}_V$, substitution into (7.39) and (7.40) yields

$$\mathbf{c} = \frac{c}{\sqrt{\gamma}}\left(1 + \frac{\mathcal{R}}{C_e + \dfrac{C_i}{1 + j\omega\tau_M}}\right)^{1/2}$$

Considerable algebraic manipulation then reveals that the absorption coefficient α_M for molecular processes is

$$\alpha_M = \frac{1}{2}\frac{\omega}{c}\frac{\omega\tau_M}{1 + (\omega\tau_M)^2}\frac{\mathcal{R}C_i}{C_e(C_e + \mathcal{R})} \qquad (7.46)$$

and the phase speed $c_p = \omega/k$ is

$$c_p = \frac{c}{\sqrt{\gamma}}\left[1 + \mathcal{R}\frac{C_V + C_e(\omega\tau_M)^2}{C_V^2 + C_e^2(\omega\tau_M)^2}\right]^{1/2} \qquad (7.47)$$

Notice that α_M is proportional to τ_M when $\omega\tau_M \ll 1$; this means that, for α_M to be important compared to other absorption coefficients, τ_M must be sufficiently large.

For low frequencies ($\omega\tau_M \ll 1$), (7.47) simplifies to

$$c_p(0) = \frac{c}{\sqrt{\gamma}}\left(1 + \frac{\mathcal{R}}{C_V}\right)^{1/2} = c \qquad (7.48)$$

the phase speed in the absence of losses. On the other hand, for high frequencies ($\omega\tau_M \gg 1$), (7.47) becomes

$$c_p(\infty) = c\left(\frac{1 + \mathcal{R}/C_e}{1 + \mathcal{R}/C_V}\right)^{1/2} \qquad (7.49)$$

which is always greater than c since $C_V > C_e$ unless $C_i = 0$. Thus, molecular thermal relaxation not only produces attenuation of sound waves but also a dispersion in phase speed.

As previously mentioned, α_M is important compared to the α's for other absorptive mechanisms only when $\omega_M = 1/\tau_M$ is relatively small. This means that when thermal relaxation is important ω_M may fall in the kilohertz or low megahertz range.

In graphing the measured absorption caused by a molecular thermal relaxation, it is customary to plot the product $\alpha_M \lambda$, the absorption per wavelength resulting from the relaxation, against frequency. When this is done, curves similar to Fig. 7.2 are obtained. Since the peak value of $\alpha_M \lambda$ occurs at $\omega = \omega_M$, the relaxation frequency $f_M = \omega_M/2\pi$ can be found directly from such a plot. The magnitude of the maximum value of $\alpha_M \lambda$ is often called μ_{max},

$$\mu_{max} = (\alpha_M \lambda)_{f=f_M} = \frac{\pi}{2} \frac{\mathscr{R} C_i}{C_e(C_e + \mathscr{R})} \tag{7.50}$$

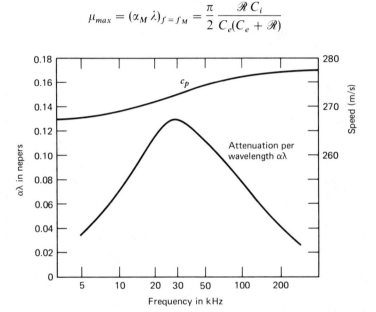

Fig. 7.2. *Variation of absorption and phase speed for* CO_2 *at 20°C.*

Through experimental measurement of μ_{max} one is able to determine a numerical relationship between C_e and C_i. Use of (7.50) allows (7.46) to be written as

$$\alpha_M \lambda = 2\mu_{max} \frac{f/f_M}{1 + (f/f_M)^2} \tag{7.51a}$$

A convenient form for the absorption coefficient follows from (7.51a),

$$\alpha_M = 2 \frac{\mu_{max}}{c} \frac{f_M f^2}{f_M^2 + f^2} \tag{7.51b}$$

(Remember that both the relaxation frequency f_M and μ_{max} are functions of the temperature.)

A classic example of a gas having only one molecular thermal relaxation is dry carbon dioxide gas at normal temperatures. Measurements have shown good agreement with theory, with $\alpha_M \lambda$ peaking at about 30 kHz and α_M about 1200 times greater than the classical absorption at this frequency.

The above equations (7.46) to (7.51b) give the contribution to the absorption coefficient of a single molecular thermal relaxation mechanism. If more than one internal degree of freedom can be excited, the total absorption coefficient is the sum (7.37) of the absorption coefficients calculated separately for each of the relaxation mechanisms.

In certain cases, small concentrations of molecules of another species can have considerable influence on the absorption coefficient of a gas. For example, the addition of water vapor to carbon dioxide gas has a profound effect. For small

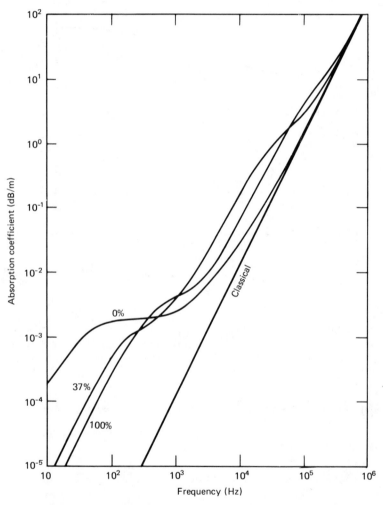

Fig. 7.3. *Absorption of sound in air at 20°C as a function of frequency. (After Bass, et al.)*

concentrations of water vapor, additional absorption mechanisms associated with the water molecules do not significantly contribute to the overall heat capacity; the absorption coefficient of the mixture remains dominated by the molecular thermal relaxation of the carbon dioxide molecules. The water vapor, however, acts as a catalyst, decreasing the relaxation time for the molecular thermal relaxation experienced by the carbon dioxide molecules. The presence of 1 percent of water vapor in carbon dioxide shifts the frequency of maximum molecular absorption per wavelength from $f_M \sim 30$ kHz to $f'_M \sim 2$ MHz. The absorption per wavelength at the relaxation frequency remains the same, but the decreased wavelength at the higher frequency greatly increases the absorption coefficient. At the relaxation frequency of the *moist* gas, the ratio of the absorption coefficients for moist (α'_M) and dry (α_M) carbon dioxide gas is about $\alpha'_M/\alpha_M \sim \frac{1}{2} f'_M/f_M \sim 33$. On the other hand, at frequencies far below the relaxation frequency of *dry* carbon dioxide, $\alpha'_M/\alpha_M \sim f_M/f'_M$, and the absorption coefficient of the moist carbon dioxide is only 0.015 times that of the dry gas.

Another polyatomic gas that has been extensively studied is air. Air consists of molecular oxygen and nitrogen with traces of other gases, the most important acoustically being carbon dioxide and water vapor. Figure 7.3 shows the absorption coefficients for different relative humidities for air, calculated from measured and

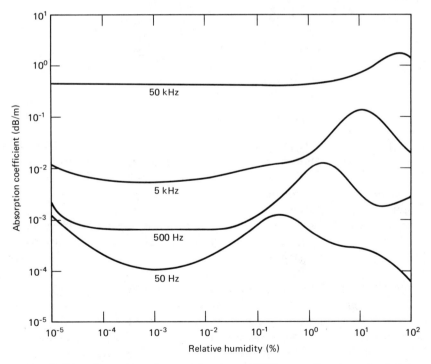

Fig. 7.4(a). *Absorption of sound in air at 20°C as a function of relative humidity. (After Bass, et al.)*

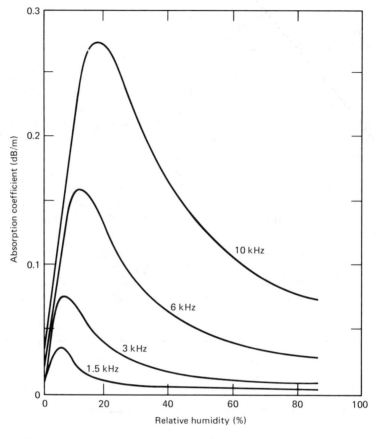

Fig. 7.4(b)

assumed relaxation times and interaction rates.[7] The appreciable increase in the absorption coefficient above that of the classical prediction for all frequencies below 1 MHz is a consequence of molecular thermal relaxations. This excess absorption increases rapidly with temperature.

Figures 7.3 and 7.4a show that extremely small concentrations of water molecules strongly affect the absorption coefficient for air, especially for frequencies below about 5 kHz. As in moist carbon dioxide, the water vapor appears to act as a catalyst, increasing the relaxation frequencies associated with the interactions of the molecules.

Figure 7.4b shows the effects of naturally occurring values of humidity on the absorption coefficients of air at audio frequencies. For humidities in excess of 20 percent but below 60 percent, the higher audio frequencies are much more strongly absorbed than are the lower frequencies. Therefore, the absorption of high-frequency sounds coming from a distant band, or the consonant sounds from a speaker in a

[7] Bass, Bauer, and Evans, *J. Acoust. Soc. Am.*, **52**, 821 (1972).

large auditorium, will be most pronounced for relative humidities around 40 percent.

In moist air, sound absorption at audio frequencies is dominated by three types of collisions: nitrogen and carbon dioxide, oxygen and carbon dioxide, and oxygen and water vapor. At very low humidities (less than one-tenth of a percent) the collisions of oxygen and carbon dioxide molecules control the relaxation frequency.[7] At higher humidities, the interaction of oxygen and carbon dioxide continues to dominate above 500 Hz, while the interaction of nitrogen and carbon dioxide assumes considerable importance below 500 Hz, particularly for relative humidities in excess of about 25 percent.[8]

7.6 ABSORPTION PHENOMENA IN LIQUIDS.

One type of excess absorption occurring in liquids is that associated with thermal relaxation of energy between external degrees of translational freedom and internal vibrations. As in the case of gases, it requires that the acoustic waves produce periodic fluctuations in temperature. Thermal relaxation has been applied successfully to explain the excess absorption observed in many nonassociated nonpolar liquids such as carbon disulfide, benzene, and acetone. For instance, it explains the behavior in acetone where the measured absorption is some 4.3 times the value predicted by classical theory. (See Table 7.1.)

Thermal relaxation, however, has not been successful in accounting for the observed excess absorption in associated polar liquids, such as the alcohols and water. It appears that in these liquids the intermolecular forces are so strong that they cause any existing thermal relaxation time to be very short. Since, in accordance with (7.46), the magnitude of the absorption coefficient is proportional to the relaxation time, the resulting absorption associated with the process is small. The fact that the excess absorption in water is not due to thermal relaxation has been demonstrated in a striking manner through measurements made in the vicinity of $4°C$.[10] If the measured excess absorption in water were caused by thermal relaxation, then this excess should vanish at $4°C$, where the coefficient of thermal expansion is zero. At this temperature, compression or rarefaction will not change the temperature and thus thermal relaxation cannot take place. Measurements in water in the vicinity of $4°C$ give no evidence of any decrease in absorption at this temperature. Therefore, it is necessary to find some other type of relaxation mechanism in order to explain the measured excess absorption in water, which is some three times that predicted by classical theory. (See Table 7.1.) One such explanation is offered by a theory of *structural relaxation*, as applied to water with some success by Hall.[11] This theory assumes the excess absorption in water results from a structural relaxation directly related to volume change and not to temperature change. In this structural relaxation theory, water is assumed to be a two-state liquid. The state of

[8] Sivian, *J. Acoust. Soc. Am.*, **19**, 914 (1947).
[9] Piercy, *J. Acoust. Soc. Am.*, **46**, 602 (1969).
[10] Fox and Rock, *Phys. Rev.*, **70**, 68 (1946).
[11] Hall, *Phys. Rev.*, **73**, 775 (1948).

lower energy is the normal state and the state of higher energy is one in which the molecules have a more closely packed structure. Under ordinary static conditions of equilibrium, most of the molecules are in the first energy state. However, the passage of a compressional wave is assumed to promote the transfer of molecules from the more open first state to the more closely packed second state. The time delays in this process and in its reversal lead to a relaxational dissipation of acoustic energy.

A detailed analysis of the influence of structural relaxation on the propagation of acoustic waves in water indicates that it may be taken into consideration by assuming the existence of a nonvanishing bulk coefficient of viscosity η_B. The resulting expression for total absorption in water becomes

$$\alpha = \frac{\omega^2}{2\rho_0 c^3} \left(\tfrac{4}{3}\eta + \eta_B\right) \tag{7.52}$$

Direct measurement of the coefficient of volume viscosity by Liebermann[12] indicates that η_B in water is approximately three times the coefficient of shear viscosity η. If this measured value of η_B is substituted into (7.52) and α/f^2 computed, the resulting constant is found to be in satisfactory agreement with the measured value listed in Table 7.1.

Figure 7.5 displays the absorption of acoustic waves in freshwater and seawater at 5 C. The pronounced difference between the two curves at frequencies below 500 kHz is evidence of additional absorptive processes active in seawater. It is natural to attribute these additional losses to the presence of dissolved salts. Laboratory measurements by Leonard[13] and co-workers have shown that the excess acoustic absorption in the midfrequency range is caused by the presence of dissolved magnesium sulfate ($MgSO_4$); the acoustic pressure associates and dissociates the $MgSO_4$ ions, and the relaxation time for this process leads to absorption. This type of process is referred to as a *chemical relaxation*.

Measurement of low-frequency absorption in seawater is extremely difficult because of the small value of the absorption coefficient at these frequencies (0.001 dB/km at 100 Hz), but recent measurements have succeeded in revealing a second relaxation mechanism active below about 1 kHz. This absorption has been shown to be mainly caused by a chemical relaxation involving boric acid. Although the boric acid concentration in the ocean is only about 4 ppm (parts per million), the associated absorption at low frequencies is nearly 300 times greater than that of freshwater and 20 times greater than that of seawater with the boric acid removed.

With two chemical relaxation mechanisms operative in saltwater in addition to the absorption mechanisms active in freshwater, we would expect from (7.37), (7.51b), and (7.52) the absorption coefficient for seawater in decibels per meter to be

$$a = \frac{A f_1 f^2}{f_1^2 + f^2} + \frac{B f_2 f^2}{f_2^2 + f^2} + C f^2 \tag{7.53a}$$

$$= a(\text{boric acid}) + a(MgSO_4) + a(H_2O) \tag{7.53b}$$

[12] Liebermann, *Phys. Rev.*, **75**, 1415 (1949).
[13] Leonard, Combs, and Skidmore, *J. Acoust. Soc. Am.*, **21**, 63 (1949).

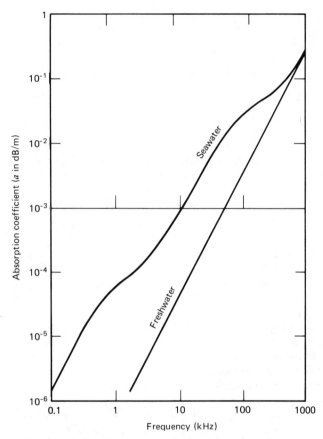

Fig. 7.5. *Sound absorption in freshwater and in seawater (35 ppt salinity) at 5°C and 1 atm. (According to Fisher and Simmons.)*

where f_1 and f_2 are the relaxation frequencies associated with the boric acid and $MgSO_4$ relaxations, respectively, and are temperature dependent. The constants A, B, and C depend on the temperature and on the hydrostatic pressure. In addition, A and B must go to zero for freshwater. Fisher and Simmons[14] have analyzed an immense amount of experimental data to arrive at empirical values for the temperature and pressure dependences of f_1, f_2, A, B, and C for seawater of salinity 35 ppt (parts per thousand) and pH = 8.0. The relaxation frequencies in hertz are

$$f_1 = 1.32 \times 10^3 (T + 273)e^{-1700/(T + 273)} \tag{7.54a}$$

and

$$f_2 = 1.55 \times 10^7 (T + 273)e^{-3052/(T + 273)} \tag{7.54b}$$

where T is in °C. The values for A, B, and C are complicated but can be simplified

[14] Fisher and Simmons, *J. Acoust. Soc. Am.*, **62**, 558 (1977).

somewhat for temperatures and pressures in the range from 0 to 30°C and 1 to 400 atm. These simpler forms, accurate to within 2 percent, are

$$A = 8.95 \times 10^{-8}(1 + 2.3 \times 10^{-2}T - 5.1 \times 10^{-4}T^2)$$

$$B = 4.88 \times 10^{-7}(1 + 1.3 \times 10^{-2}T)(1 - 0.9 \times 10^{-3}\mathscr{P}_0)$$

$$C = 4.76 \times 10^{-13}(1 - 4.0 \times 10^{-2}T + 5.9 \times 10^{-4}T^2)$$

$$\times (1 - 3.8 \times 10^{-4}\mathscr{P}_0) \tag{7.54c}$$

where $\mathscr{P}_0$ must be in atmospheres if a is to be in decibels per meter. (The pressure dependence of A has not yet been measured; the pressure dependences of B and C were measured at one temperature and assumed to apply to the entire range of valid temperatures.)

Note that the relaxation frequencies for boric acid and $MgSO_4$ are 813 Hz and 73.5 kHz at 5°C, respectively, rising to 1310 Hz and 165 kHz at 25°C.

7.7 ATTENUATION IN INHOMOGENEOUS FLUIDS. *Attenuation* is the loss of acoustic energy from a sound beam. Attenuation can be divided into two parts: absorption mechanisms that convert acoustic energy into thermal energy and other mechanisms that deflect or *scatter* acoustic energy out of the beam. When a fluid contains inhomogeneities such as suspended particles, thermal microcells of different temperatures, or regions of turbulence, acoustic energy is lost from a sound beam faster than in the homogeneous medium. Fog and smoke particles produce a decided effect on sound propagation through the atmosphere. In the immediate neighborhood of suspended particles additional viscous and heat conduction losses take place over those effective in the body of a homogeneous fluid. Furthermore, a relaxation process associated with the evaporation of the water in a fog droplet is influenced by the passage of a sound wave. The normal equilibrium between the saturated vapor near the droplet and the surrounding air is disturbed by the sound wave which is followed by a lag in its restoration. Both of these processes lead to losses that increase gradually with frequency. At 1 kHz, for example, the measured[15] contribution to attenuation from these sources in a fog with some 400 droplets/cm³ having an average radius of 6×10^{-4} cm is about 2.4×10^{-3} Np/m. This value is some hundred times greater than that measured in dry air and some ten times that measured in humid air. The presence of fog droplets in air also lowers the speed of sound in the air by amounts down to 0.9c, where c is the speed of sound in dry air at the same temperature.

Extremely high attenuations are also produced in water containing suspended gas bubbles. For instance, viscous forces and heat conduction losses associated with the compression and expansion of small gas bubbles by a passing sound wave result in a loss of energy by the sound wave. A further effect of such inhomogeneities, of particular importance in the transmission of directed sonar beams of sound energy, is *scattering* (the removal of a small amount of energy from the directed beam by each bubble and its subsequent reradiation in all directions). The presence of gas bubbles also affects the nature of the medium through which the wave is progressing, altering both its density and compressibility and thus changing the speed of sound. Such changes in speed and density may result in a considerable amount of acoustic energy being reflected and refracted away from the direction of the initial sound beam. Thus, a beam of sound waves can be attenuated by reflection, refraction, absorption, and scattering as it enters water containing a high concentration of gas bubbles.

---

[15] Knudsen, Wilson, and Anderson, *J. Acoust. Soc. Am.*, **20**, 849 (1948).

Although gas bubbles do not occur in large numbers in the main body of the ocean, high concentrations of bubbles do occur in the wakes of ships and submarines traveling on or near the surface and to modest depths when waves are breaking at the surface. Consequently, there is a tendency for bubbles to be located in specific regions and at high concentrations when they do occur. While a single bubble has little effect on the transmission of sound, the cumulative effect of many bubbles is quite pronounced. Since large bubbles rise to the surface very rapidly, their life in the medium is too short to be of much significance. However, a large number of bubbles with radii small in comparison with the wavelengths being transmitted produce significant attenuations.

Let us assume that each bubble has an effective attenuating cross section σ expressed in square meters. This effective cross section is a measure of the fraction of energy that the bubble will extract from a sound beam of 1 square meter cross section. Depending on frequency, this cross section may be either equal to, less than, or greater than the actual cross section πa^2, where a is the radius of the bubble. For instance, if $\mathscr{P}$ is the hydrostatic pressure at the depth of the bubble, ρ_0 the density of the water, and γ the ratio of specific heats for the gas within the bubble, at the resonance frequency of a gas bubble

$$f_r = \frac{1}{2\pi a} \sqrt{\frac{3\gamma\mathscr{P}}{\rho_0}} \tag{7.55}$$

(which will be derived in Chapter 10), the effective cross section may be more than a thousand times greater than the actual cross section. When the frequency of the sound wave is higher than this resonant frequency, the effective and actual cross sections are nearly equal. On the other hand, at lower frequencies the effective cross section is much less than the actual cross section. Thus, in a distributed population of bubbles, the bubbles whose resonance frequency equals that of the acoustic wave will have the greatest influence on the scattering from the sound beam.

If we now consider nearly resonant bubbles with cross section σ having a concentration N bubbles/m^3, then the loss in intensity experienced by a plane wave in traveling an infinitismal distance dx is $dI = -N\sigma I\,dx$. Integration yields

$$I = I_0 e^{-N\sigma x} \tag{7.56}$$

(While summation over all bubble sizes must be carried out to compute the total attenuation, in practice only those bubbles having resonant frequencies near that of the sound wave need be taken into consideration.) Since the intensity is proportional to the square of the pressure amplitude P, we have

$$P = P_0 e^{-(N\sigma/2)x} \tag{7.57}$$

and the attenuation is $\alpha = N\sigma/2$ Np/m or $a = 8.7\alpha = 4.35N\sigma$ dB/m.

The concentration of bubbles occurring naturally in the main body of the ocean is so small that any attenuation resulting from this source is negligible compared to that caused by viscous forces and other relaxation phenomena. However, the agitation of the ocean surface wave action produces bubbles of various sizes that may influence the propagation of sound near the surface. Also, because of scattering from their internal air sacs, fish in large schools will produce measurable attenuation at the resonant frequency of the sac. Another situation where bubbles may be important is in the wakes of ships. For example, observed attenuation coefficients for the relatively fresh wake, existing 500 m astern of a destroyer making 15 kt (equivalent to 30 km/h = 8.3 m/s), range from 0.8 dB/m at 8 kHz to 1.8 dB/m at 40 kHz.[16]

---

[16] NDRC, *Physics of Sound in the Sea*, U.S. Government Printing Office (1969).

PROBLEMS

7.1.   Using the data in the appendix, (a) compute the viscous relaxation time for glycerin. (b) For what frequency in hertz is $\omega\tau = 1$? (c) Compute and plot α/f^2 over the frequency range between 10 MHz and that determined in part (b).

7.2.   (a) Assuming the coefficient of viscosity in air to be independent of pressure, what is the viscous relaxation time in air at a pressure of 0.1 atmosphere? (b) At what frequency will the product $\omega\tau = 1$ in such air? (c) What is the theoretical viscous absorption coefficient and phase speed in such air at this frequency? (d) What is the corresponding absorption and speed in air at 1-atm pressure at this same frequency assuming viscosity to be the dominant loss mechanism?

7.3.   Calculate the relaxation time due to heat conduction in air. Compare with the corresponding value due to viscosity.

7.4.   Show that the transverse vibrations $\mathbf{u}_0 = U_0 \exp(j\omega t)$ of a plane surface along the x direction will produce transverse viscosity waves in an adjacent fluid medium whose wave equation is $\partial^2 u/\partial z^2 = (\rho_0/\eta)\,\partial u/\partial t$ in which z is the direction of propagation of the waves at right angles to the plane surface, η is the coefficient of viscosity of the medium, and ρ_0 its density. Derive the resulting harmonic solution of this wave equation, with the assumed boundary condition that the medium immediately adjacent to the vibrating plane has an identical transverse velocity. Show that these transverse waves are attenuated by $\sqrt{\rho_0\omega/2\eta}$ Np/m. In the case of air, give a formula for the thickness of the layer of air in which 1 Np of attenuation takes place. Evaluate at 10 Hz and 1 kHz.

7.5.   Given air of 13 percent relative humidity to have its maximum excess molecular absorption per wavelength at 50 kHz and assuming a single relaxation involving oxygen and carbon dioxide, (a) what is the relaxation time? (b) If the measured excess absorption at 5 kHz is 0.14 dB/m, compute and plot the excess molecular absorption per meter over the frequency range 1 to 10 kHz. (c) Combine these results with the *classical* absorption coefficients for dry air to obtain a predicted total absorption. (d) By comparing these results with the values obtained from Fig. 7.4, comment on the importance of the nitrogen-carbon dioxide relaxation mechanism for this relative humidity and range of frequencies.

7.6.   A siren is to operate in air at a frequency of 500 Hz and at a small height above the ground. Assuming hemispherical divergence and no absorption by the ground, what is the absorption coefficient and what must be the acoustic output of the siren in watts if it is to produce an intensity level of 60 dB *re* 20 μPa at a distance of 1000 ft for each of the following conditions? (a) No absorption by the air. (b) According to the classical absorption coefficient. (c) Completely dry air. (d) Air of very high relative humidity.

7.7.   Given that the ratio of heat capacities for CO_2 gas is $\gamma = 1.31$ and that its gas constant is $r = 189$ J/kg C, use the data of Fig. 7.2 and the equations of Sect. 7.5 to compute values for C_p, C_V, C_c, and C_i.

7.8.   (a) Show that the attenuation constant α as given by (7.52) has the dimensions of a reciprocal length. (b) What is the predicted absorption in decibels for sounds of 40 kHz in traversing a path length of 4 km in freshwater at 5°C? (c) In seawater at 5°C?

7.9.   A 1 kHz plane wave traverses freshwater at 15°C. (a) In what distance will it be attenuated by 10 dB? (b) Work the same problem for a frequency of 20 kHz. (c) What are the corresponding distances in seawater at 15°C? (d) In dry air? (e) In air of 37 percent relative humidity?

7.10.   (a) What is the resonance frequency of radial vibrations for an air bubble of 0.01-cm radius in water at a depth of 10 m. (b) If the effective attenuating cross section for such a bubble is 2000 times its actual cross section, how many bubbles per cubic meter will be required in order to produce an attenuation of 0.01 dB/m? (c) How does this attenuation compare with that of bubble-free seawater at the same frequency? Assume a temperature of 15 C.

CHAPTER 8

RADIATION AND RECEPTION OF ACOUSTIC WAVES

8.1 RADIATION FROM A PULSATING SPHERE. The simplest source for generating acoustic waves is a pulsating sphere—a sphere whose radius varies sinusoidally with time. From symmetry such a source will produce outgoing, harmonic, spherical waves into a medium that is infinite, homogeneous, and isotropic. (See Sect. 5.11.) The acoustic wave radiated by a pulsating sphere must be of the form (5.48)

$$\mathbf{p}(r, t) = \frac{\mathbf{A}}{r} e^{j(\omega t - kr)} \tag{8.1}$$

The boundary condition at infinity is that there is no reflection, so that only an outgoing wave exists. $\mathbf{A}$ can be evaluated from the boundary condition at the surface of the sphere: the radial component of the particle velocity must equal that of the sphere. If the surface, of average radius a, is vibrating radially with speed amplitude U_0 and frequency ω then

$$\mathbf{u}(a, t) = U_0 e^{j\omega t} \tag{8.2}$$

where it is assumed that $U_0/\omega \ll a$, so that the displacement of the surface is much less than the radius and $\mathbf{u}$ can be evaluated at $r = a$. (This is consistent with the small-amplitude approximation of linear acoustics.) The pressure on the surface is found by evaluating the specific acoustic impedance for the spherical wave at $r = a$. From (5.52)

$$\mathbf{z}(a) = \rho_0 c \cos \theta_a e^{j\theta_a} \tag{8.3}$$

where $\cot \theta_a = ka$. The pressure at the surface of the source is then

$$\mathbf{p}(a, t) = \rho_0 c U_0 \cos \theta_a e^{j(\omega t + \theta_a)} \tag{8.4}$$

and, since this must be the value of (8.1) at $r = a$, solution for $\mathbf{A}$ gives the pressure at any distance $r \geq a$

$$\mathbf{p}(r, t) = \rho_0 c U_0 \frac{a}{r} \cos \theta_a e^{j[\omega t - k(r - a) + \theta_a]} \tag{8.5}$$

The acoustic intensity

$$I = \frac{1}{2} \rho_0 c U_0^2 \left(\frac{a}{r}\right)^2 \cos^2 \theta_a \tag{8.6}$$

shows a dependence on ka through the angle θ_a.

If the radius of the source is small compared to a wavelength so that $ka \ll 1$, the specific acoustic impedance at the surface of the pulsating sphere,

$$\mathbf{z}(a) \doteq \rho_0 cka\,(j + ka) \tag{8.7}$$

is strongly reactive. This reactance is a symptom of the strong radial divergence of the acoustic wave at the source and represents the storage and release of energy because successive layers of the fluid must stretch and shrink circumferentially, altering the outward displacement. This inertial effect manifests itself in the masslike reactance of the specific acoustic impedance. In this long-wavelength limit the pressure is

$$\mathbf{p}(r, t) \doteq j\rho_0 cU_0\, \frac{a}{r}\, ka\, e^{j(\omega t - kr)} \qquad ka \ll 1 \tag{8.8}$$

and is close to $\pi/2$ out of phase with the particle speed. (Pressure and particle speed are not *exactly* $\pi/2$ out of phase, since that would lead to a vanishing intensity.) The acoustic intensity is

$$I \doteq \frac{1}{2} \rho_0 c U_0^2 \left(\frac{a}{r}\right)^2 (ka)^2 \qquad ka \ll 1 \tag{8.9}$$

Notice that for constant U_0 this intensity is proportional to the square of the frequency and depends on the fourth power of the source radius. Thus, we see that small sources (with respect to a wavelength) are inherently poor radiators of acoustic energy.

While pulsating spheres are difficult to construct and are of little practical importance, their theoretical import is great since they serve as the prototype for an important class of sources referred to as *simple sources*. A *simple source* is a closed surface, vibrating with arbitrary velocity distribution, but of such size that all dimensions are much smaller than the wavelength of the emitted sound.

8.2 SOURCE STRENGTH. Consider a source of sound of arbitrary size and shape whose surface oscillates with a single frequency but in such a manner that the velocity amplitude $\vec{U}$ and phase ϕ vary from point to point on the surface. The instantaneous velocity of a point on the surface is

$$\dot{\mathbf{u}}(t, \phi) = \vec{U} e^{j(\omega t + \phi)}$$

The source displaces a volume of fluid at the rate

$$Q e^{j\omega t} = \int_S \dot{\mathbf{u}} \cdot \hat{n}\, dS \tag{8.10}$$

where $\hat{n}$ is the unit outward normal to the surface element dS and the integral is taken over the entire surface of the source. The constant $\mathbf{Q}$ is the *complex source strength*. For example, the complex source strength of a pulsating sphere is real and is

$$Q = 4\pi a^2 U_0 \tag{8.11}$$

The importance of the source strength lies in the fact that if the dimensions of a source are much smaller than the wavelength of the sound being radiated, so that it is a *simple source*, then the details of the surface motion are not important and it will radiate exactly the same sound as any other simple source with the same source strength. The beauty of this is that once the sound field due to *one* simple source is found in terms of its source strength then the sound field of *any other* simple source is known.

8.3 ACOUSTIC RECIPROCITY. The concept of *acoustic reciprocity* can be expressed mathematically as a powerful formula which is useful in obtaining very general results, including relating sources to the fields they produce, and relating properties of sources and receivers. Let us now obtain one of the more commonly encountered statements of acoustic reciprocity.

Consider a space occupied by sources, as suggested by Fig. 8.1. By changing which sources are active and which are passive, it is possible to set up different sound fields. Choose two situations having the *same frequency* and denote them as 1 and 2, respectively. Now, establish a fixed volume V of space that does not itself contain any sources but that bounds all the sources. Let the surface of this volume be S. The volume V and the surface S remain the same for both situations. Let the velocity potential for situation 1 be $\mathbf{\Phi}_1$ and that for situation 2 be $\mathbf{\Phi}_2$. Green's theorem (see Appendix A8) thus gives the general relation

$$\oint_S (\mathbf{\Phi}_1 \nabla \mathbf{\Phi}_2 - \mathbf{\Phi}_2 \nabla \mathbf{\Phi}_1) \cdot \hat{n}\, dS = \int_V (\mathbf{\Phi}_1 \nabla^2 \mathbf{\Phi}_2 - \mathbf{\Phi}_2 \nabla^2 \mathbf{\Phi}_1)\, dV \tag{8.12}$$

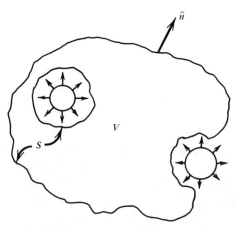

Fig. 8.1. Geometry used in deriving the theorem of acoustic reciprocity.

where $\hat{n}$ is the unit outward normal to S. Since the volume does not include any sources, and since both velocity potentials are for acoustical excitations of the same frequency, the wave equation yields

$$\nabla^2 \Phi_1 = -\mathbf{k}^2 \Phi_1$$

and

$$\nabla^2 \Phi_2 = -\mathbf{k}^2 \Phi_2$$

where $\mathbf{k} = k - j\alpha$, so that the right-hand side of (8.12) vanishes identically throughout V. Furthermore, recall that the pressure is

$$\mathbf{p} \doteq -j\omega\rho_0 \Phi$$

and the particle velocity for irrotational motion is

$$\hat{\mathbf{u}} = \nabla\Phi$$

Substitution of these expressions into the left-hand side of (8.12) gives

$$\oint_S (\mathbf{p}_1\hat{\mathbf{u}}_2 - \mathbf{p}_2\hat{\mathbf{u}}_1) \cdot \hat{n}\, dS = 0 \tag{8.13}$$

This is one form of the *principle of acoustic reciprocity*. This principle states that, for example, if in an unchanging environment the locations of a small source and a small receiver are interchanged, the received signal will remain the same.

For the purposes of this discussion, let us now develop a more restrictive but simpler form of (8.13). Assume that some portion of S is a great distance R from the acoustical sources. This portion, designated S_∞, subtends a solid angle Ω as measured from some point near the sources. Near S_∞ we must have

$$|\mathbf{p}| \rightarrow \frac{A}{R} e^{-\alpha R}$$

and

$$|\hat{\mathbf{u}}| \rightarrow \frac{A}{\rho_0 c R} e^{-\alpha R} \hat{R}$$

Therefore,

$$\int_{S_\infty} \mathbf{p}_1\hat{\mathbf{u}}_2 \cdot \hat{n}\, dS \lesssim \frac{\Omega A_1 A_2}{\rho_0 c} e^{-2\alpha R}$$

which vanishes in the limit $R \rightarrow \infty$. Thus, portions of the surface S which are at great distances from the acoustic sources make vanishing contributions to (8.13). Furthermore, let all other portions of S be perfectly rigid so that $\hat{\mathbf{u}} \cdot \hat{n} = 0$ and the integrals must vanish over those portions. [Analogously, any part of S which is pressure release would also give no contribution to (8.13) because $\mathbf{p} = 0$ on that portion.] Under this condition, (8.13) reduces to an integral over only those portions of S that correspond to sources active in situations 1 or 2.

$$\int_{sources} (\mathbf{p}_1\mathbf{\mathring{u}}_2 - \mathbf{p}_2\,\mathbf{\mathring{u}}_1) \cdot \hat{n}\ dS = 0 \qquad (8.14)$$

This simple result will now be applied to develop some important general properties of sources that are small compared to a wavelength.

8.4 SIMPLE SOURCES. Consider a region of space in which there are two irregularly shaped sources, as shown in Fig. 8.2. Let source A be active and source B be perfectly rigid in situation 1, and vice versa in situation 2. Then, if we define $\mathbf{p}_1$ as the pressure at B when source A is active, $\mathbf{\mathring{u}}_1$ as the velocity of the radiating element of source A, $\mathbf{p}_2$ as the pressure at A when source B is active, and $\mathbf{\mathring{u}}_2$ as the velocity of the radiating element of source B, application of (8.14) yields

$$\int_{S_A} \mathbf{p}_2\,\mathbf{\mathring{u}}_1 \cdot \hat{n}\ dS = \int_{S_B} \mathbf{p}_1\mathbf{\mathring{u}}_2 \cdot \hat{n}\ dS \qquad (8.15)$$

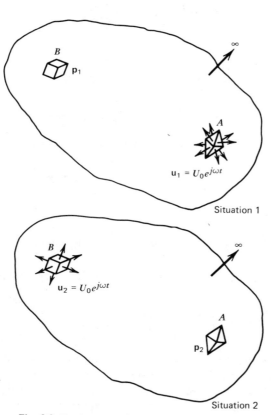

Fig. 8.2. *Reciprocity theorem applied to simple sources.*

If the sources are small with respect to a wavelength and several wavelengths apart, then the pressure is uniform over each source so that

$$\frac{1}{P_1} \int_{S_A} \hat{u}_1 \cdot \hat{n} \, dS = \frac{1}{P_2} \int_{S_B} \hat{u}_2 \cdot \hat{n} \, dS \qquad (8.16a)$$

Using (8.10) and $\mathbf{p} = \mathbf{P}(r)\exp(j\omega t)$, and taking the magnitude of (8.16a), we have

$$\frac{Q_1}{P_1(r)} = \frac{Q_2}{P_2(r)} \qquad (8.16b)$$

This shows that the ratio of the source strength to pressure amplitude at distance r is the same for all *simple sources* (at the same frequency) in the same surroundings. This allows us to calculate the pressure field of any irregular simple source since it must be identical with the pressure field produced by a pulsating sphere of the same source strength. If the simple sources are in free space, (8.8) and (8.11) show that the ratio of (8.16b) has magnitude

$$\boxed{\frac{Q}{P} = \frac{2\lambda r}{\rho_0 c}} \qquad (8.16c)$$

This is the *free-field reciprocity factor*.

Thus, rewriting (8.8) with the help of (8.11) results in

$$\mathbf{p}(r, t) = j\rho_0 c \, \frac{Qk}{4\pi r} \, e^{j(\omega t - kr)} \qquad (8.17a)$$

which must be true for *all* simple sources. The pressure amplitude is

$$P = \rho_0 c \, \frac{Qk}{4\pi r} = \frac{1}{2} \rho_0 c \, \frac{Q}{\lambda r} \qquad \text{(simple source)} \qquad (8.17b)$$

so that the intensity is

$$I(r) = \frac{1}{8} \rho_0 c \left(\frac{Q}{\lambda r} \right)^2 \qquad \text{(simple source)} \qquad (8.17c)$$

Integration of the intensity over a sphere centered at the source gives the power radiated

$$\Pi = \frac{\pi}{2} \rho_0 c \left(\frac{Q}{\lambda} \right)^2 \qquad \text{(simple source)} \qquad (8.18)$$

Another case of practical interest is that of a simple source mounted on or very close to a *rigid plane boundary*, as suggested in Fig. 8.3. If the dimensions of the boundary are much greater than a wavelength of sound, then the boundary can be considered a plane of infinite extent. This kind of boundary is often termed a *baffle*. Since the baffle is perfectly rigid, the acoustic waves generated by the simple source will reflect from this surface so that $\hat{u} \cdot \hat{n} = 0$, where $\hat{n}$ is the normal to the rigid plane. Thus, the pressure of the reflected wave and that of the incident wave are in phase and identical at the wall. Under the additional condition that the source is very close to (if not actually mounted on) the baffle, so that there is much less than a wavelength separating the furthest portions of the source from the baffle, then the wave reflected from the baffle is virtually in phase everywhere with that radiated by the source. The pressure field in the half-space occupied by the source will thus be twice that generated by the source (with the same source strength) in free space,

$$\mathbf{p}(r, t) = j\rho_0 c \frac{Qk}{2\pi r} e^{j(\omega t - kr)}$$

(baffled simple source) (8.19)

The intensity is increased by four

$$I(r) = \frac{1}{2} \rho_0 c \left(\frac{Q}{\lambda r}\right)^2$$

(baffled simple source) (8.20)

and integration of the intensity over a hemisphere (there is no acoustic penetration of the space behind the baffle) gives twice the power radiated

$$\Pi = \pi \rho_0 c \left(\frac{Q}{\lambda}\right)^2$$

(baffled simple source) (8.21)

A doubling of the power output of the source may seem surprising but results from the fact that the source has the same source strength in both cases: the source face is moving with the same velocity in both cases but, in the baffled case, it is working into twice the force and, therefore, must expend twice the power to maintain its own motion in the presence of the doubled pressure.

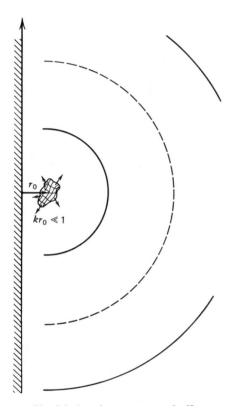

Fig. 8.3. *Simple source on a baffle.*

8.5 DIPOLE RADIATION.
A spatial configuration of simple sources, either discrete or distributed, each with its own complex source strength, can be used to represent more complicated sources. The pressure at a field point is the sum of the pressures produced by the individual sources.

As a simple example, consider the *acoustic doublet* which consists of two simple sources of equal strengths, separated by a distance d, vibrating with the same frequency but 180° out of phase with each other (Fig. 8.4). The pressure at point (r, θ) due to source 1 is

$$\mathbf{p}_1 = \frac{A}{r + \Delta r_1} e^{j[\omega t - k(r + \Delta r_1)]}$$

and that due to source 2 is

$$\mathbf{p}_2 = -\frac{A}{r - \Delta r_2}\, e^{j[\omega t - k(r - \Delta r_2)]}$$

where r is the distance from the field point to the midpoint between the sources, Δr_1 and Δr_2 are as shown, and the minus sign accounts for the phase difference of the sources. The acoustic pressure at (r, θ) is then

$$\mathbf{p} = \frac{A}{r}\left(\frac{e^{-jk\Delta r_1}}{1 + \Delta r_1/r} - \frac{e^{jk\Delta r_2}}{1 - \Delta r_2/r}\right) e^{j(\omega t - kr)} \tag{8.22}$$

The process of replacing Δr_1 and Δr_2 by r and θ is straightforward but tedious, and the resulting expression for $\mathbf{p}(r, \theta, t)$ is too complicated for efficient analysis. However, in the most frequently encountered cases, the observation of the pressure is

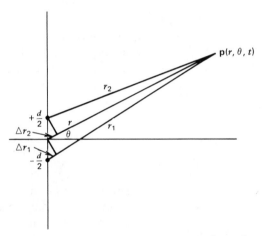

Fig. 8.4. Geometry used in deriving the radiation characteristics of an acoustic dipole.

made at distances great compared to the separation of the sources. Therefore, it will be useful to derive the form of (8.22) suitable in the limit $r \gg d$. This is referred to as the *far-field approximation*. If $r \gg d$, $\Delta r_1 \approx \Delta r_2 \approx (d/2)\sin\theta$ and $\Delta r_1/r \ll 1$, so that (8.22) becomes

$$\mathbf{p}(r, \theta, t) = -j\,\frac{2A}{r}\,\sin(\tfrac{1}{2}kd\,\sin\theta)e^{j(\omega t - kr)}$$

For frequencies such that $kd \ll 1$, this equation further reduces to

$$\mathbf{p}(r, \theta, t) = -j\,\frac{Akd}{r}\,\sin\theta\; e^{j(\omega t - kr)} \tag{8.23}$$

This pressure field is referred to as *acoustic dipole radiation*.

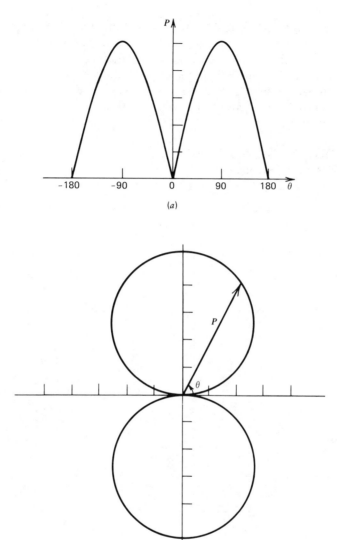

Fig. 8.5. *Pressure amplitude distribution of dipole radiation.*
(a) Cartesian representation. (b) Polar representation.

Equation (8.23) describes a spherically diverging wave with pressure amplitude

$$P(r, \theta) = \frac{Akd}{r} \left| \sin \theta \right| \tag{8.24}$$

that decreases as $1/r$, and at fixed r has a dependence on θ, as shown in Fig. 8.5.

Acoustic dipole radiation is encountered frequently: a simple source close to a pressure-release plane surface will produce an out-of-phase *image* on the other side of the boundary. Working together, the source and its image produce a dipole radiation field in the fluid half-space. Another example is any oscillating body that is free to radiate from both sides, such as an oscillating sphere or an unenclosed loudspeaker. The positive acoustic pressure produced on the advancing side of the source is accompanied by a negative pressure on the receding side which, if the wavelength is large compared to the dimensions of the source, will produce a dipolelike radiation pattern.

8.6   THE CONTINUOUS-LINE SOURCE.   As an example of a distribution of point sources used to describe an extended source, consider a long, thin cylindrical source of length L and radius a. This configuration, suggested in Fig. 8.6, is termed

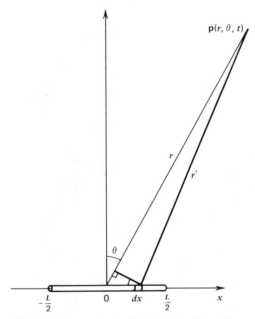

Fig. 8.6. *Geometry used in deriving the radiation characteristics of a continuous-line source.*

a *continuous-line source*. Let the surface vibrate radially with particle speed $U_0 \exp(j\omega t)$. While the acoustic field established by this source is rather complicated, a simple approximate expression can be obtained for the *far-field approximation*.

Consider the source to be made up of a large number of cylinders of length dx that are placed adjacent with a common axis. Each of these elements can be considered a simple source of strength

$$dQ = U_0 2\pi a \ dx$$

and each generates the increment of pressure given by (8.17a),

$$dp = j\frac{\rho_0 ck}{4\pi r'} U_0 2\pi a \, dx \, e^{j(\omega t - kr')}$$

at the field point (r, θ), which is related to r' by

$$r' \doteq r - x \sin\theta \qquad r \gg L \tag{8.25}$$

The total pressure is found by integrating dp over the length of the source,

$$p(r, \theta, t) = j\frac{\rho_0 cU_0 ka}{2}\int_{-L/2}^{L/2}\frac{1}{r'} e^{j(\omega t - kr')} \, dx$$

Under the stated assumption $r \gg L$, the denominator of the integrand can be re-placed by its approximate value r. This amounts to making a very small error in the amplitude of the acoustic fields generated by each of the simple sources at distance r; for very large r, it is seen that all these amplitudes become essentially identical. In the exponent, however, this simplification cannot be made because the relative phases of the elements are independent of r and can be very strong functions of angle. The integral now takes the form

$$p(r, \theta, t) = j\frac{\rho_0 cU_0 ka}{2r} e^{-jkr}\int_{-L/2}^{L/2} e^{j(\omega t + kx \sin\theta)} \, dx$$

and evaluation is immediate,

$$p(r, \theta, t) = \frac{1}{2} j\rho_0 cU_0 \frac{a}{r} kL \, e^{j(\omega t - kr)}\left[\frac{\sin(\frac{1}{2}kL \sin\theta)}{\frac{1}{2}kL \sin\theta}\right] \tag{8.26}$$

The acoustic pressure amplitude in the far field can be written

$$\boxed{P(r, \theta) = P_{ax}(r)H(\theta)} \tag{8.27}$$

where

$$H(\theta) = \left|\frac{\sin v}{v}\right| \qquad v = \frac{1}{2}kL \sin\theta \tag{8.28a}$$

and

$$P_{ax}(r) = \frac{1}{2}\rho_0 cU_0 \frac{a}{r} kL \tag{8.28b}$$

This procedure of separating the expression for the far-field pressure into one factor that depends only on the angle and another that depends only on the distance from the source is common practice in describing the sound fields of complicated sources. Note that in the far field the axial pressure is proportional to r^{-1} just as for a simple source.

The functional behavior of $(\sin v)/v$ is shown in Fig. 8.7.* The corresponding *beam pattern* $b(\theta) = 20 \log H(\theta)$ is plotted in Fig. 8.8 for the case $kL = 24$. Most of the acoustic energy is projected in the *major lobe*, which is contained within the angles given by $\frac{1}{2}kL \sin \theta = \pm\pi$ or $|\sin \theta| = \lambda/L$. There are *nodal surfaces* (cones in the present case) determined by $|\sin \theta| = n\lambda/L$, where $n = 1, 2, 3, \ldots, [L/\lambda]$ which demark the angular boundaries of lobes of lesser strength which are called *minor lobes*. Clearly, the larger the value of kL the more narrowly directed will be the major lobe and the greater the number of minor lobes.

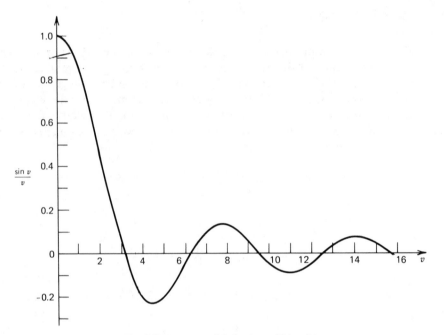

Fig. 8.7. *Functional behavior of* $(\sin v)/v$.

Notice that the pressure, when expressed in terms of the source strength $Q = U_0 2\pi aL$, becomes

$$\mathbf{p}(r, \theta, t) = \frac{1}{2} j\rho_0 c \, \frac{Qk}{2\pi r} \frac{\sin v}{v} e^{j(\omega t - kr)} \tag{8.29a}$$

where v is given by (8.28a), and

$$P(r, \theta) = \frac{1}{2} \rho_0 c \, \frac{Qk}{2\pi r} H(\theta) \tag{8.29b}$$

* The function $(\sin v)/v$ is also known as the zeroth order spherical Bessel function of the first kind, $j_0(v)$.

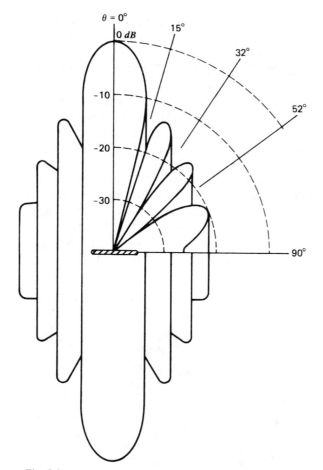

Fig. 8.8. *Beam pattern* $b(\theta)$ *for a continuous-line source with* $kL = 24$.

8.7 DIRECTIONAL FACTOR AND BEAM PATTERNS.

In the previous section, it was seen that in the far field the radiation from a continuous-line source can be expressed as a product of an on-axis pressure $P_{ax}(r)$ which depends only on r and a term $H(\theta)$, which depends only on angle. For sources of lower symmetry, this same separation is possible as long as attention is restricted to the far field.

The term that depends on the angle, denoted more generally by $H(\theta, \phi)$ and called the *directional factor*, is always normalized so that its maximum value is unity, as illustrated by (8.28a). The directions for which $H = 1$ determine the *acoustic axes*. In cases with high degrees of symmetry, the acoustic "axis" becomes a *plane* (as for the continuous-line source) or a *line* (as will be seen for the piston source). The variation of intensity level with angle is the *beam pattern*

$$b(\theta, \phi) = 10 \log \frac{I(r, \theta, \phi)}{I_{ax}(r)} = 20 \log H(\theta, \phi)$$

The term $P_{ax}(r)$ is the far-field pressure on the acoustic axis ; the pressure along any other radial line is simply $P_{ax}(r)$ reduced by the factor $H(\theta, \phi)$.

8.8 RADIATION FROM A PLANE CIRCULAR PISTON.

Another type of extended radiator of particular interest is the rigid circular piston mounted flush with the surface of an infinite baffle and vibrating with simple harmonic motion. The solution in this example is applicable to a number of related problems, including the radiation from the open end of a flanged organ pipe. It makes no difference whether the vibrating surface is a piston or a circular layer of air, provided only that all portions of the layer vibrate in phase with the same amplitude. [In actual practice, this limitation is not strictly obeyed for an organ pipe (the amplitude of the air motion near the edges of the pipe is somewhat less than at its center), but the results obtained by assuming that the radiation is similar to that from a rigid piston are in reasonably good agreement with observed values.]

Assume a piston of radius a is mounted on a flat rigid baffle of infinite extent.

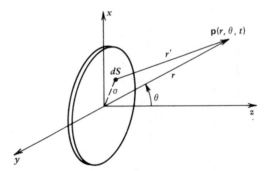

Fig. 8.9. *Geometry used in deriving the radiation characteristics of a flat piston.*

Let the radiating surface of the piston move uniformly with speed $U_0 \exp(j\omega t)$. normal to the baffle. The geometry and coordinates are sketched in Fig. 8.9. The pressure at the field point can be obtained by dividing the surface of the piston into infinitesimal elements, each of which acts like a baffled simple source of strength $dQ = U_0\, dS$. The pressure generated by one of these sources is given by (8.19), and the total pressure is

$$\mathbf{p}(r, \theta, t) = j\,\frac{\rho_0 c U_0 k}{2\pi} \int_S \frac{e^{j(\omega t - kr')}}{r'}\, dS \tag{8.30}$$

where the surface integral is taken over the region $\sigma \le a$.

(a) Axial Response. The field along the acoustic axis (the z axis) is relatively simple to calculate. With reference to Fig. 8.9, we have

$$\mathbf{p}(r, 0, t) = j\,\frac{\rho_0 c U_0 k}{2\pi}\, e^{j\omega t} \int_0^a \frac{\exp\left(-jk\sqrt{r^2 + \sigma^2}\right)}{\sqrt{r^2 + \sigma^2}}\, 2\pi\sigma\, d\sigma$$

The integrand is a perfect differential since

$$\frac{\sigma}{\sqrt{r^2 + \sigma^2}} \exp(-jk\sqrt{r^2 + \sigma^2}) = -\frac{d}{d\sigma}\left(\frac{\exp(-jk\sqrt{r^2 + \sigma^2})}{jk}\right)$$

and the complex acoustic pressure is

$$\mathbf{p}(r, 0, t) = \rho_0 c U_0\, e^{j\omega t}[e^{-jkr} - \exp(-jk\sqrt{r^2 + a^2})] \qquad (8.31a)$$

The pressure amplitude on the axis of the piston is the magnitude of the above expression,

$$P(r, 0) = 2\rho_0 c U_0 \left| \sin\left\{\frac{1}{2} kr\left[\sqrt{1 + \left(\frac{a}{r}\right)^2} - 1\right]\right\}\right| \qquad (8.31b)$$

For $r/a \gg 1$, the square root can be simplified to

$$\sqrt{1 + \left(\frac{a}{r}\right)^2} \doteq 1 + \frac{1}{2}\left(\frac{a}{r}\right)^2$$

If also $r/a \gg ka$ so that the field point is distant compared to both the radius and wavelength, then the pressure amplitude on the axis has *asymptotic* form

$$P_{ax}(r) = \frac{1}{2}\rho_0 c U_0 \frac{a}{r} ka \qquad (8.32)$$

which reveals the expected spherical divergence at distances satisfying $r/a \gg 1$ and $r/a \gg ka$.

Study of (8.31b) reveals that the axial pressure exhibits strong interference effects, fluctuating between 0 and $2\rho_0 c U_0$ as r ranges between 0 and ∞. These extremes of pressure occur for values of r satisfying

$$\frac{1}{2} kr\left[\sqrt{1 + \left(\frac{a}{r}\right)^2} - 1\right] = m\frac{\pi}{2} \qquad \begin{matrix} \text{maxima:} & m \text{ odd} \\ \text{minima:} & m \text{ even} \end{matrix}\Bigg\} \qquad (8.33)$$

where $m = 0, 1, 2, 3, \ldots$. Solution of the above for the values of r at the extrema yields

$$\frac{r_m}{a} = \frac{1}{m}\frac{a}{\lambda} - \frac{m}{4}\frac{\lambda}{a} \qquad (8.34)$$

Moving in toward the source from large r, one encounters the first local maximum in axial pressure at $m = 1$, for which

$$\frac{r_1}{a} = \frac{a}{\lambda} - \frac{1}{4}\frac{\lambda}{a}$$

For still smaller r, the pressure amplitude falls to another minimum at $m = 2$, for which

$$\frac{r_2}{a} = \frac{1}{2}\left(\frac{a}{\lambda} - \frac{\lambda}{a}\right)$$

and then continues to fluctuate between 0 and $2\rho_0 cU_0$ until the face of the piston is reached. A sketch of this behavior is shown in Fig. 8.10.

Thus, for values of r greater than r_1, the axial pressure displays a monotonically decreasing behavior going asymptotically to a $1/r$ dependence. For values of r less than r_1, the axial pressure displays a strong interference effects, suggesting that the acoustic field close to the piston is complicated. The distance r_1 thus serves as a convenient demarcation between the complicated *near field* found close to the source and the simpler *far field* found at large distances from the source. The quantity r_1 has physical meaning only if the ratio a/λ is large enough so that r_1 is positive. Indeed, if $a = \lambda/2$ then r_1 is zero and there is no near field.

(b) Far Field. To aid in the evaluation of the far field, introduce additional coordinates, as indicated in Fig. 8.11. Given the position (r, θ) at which the pressure

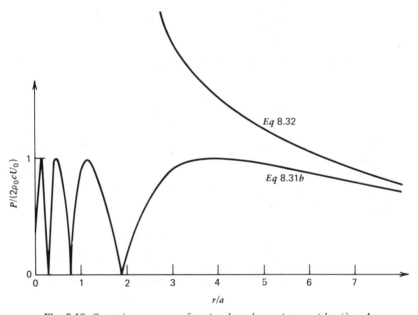

Fig. 8.10. *On-axis response of a circular plane piston with $a/\lambda = 4$.*

is desired, let the x and y axes be oriented so that the field point lies in the xz plane. This allows us to break the piston surface into a column of continuous line sources of differing lengths, each parallel to the y axis so that we are on the acoustical axis of each line source. Since we wish the far-field limit of the piston radiation pattern, we impose the restriction $r \gg a$ so that the contributions to the field point from each of the line sources is simply its far-field axial pressure. Since each line is of length $2a \sin \phi$ and width dx, the source strength from one such source is

$$dQ = 2U_0 a \sin \phi \, dx$$

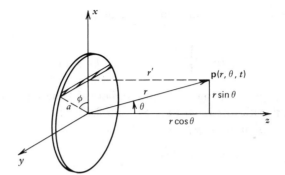

Fig. 8.11. *Geometry used in deriving the far-field radiation characteristics of a circular plane piston.*

and the incremental pressure $d\mathbf{p}$ for this baffled source is from (8.29a)

$$d\mathbf{p} = j\rho_0 c \, \frac{U_0}{\pi r'} \, ka \sin \phi \, e^{j(\omega t - kr')} \, dx$$

For $r \gg a$, r' has the approximate form

$$r' \doteq r\left(1 - \frac{a}{r} \sin \theta \cos \phi\right) = r + \Delta r$$

and the acoustic pressure is

$$\mathbf{p}(r, \, \theta, \, t) = j\rho_0 c \, \frac{U_0}{\pi r} \, ka \, e^{j(\omega t - kr)} \int_{-a}^{a} e^{jka \sin \theta \cos \phi} \sin \phi \, dx$$

where $r' \to r$ in the denominator, but $r' = r + \Delta r$ in the phase in accordance with the far-field approximation. Using the relationship $x = a \cos \phi$, we can convert the integration from dx to $d\phi$

$$\mathbf{p}(r, \, \theta, \, t) = j\rho_0 c \, \frac{U_0}{\pi} \, \frac{a}{r} \, ka \, e^{j(\omega t - kr)} \int_{0}^{\pi} e^{jka \sin \theta \cos \phi} \sin^2 \phi \, d\phi$$

By symmetry, the imaginary part of the integral vanishes. The real part is tabulated in terms of Bessel functions,

$$\int_{0}^{\pi} \cos(z \cos \phi) \sin^2 \phi \, d\phi = \pi \, \frac{J_1(z)}{z}$$

so that

$$\mathbf{p}(r, \, \theta, \, t) = j \, \frac{\rho_0 c}{2} \, U_0 \, \frac{a}{r} \, ka \, e^{j(\omega t - kr)} \left[\frac{2J_1(ka \sin \theta)}{ka \sin \theta} \right] \qquad (8.35)$$

All of the angular dependance of $\mathbf{p}$ is in the bracketed term, and this term goes to unity as θ goes to 0. Thus, we can identify the directional factor $H(\theta)$ for the piston as

$$H(\theta) = \left| \frac{2J_1(v)}{v} \right| \qquad v = ka \sin \theta \qquad (8.36)$$

A plot of $2J_1(v)/v$ is given in Fig. 8.12, and numerical values are given in Appendix A6. Notice the similarities between Figs. 8.7 and 8.12. The axial pressure amplitude is identical with the asymptotic expression (8.32).

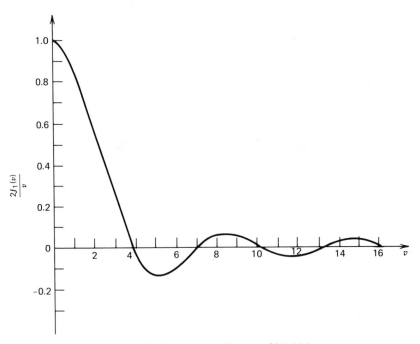

Fig. 8.12. *Functional behavior of $2J_1(v)/v$.*

The angular dependence of $H(\theta)$ reveals that there are pressure nodes along angles θ_m such that

$$ka \sin \theta_m = j_{1m}$$

where j_{1m} designates the values of the argument of J_1 that reduce this Bessel function to zero, $J_1(j_{1m}) = 0$. (See Appendix A5.) Notice that the form of $H(\theta)$ yields a maximum along $\theta = 0$. The angles θ_m define conical nodal surfaces with vertices at $r = 0$. Between these surfaces lie pressure lobes, as suggested in Fig. 8.13. The relative strengths and angular locations of the acoustic pressure maxima in the lobes are given by the relative maxima of $H(\theta)$. Thus, for example, if the intensity level on the axis is set at 0 dB, then the level of the maximum signal in the first side lobe (between θ_1 and θ_2) is about -17 dB. The main lobe is thus of considerable relative

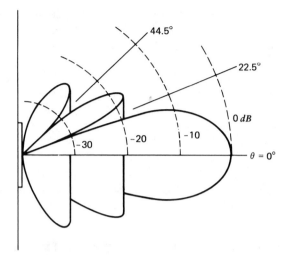

Fig. 8.13. *Beam pattern b(θ) for a circular plane piston with kL = 10.*

strength, so that θ_1 is a measure of the solid angle of the space strongly ensonified by the piston.

When the radius of the piston is large in comparison with the wavelength of the sound, $ka \gg 1$, the radiation pattern has many side lobes and the angular width of the major lobe is small. On the other hand, if the wavelength is much greater than the radius, then $ka \ll 1$ and only the major lobe will be present. For very small values of ka, the directional factor is nearly equal to unity for all angles, so that the pressure field is the same as for a baffled simple source of the same strength. This is not unexpected since for very small values of ka a piston becomes a simple source.

The radiation patterns produced by a piston-type loudspeaker differ to some extent from these idealized patterns. One reason for this discrepancy is that the area of the baffle in which the speaker is mounted is necessarily finite. At high frequencies even a baffle of small linear dimensions corresponds quite closely to an ideal infinite baffle but, at low frequencies, where the wavelength of the sound may be the same as or greater than the linear dimensions of the baffle, the assumption that each element of the piston radiates with hemispherical divergence will be in error. In addition, the radiation from the back of the speaker may propagate into the region in front of the speaker so that the resulting radiation pattern approximates that of an acoustic doublet rather than that of a piston in an infinite baffle. Another reason for this discrepancy is that the material of an actual speaker cone is not perfectly rigid. Driving the speaker at its center establishes higher velocity amplitude in the inner parts of the cone than near its rim, and at high frequencies the cone may even vibrate in normal modes similar to those of the circular membrane. Under these circumstances, U_0 may become a complex function $\mathbf{U}_0$ of the radial distance and angle and, hence, must be considered as a variable in the integration. (For example, if U_0 decreases with increasing σ, this produces a broadening of the major lobe with an accompanying decrease in the intensity of the side lobes.) By a suitable choice of the relation between U_0 and σ, we may obtain a wide variety of radiation patterns.

The increase in directionality that accompanies an increase in frequency some-times produces wide variations in the relative intensities of the low-frequency and the high-frequency sounds generated by a single speaker, the relative intensity of the higher frequency components being much greater at points on or near the axis of the speaker than at larger angles. In rooms of moderate dimensions this effect is not too important unless the walls have a high absorption coefficient, since reflections from the walls scatter the radiated sound to such an extent that the intensity distribution is nearly uniform. When public address systems are used outdoors or in large auditoriums, however, the scattering is negligible, and uniform distribution of the high-frequency components can be obtained only by employing several speakers aimed in different directions.

8.9   BEAM WIDTH, SOURCE LEVEL, DIRECTIVITY, AND DIRECTIVITY INDEX.

From the preceding sections, it is clear that the pressure field radiated by a real acoustic source may be rather complicated. Several definitions are in general use to describe quantitatively the more important aspects of the radiation from a source without the necessity of displaying the entire radiation pattern.

(a) Beam Width.   No standard value of the ratio $I(\theta, \phi)/I_{ax}$ has been agreed on for measuring or calculating the angles θ and ϕ that mark the effective extremity of the major lobe. Hence, the particular value employed must be clearly stated when beam widths are specified in this manner. The ratios used by various authors and experi-menters range from a maximum of 0.5 (down 3 dB), through 0.25 (down 6 dB), to a minimum of 0.1 (down 10 dB). As an illustration of the ambiguity that arises if the ratio is not specified, consider a piston that is radiating sound of wavelength $\lambda = a/4$ so that $ka = 8\pi$. The calculated beam widths corresponding to the three ratios given above are then 7.4° (down 3 dB), 10.1° (down 6 dB), and 12.9° (down 10 dB), whereas the beam width corresponding to the first *null* of intensity is 17.3°. It should be noted that, even when the outer limits of the beam are defined as being down 10 dB relative to the axial level, they still are some 7.5 dB higher than the maximum level in the first minor lobe.

(b) Source Level.   In most applications it is important to relate the pressure on the acoustic axis in the far field to the available acoustic power. The usual measure of the output of a source is the *source level SL* which expresses the axial response of the source in decibels. Assume that the acoustic axis of the source has been deter-mined and the pressure amplitude along that line is measured in the far field (where the pressure is inversely proportional to the distance from the source). The curve of $P_{ax}(r)$ versus r can be *extrapolated back to a position* 1 m *from the source*. Let us call this pressure amplitude $P_{ax}(1)$ defined by

$$P_{ax}(1) = \lim_{r \downarrow 1} P_{ax}(r) \qquad (8.37)$$

This extrapolated value provides the means of comparing the strengths of different sources on the axes of their major lobes. To convert this to a decibel rating, we must reference to an effective pressure of 1 μPa, 20 μPa, or 1 μbar as discussed in Sect. 5.12. Since $P_{ax}(1)$ is a peak pressure amplitude, it must be reduced to an effective (or rms) value $P_e(1)$, given by

$$P_e(1) = \frac{1}{\sqrt{2}} P_{ax}(1) \tag{8.38}$$

The *source level* is then

$$SL(P_{ref}) = 20 \log \frac{P_e(1)}{P_{ref}} \tag{8.39}$$

(c) Directivity.   Given the pressure field with amplitude $P(r, \theta, \phi)$ established by some source, the intensity distribution in the far field is

$$I(r, \theta, \phi) = \frac{1}{2\rho_0 c} P^2(r, \theta, \phi)$$

The total radiated power is obtained by integrating the intensity over a surface enclosing the source,

$$\Pi = \frac{1}{2\rho_0 c} \int_{4\pi} P^2(r, \theta, \phi) r^2 \, d\Omega$$

where a sphere of radius r has been chosen for convenience. Recalling that $P(r, \theta, \phi) = P_{ax}(r)H(\theta, \phi)$ and noticing that r is constant for the integration, we can write

$$\Pi = \frac{1}{2\rho_0 c} r^2 P_{ax}^2(r) \int_{4\pi} H^2(\theta, \phi) \, d\Omega \tag{8.40}$$

For a *simple source* that generates the *same acoustic power*, the pressure amplitude $P_s(r)$ to be found at the distance r is given by

$$\Pi = \frac{1}{2\rho_0 c} 4\pi r^2 P_s^2(r) \tag{8.41}$$

Clearly, for the same acoustic power, the directional source will have greater intensity at a distance r on the acoustic axis than will the simple source. The ratio of these intensities therefore reveals how much more efficiently a directional source concentrates the available acoustic power into a preferred direction. This ratio is the *directivity D*. Thus,

$$D = \frac{I_{ax}(r)[directional]}{I_s(r)[spherical]} = \frac{P_{ax}^2(r)}{P_s^2(r)} \tag{8.42}$$

(Both sources generate the *same power*, and both intensities are evaluated at the *same distance* from the sources in the far field.) Substitution of (8.40) and (8.41) into the definition of D results in

$$D = \frac{4\pi}{\int_{4\pi} H^2(\theta, \phi) \, d\Omega} \tag{8.43}$$

Thus, the directivity D of a source is the reciprocal of the average of $H^2(\theta, \phi)$ over solid angle.

We can now express the source level in terms of the acoustic power output and the directivity of the source. Combination of (8.38), (8.40), and (8.43) results in

$$\Pi = \frac{4\pi}{D} \frac{P_e^2(1)}{\rho_0 c}$$

Substitution for $P_e(1)$ into (8.39) then gives

$$SL(P_{ref}) = 10 \log \frac{D\rho_0 c\Pi}{4\pi P_{ref}^2} \tag{8.44}$$

(d) Directivity Index. It is convenient to define the *directivity index DI* as

$$DI = 10 \log D \tag{8.45}$$

which is the decibel equivalent of the directivity. For the pulsating sphere, $H(\theta, \phi) = 1$ and $D = 1$ so that $DI = 0$ dB. If the source is a hemisphere mounted on a rigid baffle, $H(\theta, \phi) = 1$ for $\Omega \leq 2\pi$ and $H(\theta, \phi) = 0$ for $\Omega > 2\pi$ so that $D = 2$ and $DI = 3$ dB.

If the reference pressure is 1 μbar $= 0.1$ Pa, then (8.44) becomes

$$SL(1 \ \mu b) = 10 \log \Pi + DI + 10 \log\left(\frac{25}{\pi} \rho_0 c\right)$$

where power, density, and speed of sound *must* be in MKS units. If we apply this formula to water and evaluate $\rho_0 c$, resulting expressions for SL become

$$SL(1 \ \mu b) = 10 \log \Pi + DI + 71 \tag{8.46}$$

and

$$SL(1 \ \mu Pa) = 10 \log \Pi + DI + 171 \tag{8.47}$$

In air the conventional reference pressure is 20 μPa, and the source level becomes

$$SL(20 \ \mu Pa) = 10 \log \Pi + DI + 109 \tag{8.48}$$

The directional factor for a continuous-line source is (8.28a). A study of the cylindrical geometry reveals

$$D = 4\pi \bigg/ 2 \int_0^{\pi/2} H^2(\theta) 2\pi \cos \theta \, d\theta$$

and the change of variable $v = (\frac{1}{2})kL \sin \theta$ gives

$$D = \tfrac{1}{2}kL \Big/ \int_0^{kL/2} \left(\frac{\sin v}{v}\right)^2 dv$$

If the line is long so that $kL \gg 1$, then the upper limit can be taken arbitrarily large with little loss in accuracy. Then, since

$$\int_0^\infty \left(\frac{\sin v}{v}\right)^2 dv = \frac{\pi}{2}$$

the directivity of the continuous-line source can be approximated as

$$D \doteq kL/\pi = 2L/\lambda \qquad kL \gg 1 \tag{8.49}$$

The directivity of a piston is determined by (8.43) and the directional factor (8.36),

$$D = \frac{4\pi}{\displaystyle\int_0^{\pi/2} \left[\frac{2J_1(ka \sin \theta)}{ka \sin \theta}\right]^2 2\pi \sin \theta \, d\theta}$$

where $2\pi \sin \theta \, d\theta$ is the increment of the solid angle $d\Omega$ for this axisymmetric case. Direct integration yields

$$D = \frac{(ka)^2}{1 - J_1(2ka)/ka} \tag{8.50}$$

For low frequencies such that $ka \to 0$, the Bessel function can be replaced by the first two terms of its series expansion, and in this limit we have $D \to 2$, which is the same as a hemispherical source on an infinite baffle. For the more interesting case $ka \gg 1$, the Bessel function becomes small and we have

$$D \doteq (ka)^2 \qquad ka \gg 1 \tag{8.51}$$

which shows that the piston is highly directive at higher frequencies.

8.10 DIRECTIONAL FACTORS OF REVERSIBLE TRANSDUCERS. While the details of operation of a few of the more common acoustic sources and receivers will be developed in Chapter 14, it is appropriate here to develop an important relationship between the transmitting and receiving directional properties of a *reversible transducer*. A reversible transducer is one that can be used either as a source or a detector of acoustic energy. The common office intercom incorporates such devices. The acoustic element, usually a small loudspeaker, can be switched from acting as an acoustic source (to generate a message) to acting as a receiver (to detect the response to the message).

If a reversible transducer exhibits directionality as a source, it is not unexpected that it will also be directional as a receiver. For example, a plane wave falling obliquely on the surface of a large plane piston will cause the piston to move with some normal component of velocity which is related to the *spatially averaged* pressure exerted on the piston. Thus, the response of the piston to the incident plane wave must depend on the angle of arrival of the wave if the wavelength of the sound is comparable to or smaller than the dimensions of the piston. The measure of this response is the *receiving directional factor* $H_r(\theta, \phi)$. We will show in what follows that the transmitting and receiving directional factors for a reversible transducer are identical.

The receiving directional factor is defined quite explicitly: consider plane waves incident on the receiver from a direction specified by θ and ϕ. Let $\langle \mathbf{p}_B \rangle_S$ be the average of the incident sound pressure over the active face of the receiver, measured with the active face (*diaphragm*) of the receiver held perfectly still (*blocked*). The receiving directional factor is then *defined* as

$$H_r(\theta, \phi) = \left| \frac{\langle \mathbf{p}_B(\theta, \phi) \rangle_S}{\langle \mathbf{p}_{Bax} \rangle_S} \right| \tag{8.52}$$

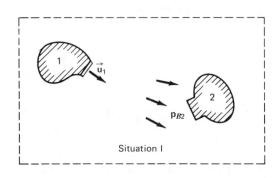

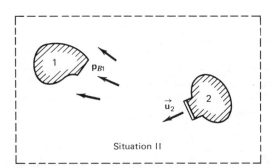

Fig. 8.14. *The reciprocity theorem applied to reversible transducers.*

This measures the phase cancellation of the incident wave over the *blocked* active surface of the receiver as a function of θ and ϕ and thus gives the directional sensitivity of the receiver. (The receiving directional factor is defined with the diaphragm blocked to eliminate any reradiated sound field; more on this in Chapter 14.)

With the help of the reciprocity theorem a relationship can easily be established between H_r and H for reversible transducers. Consider the situation represented in Fig. 8.14. There are two reversible transducers (with surfaces, other than their diaphragms, perfectly rigid) situated a large distance r apart in otherwise free space. (The requirement of large r ensures that near-field effects are avoided.) Situation I requires one of the transducers to be active and the other passive with its diaphragm blocked. Situation II reverses the roles of the two transducers. For this case, (8.14) yields

$$\int_{A_2} \mathbf{p}_{B2}\, \hat{\mathbf{u}}_2 \cdot \hat{n}\, dS = \int_{A_1} \mathbf{p}_{B1}\hat{\mathbf{u}}_1 \cdot \hat{n}\, dS$$

where $\mathbf{p}_B$ is the pressure distribution over each blocked diaphragm and A is the area of the diaphragm of each of transducers 1 and 2. If each diaphragm moves as a unit, so that $\hat{\mathbf{u}}_1$ and $\hat{\mathbf{u}}_2$ are constant over A_1 and A_2, then this simplifies to

$$\mathbf{u}_2\langle \mathbf{p}_{B2}\rangle A_2 = \mathbf{u}_1\langle \mathbf{p}_{B1}\rangle A_1 \tag{8.53}$$

where $\mathbf{u}_1$ and $\mathbf{u}_2$ are the components of the particle velocities perpendicular to the diaphragms.

Now, if transducer 2 is vanishing small, it does not appreciably disturb the pressure field $\mathbf{p}_1$ which is radiated by transducer 1, so that $\mathbf{p}_{B2} = \mathbf{p}_1(r, \theta, \phi, t)$. Furthermore, the pressure $\mathbf{p}_{B2}$ is uniform over the active surface of transducer 2, so that (8.53) becomes

$$\mathbf{u}_2\, \mathbf{p}_1(r, \theta, \phi, t)A_2 = \mathbf{u}_1\langle \mathbf{p}_{B1}(\theta, \phi, t)\rangle_{A_1} A_1 \tag{8.54}$$

Now, if transducer 1 and 2 are rotated so that they are on each others acoustic axis, (8.54) gives the additional equality

$$\mathbf{u}_2\, \mathbf{p}_{1ax}(r, t)A_2 = \mathbf{u}_1\langle \mathbf{p}_{B1ax}(t)\rangle_{A_1} A_1 \tag{8.55}$$

The magnitude of the ratio of the above pair of equations yields

$$\left| \frac{\mathbf{p}_1(r, \theta, \phi, t)}{\mathbf{p}_{1ax}(r, t)} \right| = \left| \frac{\langle \mathbf{p}_{B1}(\theta, \phi, t)\rangle_{A_1}}{\langle \mathbf{p}_{B1ax}(t)\rangle_{A_1}} \right| \tag{8.56}$$

The left-hand side of (8.56) is $H(\theta, \phi)$ and the right-hand side is $H_r(\theta, \phi)$. Thus,

$$H(\theta, \phi) = H_r(\theta, \phi) \tag{8.57}$$

and a reversible acoustic transducer has the same directional properties whether it is transmitting or receiving.

8.11 ROUGH ESTIMATES OF RADIATION PATTERNS FOR SOURCES OF SIMPLE GEOMETRY.

For reasonably directive sources of simple geometry, most of the important properties of the radiation fields can be estimated from the size and geometry of the source and the frequency of the excitation. The source considered may be like one of those we have quantitatively discussed or may be a mosaic or array of such sources. The requirement that the source be reasonably directive can be stated by the condition $kL \gg 1$ or $\lambda \ll 2\pi L$ where L is the greatest dimension of the source (active face).

(a) **Extent of the Near Field.** Obtaining a criterion for the extent of the near field of complicated sources is not straightforward. However, for linelike or flat pistonlike sources, a criterion for the inner boundary of the far field is that the distances from the various portions of the source to a point on the acoustic axis must not differ by more than *about* one half-wavelength. This yields, from Fig. 8.15,

$$r_{max} - r_{min} \sim \lambda/2$$

where r_{max} is the maximum distance from the furthest portion of the source to the field point and r_{min} is the distance from the nearest portion of the source to the field point. If the greatest extent of the source is L, the value of r_{min} estimating the beginning of the far field is roughly

$$\frac{r_{min}}{L} \sim \frac{1}{4}\frac{L}{\lambda} \qquad (8.58)$$

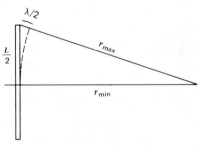

Fig. 8.15. Geometry used in estimating the extent of the near field.

(b) Behavior of the Far Field. The major lobe corresponds to that portion of the radiation pattern for which the source elements are essentially in phase. In Fig. 8.16, as the angle θ is increased more interference is introduced and the edge of the major lobe is approached. Very approximately, when θ increases, until half of the incremental elements are out of phase with the other half, a nodal surface will be encountered. From Fig. 8.16 it can be seen that this occurs at an angle of *about* λ/L. Thus, the half angle subtended by the major lobe can be estimated by

$$\sin\theta_1 \sim \frac{\lambda}{L} \qquad (8.59)$$

The reader should verify that (8.58) and (8.59) are in agreement with the quantitative predictions for the circular piston and that (8.59) agrees with the major-lobe width calculated for the continuous-line source.

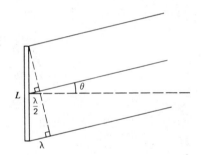

Fig. 8.16. Geometry used for estimating beam width.

Now consider the more complicated pistonlike source shown in Fig. 8.17 which is rectangular with dimensions L_1 and L_2. The major lobe will have different beam widths depending on whether the angle is measured with respect to the height or length of the piston. The two angular widths are

$$2\theta_{11} \sim 2\lambda/L_1$$

in the one direction and

$$2\theta_{12} \sim 2\lambda/L_2$$

in the other. Since the near field extends further outward as the source size increases, the limiting distance to the far field is determined by the larger of the two sides, so that, for $L_2 > L_1$,

$$\frac{r_{min}}{L_2} \sim \frac{1}{4}\frac{L_2}{\lambda}$$

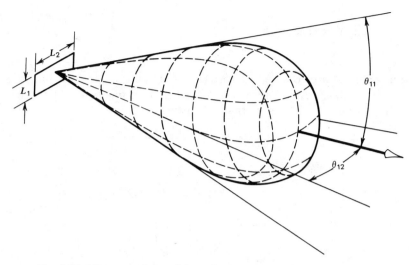

Fig. 8.17. *The major lobe of the radiation pattern for a plane piston.*

(c) Estimation of Directivity.   It is useful to be able to estimate the directivity D since an exact evaluation of the integral expression (8.43) may be too difficult or more accurate than required by the problem at hand.

If the source is reasonably directive and is designed so that the side lobes are considerably weaker than the major lobes, then D can be estimated by setting the integrand to unity over the strong central portion of the major lobe and to zero otherwise. The expression for D now has the approximate form

$$D \sim 4\pi \left/ \int_{\substack{central \\ portion}} d\Omega \right.$$

or

$$\boxed{D \sim \frac{4\pi}{\Omega_{eff}}} \tag{8.60}$$

Evaluation of D is thus reduced to the *geometrical* problem of obtaining the size of the effective solid angle Ω_{eff} subtended by the central portion of the major lobe. A first estimate would be to base Ω_{eff} on θ_1. However, since for highly directive sources the angular beam width $2\theta_1$ tends to overestimate Ω_{eff}, a better approximation would be to take that portion of the major lobe over which the directional factor H falls from its maximum value of 1 to a value of 0.5. (This corresponds to retaining just that portion of the beam pattern that is within $-$ 6 dB of unity and then setting all the retained portion to unity.) Since $(\sin v)/v$ is a good approximation for most major lobes, this determines an effective angular beam width $2\theta'$ where θ' is such that

$$\frac{\sin\left(\frac{1}{2}kL \sin \theta'\right)}{\frac{1}{2}kL \sin \theta'} = \frac{1}{2}$$

Numerical trial and error shows that $\frac{1}{2}kL \sin \theta' \approx 2$, and for highly directive beams where θ_1 is small, this yields

$$\theta' \sim 2\theta_1/\pi \tag{8.61}$$

For a linelike source the central portion of the major lobe is distributed over the surface of a sphere, as shown in Fig. 8.18. The height of this belt is approximated by $2\theta'$, and the circumference is 2π, so that we have $\Omega_{eff} \approx 4\pi\theta'$ or

$$D \sim \tfrac{1}{4}kL \quad \text{(linelike source)} \tag{8.62}$$

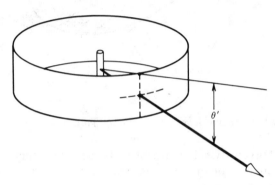

Fig. 8.18. *Area of unit sphere radiated by a continuous-line source.*

For a pistonlike source the central portion of the major lobe is the roughly elliptical patch shown in Fig. 8.19. Again, on a unit sphere the area of this patch is well-approximated by $\Omega_{eff} = \pi\theta'_1\theta'_2$, where $2\theta'_1$ is the effective angular beam width pertinent to the length L_1 and $2\theta'_2$ is the effective angular beam width pertinent to L_2. The resultant directivity is

$$D \sim \tfrac{1}{4}kL_1kL_2 \quad \text{(pistonlike source)} \tag{8.63}$$

Comparison of these estimates with the high-frequency limits for the continuous-line source (8.49) and the circular piston (8.51) reveals that these estimates are not far from being correct.

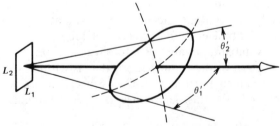

Fig. 8.19. *Area of unit sphere radiated by a plane piston.*

8.12   RADIATION IMPEDANCE.   In Chapter 2 it was found useful to define the input mechanical impedance as the force applied to the string divided by the resulting speed of the string at the point where the force is applied. If the force is not applied directly to the string, but to some device attached to the string, then it was shown in a problem that the force applied to the device divided by the speed of the device was equal to the mechanical impedance of the device plus the input mechanical impedance of the string as seen by the device. Similarly, in the discussion of acoustic sources it will be useful to express the *input mechanical impedance* of the source in terms of the *mechanical impedance of the source radiating into a vacuum* and the *radiation impedance of the fluid.*

Consider a transmitter whose active face is driven with a velocity which may be a function of position on the face. If $d\mathbf{f}_s$ is the component of force in the direction of motion on the fluid by an element of the active face, the radiation impedance is

$$\mathbf{Z}_r = \int \frac{d\mathbf{f}_s}{\mathbf{u}} \tag{8.64}$$

If the face is rigid with mass m, mechanical resistance R_m, and stiffness s and is driven in rectilinear motion with a speed $\mathbf{u}_0 = U_0 \exp(j\omega t) = j\omega\xi_0$ by an externally applied force $\mathbf{f} = F \exp(j\omega t)$, Newton's law of motion yields

$$\mathbf{f} - \mathbf{f}_s - R_m \frac{d\xi_0}{dt} - s\xi_0 = m \frac{d^2\xi_0}{dt^2}$$

where the force of the fluid on the diaphragm is $-\mathbf{f}_s = -\mathbf{Z}_r\mathbf{u}_0$. Recalling that $\mathbf{Z}_m = R_m + j(\omega m - s/\omega)$ and solving for $\mathbf{u}_0$ gives

$$\mathbf{u}_0 = \frac{\mathbf{f}}{\mathbf{Z}_m + \mathbf{Z}_r} \tag{8.65}$$

or

$$U_0 = \frac{F}{\mathbf{Z}_m + \mathbf{Z}_r}$$

Thus, in the presence of fluid loading, the applied force encounters the sum of the mechanical impedance of the source and the radiation impedance.

The radiation impedance can be expressed as

$$\mathbf{Z}_r = Z_r e^{j\theta} = R_r + jX_r \tag{8.66}$$

where R_r and X_r are the *radiation resistance* and the *radiation reactance.* A positive R_r will increase the mechanical resistance, thereby increasing the power dissipated by the source. This additional power dissipation is that radiated away by the acoustic field. In fact, the radiation resistance R_r can be found directly from the power radiated into the fluid by the source

$$\Pi = \frac{1}{T} \int_0^T \mathrm{Re}\{\mathbf{f}_s\}\mathrm{Re}\{\mathbf{u}_0\}dt$$

which with the help of (8.66) becomes

$$\Pi = \tfrac{1}{2}U_0^2 Z_r \cos\theta = \tfrac{1}{2}U_0^2 R_r \tag{8.67}$$

A positive X_r will manifest itself as a mass loading which decreases the resonance frequency ω_0 of the oscillator from $\sqrt{s/m}$ to $\sqrt{s/(m+m_r)}$, where $m_r = X_r/\omega$ is the *radiation mass*. The effect of the radiation mass is usually negligible for sources operating in light media such as air, but for a dense medium such as water the decrease in resonance frequency resulting from the presence of the medium may be quite marked.

The calculation of the radiation impedance Z_r of a baffled piston is not easy and the details of the calculation will not be given. The results[1] are

$$R_r = \pi a^2 \rho_0 c R_1(2ka)$$
$$X_r = \pi a^2 \rho_0 c X_1(2ka) \tag{8.68a}$$

where

$$R_1(x) = 1 - \frac{2J_1(x)}{x} = \frac{x^2}{2^2 1!\,2!} - \frac{x^4}{2^4 2!\,3!} + \cdots$$

$$X_1(x) = \frac{4}{\pi}\left[\frac{x}{3} - \frac{(x)^3}{3^2 \cdot 5} + \frac{(x)^5}{3^2 \cdot 5^2 \cdot 7} - \cdots\right] \tag{8.68b}$$

Sketches of R_1 and X_1 are shown in Fig. 8.20 and numerically tabulated in Appendix A6.

In the low-frequency limit, $ka \ll 1$, the radiation impedance of a baffled, circular piston can be approximated by the first terms of the power expansions. The radiation resistance becomes

$$R_r = \tfrac{1}{2}\pi a^2 \rho_0 c(ka)^2$$

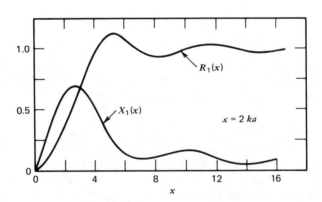

Fig. 8.20. *Piston impedance functions.*

[1] Kinsler and Frey, *Fundamentals of Acoustics*, 2 ed., Wiley (1962).

and the radiation reactance becomes

$$X_r = \pi a^2 \rho_0 c \, \frac{8}{3\pi} \, (ka)$$

The low-frequency reactance is like that of a mass of value

$$m_r = \frac{X_r}{\omega} = \pi a^2 \rho_0 \, \frac{8a}{3\pi}$$

Thus, the piston appears to be loaded with a cylindrical volume of fluid whose cross-sectional area is that of the piston and whose effective height is $8a/(3\pi) \approx 0.85a$. For large pistons or for high frequencies ($ka \gg 1$), $X_1(2ka) \doteq 0$, $R_1(2ka) \doteq 1$, and $R_r \doteq \pi a^2 \rho_0 c$. This yields

$$\mathbf{Z}_r \doteq \pi a^2 \rho_0 c$$

and

$$\Pi \doteq \tfrac{1}{2}\rho_0 \, c\pi a^2 U_0^2 = \tfrac{1}{2}\rho_0 \, cS U_0^2 \tag{8.69}$$

which is the same as the power that would be carried by a plane wave of particle speed amplitude U_0 in a fluid of characteristic impedance $\rho_0 c$ through a cross-sectional area S. Such behavior is to be expected, for when $ka \gg 1$ the radiation resistance is simply $S\rho_0 c$ (the cross-sectional area of the piston multiplied by the characteristic impedance $\rho_0 c$) and $X_r \to 0$.

The radiation impedance of the pulsating sphere is

$$\mathbf{Z}_r = 4\pi a^2 \rho_0 c \, \cos \theta_a \, e^{j\theta_a}$$

For high frequencies ($ka \gg 1$), this reduces to a pure radiation resistance $\mathbf{Z}_r = R_r$ where

$$R_r \doteq 4\pi a^2 \rho_0 c$$

which is $S\rho_0 c$, where S is the surface area of the sphere. For low frequencies ($ka \ll 1$), $\mathbf{Z}_r$ becomes

$$\mathbf{Z}_r \doteq 4\pi a^2 \rho_0 c(ka)^2 + j4\pi a^2 \rho_0 cka$$

The radiation resistance is much less than the radiation reactance, and the radiation reactance is again masslike

$$m_r = \frac{X_r}{\omega} = 3\rho_0(\tfrac{4}{3}\pi a^3)$$

which reveals that in the low-frequency limit the radiation mass is three times the mass of the fluid displaced by the sphere.

8.13   THE SIMPLE LINE ARRAY.   Consider a line of N simple sources with adjacent elements spaced distance d apart, as shown in Fig. 8.21. If all sources have the same source

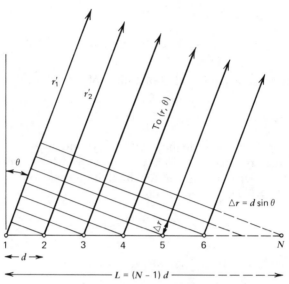

Fig. 8.21. *Geometry used in deriving the radiation characteristics of a line array.*

strength and radiate waves with the same phase, then the ith source generates a pressure wave of the form

$$\frac{A}{r_i'} e^{j(\omega t - kr_i')}$$

where r_i' is the distance from this source to (r, θ). The resultant pressure at the field point is the summation

$$\mathbf{p}(r, \theta, t) = \sum_{i=1}^{N} \frac{A}{r_i'} e^{j(\omega t - kr_i')} \tag{8.70}$$

If we restrict attention to the far field [specified by the condition $r \gg L$ where $L = (N - 1)d$ is the length of the array], we can approximate by assuming all r_i' are parallel. Then $r_i' = r_1 - (i - 1)\Delta r$ where $\Delta r = d \sin \theta$. The distance to the center of the array can be expressed as $r = r_1 - \frac{1}{2}(L/d)\Delta r$. In the far field, r_i' in the denominator of (8.70) can be replaced with r and (8.70) takes the form

$$\mathbf{p}(r, \theta, t) = \frac{A}{r} e^{-j\frac{1}{2}(L/d)k\Delta r} e^{j(\omega t - kr)} \sum_{i=1}^{N} e^{j(i-1)k\Delta r}$$

Use of the trignometric identities in the Appendix A3 results in

$$\mathbf{p}(r, \theta, t) = \frac{A}{r} e^{j(\omega t - kr)} \left[\frac{\sin\left(\dfrac{N}{2} k\Delta r\right)}{\sin\left(\dfrac{1}{2} k\Delta r\right)} \right]$$

The pressure on the axis ($\theta = 0$) is

$$\mathbf{p}(r, 0, t) = N \frac{A}{r} e^{j(\omega t - kr)}$$

and has the maximum possible pressure amplitude

$$P_{ax}(r) = NA/r \qquad (8.71a)$$

Identification of the directional factor

$$H(\theta) = \left| \frac{1}{N} \frac{\sin\left(\dfrac{N}{2} kd \sin \theta\right)}{\sin\left(\dfrac{1}{2} kd \sin \theta\right)} \right| \qquad (8.71b)$$

allows us to write the amplitude of the pressure in the familiar form

$$P(r, \theta) = P_{ax}(r)H(\theta) \qquad (8.71c)$$

Notice that the denominator of $H(\theta)$ may vanish if $\frac{1}{2}kd|\sin \theta| = m\pi$, but the numerator vanishes also, and the pressure amplitude becomes $P_{ax}(r)$. Thus, we can have more than one major lobe. The angles of these occur for

$$|\sin \theta| = m \frac{\lambda}{d} \qquad m = 0, 1, 2, \ldots, \left[\frac{d}{\lambda}\right] \qquad (8.72)$$

(This result can be restated as $|\Delta r| = m\lambda$, which reveals that the radiated pressure is maximized at those angles for which the distances from the field point to the adjacent array elements differ by integral numbers of wavelengths.)

There are additional zeros in the numerator at angles given by

$$|\sin \theta| = \frac{n}{N} \frac{\lambda}{d} \qquad \frac{n}{N} \neq m \qquad (8.73)$$

where the integer n is not zero or a multiple of N. Since the denominator is not zero, the pressure vanishes and these values of θ determine the nodal surfaces in the far field. There are also secondary maxima of $H(\theta)$ that designate the directions and magnitudes of the minor lobes. The directions of these side lobes are given approximately by

$$|\sin \theta| \approx \frac{n + \frac{1}{2}}{N} \frac{\lambda}{d} \qquad \begin{array}{c} n/N \neq m \\ (n + 1)/N \neq m \end{array} \qquad (8.74)$$

and the amplitudes by

$$P_n(0) \approx \frac{P_{ax}}{N \sin\left(\dfrac{n + \frac{1}{2}}{N} \pi\right)} \qquad (8.75)$$

A sketch of a representative beam pattern for a linear array is given in Fig. 8.22.

Certain high-fidelity loudspeakers contain such line arrays, mounted vertically so that vertical directivity is large and horizontal directivity small.

In some applications it is desired to have a single major lobe, as narrow as possible. A simple requirement which results in one major lobe *almost* as narrow as possible is to have

$\theta = \pi/2$ when $n = N - 1$. This gives

$$\frac{\lambda}{d} = \frac{N}{N-1} \tag{8.76}$$

and the beam pattern terminates just short of a second major lobe. The major lobe is contained within angles θ_1 where

$$\sin \theta_1 = 1/(N-1) \tag{8.77}$$

For simple line arrays of many elements this equation reveals that if only one major lobe is to occur the minimal angular width of the major lobe is approximately $2/N$. The directivity is then approximately equal to the number of elements.

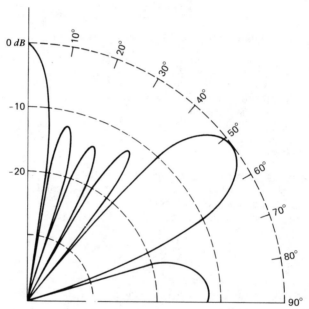

Fig. 8.22. Beam pattern $b(\theta)$ for a line array with $kd = 8$ and $N = 5$.

For very large arrays it is often desired to be able to transmit or receive in various directions without physically rotating the array. This can be accomplished by *electronic steering*: if a *time delay* $i\tau$ is inserted into the electronic signal for the ith element of the array, the expression before (8.70) is replaced with

$$\frac{A}{r_i'} e^{j[\omega(t + i\tau) - kr_i']}$$

and the directional factor becomes

$$H(\theta) = \left| \frac{1}{N} \frac{\sin\left[\frac{N}{2} kd\left(\sin\theta - \frac{c\tau}{d}\right)\right]}{\sin\left[\frac{1}{2} kd\left(\sin\theta - \frac{c\tau}{d}\right)\right]} \right| \tag{8.78}$$

The major lobe now points in the direction θ_0 given by

$$\sin \theta_0 = c\tau/d \tag{8.79}$$

Thus, the introduction of a progressive time delay across the array *steers* the major lobe off the $\theta = 0$ plane into a cone determined by θ_0. Notice that (8.79) is independent of frequency.

Another technique for tailoring the properties of the beam is to *shade* the response of the array by applying different gains to the individual elements of the array. (See Urick[2] for a good bibliography on this subject.)

8.14   THE PRODUCT THEOREM.   In the preceding discussion of an array, it was assumed that each element is a simple source so that the individual pressure waveforms were spherically symmetric. It is now straightforward to generalize the results for an array of simple sources to an array of identical sources (all oriented in the same direction) which are not simple. If attention is restricted to the far field, so that all rays from the array to the field point (r, θ, ϕ) can be considered parallel, the pressure generated at (r, θ, ϕ) by each element must therefore contain the factor $H_e(\theta, \phi)$ which describes the directionality of each single element of the array, and since all rays are parallel, this factor must be the same in each term of (8.70). Given this, (8.71c) can be modified and generalized to

$$P(r, \theta, \phi) = P_{ax}(r) \, | \, H_e(\theta, \phi) H(\theta, \phi) | \tag{8.80}$$

where $H(\theta, \phi)$ is the directional factor for the array with simple sources at the position of each element and $H_e(\theta, \phi)$ is the directional factor for a single element. This reveals that the directional factor of an array of identical sources is the product of the directional factor of an array (with identical geometry but with simple sources) and the directional factor of a single element of the array.

PROBLEMS

8.1.   A pulsating sphere radiates spherical waves into air so as to produce an intensity of 50 mW/m² at a distance of 1.0 m from the center of the sphere. (a) What is the acoustic power radiated? Assume the sphere has a radius of 0.1 m. (b) If the frequency is 100 Hz, compute the intensity, the pressure amplitude, and the particle speed amplitude at the surface of the sphere. (c) Repeat part (b) at a distance of 0.5 m from the center of the sphere.

8.2.   A pulsating sphere of radius a vibrates with a surface velocity amplitude U_0 and at such a high frequency that $ka \gg 1$. Derive expressions for the pressure amplitude, the particle velocity amplitude, the intensity, and the total acoustic power radiated in the resulting acoustic wave.

8.3.   (a) A spherical source of radius a is operated in water at a frequency for which $ka = 1$. Evaluate the specific acoustic impedance at the source radius for this frequency. Find the error in calculating the acoustic intensity by the formula valid for $ka \ll 1$. (b) If the source strength of a small ($ka \ll 1$) spherical source is kept constant, find the frequency dependence of the radiated power. If this small source is operated with constant acceleration amplitude, find the frequency dependence of the radiated power.

8.4.   A simple source of sound in air radiates spherical waves at 400 Hz and at an acoustic power of 10 mW. Compute (a) the intensity at 0.5 m from the source, (b) the pressure amplitude at this distance, (c) the particle speed amplitude at this distance, (d) the particle displacement amplitude at this distance, and (e) the condensation amplitude at this distance.

[2] Urick, *Principles of Underwater Sound*, McGraw-Hill (1975).

8.5. A hemisphere of radius a and a piston of radius a are each mounted so that they radiate on one side of an infinite baffle. They are both vibrating with the same maximum speed amplitude U_0 and at the same frequency so that $ka \ll 1$. (a) For a distance such that $r \gg a$, what is the ratio of the axial intensity of the piston to that of the hemisphere? (b) What is the ratio of the total power radiated by the hemisphere to that radiated by the piston?

8.6. A simple line source is designed so that $kL = 50$. (a) How many major lobes are there? (b) Find the total number of nodal surfaces. (c) Find the angular width in degrees of the major lobe that is centered at $\theta = 0$. (d) Estimate the relative strength in decibels of the first side lobe.

8.7. For a baffled piston of radius a, (a) find the smallest angle θ_1 for which the pressure is zero in the far field, (b) find the greatest finite distance for which the pressure is zero on the acoustic axis, and (c) discuss the possibility of obtaining $\theta_1 \ll 1$ and $r_1/a \ll 1$ simultaneously.

8.8. A plane circular piston in an infinite baffle operates into water. The radius of the piston is 1 m. At a frequency of $(6/\pi)$ kHz the sound pressure level on axis at 1 km is 100 dB re 1 μbar. (a) Find all angles at which the pressure amplitude in the far field is zero. (b) Find the rms speed of the piston. (c) If the frequency were doubled while keeping the speed amplitude of the piston constant, what would be the decibel change in the sound pressure level on the axis in the far field and what would be the decibel change in the directivity index?

8.9. A circular piston-type sonar transducer of 0.5-m radius radiates 5000 W of acoustic power into water at a frequency of 10 kHz. (a) What is its beam width at the down 10-dB direction? (b) What is the axial pressure level in dB re 1 μbar at a distance of 10 m from the face of the transducer?

8.10. A piston of radius a is mounted so as to radiate on one side of an infinite baffle into air. The piston is driven at a frequency such that the wavelength of the radiated sound equals πa. (a) Compute and plot the relative axial intensities produced by the piston from its surface to a distance of 0.5 m. (b) Over what range of distances is the divergence approximately spherical?

8.11. By expanding $e^{jka \sin \theta \cos \phi}$ as a power series show that

$$\int_0^\pi e^{jka \sin \theta \cos \phi} \sin^2 \phi \, d\phi = \pi \frac{J_1(ka \sin \theta)}{ka \sin \theta}$$

8.12. Show that the nodal angles of the piston can be approximated by $\sin \theta_m \doteq (m + \frac{1}{4})\pi/(ka)$. Estimate the error in θ_m for the first nodal surface given by $m = 1$.

8.13. A continuous-line source is designed so that $kL = 50$. (a) If the length of the array is 100 m, estimate the distance to the far field. (b) Estimate the directivity index.

8.14. A piston is mounted so as to radiate on one side of an infinite baffle into air. The radius of the piston is a, and it is driven at a frequency such that the wavelength of the radiated sound equals πa. (a) If $a = 0.1$ m, and the maximum displacement amplitude of the piston is 0.0002 m; how much acoustic power is radiated? (b) What is the axial intensity at a distance of 2.0 m? (c) What is the directivity index of the radiated beam? (d) What is the radiation mass loading action on the piston?

8.15. A flat piston of 0.2-m radius radiates 100 W of acoustic power at 20 kHz in water. (a) Assuming the radiation to be equivalent to that of a piston mounted in an infinite baffle and radiating on only one side, what is the velocity amplitude of the piston? (b) What is the radiation mass loading of the piston? (c) What is the beam width at the down 10-dB direction? (d) What is the directivity index of the beam?

8.16. A flat piston of 0.15-m radius is mounted so as to radiate on one side of an infinite baffle into air. The frequency is 330 Hz. (a) What must be the speed amplitude of the piston, if it is to radiate 0.5 W of acoustic power? (b) If the piston has a mass of 0.015 kg, a stiffness constant of 2000 N/m, and negligible internal mechanical resistance, what force amplitude is required in order to produce this velocity amplitude?

8.17. (a) Find the resonance frequency in hertz of a piston transducer with the parameters m, s, and R_m radiating into a fluid with specific acoustic impedance $\rho_0 c$. Assume $ka \gg 2$. (b) Sketch the frequency dependence of the radiated power if the transducer is driven with a force of constant amplitude. Assume that the resonance frequency occurs well above the lower limit of the approximations implicit in $ka \gg 2$. Indicate where the transducer is mass controlled and where it is compliance controlled.

8.18. It is desired to design a highly directive piston transducer that will produce a given acoustic pressure amplitude P on axis at a specified range r. The operating frequency must be f, and the total acoustic power output is fixed. Find the radius and speed amplitude of this transducer.

8.19. A baffled piston transducer with radius 10 cm is normally operated at a frequency of 15 kHz. If it is desired to operate this same transducer at 3.5 kHz while maintaining the same acoustic pressure at each point on the axis in the far field, calculate the ratio of the total acoustic power output at 3.5 kHz to that at 15 kHz. Assume operation in water.

8.20. It is desired to design an underwater linear array of 30 equally spaced elements. (The array is not steered or shaded.) (a) If the major lobe is to be (nearly) as narrow as possible at 300 Hz, find the spacing between elements. (b) What is the angular width in degrees of the major lobe? (c) What is the geometrical shape of the axis of the major lobe? (d) Estimate the directivity index of the array.

8.21. Show that if the time delay introduced between adjacent elements of a linear array is $\tau = d/c$, where d is the spacing of the elements, then the steered array has a major lobe centered on the line $\theta = 90°$. This is called an *end-fired* array.

8.22. For a steered line array, show that the condition $\lambda/d \geq (1 + |c\tau/d|)N/(N - 1)$ guarantees that there will be only one major lobe.

8.23. It is desired to design an end-fired linear array of 30 equally spaced elements to be operated in water at 300 Hz. (a) Use the results of Problems 8.21 and 8.22 to find the spacing between elements if there is to be only one major lobe. (b) What is the geometric shape of the major lobe axis? (c) Find the angular width in degrees of the major lobe. (d) Estimate the directivity index of this end-fired array.

8.24. Write the directional factor $H(\theta, \phi)$ for a rectangular piston transducer of dimensions L_x and L_y in terms of the dimensions of the piston and k.

CHAPTER 9

PIPES, CAVITIES, AND WAVEGUIDES

9.1 INTRODUCTION. Thus far, we have considered the behavior of acoustic waves under relatively simple geometric conditions. There are, however, many more complicated situations in acoustics for which the geometry of the boundaries confines the wave to a limited region of space. These situations will occupy our attention for most of the remainder of this book. In this and the next chapter we will discuss the basic properties of sound in pipes, cavities, and waveguides.

When sound propagates in a rigid-walled pipe with wavelength larger than the radius, the acoustic motion is essentially planar, much as longitudinal waves in a bar. The resonance properties of pipes driven at one end and terminated at the other have important application in the laboratory for measurement of acoustical impedances and of the absorptive properties of materials. This study of pipes will also reveal many properties of brasses, woodwinds, and organ pipes. Pipes also serve as models for ventilation ducts.

For larger spaces in which the dimensions are not smaller than a wavelength, two- and three-dimensional standing waves can be stimulated. The basic properties of the normal modes describing these standing waves in rigid-walled volumes offer some simple explanations for the behavior of lower-frequency sound in rooms, auditoriums, concert halls, and so forth.

Finally, we will study the simple waveguide with uniform cross section and develop the concepts of the group speed and the phase speed associated with a sound wave propagating in a waveguide. Application of acoustic waveguides is found in surface-wave delay lines, in high-frequency electronic systems, and in the propagation of sound in the oceans and the atmosphere.

9.2 RESONANCE IN PIPES. Assume that the fluid in a pipe of cross-sectional area S and length L is driven by a piston at $x = 0$ and that the pipe is terminated at $x = L$ in a mechanical impedance $\mathbf{Z}_{mL}$. If the piston vibrates harmonically at a frequency sufficiently low that only plane waves propagate (see Sect. 9.8), the wave in the pipe will be of the form

$$\mathbf{p} = \mathbf{A}e^{j[\omega t + k(L - x)]} + \mathbf{B}e^{j[\omega t - k(L - x)]} \tag{9.1}$$

where $\mathbf{A}$ and $\mathbf{B}$ are determined by the boundary conditions at $x = 0$ and $x = L$.

At $x = L$, the continuities of force and particle speed require that the mechanical impedance of the wave at $x = L$ equals the mechanical impedance of the termination, $\mathbf{Z}_{mL}$. Since the force of the fluid on the termination is $\mathbf{p}(L, t)S$ and the particle speed is $\mathbf{u}(L, t) = -(1/\rho_0)\int(\partial p/\partial x)\,dt$,

$$\mathbf{Z}_{mL} = \rho_0\,cS\,\frac{\mathbf{A} + \mathbf{B}}{\mathbf{A} - \mathbf{B}} \tag{9.2}$$

The input mechanical impedance $\mathbf{Z}_{m0}$ at $x = 0$ is correspondingly given by

$$\mathbf{Z}_{m0} = \rho_0\,cS\,\frac{\mathbf{A}e^{jkL} + \mathbf{B}e^{-jkL}}{\mathbf{A}e^{jkL} - \mathbf{B}e^{-jkL}} \tag{9.3}$$

Combining these equations to eliminate $\mathbf{A}$ and $\mathbf{B}$, we obtain

$$\frac{\mathbf{Z}_{m0}}{\rho_0\,cS} = \frac{\dfrac{\mathbf{Z}_{mL}}{\rho_0\,cS} + j\,\tan kL}{1 + j\,\dfrac{\mathbf{Z}_{mL}}{\rho_0\,cS}\,\tan kL} \tag{9.4}$$

This equation is identical with (3.33) with the replacement of $\rho_L c$ with $\rho_0\,cS$, and the substitution

$$\frac{\mathbf{Z}_{mL}}{\rho_0\,cS} = r + jx \tag{9.5}$$

leads directly to (3.34). Recalling the discussion following that equation, we see that the frequencies of resonance and antiresonance are determined by the vanishing of the reactance,

$$-j\,\frac{x\,\tan^2 kL + (r^2 + x^2 - 1)\tan kL - x}{(r^2 + x^2)\tan^2 kL - 2x\,\tan kL + 1} = 0 \tag{9.6}$$

The solution identified with *small* input resistance denotes *resonance*, and that identified with *large* input resistance denotes *antiresonance*. (In the limiting case $r = 0$, there is only one solution, corresponding to resonance.)

Let the pipe be driven at $x = 0$ and *closed* at $x = L$ by a rigid cap. To obtain the condition of resonance most simply, let $|\mathbf{Z}_{mL}/(\rho_0\,cS)| \to \infty$ in (9.4). This yields

$$\frac{\mathbf{Z}_{m0}}{\rho_0\,cS} = -j\,\cot kL \tag{9.7}$$

The reactance is zero when $\cot kL = 0$,

$$k_n L = (2n - 1)\pi/2 \qquad n = 1, 2, 3, \ldots \tag{9.8a}$$

or

$$f_n = \frac{2n - 1}{4}\,\frac{c}{L} \tag{9.8b}$$

This formula is identical with (2.22) for the forced-fixed string. The resonance frequencies are the odd harmonics of the fundamental. The driven, closed pipe has a pressure antinode at $x = L$ and a pressure node at $x = 0$. Notice that this requires that the driver presents a vanishing mechanical impedance to the tube. The implication of this, and the effects of the mechanical properties of the driver on the behavior of the driver-pipe system, will be discussed in Sect. 9.6.

Now, consider a pipe driven at $x = 0$ and *open-ended* at $x = L$. On first examination, it might be thought that this will lead to $Z_{mL} = 0$ for which $Z_{m0}/(\rho_0 cS) = j \tan kL$ with resonance occurring at $f_n = (n/2)c/L$, $n = 1, 2, 3, \ldots$. However, this is not the case, most elementary physics textbooks notwithstanding. The condition at $x = L$ is not $Z_{mL} = 0$ since the open end of the tube radiates sound into the surrounding medium. The appropriate value for Z_{mL} is therefore

$$Z_{mL} = Z_r \tag{9.9}$$

where Z_r is the radiation impedance of the open end of the pipe.

For example, assume that the open end of a circular pipe of radius a is surrounded by a *flange* large with respect to the wavelength of the sound. Consistent with the assumption that the wavelength is large compared to the transverse dimensions of the tube ($\lambda \gg a$), the opening resembles a baffled piston in the low-frequency limit. We have, therefore, from (8.68) *et seq.*

$$\frac{Z_{mL}}{\rho_0 cS} = \frac{1}{2}(ka)^2 + j\frac{8}{3\pi}ka \qquad \text{(flanged)} \tag{9.10}$$

where both $r = (ka)^2/2$ and $x = 8ka/(3\pi)$ are much less than unity and $r \ll x$. Solution of (9.6) under these conditions gives $\tan kL = -x$ for the resonance frequencies. Since $x \ll 1$, this yields

$$\tan(n\pi - k_n L) = \frac{8}{3\pi}ka \doteq \tan\left(\frac{8}{3\pi}ka\right)$$

where $n = 1, 2, 3, \ldots$. Therefore,

$$n\pi = k_n L + \frac{8}{3\pi}k_n a \tag{9.11a}$$

and the resonance frequencies are

$$f_n = \frac{n}{2}\frac{c}{L + \frac{8}{3\pi}a} \tag{9.11b}$$

These resonance frequencies are all harmonics of the fundamental, and it is apparent that the *effective length* L_{eff} of such a pipe is not L but rather $L + 8a/(3\pi)$. This predicted *end correction* for a flanged pipe is in reasonable agreement with experimentally measured values of around $0.82a$.

For an *unflanged*, open pipe, both experiments and theory indicate that the radiation impedance is approximately

$$\frac{Z_{mL}}{\rho_0 cS} = \frac{1}{4}(ka)^2 + j0.6ka \qquad \text{(unflanged)} \tag{9.12}$$

The end correction for an unflanged, open pipe is therefore $0.6a$, so that $L_{eff} = L + 0.6a$.

In both cases, the end corrections are independent of frequency, so that the resonance frequencies of flanged and unflanged open pipes are harmonics of the fundamental (so long as $\lambda_n \gg a$).

These considerations reveal that the resonances of a suitably driven, open-ended organ pipe correspond to the harmonics of the driving frequency. It should be noted that this result has been obtained only for pipes of constant cross section. The presence of any *flare* in the pipe, as found in many wind instruments and some organ pipes, modifies these results. In particular, the resonance frequencies may no longer be harmonics of the fundamental. Indeed, the design of the flare is very important in emphasizing or reducing certain of the harmonics present in the forcing function and therefore in controlling the quality or *timbre* of the sound radiated by the pipe.

9.3 POWER RADIATION FROM OPEN-ENDED PIPES.

Solution of (9.2) for **B/A** yields

$$\frac{\mathbf{B}}{\mathbf{A}} = \frac{\mathbf{Z}_{mL}/(\rho_0 cS) - 1}{\mathbf{Z}_{mL}/(\rho_0 cS) + 1} \tag{9.13}$$

and the power transmission coefficient can be found from

$$T_\pi = 1 - |\mathbf{B/A}|^2 \tag{9.14}$$

once the termination impedance $\mathbf{Z}_{mL}$ is known.

For an open-ended pipe terminated in a flange, $\mathbf{Z}_{mL}$ is given by (9.10), and (9.13) becomes

$$\frac{\mathbf{B}}{\mathbf{A}} = - \frac{[1 - \frac{1}{2}(ka)^2] - j\dfrac{8}{3\pi}ka}{[1 + \frac{1}{2}(ka)^2] + j\dfrac{8}{3\pi}ka} \tag{9.15}$$

This, in turn, yields

$$T_\pi = \frac{2(ka)^2}{[1 + \frac{1}{2}(ka)^2]^2 + \left(\dfrac{8}{3\pi}\right)^2(ka)^2} \tag{9.16a}$$

Since $ka \ll 1$, the power transmission coefficient is extremely small and can be further simplified,

$$T_\pi \doteq 2(ka)^2 \quad \text{(flanged)} \tag{9.16b}$$

and (9.15) shows that **B/A** is very nearly -1. The pressure amplitude of the reflected wave is only slightly less than that of the incident wave, and at $x = L$ its pressure differs in phase by nearly $180°$; a condensation is reflected as a rarefaction. In contrast, the incident and reflected particle speeds are nearly in phase at the orifice of the pipe, so that this position is approximately an antinode of particle speed. Thus, in spite of the fact that the amplitude of the particle speed at the orifice is almost twice that of the incident wave alone, only a small percentage of the incident power is transmitted out of a flanged pipe. This is another statement of the fact that sources whose dimensions are small compared with the wavelength of the sound are very inefficient as radiators of acoustic energy.

For an unflanged pipe, Z_{mL} is given by (9.12), and the transmission coefficient becomes

$$T_\pi = \frac{(ka)^2}{[1 + \tfrac{1}{4}(ka)^2]^2 + (0.6ka)^2} \tag{9.17a}$$

or

$$T_\pi \doteq (ka)^2 \qquad \text{(unflanged)} \tag{9.17b}$$

so that the presence of a wide flange at the end of a pipe approximately doubles the radiation of sound at low frequencies.

Note that when a pipe is terminated in a gradual flare the low-frequency power transmission is still further increased.

In the vicinity of resonance we can write $\omega = \omega_n + \Delta\omega$, and the input impedance of the unflanged pipe (9.7) is then well-approximated by

$$\frac{Z_{m0}}{\rho_0 cS} \approx \tfrac{1}{4}(k_n a)^2 + j\,\Delta\omega\,\frac{L}{c}$$

The half-power points are

$$\omega_{u,l} = \omega_n \pm \tfrac{1}{4}(k_n a)^2\,\frac{c}{L}$$

and the Q for the nth resonance is

$$Q_n = \frac{\omega_n}{\omega_u - \omega_l} = \frac{2}{n\pi}\frac{L}{a}\frac{L + 0.6a}{a}$$

The radiated power, $\Pi = F^2 R_{m0}/(2Z_{m0}^2)$ where $R_{m0} = \mathrm{Re}\{Z_{m0}\}$ and F is the force amplitude, has the value

$$\Pi_n = \frac{F^2}{\rho_0 cS}\frac{2}{(k_n a)^2} = \frac{2}{(n\pi)^2}\frac{F^2}{\rho_0 cS}\left(\frac{L + 0.6a}{a}\right)^2$$

Thus, we see that the Q's of the resonances decrease as $1/n$ and that the power radiated at resonance decreases as $1/n^2$ in the low-frequency region for constant applied force amplitude.

9.4   STANDING WAVE PATTERNS.   The phase interference between the transmitted and reflected waves in a terminated pipe will result in the formation of a standing wave pattern. In this section we will show how the measured properties of the standing wave can be used to determine the load impedance.

Let us choose to write

$$\mathbf{A} = A \tag{9.18a}$$

and

$$\mathbf{B} = Be^{j\theta} \tag{9.18b}$$

where A and B are real and positive. Substituting (9.18) into (9.2) gives

$$\frac{\mathbf{Z}_{mL}}{\rho_0 cS} = \frac{1 + \dfrac{B}{A}\, e^{j\theta}}{1 - \dfrac{B}{A}\, e^{j\theta}} \tag{9.19}$$

Thus, given B/A and θ, $\mathbf{Z}_{mL}$ can be determined. By substituting (9.18) into (9.1) and solving for the amplitude $P = |\mathbf{p}|$ of the wave, we obtain

$$P = \{(A + B)^2 \cos^2[k(L - x) - \theta/2] + (A - B)^2 \sin^2[k(L - x) - \theta/2]\}^{1/2} \tag{9.20}$$

The amplitude at a pressure antinode is $A + B$, and the amplitude at a node is $A - B$. The ratio of pressure amplitude at an antinode to that at a node is the *standing wave ratio*

$$\mathrm{SWR} = \frac{A + B}{A - B} \tag{9.21}$$

which can be rearranged at once to provide

$$\frac{B}{A} = \frac{\mathrm{SWR} - 1}{\mathrm{SWR} + 1} \tag{9.22}$$

Thus, measurement of the SWR by probing the sound field in the pipe with a small microphone yields a value for B/A. The phase angle θ can be evaluated from the distance of the first node from the end at $x = L$. From (9.20), these nodes are located at $k(L - x_n) - \theta/2 = (n - \tfrac{1}{2})\pi$, so that for the first node

$$\theta = 2k(L - x_1) - \pi \tag{9.23}$$

For example, let us assume that in some terminated tube the standing wave ratio is $\mathrm{SWR} = 2$ and the first node is $3/8$ of a wavelength from the end. Then $L - x = 3\lambda/8$, and $\theta = 2(2\pi/\lambda)(3\lambda/8) - \pi = \pi/2$. Furthermore, $B/A = (2 - 1)/(2 + 1) = 1/3$ and

$$\frac{\mathbf{Z}_{mL}}{\rho_0 cS} = \frac{1 + \tfrac{1}{3}e^{j\pi/2}}{1 - \tfrac{1}{3}e^{j\pi/2}} \approx 0.80 + j0.60$$

Since the mechanical impedance of the termination can be a complicated function of frequency, it may be necessary to repeat the above measurements over the range of frequencies of interest. When a large number of frequencies are measured, it is inconvenient and unnecessary to carry out the calculation as outlined above. Instead, use of a "Smith chart," a nomographic chart, enables rapid determination of r and x in (9.5) directly from measurements of the standing wave ratio and the position of the node nearest the end.[1]

The reflective and absorptive properties, at normal incidence, of such materials as acoustic tiles and other sound control materials can be determined by mounting

---

[1] Beranek, *Acoustic Measurements*, 317–321, Wiley (1949).

a small section of the material at the end of a standing wave tube and making the measurements and calculations described in this section.

9.5 ABSORPTION OF SOUND IN PIPES. If absorptive processes within the fluid and at the walls of the pipe are considered, the solution for the pipe driven at $x = 0$ is found (see Chapter 7) by substituting the complex propagation constant

$$\mathbf{k} = k - j\alpha \tag{9.24}$$

into the lossless solutions obtained in the previous sections.

As an example, for a rigid termination at $x = L$ the pressure is

$$\mathbf{p}(x, t) = \frac{F}{S} \frac{\cos[\mathbf{k}(L - x)]}{\cos \mathbf{k}L} e^{j\omega t} \tag{9.25}$$

and the input impedance is

$$\mathbf{Z}_{m0} = -j\rho_0 \frac{\omega}{\mathbf{k}} S \cot \mathbf{k}L \tag{9.26}$$

With the help of the expansions of sines and cosines of complex argument, the above expression becomes

$$\frac{\mathbf{Z}_{m0}}{\rho_0 cS} = -j \frac{1 + j\alpha/k}{1 + (\alpha/k)^2} \frac{\cos kL \sin kL + j \sinh \alpha L \cosh \alpha L}{\sin^2 kL \cosh^2 \alpha L + \cos^2 kL \sinh^2 \alpha L}$$

The terms in α/k introduce a phase angle $\tan^{-1}(\alpha/k)$ which can be ignored with no significant loss of accuracy when $\alpha/k \ll 1$. If we further assume that the pipe is of reasonable length so that $\alpha L \ll 1$, then the input impedance assumes a simpler form

$$\frac{\mathbf{Z}_{m0}}{\rho_0 cS} \doteq \frac{\alpha L - j \cos kL \sin kL}{\sin^2 kL + (\alpha L)^2 \cos^2 kL} \tag{9.27}$$

In the lossless limit, the resistance becomes zero and the reactance becomes proportional to $\cot kL$. The major effects of absorption are to introduce a small resistance which maximizes when $k_n L = n\pi$ and to alter the behavior of the reactance in this same region so that it no longer becomes infinite in magnitude but rather remains bounded and changes from positive to negative value very rapidly. These effects are shown in Fig. 9.1.

The power dissipated by the pipe is that delivered by the source, $\Pi = F^2 R_{m0}/(2Z_{m0}^2)$, where $\mathbf{Z}_{m0} = R_{m0} + jX_{m0}$. This becomes

$$\Pi = \frac{1}{2} \frac{F^2}{\rho_0 cS} \alpha L \frac{\sin^2 kL + (\alpha L)^2 \cos^2 kL}{(\alpha L)^2 + \cos^2 kL \sin^2 kL} \tag{9.28}$$

At mechanical *resonance* $\cos kL = 0$, and the power consumption is

$$\Pi_r = \frac{1}{2} \frac{F^2}{\rho_0 cS} \frac{1}{\alpha L}$$

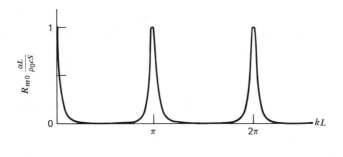

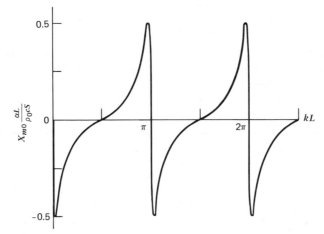

Fig. 9.1. *The input mechanical resistance R_{m0} and reactance X_{m0} for a rigidly terminated pipe of length L with $\alpha L = 0.1$.*

whereas at *antiresonance* $\sin kL = 0$ and we have

$$\Pi_a = \frac{1}{2}\frac{F^2}{\rho_0 cS}\,\alpha L$$

Notice that the frequencies of resonance and antiresonance are close to the natural frequencies of the undamped open, rigid and rigid, rigid pipes, respectively.

If we examine the input impedance in the immediate frequency range of resonance, and define the deviation from resonance by the incremental angular frequency $\Delta\omega$, then

$$kL = \frac{\omega_n + \Delta\omega}{c}L = (2n-1)\frac{\pi}{2} + \Delta\omega\,\frac{L}{c} \tag{9.29a}$$

and

$$\frac{\mathbf{Z}_{m0}}{\rho_0 cS} \doteq \alpha L + j\,\Delta\omega\,\frac{L}{c} \tag{9.29b}$$

The resistance is constant over this range, and the half-power points are determined from the frequencies for which reactance equals resistance, $\Delta\omega = \pm\alpha c$. The frequency interval between upper and lower half-power points is $2\alpha c$, so that $Q = \omega_n/(2\alpha c)$ or

$$Q_n = \frac{1}{2}\frac{1}{\alpha/k_n} \tag{9.30}$$

Laboratory measurements of acoustic absorption in fluids are frequently made on fluids contained within cylindrical pipes. In one method, a probe microphone is used to measure pressure amplitudes of a plane progressive wave at two or more positions along the length of the pipe. If P_1 is the pressure amplitude at x_1 and P_2 that at x_2, then the attenuation constant may be determined from

$$P_2 = P_1 e^{-\alpha(x_2 - x_1)}$$

When this equation is used, steps must be taken to eliminate reflected waves either through the use of a nonreflecting termination at the end of the pipe or through the use of short pulses or long pipes so that measurements at x_1 and x_2 may be made before a reflected pulse is returned.

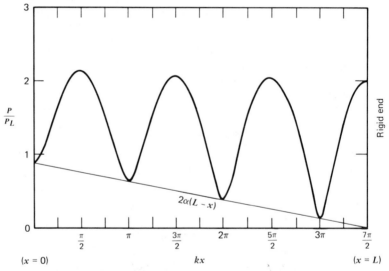

Fig. 9.2. *Spatial dependence of the pressure amplitude of a damped standing wave in a driven pipe rigidly terminated at $x = L$ with $\alpha/k = 0.04$.*

A second method utilizes the standing wave (Fig. 9.2). Let us assume that the termination at $x = L$ is rigid. Then at $x = L$ the amplitude P_L of the reflected wave will equal that of the incident wave. The resulting pressure amplitude P at any position along the pipe may be shown from (9.25) to be

$$P = 2P_L\{\cos^2[k(L - x)]\cosh^2[\alpha(L - x)] + \sin^2[k(L - x)]\sinh^2[\alpha(L - x)]\}^{1/2} \tag{9.31}$$

The nodes occur at

$$k(L - x) = (2n - 1)\pi/2 \qquad n = 1, 2, 3, \ldots$$

and have relative amplitudes

$$P_{min}/P_L = 2 \sinh[\alpha(L - x)] \approx 2\alpha(L - x) \tag{9.32}$$

The pressure amplitudes at successive nodes can be measured by means of a probe microphone. The value of α may then be determined by drawing a smooth curve through these points, as indicated in Fig. 9.2. The antinodes occur at $k(L - x) = n\pi$ where $n = 0, 1, 2, \ldots$ and give maximum pressure amplitudes

$$P_{max}/P_L = 2 \cosh[\alpha(L - x)] \approx 2 + [\alpha(L - x)]^2 \tag{9.33}$$

Acoustic absorption determined by either of the above methods is always higher than that measured in large volumes of the fluid because of the losses taking place at the walls of the pipe. One source of this increased absorption, studied by Helmholtz in 1863, is associated with viscous resistance offered to fluid motion at the walls of the pipe. This results in laminar motion, with the amplitude of the particle speed increasing rapidly from zero at the wall to nearly its maximum value at a distance $2.28\sqrt{2\eta/(\rho_0 \omega)}$ from the wall. Another source is the exchange of heat energy between the fluid and the wall. The generation of heat, and its communication by conduction to and from the walls of the pipe, was first studied by Kirchhoff in 1868. He assumed that the layer of fluid in contact with the wall can have no velocity and is isothermal. This analysis, as presented by Rayleigh,[2] shows that the true coefficient of shear viscosity η must be replaced by an *effective* coefficient

$$\eta_e = \eta\left[1 + (\gamma - 1)\sqrt{\frac{\kappa}{C_{\mathscr{P}}\eta}}\right]^2 \tag{9.34}$$

where κ is the thermal conductivity of the fluid, γ is the ratio of specific heats, and $C_{\mathscr{P}}$ is its specific heat at constant pressure. Experimentally measured values for air at 20°C are $\eta = 1.81 \times 10^{-5}$ Pa $\cdot$ s, $\rho_0 = 1.21$ kg/m³, $\gamma = 1.402$, and $\kappa/(C_{\mathscr{P}}\eta) = 1.35$, so that

$$\eta_e = 2.15\eta = 3.9 \times 10^{-5} \text{ Pa} \cdot \text{s}$$

The effect of heat conduction on the absorption of sound waves is equivalent to a 115 percent increase in the coefficient of shear viscosity.

When wall effects dominate, the absorption coefficient and phase speed of an acoustic wave propagating in a circular pipe of radius a are[3]

$$\alpha = \frac{1}{ac}\sqrt{\frac{\eta_e \omega}{2\rho_0}} \tag{9.35a}$$

$$c_p = c\left(1 - \frac{1}{2a}\sqrt{\frac{2\eta_e}{\rho_0 \omega}}\right) \tag{9.35b}$$

[2] Rayleigh, *Theory of Sound*, Sect. 348–350, Macmillan and Company, Ltd. (1929).

[3] Weston, *Proc. Phys. Soc.* (London), **B66**, 695 (1953).

Substitution into these equations of the values for air at 20°C yields

$$\alpha = 2.93 \times 10^{-5} f^{1/2}/a$$

$$c_p = c[1 - 2.84 \times 10^{-3}/(af^{1/2})]$$

The absorption coefficient for a frequency of 10 kHz in air is $\alpha = 2.93 \times 10^{-3}/a$ Np/m. This absorption is so small that it indicates that viscous and heat-conduction losses at the walls are negligible in absorbing sound in ventilating ducts. However, small as this absorption may be, it must be considered when acoustic measurements are made at high frequencies in small pipes. For instance, at a frequency of 10 kHz in a pipe of 0.01-m radius, it is much larger than the absorption taking place within the body of the fluid, which is $\alpha \sim 5 \times 10^{-11} f^2$. However, as frequency increases, absorption of acoustic energy within the body of the fluid increases more rapidly than at the wall, and at frequencies above 1 MHz it becomes dominant in all but the most narrow of pipes.

In the treatment of wave propagation through *liquids* confined within pipes, the influence of heat conduction at the walls is usually negligible, so that only the ordinary shear coefficient of viscosity needs to be considered.

9.6 BEHAVIOR OF THE COMBINED DRIVER-PIPE SYSTEM. We have so far considered the resonance properties of the pipe. A more realistic investigation of resonating pipes must account for the properties of the mechanical driver, which itself is driven by the applied force. The driver has its own mechanical impedance, so that when a force is applied to the *driver-pipe system* the mechanical resonances of the *combined system* involve the mechanical behavior of the driver as well as that of the pipe.

For example, let the driver be a damped, harmonic oscillator, as suggested in Fig. 9.3, excited with the externally applied force $\mathbf{f} = F \exp(j\omega t)$. Newton's second law for the motion of the mass is

$$m \frac{d^2\xi}{dt^2} = -R_m \frac{d\xi}{dt} - s\xi - S\mathbf{p}(0, t) + \mathbf{f} \tag{9.36}$$

where ξ is the displacement of the mass to the right and $\mathbf{p}(0, t)$ is the pressure in the

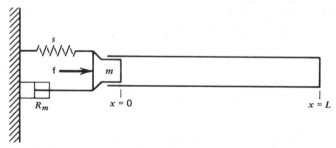

Fig. 9.3. *Schematic representation of a driver-tube system.*

pipe at $x = 0$. The complex speed of the mass is $u(0, t) = d\xi/dt$, the particle speed of the fluid in the pipe at $x = 0$, so that (9.36) becomes

$$\left[R_m + j\left(\omega m - \frac{s}{\omega} \right) + \frac{S\mathbf{p}(0, t)}{\mathbf{u}(0, t)} \right] \mathbf{u}(0, t) = \mathbf{f} \tag{9.37}$$

The input mechanical impedance $\mathbf{Z}_{md}$ of the *driver* is

$$\mathbf{Z}_{md} = R_m + j(\omega m - s/\omega) \tag{9.38}$$

and the input mechanical impedance of the *pipe* is

$$\mathbf{Z}_{m0} = \frac{S\mathbf{p}(0, t)}{\mathbf{u}(0, t)} \tag{9.39}$$

Thus, (9.37) shows that the input mechanical impedance $\mathbf{Z}_m$ of this *system* is the series combination of $\mathbf{Z}_{md}$ and $\mathbf{Z}_{m0}$, so that

$$\mathbf{f} = \mathbf{Z}_m \mathbf{u}(0, t) = (\mathbf{Z}_{md} + \mathbf{Z}_{m0})\mathbf{u}(0, t) \tag{9.40}$$

Driver and pipe have their own frequencies of mechanical resonance: the driver resonates when its reactance vanishes, $\omega_0 = \sqrt{s/m}$, and the pipe resonates when $\text{Im}\{\mathbf{Z}_{m0}\} = 0$. Now, when the combined system is driven, the input impedance seen by the applied force is the sum of the impedances of the source and pipe, so that the frequencies of mechanical resonance of the combined system are found from

$$\text{Im}\{\mathbf{Z}_{md} + \mathbf{Z}_{m0}\} = 0 \tag{9.41}$$

Let us assume that the driven pipe has a *rigid* termination at $x = L$. With the results of the previous section, (9.41) becomes

$$\omega m - \frac{s}{\omega} - \frac{S\rho_0 c \cos kL \sin kL}{\sin^2 kL + (\alpha L)^2 \cos^2 kL} = 0$$

Use of $\omega/k = c$ and rearrangement yields

$$\frac{\cos kL \sin kL}{\sin^2 kL + (\alpha L)^2 \cos^2 kL} = akL - \frac{b}{kL} \tag{9.42a}$$

where

$$a = \frac{m}{S\rho_0 L} \quad \text{and} \quad b = \frac{sL}{S\rho_0 c^2} \tag{9.42b}$$

Notice that a is the ratio of the mass of the moving element of the driver to the mass of the fluid in the tube and b is the ratio of the stiffness of the suspension of the moving element of the driver to the stiffness of the compressible fluid filling the pipe. Plotting both sides of (9.42) against kL on the same set of axes provides the frequencies of mechanical resonance from the values of kL for which the two curves intersect. Two examples are given in Fig. 9.4. The examples illustrate the effects of two different driver conditions. (1) For a light, flexible driver with small values of a and b (Fig. 9.4a), the curves tend to intersect for $k_n L \sim (2n - 1)\pi/2$, so that

$L \sim (2n - 1)\lambda/4$ and there is nearly a pressure *node* at $x = 0$. (2) On the other hand, if the driver is heavy and stiff so that a and b are large, Fig. 9.4b shows that most of the resonances occur for $k_n L \sim n\pi$; there is nearly a pressure *antinode* at $x = 0$. However, in the vicinity of the driver resonance, $kL = 3.6\pi$ in Fig. 9.4b, the system resonances tend toward values of $k_n L$ corresponding to a pressure node at $x = 0$.

Since there will always be a pressure antinode at $x = L$, if we obtain the pressure amplitude at this end in terms of the applied force and the mechanical impedance, then we can determine the behavior of the antinodal pressure amplitude as a function of frequency. From (9.25) we have

$$\mathbf{p}(x, t) = \mathbf{p}(0, t) \frac{\cos[\mathbf{k}(L - x)]}{\cos kL}$$

Evaluation of the above equation at $x = L$ and then use of (9.39) and (9.40) results in

$$\mathbf{p}(L, t) = \frac{F}{S} \frac{\mathbf{Z}_{m0}}{\mathbf{Z}_m} \frac{e^{j\omega t}}{\cos kL} \tag{9.43}$$

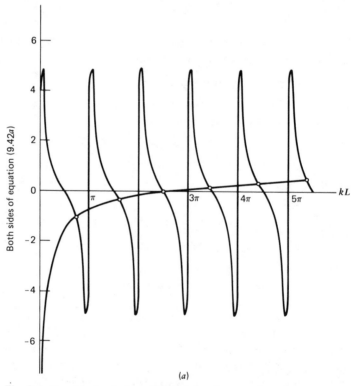

(a)

Fig. 9.4. *Graphical solution of (9.42a) for the resonance frequencies of a driver-pipe system rigidly terminated at $x = L$ with $\alpha L = 0.1$. (a) A light, flexible driver with $a = 0.04$ and $b = 2.67$. The driver is resonant at $kL = 2.6\pi$. (b) A heavy, stiff driver with $a = 0.25$ and $b = 32$. The driver is resonant at $kL = 3.6\pi$.*

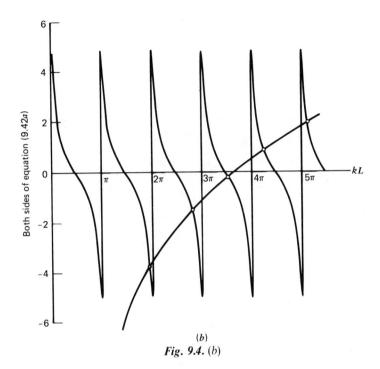

(b)

Fig. 9.4. (b)

and the pressure amplitude $P(L)$ at the rigid end is, for $\alpha L \ll 1$,

$$P(L) = \rho_0 c \, \frac{F}{Z_m} \, \frac{1}{\sqrt{\sin^2 kL + (\alpha L)^2 \cos^2 kL}} \tag{9.44}$$

Figure 9.5 displays the pressure amplitude at the rigid end for the same driver-pipe systems illustrated in Fig. 9.4 (1) The system with the light, flexible driver (Fig. 9.5a) displays resonances of nearly constant spacing in frequency and with nearly equal maximum pressure amplitudes. (2) The heavy, stiff driver (Fig. 9.4b) produces pressure amplitudes at the resonances that are much stronger for frequencies close to the resonance frequency of the driver. Furthermore, the driver resonance introduces an "extra" resonance for kL between 3π and 4π. (3) Finally, note that the two curves in Fig. 9.4b do not intersect near $kL = \pi$. Even though this is not, strictly speaking, a true resonance, since the reactance of the system does not go to zero, there is a relative minimum in the reactance so that the response $P(L)$ of the driver-pipe system peaks at this frequency.

This interaction of the driver with the pipe in determining the resonances of the system is displayed prominently in many musical instruments. In the brasses, for example, the player can, by altering the tension of his lips, influence the reactance of the driver (himself) and, therefore, the resonance frequencies of the system. The player may thus be able to "lip" the desired note about a semitone* away from the pertinent resonance frequency of the instrument.

---

* In the even-tempered musical scale, two frequencies f_2 and f_1 a semitone apart are related by $f_2/f_1 = 2^{1/12}$.

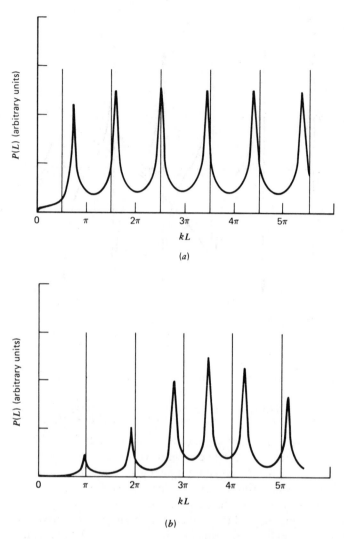

Fig. 9.5. *The antinodal pressure amplitude for a rigidly terminated driver-pipe system excited by a force of constant amplitude. (a) For the light, flexible driver of Fig. 9.4a. (b) For the heavy, stiff driver of Fig. 9.4b. For both drivers, $Rx/(S\rho_0 c)$ = 0.0715.*

9.7   THE RECTANGULAR CAVITY. Consider a rectangular cavity of dimensions L_x, L_y, L_z, as indicated in Fig. 9.6. This box could represent a living room or auditorium, a simple model of a concert hall, or any other rectangular space that has few windows or other openings and fairly rigid walls. Such applications will be encountered in Sect. 13.9.

Assume that all surfaces of the cavity are perfectly rigid so that $\hat{n} \cdot \vec{u} = 0$ at all boundaries. Then $\hat{n} \cdot \nabla p = 0$ there and thus

$$\left(\frac{\partial p}{\partial x}\right)_{x=0} = \left(\frac{\partial p}{\partial x}\right)_{x=L_x} = 0$$

$$\left(\frac{\partial p}{\partial y}\right)_{y=0} = \left(\frac{\partial p}{\partial y}\right)_{y=L_y} = 0 \tag{9.45}$$

$$\left(\frac{\partial p}{\partial z}\right)_{z=0} = \left(\frac{\partial p}{\partial z}\right)_{z=L_z} = 0$$

Since acoustic energy cannot escape from a closed cavity with rigid boundaries, appropriate solutions of the wave equation are standing waves. Substitution of

$$\mathbf{p}(x, y, z, t) = \mathbf{X}(x)\mathbf{Y}(y)\mathbf{Z}(z)e^{j\omega t} \tag{9.46}$$

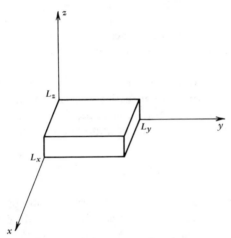

Fig. 9.6. *The rectangular cavity.*

into the wave equation and separation of variables (as performed in Chapter 4) results in the set of equations

$$\left(\frac{d^2}{dx^2} + k_x^2\right)\mathbf{X} = 0$$

$$\left(\frac{d^2}{dy^2} + k_y^2\right)\mathbf{Y} = 0 \tag{9.47}$$

$$\left(\frac{d^2}{dz^2} + k_z^2\right)\mathbf{Z} = 0$$

where the separation constants must be related by

$$(\omega/c)^2 = k^2 = k_x^2 + k_y^2 + k_z^2 \tag{9.48}$$

Application of the boundary conditions (9.45) shows that cosines are appropriate solutions, and (9.46) becomes

$$\mathbf{p}_{lmn} = \mathbf{A}_{lmn} \cos k_{xl} x \cos k_{ym} y \cos k_{zn} z \; e^{j\omega lmn} \tag{9.49}$$

where the components of $\vec{k}$ are

$$
\begin{aligned}
k_{xl} &= l\pi/L_x & l &= 0, 1, 2, \cdots \\
k_{ym} &= m\pi/L_y & m &= 0, 1, 2, \cdots \\
k_{zn} &= n\pi/L_z & n &= 0, 1, 2, \cdots
\end{aligned}
\tag{9.50}
$$

Thus, the allowed frequencies of vibration are *quantized*,

$$\omega_{lmn} = c\sqrt{(l\pi/L_x)^2 + (m\pi/L_y)^2 + (n\pi/L_z)^2} \tag{9.51}$$

Each *eigenfunction* given by (9.49) has its own characteristic *eigenfrequency* (9.51) and can be specified by the ordered integers (l, m, n).

The form (9.49) gives three-dimensional standing waves in the cavity with nodal planes parallel to the walls. Between these nodal planes the pressure varies sinusoidally, with the pressure within a given loop in phase, and with adjacent loops 180° out of phase.

However, just as a standing wave on a string could be considered as two traveling waves moving in opposite directions, the eigenfunctions in the rectangular cavity can be decomposed into traveling plane waves. If the solutions (9.49) are represented in complex exponential form and expanded as a sum of products, it is seen that

$$\mathbf{p}_{lmn} = \tfrac{1}{8}\mathbf{A}_{lmn} \sum_{\pm} e^{j(\omega lmn \pm k_{xl}x \pm k_{ym}y \pm k_{zn}z)} \tag{9.52}$$

where the summation is taken over all the permutations of plus and minus signs. Each of these eight terms represents a plane wave traveling in the direction of its propagation vector $\vec{k}_{lmn}$ whose projections on the coordinate axes are $\pm k_{xl}$, $\pm k_{ym}$, and $\pm k_{zn}$. Thus, the standing wave solution can be viewed as a superposition of eight traveling waves (one into each quadrant) whose directions of propagation are fixed by the boundary conditions. It is the coherent combination of these waves, each traveling obliquely with respect to the coordinate axes, that results in the three-dimensional standing wave (9.49).

In a similar fashion, separation of variables can be used to solve for the standing waves in other cavities of simple geometry, such as cylindrical and spherical cavities. In these cases the eigenfunctions may involve Bessel and Legendre functions. Treatment of these problems is straightforward but outside the scope of this book.[4]

9.8 THE WAVEGUIDE OF CONSTANT CROSS SECTION. Consider a waveguide with rectangular cross section, as shown in Fig. 9.7. Assume the side walls to be rigid and the boundary at $z = 0$ to be a source of acoustic energy. The absence of another boundary on the z axis allows energy to propagate down the waveguide. This suggests a wave pattern consis-

---

[4] Morse and Ingard, *Theoretical Acoustics*, McGraw-Hill (1968).

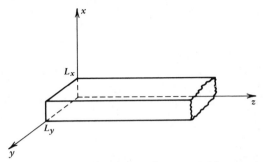

Fig. 9.7. *The rectangular waveguide.*

ting of standing waves in the transverse directions (x and y) and a traveling wave in the z direction.

Since the cross section is rectangular and the boundaries are rigid, it can be seen intuitively and verified mathematically that acceptable eigenfunctions are

$$\mathbf{p}_{lm} = \mathbf{A}_{lm} \cos k_{xl} x \cos k_{ym} y \; e^{j(\omega t - k_z z)} \tag{9.53}$$

where substitution into the wave equation shows that

$$(\omega/c)^2 = k^2 = k_{xl}^2 + k_{ym}^2 + k_z^2 \tag{9.54a}$$

The allowed values of k_{xl} and k_{ym} are found from the rigid boundary conditions to be

$$k_{xl} = l\pi/L_x \qquad l = 0, 1, 2, \cdots$$
$$k_{ym} = m\pi/L_y \qquad m = 0, 1, 2, \cdots \tag{9.54b}$$

and k_z is therefore given by

$$k_z = \sqrt{(\omega/c)^2 - k_{xl}^2 - k_{ym}^2} \tag{9.55}$$

Since ω can now have any value, $\mathbf{p}_{lm}$ is a solution for *all* values of ω, in contrast with the cavity for which the allowed frequencies were quantized. Thus $\check{k}$ is *not* fixed, and neither is k_z. It is convenient to define the *transverse component* k_{lm} of the propagation vector $\check{k}$. For this rectangular cross section

$$k_{lm} = \sqrt{k_{xl}^2 + k_{ym}^2} \tag{9.56a}$$

so that (9.55) can be written more succinctly as

$$k_z = \sqrt{(\omega/c)^2 - k_{lm}^2} \tag{9.56b}$$

When $\omega/c > k_{lm}$, then k_z is real. The wave advances in the $+z$ direction and the eigenfunction is called a *propagating mode*. The limiting value of ω/c for which k_{lm} remains real is given by $\omega/c = k_{lm}$, and this defines the *cutoff frequency*

$$\boxed{\omega_{lm} = ck_{lm}} \tag{9.57}$$

for the (l, m) mode. If the input frequency is lowered below cutoff, the argument of the square root in (9.56b) becomes negative and k_z must be pure imaginary

$$k_z = \pm j\sqrt{k_{lm}^2 - (\omega/c)^2}$$

The minus sign must be taken on physical grounds so that $\mathbf{p} \to 0$ as $z \to \infty$, and the eigenfunctions have the form

$$\mathbf{p}_{lm} = \mathbf{A}_{lm} \cos k_{xl} x \cos k_{ym} y \exp\left[-\sqrt{k_{lm}^2 - (\omega/c)^2} z \right] e^{j\omega t} \tag{9.58}$$

which represents a standing wave that decays exponentially with z. In this form, the eigenfunction is termed an *evanescent mode*: no energy is propagated down the waveguide. If the waveguide is excited with a frequency just below the cutoff frequency of some particular mode, then this and higher modes are evanescent and not important at appreciable distances from the source. All modes having cutoff frequencies below the driving frequency may propagate energy and may be seen at large distances.

In a rigid-walled waveguide, only plane waves propagate if the frequency of the sound is sufficiently low. For a waveguide of rectangular cross section of greatest dimension L, this frequency is easily shown to be $f_c = \frac{1}{2}c/L$.

The *phase speed* of a mode is seen from (9.53) to be

$$c_p = \frac{\omega}{k_z} = \frac{c}{\sqrt{1 - (k_{lm}/k)^2}} = \frac{c}{\sqrt{1 - (\omega_{lm}/\omega)^2}} \tag{9.59}$$

and is clearly not equal to c. A physical understanding of this important result can be obtained if we write the cosines in (9.53) in complex exponential form. The solution then consists of the sum

$$\mathbf{p}_{lm} = \frac{1}{4}\mathbf{A}_{lm} \sum_{\pm} e^{j(\omega t \pm k_{xl}x \pm k_{ym}y - k_z z)}$$

where all permutations of the plus and minus signs are necessary. (Recall that the absence of a reflecting boundary at large z requires that only the minus sign appear before k_z.) It is easy to see that the propagation vector $\check{k}$ for each of the four traveling waves makes an angle with the z axis given by

$$\cos \theta = k_z/k \tag{9.60}$$

which can be rewritten

$$\cos \theta = \sqrt{1 - (\omega_{lm}/\omega)^2} \tag{9.61a}$$

Then, the phase speed (9.59) is

$$c_p = c/\cos \theta \tag{9.61b}$$

Figure 9.8 gives the surfaces of constant phase for the two component waves that represent the $(0, 1)$ mode of a rigid-walled waveguide. The waves exactly cancel each other for $y = L_y/2$, so that there is a nodal plane midway between the walls. At the upper and lower walls the waves are always in phase so that the pressure amplitude is maximized at these rigid boundaries. Notice that the *apparent wavelength* λ_z measured in the z direction is

$$\lambda_z = \lambda/\cos \theta$$

The lowest mode for a rigid-walled waveguide is the $(0, 0)$ mode. For this case, $k_z = k$ and the four component waves collapse into a single plane wave that travels down the axis of the waveguide with phase speed c.

For all other modes, the propagation vectors of the component waves can be at angles to the waveguide axis, one pointing into each of the four forward quadrants. From (9.59), at frequencies far above the cutoff of the (l, m) mode, we have $\omega \gg \omega_{lm}$ so that θ tends to zero and the waves are traveling almost straight down the waveguide. As the input frequency is decreased toward that of cutoff ω_{lm}, the angle θ increases so that the component waves travel

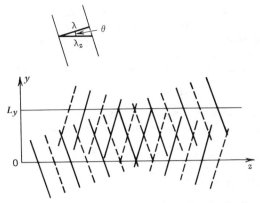

Fig. 9.8. *Component plane waves for the* (0, 1)
mode in a rigid-walled, rectangular waveguide.

in increasingly transverse directions; indeed, when the frequency is lowered to a value such that $\omega = \omega_{lm}$, the component waves are traveling transversely to the axis of the waveguide. Now, if we imagine that each component wave carries energy down the waveguide by a process of continual reflection from the walls (much like a bullet ricocheting down a hard-walled corridor), and remember that the energy of a wave is propagated with speed c in the direction $\vec{k}$, then we see that the speed c_g with which energy moves in the z direction is given by the component of the plane wave velocity $\tilde{c}$ along the waveguide axis,

$$c_g = c \cos \theta = c\sqrt{1 - (\omega_{lm}/\omega)^2} \tag{9.62}$$

The speed c_g is called the *group speed*.

For a given driving frequency ω, each normal mode for which $\omega_{lm} < \omega$ has its own individual values of c_p and c_g. The behavior of the group and phase speeds as functions of frequency for three modes [including the (0, 0) mode] in a rigid-walled waveguide are shown in Fig. 9.9. [The (0, 0) mode is the plane wave solution so that its group and phase speeds are identical and equal to c for all frequencies.]

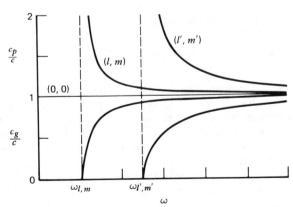

Fig. 9.9. *Group and phase speeds for three normal
modes in a rigid-walled waveguide.*

Concerning the boundary condition at $z = 0$, it is clear that we must match the allowed mode solutions to the acoustic behavior of the active surface at this end of the waveguide. If we know the pressure or velocity distribution of the source, then these can be related to the behavior of $\mathbf{p}(x, y, 0, t)$ for the pressure or $\hat{z} \cdot \hat{\mathbf{u}}(x, y, 0, t)$ for the velocity. For example, if the pressure distribution of the source is known, then the boundary condition is

$$\mathbf{p}(x, y, 0, t) = \mathbf{P}(x, y)e^{j\omega t} \tag{9.63}$$

But $\mathbf{p}$ can also be written as a superposition of the normal modes of the waveguide. For the case of a waveguide with rigid walls and rectangular cross section, we have

$$\mathbf{p}(x, y, z, t) = \sum_{l, m} \mathbf{A}_{lm} \cos k_{xl} x \cos k_{ym} y \; e^{j(\omega t - k_z z)} \tag{9.64a}$$

Evaluation at $z = 0$ and use of (9.63) gives

$$\mathbf{P}(x, y) = \sum_{l, m} \mathbf{A}_{lm} \cos k_{xl} x \cos k_{ym} y \tag{9.64b}$$

from which we can determine the required values of $\mathbf{A}_{lm}$ by inverting (9.64b) first for the x dimension and then for the y direction. This is merely a two-stage application of the analogous inversions done in Sects. 1.14 and 2.10. Further details will not be developed here.

The existence of three speeds c_p, c_g, and c in the description of each normal mode of a waveguide serves to elucidate the propagation of transient signals. Consider the case of a well-defined pulse in a single mode that is allowed to propagate down the waveguide. Recalling the elements of Fourier superposition set forth in Sect. 1.15, we can write the dependence of the pulse on distance and time in the form of a weighted superposition of monofrequency components. The spectral density $\mathbf{g}(\omega)$ can be found from the behavior of the source at $z = 0$. If, instead of $\exp(j\omega t)$, the source generates a known signature $f(t)$, then

$$f(t) = \int_{-\infty}^{\infty} \mathbf{g}(\omega)e^{j\omega t} \, d\omega \tag{9.65a}$$

and inversion yields

$$\mathbf{g}(\omega) = \frac{1}{2\pi} \int_{-\infty}^{\infty} f(t)e^{-j\omega t} \, dt \tag{9.65b}$$

The extension of (9.64a) then becomes

$$\mathbf{p} = \sum_{l, m} \mathbf{A}_{lm} \cos k_{xl} x \cos k_{ym} y \int_{-\infty}^{\infty} \mathbf{g}(\omega)e^{j[\omega t - k_z(\omega)z]} \, d\omega \tag{9.66}$$

Because of the distance-dependent phase in the integrand of (9.66), it is clear that the pulse will evolve in shape as it travels along the z axis. If the pulse is initially well-defined, then $\mathbf{g}(\omega)$ is a smoothly varying function of frequency, strong over a wide range of frequencies. Under this condition, the portion of the integrand that contributes the most to the pulse is that for which the exponent is nearly constant as a function of frequency—for other values of frequency the phase of the integrand is rapidly varying so that adjacent cycles of the integrand tend to cancel. Thus, the major portion of the pulse will begin near the time for which the phase is stationary, and for each mode this time is found from

$$\frac{d}{d\omega} [\omega t - k_z(\omega)z] = 0$$

This results in

$$t - \frac{dk_z}{d\omega} z = 0$$

and it is clear that the most significant portion of the pulse travels down the waveguide with a speed c_g given by

$$c_g = \frac{d\omega}{dk_z} \qquad (9.67)$$

[With a little manipulation, it can be seen that this is identical with (9.62) for the waveguide of rectangular cross section; (9.67) is more general, however, and can be applied to any lossless dispersive medium.] The phase speed of each monofrequency component of the signal is, of course, still given by

$$c_p = \frac{\omega}{k_z} \qquad (9.68)$$

Consider qualitatively what happens when $f(t)$ is a turned-on sine wave of frequency ω as illustrated in Fig. 9.10a. Assume that the source is specially designed to excite only one mode. The Fourier spectrum of this transient will contain a continuous range of frequencies, each of which will travel down the waveguide, interfering with the others to produce the observed waveform. The waveform at a distance z down the waveguide is sketched in Fig. 9.10b. The various aspects of this waveform that should be noted are the following. (1) The first signal arrives at $t = z/c$. Since the basic mechanism for sound propagation (the transfer of information from one molecule to another at the thermodynamic speed c) is not changed by the presence of boundaries, the first information that the source has been turned on must arrive at z by the shortest path (directly down the z-axis) and must therefore be of the highest frequency components. (2) At very long elapsed time, the waveguide attains steady-state conditions corresponding to monofrequency excitation at the *carrier frequency* ω. (3) For intermediate times, the waveform is characterized by a signal with increasing amplitude and period. Detailed analysis of the waveform reveals that the time for the envelope to attain one-half of its steady-state value is given by $t = z/c_g$ where c_g is the group speed of the carrier frequency.

It is straightforward to derive the behavior of a rigid-walled, *circular* waveguide of radius $r = a$. Separation of variables results in

$$\mathbf{p}_{ml} = A_{ml} J_m(k_{ml} r)\cos(m\theta)\ e^{j(\omega t - k_z z)} \qquad (9.69)$$

where r, θ, and z are the usual cylindrical coordinates, J_m is the mth order Bessel function, and

$$k_z^2 = (\omega/c)^2 - k_{ml}^2 \qquad (9.70a)$$

where k_{ml} is determined by the boundary conditions. Since $\hat{r} \cdot \nabla p = 0$ at $r = a$,

$$k_{ml} = j'_{ml}/a \qquad (9.70b)$$

where j'_{ml} are the zeros of $dJ_m(z)/dz$. These values are tabulated in Appendix A5.

Once the values of k_{ml} are found, all the salient results developed for rectangular waveguides can be applied simply by substituting the values of k_{ml} for a circular waveguide. For example, the (0, 0) mode is a plane wave that propagates with $c_p = c$ for all $\omega > 0$. The

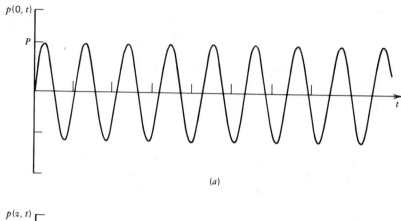

(a)

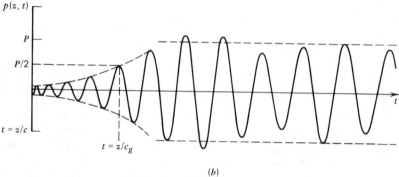

(b)

Fig. 9.10. *Effects of dispersion on the propagation of a pulse in a waveguide. (a) A gated sine wave at the source, where $p = 0$ for $t < 0$, $p = P \sin \omega t$ for $t > 0$. (b) The pressure at a distance z down the axis of the waveguide, where $p = 0$ for $t < z/c$.*

nonplanar mode with the lowest cutoff frequency is the (1, 1) mode (the first "sloshing" mode) with cutoff frequency $\omega_{1,1} = 1.84c/a$ or

$$f_{1,1} \approx 101/a \qquad \text{for air} \tag{9.71}$$

It is of great practical importance that for frequencies below $f_{1,1}$ only plane waves can propagate in a rigid-walled, circular waveguide.

PROBLEMS

9.1.   Find the shortest length pipe for which the input acoustic resistance equals the input acoustic reactance at a frequency of 500 Hz when the terminating load impedance is three times the acoustic impedance of the air in the pipe.

9.2.   Show that for $kL \ll 1$ the input impedance of the forced, rigid tube is springlike. Explain the physical significance of this stiffness for a perfect gas.

9.3.   A condenser-microphone diaphragm is stretched across one end of a pipe of 0.02-m radius and 0.01-m length, open at the other end. Compute and plot the ratio between the pressure at the diaphragm, considered rigid, and that at the open end as a function of frequency from 100 to 2000 Hz.

9.4. The air in a pipe of 1.0-cm length and 0.05-m radius is being driven at one end by a piston of negligible mass. The far end of the pipe is open and has a large flange attached to it. (a) What is the fundamental resonance frequency of the system? (b) If the displacement amplitude of the piston is 0.01 m when driven at the above frequency, what acoustic power is being transmitted by the plane waves moving toward the open end of the pipe? (c) What acoustic power is being transmitted out through the open end of the pipe?

9.5. The air in a pipe of 0.05-m radius and 1.0-m length is being driven by a piston of 0.015-kg mass and 0.05-m radius inserted in one end of the pipe. The other end of the pipe is terminated in an infinite baffle. (a) At a frequency of 150 Hz, what is the mechanical impedance of the piston, including the loading effect of the air in the pipe? (b) What is the amplitude of the force required to drive the piston with a displacement amplitude of 0.005 m at this frequency? (c) How much acoustic power in watts will be radiated out through the open end of the pipe?

9.6. A 500-Hz sound wave in air having a pressure level of 60 dB re 0.0002 μbar is normally incident on a boundary between the air and a second medium having a characteristic impedance of 830 Pa · s/m. (a) What is the effective (rms) pressure amplitude of the reflected waves? (b) Of the transmitted waves? (c) At what distance from the boundary is the pressure amplitude in the pattern of standing waves equal to that of the incident wave?

9.7. Plane waves in water are normally incident on a flat concrete wall that may be considered to absorb all of the sound energy transmitted into it. The speed of sound in the water is 1480 m/s and the frequency of the waves is 1480 Hz. The pattern of standing waves has a peak pressure amplitude of 15 Pa at the wall and a pressure amplitude of 5 Pa at the nearest pressure node at a distance of 0.25 m from the wall. (a) What is the ratio of the intensity of the reflected wave to that of the incident wave? (b) What is the specific acoustic impedance of the wall?

9.8. Plane waves in water of 1 kHz are normally incident on a concrete wall. (a) What is the resulting standing wave ratio? (b) To what difference in pressure levels, in decibels, is this equivalent? (c) Where are the first three nodes located?

9.9. Given that 200-Hz plane waves in air are normally incident on an acoustic tile panel having a normal specific acoustic impedance of $1000 - j2000$ Pa · s/m. (a) What is the standing wave ratio in the resulting pattern of standing waves? (b) Where are the first two nodes located?

9.10. (a) Calculate the value of $\sqrt{2\eta/\rho_0\,\omega}$ for acoustic waves of 200 Hz in air. (b) Including the effects of both viscous and heat-conduction losses at the walls of the pipe, calculate the phase speed of 200-Hz plane waves through air in a pipe of 1-cm radius. (c) What is the corresponding absorption coefficient in nepers per meter? (d) What is the attenuation in intensity level produced in a 2-m length of this pipe?

9.11. Calculate and compare the absorption coefficient in decibels per meter in dry air for plane waves in a pipe of 1.0-cm radius with that for plane waves in the unbounded medium at frequencies of 1, 10, and 100 kHz.

9.12. Calculate a value for the absorption in decibels per meter at a frequency of 20 kHz, (a) in freshwater contained in a pipe of 1.0-cm radius, (b) in a large body of freshwater at 15°C, (c) in a large body of seawater at 15°C.

9.13. Plane waves of sound are being propagated in an air-filled pipe of 0.1-m radius by means of a loudspeaker fitted into one end of the pipe. The far end of the pipe is closed by means of a rigid cap. The frequency radiated by the loudspeaker is 6 kHz. The measured standing wave ratio of pressure at one position in the pipe is 8. At a second position 0.5 m further down the pipe, the measured standing wave ratio is 9. (a) Derive an equation, involving these ratios and the distance between them, that may be used in calculating the absorption constant for waves being propagated within the pipe. Simplify your equation for $\alpha \ll 1$. (b) What is the numerical value of α corresponding to the above data? (c) Calculate the absorption constant to be anticipated if there were only viscous and heat-conduction losses at the walls of the pipe. (d) Assuming the remainder of the measured absorption constant to result from the presence of water vapor, use Fig. 7.4 to estimate the relative humidity of the air in the pipe.

9.14. A piston of mass m and radius a is mounted in one end of a pipe of length L and radius a, where $a \ll L$. The other end of the pipe opens into an infinite plane flange. (a) Derive an approximate equation giving the acoustic power radiated out through the open end of the pipe when the piston is driven by a force $F \cos \omega t$ at such high frequencies that $ka \gg 1$. (b) Also derive an approximate equation that is valid for such low frequencies that both $ka \ll 1$ and $kL \ll 1$.

9.15. Verify that the frequencies of resonance and those near antiresonance of the forced-open tube with damping correspond to the frequencies of maximum and minimum power dissipation, respectively. Assume $ka \ll 1$.

9.16. Find the lowest normal mode frequency of a fluid-filled cubic cavity (L on a side) which has five rigid sides and one pressure release side. Sketch the pressure distribution associated with this mode.

9.17. Calculate the lowest 10 eigenfrequencies for a rigid-walled rectangular room of dimensions 3.12 m × 4.69 m × 6.24 m. Assume $c = 345$ m/s.

9.18. Consider a fluid in an infinitely long rigid-walled pipe of rectangular cross section (height H and width W). For the lowest mode, (a) sketch the pressure distribution across the pipe. (b) Find the cutoff frequency and the phase speed.

9.19. An irrigation canal measures 30-m wide and 10-m deep. If it is completely filled with water, calculate the cutoff frequency of the lowest mode of propagation. Assume the concrete bed is perfectly rigid.

9.20. Show from (9.67) that $c_g/c = \sqrt{1 - (\omega_{lm}/\omega)^2}$ for a rigid-walled waveguide.

CHAPTER 10

RESONATORS, DUCTS, AND FILTERS

10.1 THE LONG-WAVELENGTH LIMIT. Analysis of many acoustic devices becomes simple if the wavelength in the fluid is much longer than the dimensions of the device. For example, in the last chapter it was shown that for a rigid-walled waveguide there is a frequency below which the only propagating waveform that can exist is a plane wave traveling straight down the waveguide with phase speed $c_p = c$. Thus, propagation in such a waveguide is extremely simple if the wavelength is sufficiently long compared to the cross-sectional dimensions of the waveguide.

If the wavelength is greater than *all* dimensions, further simplifications are possible. In this limit, each acoustic variable is time varying but almost independent of distance over the dimensions of the device. Thus, spatial coordinates can be ignored in the equation of motion, and such a device behaves as if it were a harmonic oscillator with one degree of freedom. In such cases, all the results of Chapter 1 can be applied. Acoustical devices in this long-wavelength limit are often termed *lumped acoustic elements*.

10.2 THE HELMHOLTZ RESONATOR. A simple example of a lumped acoustic element is the *Helmholtz resonator* shown in Fig. 10.1. This device consists of a rigid-walled cavity of volume V with a neck of area S and length L. If $\lambda \gg L$, the fluid in the neck moves as a unit to provide the *mass* element. If $\lambda \gg V^{1/3}$, the acoustic pressure within the cavity provides the *stiffness* element. If $\lambda \gg S^{1/2}$, the opening radiates sound as a simple source does, thus providing the *resistance* element. (Additional resistance is provided by the viscous losses in the neck. However, for necks of 1-cm diameter and greater, viscous losses are usually less than those associated with radiation and can be safely ignored in this development.) Let us analyze the Helmholtz resonator by calculating the values of the analogous mechanical system.

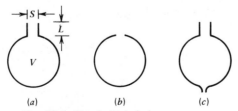

Fig. 10.1. *Simple Helmholtz resonators.*

The fluid in the neck has a total effective mass

$$m = \rho_0 SL' \tag{10.1}$$

where L', the effective length of the neck, is longer than the physical length because of its radiation-mass loading. It was seen in Sect. 9.2 that at low frequencies a circular opening of radius a is loaded with a radiation mass equal to that of the fluid contained in a cylinder of area πa^2, and length $0.85a$ if terminated in a wide flange and $0.6a$ if unflanged. If the mass loading at the inner end of the neck is assumed to be equivalent to a flanged termination, then

$$L' = L + 2(0.85a) = L + 1.7a \qquad \text{(outer end flanged)} \tag{10.2a}$$

$$L' = L + (0.85 + 0.6)a = L + 1.5a \qquad \text{(outer end unflanged)} \tag{10.2b}$$

(An opening consisting of a hole in a thin wall of a resonator will have an effective length equal to $1.7a$.)

To determine the stiffness of the system, consider the neck to be fitted with an air-tight piston. When this piston is pushed in a distance ξ, the volume of the cavity is changed by $\Delta V = -S\xi$, resulting in a condensation $\Delta\rho/\rho = -\Delta V/V = S\xi/V$. The pressure increase (in the acoustic approximation) is

$$p = \rho_0 c^2(\Delta\rho/\rho) \qquad \text{or} \qquad p = \frac{\rho_0 c^2 S}{V}\xi \tag{10.3a}$$

The force $f = pS$ required to maintain the displacement is $(\rho_0 c^2 S^2/V)\xi$ and the effective stiffness s is

$$s = \rho_0 c^2 \frac{S^2}{V} \tag{10.3b}$$

If it is assumed that the moving fluid in the neck radiates sound into the surrounding medium in the same manner as an open-ended pipe, then for $\lambda \gg a$ the radiation resistance is as given in Sect. 9.2,

$$R_r = \rho_0 c \frac{k^2 S^2}{2\pi} \qquad \text{(flanged)} \tag{10.4a}$$

or

$$R_r = \rho_0 c \frac{k^2 S^2}{4\pi} \qquad \text{(unflanged)} \tag{10.4b}$$

The instantaneous complex driving force produced by a sound wave of amplitude P impinging on the resonator opening is

$$\mathbf{f} = SPe^{j\omega t} \tag{10.5}$$

The resulting differential equation for the inward displacement ξ of the fluid in the neck is

$$m\frac{d^2\xi}{dt^2} + R_r\frac{d\xi}{dt} + s\xi = SPe^{j\omega t} \tag{10.6}$$

Since this equation is analogous to that of a driven oscillator, its solution may be obtained by analogy. In particular, the mechanical impedance of a Helmholtz resonator is

$$\mathbf{Z}_m = R_r + j(\omega m - s/\omega) \tag{10.7}$$

and resonance occurs when the reactance goes to zero:

$$\omega_0 = c\sqrt{\frac{S}{L'V}} \tag{10.8}$$

In deriving this equation, no assumption has been made restricting the shape of the resonator. For a given opening, it is the volume of the cavity, and not its shape, that is important. In fact, so long as all dimensions of the cavity are considerably less than a wavelength and the opening is not too large, the resonant frequencies of resonators having the same ratio $S/(L'V)$ but having very different shapes are found to be identical. [Helmholtz resonators have additional resonance frequencies higher than that given by (10.8). The origin of these higher frequencies is quite different from that of the fundamental, for they result from standing waves in the cavity, rather than from the oscillatory motion of the mass of fluid in the neck. The overtone frequencies therefore depend on the shape of the cavity and are not harmonically related to the fundamental. In general, the frequency of the first overtone is several times as great as that of the fundamental.]

The sharpness of resonance of a driven Helmholtz resonator, measured by its quality factor Q (see Sect. 1.10), is given by $Q = \omega_0 m/R_r$, or

$$Q = 2\pi\sqrt{V(L/S)^3} \tag{10.9}$$

This expression is derived on the assumption that there are no losses except those resulting from acoustical radiation and that this radiation is similar to that for a *flanged* termination.

In his original investigation of the frequencies present in complex musical tones, Helmholtz used a graduated series of resonators of the type pictured in Fig. 10.1c, with their individual volumes and areas of openings chosen to cover a wide range of frequencies. Whenever an incident sound wave has a frequency component corresponding to the resonance frequency of a particular resonator, greatly amplified sound pressure at this frequency will be produced within the cavity of the resonator. Such resonance may be audibly detected by connecting the small nipple opposite the neck of the resonator to the ear, either directly or through a short rubber tube.

Let us define the *pressure amplification* of the resonator as the ratio of the acoustic pressure amplitude P_c within the cavity to the external driving pressure amplitude P of the incident sound wave. The pressure amplitude P_c is obtained from (10.3a). Then, from the mechanical impedance $\mathbf{Z}_m = F/(d\xi/dt)$, we have at resonance $|\xi| = PS/(\omega_0 R_r)$ which when combined with (10.3b) yields

$$P_c/P = Q \quad \text{(at resonance)} \tag{10.10}$$

Thus, at resonance, the Helmholtz resonator acts like an amplifier of gain Q.

When a loudspeaker is mounted in a closed cabinet, the combined system may be treated as a Helmholtz resonator, in which both the reactance of the air and the mass of the speaker cone contribute to the effective mass of the system. Similarly, both the stiffness of the cavity and that of the speaker cone contribute to the effective stiffness. The effective resistance is the sum of that due to the radiation of acoustic energy and that due to the internal mechanical resistance of the speaker cone.

10.3   THE RESONANT BUBBLE.   A second example of a lumped acoustic system is an air bubble in water. As discussed in Chapter 7, such bubbles can greatly affect the attenuation of sound in the sea.

If the equilibrium radius a of the bubble is much less than the wavelength of the sound, the bubble can be driven into radial oscillations. Let us specify the *outward* displacement of the surface of the bubble by ξ. If the bubble is subjected to a static compression, the overpressure within the bubble must be given by $p = -\rho_0' c'^2 \Delta V/V$, where ρ_0' and c' are for the gaseous interior of the bubble. For a perfect gas, we have from (5.18) the relation $\rho_0' c'^2 = \gamma \mathscr{P}_0$ where $\mathscr{P}_0$ is the total hydrostatic pressure at the depth of the bubble. (We neglect the effect of surface tension which tends to increase the hydrostatic pressure within the bubble for very small bubbles; for bubbles of greatest interest, this effect can be ignored without introducing much error.) The condensation within the bubble is $\Delta V/V = -4\pi a^2 \xi/(\frac{4}{3}\pi a^3)$, and the compressive force on the surface of the bubble is $f = -pA = -4\pi a^2 p$. By combining these equations we obtain

$$f = -12\pi a\gamma \mathscr{P}_0 \xi$$

Thus, the bubble has a stiffness

$$s = 12\pi a\gamma \mathscr{P}_0 \tag{10.11}$$

When the bubble is oscillating, it must act like a very small pulsating sphere and radiate sound uniformly in all directions. It therefore has the radiation impedance for a pulsating sphere in the low frequency limit. As seen in Chapter 8, this is given by a radiation mass

$$m_r = 4\pi a^3 \rho_0 \tag{10.12}$$

and a radiation resistance

$$R_r = 4\pi a^2 \rho_0 c(ka)^2 \tag{10.13}$$

where ρ_0 is the density and c the speed of sound for water. (It is clear that the mass of the bubble is negligible compared to m_r.) Thus, the bubble acts like a harmonic oscillator with resonance angular frequency

$$\omega_0 = \frac{1}{a}\sqrt{\frac{3\gamma\mathscr{P}_0}{\rho_0}} \tag{10.14}$$

In addition to the losses given by the radiation resistance, there is absorption of energy by the bubble: because the bubble is surrounded by a fluid of relatively high thermal conductivity, there can be significant heat transfer between the outer layer of gas in the bubble and the surrounding water. The expansions and compressions of the gas are not entirely adiabatic. Analysis is quite involved,[1] but the result is that the bubble can be characterized by an

---

[1] Devin, *J. Acoust. Soc. Am.*, **31**, 1654 (1959).

additional (mechanical) resistance R_m which, at the resonance frequency ω_0, can be approximated as

$$\frac{R_m}{\omega_0 m_r} = 1.6 \times 10^{-4}\sqrt{\omega_0} \qquad \text{(resonance)} \qquad (10.15)$$

This expression is accurate to within about 10 percent for frequencies below 40 kHz. Losses from viscosity are negligible in this frequency range. Thus, the complete resistance associated with the bubble at resonance is $R = R_m + R_r$, and the total input mechanical impedance Z_m of the bubble is

$$Z_m = (R_m + R_r) + j(\omega m_r - s/\omega) \qquad (10.16)$$

The quality factor is now seen to be

$$Q = \frac{\omega_0 m_r}{R_m + R_r} = \frac{1}{ka + 1.6 \times 10^{-4}\sqrt{\omega_0}} \qquad (10.17)$$

For example, for a 0.065-cm radius air bubble just beneath the surface of the ocean, the resonance frequency calculated from (10.14) is 5 kHz. The wavelength in seawater for this frequency is about 30 cm, so that $ka = 0.0136$ and the assumption that the bubble is small is easily justified. From (10.17) we have $Q = 24$.

As noted, the bubble acts as a scatterer and an absorber of sound energy. Both of these processes remove energy from the incident sound beam. The scattering of energy out of the incident beam can be described by the *scattering cross section* σ_s and the absorption of energy by the *absorption cross section* σ_a. The combined effects of absorption and scattering are described by the *extinction cross section*

$$\sigma = \sigma_s + \sigma_a \qquad (10.18)$$

(It is this last parameter σ which was introduced in Chapter 7.)

Let us first consider the *total energy loss* from the beam of sound. The extinction cross section σ is that area transverse to the incident beam that intercepts sufficient power from the sound beam to just equal the total power loss by scattering from, and absorption within, the bubble. Thus, if Π is the absorbed and scattered power and I the incident intensity,

$$\sigma = \Pi/I \qquad (10.19)$$

The total power absorbed and scattered by the bubble is

$$\Pi = \tfrac{1}{2}U_0^2(R_m + R_r) \qquad (10.20)$$

where U_0 is the amplitude of the radial speed of the surface of the bubble. The compressive force applied to the bubble by the surrounding fluid is $F = 4\pi a^2 \mathbf{p}_i$, where $\mathbf{p}_i = P_i \exp(j\omega t)$ is the pressure applied to the bubble by the incident sound field. Thus, from $F/u(a) = Z$ we have

$$U_0 = \frac{4\pi a^2 P_i}{Z}$$

and the scattered and absorbed power is

$$\Pi = \tfrac{1}{2}(4\pi a^2 P_i)^2 \frac{R_m + R_r}{(R_m + R_r)^2 + (\omega m_r - s/\omega)^2}$$

The intensity is $P_i^2/(2\rho_0 c)$, so that the extinction cross section is

$$\sigma = 4\pi a^2 \frac{R_m + R_r}{Z^2} (4\pi a^2 \rho_0 c) \tag{10.21}$$

The scattering cross section σ_s can also be obtained relatively easily. The incident pressure $\mathbf{p}_i$, the radial speed of the surface of the bubble $\mathbf{u}(a)$, and the total mechanical impedance $\mathbf{Z}_m$ are related by

$$4\pi a^2 \mathbf{p}_i = \mathbf{Z}_m \mathbf{u}(a)$$

Because of the oscillatory motion of the bubble, there is a radiated pressure wave $\mathbf{p}_r$ that falls off with $1/r$. It is this wave that comprises the sound scattered by the bubble. From the definition of radiation impedance, we must have

$$4\pi a^2 \mathbf{p}_r = \mathbf{Z}_r \mathbf{u}(a)$$

By combining these two equations we obtain

$$\frac{\mathbf{p}_r(a)}{\mathbf{p}_i} = \frac{\mathbf{Z}_r}{\mathbf{Z}_m} = \frac{R_r + j\omega m_r}{(R_m + R_r) + j(\omega m_r - s/\omega)} \tag{10.22}$$

The power *radiated* by the bubble is

$$\Pi_r = 4\pi a^2 \frac{|\mathbf{p}_r(a)|^2}{2\rho_0 c}$$

and the incident intensity is $I = |\mathbf{p}_i|^2/(2\rho_0 c)$, so that the scattering cross section is $\sigma_s = \Pi_r/I$ or

$$\sigma_s = 4\pi a^2 \frac{R_r^2 + (\omega m_r)^2}{Z_m^2}$$

Since we are restricting attention to frequencies such that $ka \ll 1$, then $R_r \ll \omega m_r$ and the scattering cross section is well-approximated by

$$\sigma_s = 4\pi a^2 (\omega m_r/Z_m)^2 \tag{10.23}$$

Comparison of (10.21) and (10.23) reveals that

$$\sigma_s/\sigma = kaQ \tag{10.24}$$

where use has been made of (10.12) and (10.17).

At resonance, $\mathbf{Z}_m = R_m + R_r$, the scattering cross section is

$$\sigma_s = 4\pi a^2 Q^2 \qquad (\omega = \omega_0) \tag{10.25}$$

and the extinction cross section is

$$\sigma = 4\pi a^2 Q/(ka) \qquad (\omega = \omega_0) \tag{10.26}$$

Thus, at resonance the extinction cross section for the 0.065-cm radius bubble mentioned earlier is about 7000 times greater than its geometrical cross section πa^2. This demonstrates in a striking way the efficiency of a resonant bubble as an attenuator of an incident sound beam.

For frequencies well below resonance, $\mathbf{Z}_m$ is dominated by the stiffness and the scattering cross section becomes

$$\sigma_s = 4\pi a^2 (\omega/\omega_0)^4 \qquad (\omega \ll \omega_0)$$

This is the famous Rayleigh law of scattering which in optics predicts that the sky is blue; since blue light has higher frequency than red light, blue light is scattered more strongly than red and more of it reaches the earth's surface from high atmospheric scattering.

10.4 ACOUSTIC IMPEDANCE. Many other simple systems may be converted into analogous mechanical systems and solved. However, it is also possible to convert such systems into analogous electrical systems in which the motion of the fluid is equivalent to the behavior of current in an electric circuit having lumped elements of inductance, capacitance, and resistance. The electrical analogue of the pressure difference across an acoustic element is the voltage across the corresponding part of the electric circuit. One acoustic analogue of current at some point in the circuit is the *volume velocity* U of the fluid in the corresponding acoustic element. (Strictly speaking, since U is not a vector, it should not be termed a "velocity," but this is the accepted convention.)

In general, the *acoustic impedance* Z of a fluid acting on a surface of area S is the complex quotient of the acoustic pressure at the surface divided by the volume velocity at the surface

$$\mathbf{Z} = \frac{\mathbf{p}}{\mathbf{U}} \tag{10.27}$$

We have now encountered three kinds of impedances, an inexcusable redundancy if it were not for the fact that these various impedances are useful in different kinds of calculations.

1. The *specific acoustic impedance* z (pressure/particle speed) is a characteristic property of the medium and of the type of wave that is being propagated. It is useful in calculations involving the transmission of acoustic waves from one medium to another.

2. The *acoustic impedance* Z (pressure/volume velocity) is useful in discussing acoustic radiation from vibrating surfaces, and the transmission of this radiation through lumped acoustic elements or through pipes and horns. The acoustic impedance is related to the specific acoustic impedance at a surface by

$$\mathbf{Z} = \mathbf{z}/S \tag{10.28}$$

3. The *radiation impedance* Z_r (force/speed) is used in calculating the coupling between acoustic waves and a driving source or driven load. It is part of the *mechanical impedance* Z_m of a vibrating system associated with the radiation of sound. Radiation impedance is related to specific acoustic impedance at a surface by

$$\mathbf{Z}_r = S\mathbf{z} \tag{10.29}$$

(a) Lumped Acoustic Impedance. When concentrated rather than distributed impedances are considered, the impedance of a portion of the acoustic system is defined as the complex ratio of the pressure difference $\mathbf{p}$ that is driving that portion to the resultant volume velocity $\mathbf{U}$. The acoustic impedance units are $Pa \cdot s/m^3$, often termed an *acoustic ohm*.

One example of lumped acoustic impedance, which we developed in a different form earlier, is the Helmholtz resonator. If we wish to cast (10.6) into the form of acoustic impedance, divide by S and notice that $\mathbf{U} = (d\xi/dt)S$. The result is

$$\frac{\mathbf{p}}{\mathbf{U}} = \mathbf{Z} = R + j\left(\omega M - \frac{1}{\omega C}\right)$$

where

$$R = R_r/S^2 = \rho_0 ck^2/(2\pi) \qquad \text{(flanged)}$$

$$M = m/S^2 = \rho_0 L'/S$$

and

$$C = S^2/s = V/(\rho_0 c^2)$$

Electrically, this is a series RLC circuit where L is the electrical equivalent M. The analogue is presented in Fig. 10.2. Thus, the Helmholtz resonator can be represented by Fig. 10.2, where R_r, m, and s are given by (10.4), (10.1), and (10.3b).

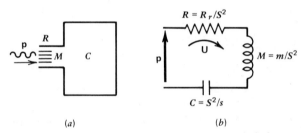

(a)   (b)

Fig. 10.2. *Schematic representation of a Helmholtz resonator.*

The three basic elements of mechanical, acoustic, and electric systems are represented schematically in Fig. 10.3. The *inertance M* of an acoustic system is represented by the fluid contained in a constriction that is short enough so that all particles may be assumed to move in phase when actuated by a sound pressure. The *compliance C* of the system is represented by an enclosed volume, with its associated stiffness. Although resistance may be contributed to an acoustic system by a number of different factors, irrespective of its origin, it is conventionally represented by narrow slits in a pipe.

(b) Distributed Acoustic Impedance. When one or more dimensions of an acoustic system are not small as compared to a wavelength, it may no longer be possible to treat the system as one having lumped constants, and it then must be considered as having distributed constants. The simplest system of this type is one in which low-frequency plane waves are propagated through a pipe. If the waves are propagated in the positive x direction, the ratio of acoustic pressure to particle

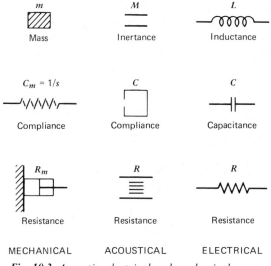

Fig. 10.3. *Acoustic, electrical and mechanical analogues.*

speed is given by the characteristic impedance $\rho_0 c$ of the medium, and hence the acoustic impedance at any cross section S in the pipe is

$$Z = \frac{\mathbf{p}}{\mathbf{U}} = \frac{1}{S}\frac{\mathbf{p}}{\mathbf{u}} = \frac{\rho_0 c}{S} \tag{10.30}$$

The propagation of plane waves in such a pipe is analogous to high-frequency currents along a transmission line. If such a transmission line possesses inductance L_e and capacitance per unit length C_e, it can be shown (see any standard textbook on electricity and magnetism) that the input electrical impedance is $\sqrt{L_e/C_e}$. We may consider the medium in the pipe to possess a *distributed inertance* M_1 per unit length and a *distributed compliance* C_1 per unit length. Each unit length of the pipe contains a mass $m_1 = \rho_0 S$ of fluid. The acoustic inertance M_1 per unit length is therefore

$$M_1 = m_1/S^2 = \rho_0/S \tag{10.31}$$

The mechanical stiffness of each unit length is found as follows. If the fluid is compressed by Δl, then $p = \rho_0 c^2 \, \Delta l/l$. The applied force to do this is pS, and the stiffness is therefore $s = pS/\Delta l$ or $s_1 = S\rho_0 c^2$ for a unit length. Then, since the mechanical compliance $C_m = 1/s$ is related to the acoustic compliance C by $C = S^2 C_m$, we have

$$C_1 = S/(\rho_0 c^2) \tag{10.32}$$

The acoustic impedance of a pipe is then, by analogy,

$$Z = \sqrt{M_1/C_1} = \rho_0 c/S$$

which is in agreement with (10.30). Furthermore, as will be shown in the following sections, the reflection and transmission of sound waves at a point where the acoustic impedance of a pipe changes are analogous to the behavior of current waves in a transmission line at a point where there is an impedance differing from the characteristic impedance of the line.

10.5 WAVES IN A PIPE. Assume that at some point $x = 0$ along a pipe the acoustic impedance changes from $\rho_0 c/S$ to $\mathbf{Z}_0$. If a wave traveling in the positive x direction and represented by

$$\mathbf{p}_i = \mathbf{A}e^{j(\omega t - kx)} \tag{10.33a}$$

is incident at this point, a reflected wave

$$\mathbf{p}_r = \mathbf{B}e^{j(\omega t + kx)} \tag{10.33b}$$

traveling in the negative x direction will be produced. Given the impedance $\mathbf{Z}_0$ observed at $x = 0$, we can solve for the power reflection and transmission coefficients. Since the acoustic impedance $\mathbf{Z}$ at any point in the tube is given by

$$\mathbf{Z} = \frac{\mathbf{p}_i + \mathbf{p}_r}{\mathbf{U}_i + \mathbf{U}_r} = \frac{\rho_0 c}{S} \frac{\mathbf{A}e^{-jkx} + \mathbf{B}e^{jkx}}{\mathbf{A}e^{-jkx} - \mathbf{B}e^{jkx}} \tag{10.34}$$

at $x = 0$ we must have

$$\mathbf{Z}_0 = \frac{\rho_0 c}{S} \frac{\mathbf{A} + \mathbf{B}}{\mathbf{A} - \mathbf{B}} \tag{10.35}$$

By solving for the ratio $\mathbf{B}/\mathbf{A}$ we obtain

$$\frac{\mathbf{B}}{\mathbf{A}} = \frac{\mathbf{Z}_0 - \rho_0 c/S}{\mathbf{Z}_0 + \rho_0 c/S}$$

The sound *power reflection coefficient* $R_\pi = |\mathbf{B}/\mathbf{A}|^2$ is

$$R_\pi = \frac{(R_0 - \rho_0 c/S)^2 + X_0^2}{(R_0 + \rho_0 c/S)^2 + X_0^2} \tag{10.36a}$$

where $\mathbf{Z}_0$ has been replaced $R_0 + jX_0$. Correspondingly, the sound *power transmission coefficient* $T_\pi = 1 - R_\pi$ is

$$T_\pi = \frac{4R_0 \rho_0 c/S}{(R_0 + \rho_0 c/S)^2 + X_0^2} \tag{10.36b}$$

Note that the above equations are identical in form to those developed for normal reflection from a plane interface between two fluids (see Sect. 6.2) if $\mathbf{Z}_0 = \rho_2 c_2/S$, where ρ_2 and c_2 characterize the system to the right of $x = 0$.

For example, let us apply these equations to plane waves in a pipe of cross-sectional area S_1 as they enter a second pipe of area S_2, as shown in Fig. 10.4. The second pipe is either of infinite length or so terminated that no reflected wave is returned from its far end. When the wavelength is large compared to the diameters of the pipes, we may assume that the spatial extent (along the axis of the pipe) of the complicated flow that accompanies the adjustment of the wave from one cross-sectional area to the other is small compared to a wavelength so that the acoustic impedance seen by the wave incident on the junction is $Z_0 = \rho_0 c/S_2$, the acoustic impedance of plane waves in the second pipe. By substituting this expression into the general equation (10.36), we have

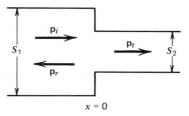

Fig. 10.4. Transmission and reflection of a plane wave at a junction between two pipes.

$$R_\pi = \frac{(S_1 - S_2)^2}{(S_1 + S_2)^2}$$

(10.37a)

and

$$T_\pi = \frac{4S_1 S_2}{(S_1 + S_2)^2}$$

(10.37b)

If the end of the pipe is closed, $S_2 = 0$. Then $Z_0 = \infty$ and $R_\pi = 1$, as to be expected from the discussion of rigidly terminated pipes in Chapter 9. If the end of the pipe is open, the impedance of the junction is not zero, but Z_{mL}/S^2 where Z_{mL} is given by (9.12).

Frequently, the termination impedance Z_0 is not as simple. As an illustration, consider a pipe that branches into two pipes each with arbitrary input impedance, as indicated in Fig. 10.5. If the junction is chosen as the origin, the pressures produced at $x = 0$ by the waves in the three pipes are

$$\mathbf{p}_i = \mathbf{A}e^{j\omega t} \qquad \mathbf{p}_r = \mathbf{B}e^{j\omega t}$$

and

$$\mathbf{p}_1 = \mathbf{Z}_1\mathbf{U}_1 e^{j\omega t} \qquad \mathbf{p}_2 = \mathbf{Z}_2 \mathbf{U}_2 e^{j\omega t}$$

where $\mathbf{A}$ and $\mathbf{B}$ are the amplitudes of the incident and reflected waves in the main pipe, and $\mathbf{Z}_1$, $\mathbf{Z}_2$ and $\mathbf{U}_1$, $\mathbf{U}_2$ are the input impedances and volume velocity complex amplitudes in the two branches. Assuming that the wavelength is long compared to

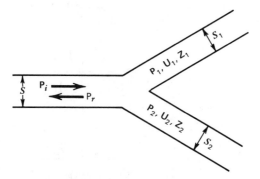

Fig. 10.5. Conditions at a branch.

the extent of the complicated flow pattern near the junction, we may apply the condition of continuity of pressure to the junction and obtain

$$\mathbf{p}_i + \mathbf{p}_r = \mathbf{p}_1 = \mathbf{p}_2 \tag{10.38}$$

The condition of continuity of volume velocity requires that

$$\mathbf{U}_i + \mathbf{U}_r = \mathbf{U}_1 + \mathbf{U}_2 \tag{10.39}$$

Dividing (10.39) by (10.38), we obtain

$$\frac{\mathbf{U}_i + \mathbf{U}_r}{\mathbf{p}_i + \mathbf{p}_r} = \frac{\mathbf{U}_1}{\mathbf{p}_1} + \frac{\mathbf{U}_2}{\mathbf{p}_2}$$

which may be written in the more revealing form

$$\frac{1}{\mathbf{Z}_0} = \frac{1}{\mathbf{Z}_1} + \frac{1}{\mathbf{Z}_2} \tag{10.40}$$

Thus, the combined *admittance* $1/\mathbf{Z}_0$ associated with the incident and reflected waves equals the sum of the admittances $1/\mathbf{Z}_1$ and $1/\mathbf{Z}_2$ of the two branches.

As a special case of branching, consider a side branch $(1 = b)$ of arbitrary acoustic impedance $\mathbf{Z}_b$ connected at $x = 0$ to an infinitely long pipe $(2 = t)$ of cross-sectional area S. In this case (10.40) may be solved for $\mathbf{B}/\mathbf{A}$

$$\frac{\mathbf{B}}{\mathbf{A}} = -\frac{\dfrac{\rho_0 c}{2S}}{\dfrac{\rho_0 c}{2S} + \mathbf{Z}_b} \tag{10.41}$$

The corresponding ratio for the pressure amplitude of the wave transmitted along the main pipe is obtained by combining (10.41) and (10.38),

$$\frac{\mathbf{A}_t}{\mathbf{A}} = \frac{\mathbf{Z}_b}{\dfrac{\rho_0 c}{2S} + \mathbf{Z}_b} \tag{10.42}$$

In deriving the expressions for the sound power reflection and transmission coefficients, it is convenient to replace the acoustic impedance of the branch by $\mathbf{Z}_b = R_b + jX_b$. The reflection coefficient is then

$$R_\pi = \left(\frac{B}{A}\right)^2 = \frac{\left(\dfrac{\rho_0 c}{2S}\right)^2}{\left(\dfrac{\rho_0 c}{2S} + R_b\right)^2 + X_b^2} \tag{10.43}$$

and the transmission coefficient along the pipe is

$$T_\pi = \left(\frac{A_t}{A}\right)^2 = \frac{R_b^2 + X_b^2}{\left(\dfrac{\rho_0 c}{2S} + R_b\right)^2 + X_b^2} \tag{10.44}$$

The ratio of the power transmitted into the branch to that in the incident wave is given by $T_{\pi b} = 1 - R_\pi - T_\pi$,

$$T_{\pi b} = \frac{\dfrac{\rho_0 c}{S} R_b}{\left(\dfrac{\rho_0 c}{2S} + R_b\right)^2 + X_b^2} \tag{10.45}$$

The power transmitted past the junction and along the main pipe is zero only when $T_\pi = 0$, which requires that both R_b and X_b be zero. A branch for which this is true does not absorb all the sound energy that reaches the junction but, on the contrary, absorbs no energy and reflects the incident energy back through the pipe toward the source. If R_b is greater than zero but not infinite, some acoustic energy is dissipated in the branch and some is transmitted beyond the junction, irrespective of the particular value of X_b. At the opposite extreme, as either R_b or X_b becomes very large compared to $\rho_0 c/S$, almost all the incident power is transmitted beyond the branch. In the limit $R_b = X_b = \infty$, which corresponds to no branch, the power transmission ratio equals unity. It should also be noted that, if $R_b = 0$, no acoustic power is dissipated in the branch.

10.6 ACOUSTIC FILTERS. The ability of a side branch to attenuate the sound energy transmitted in a pipe is the basis of a class of acoustic filters. Depending on the input impedance of the side branch, such systems can act as *low-pass, high-pass,* or *band-pass* filters. One example of each type will be considered.

(a) Low-Pass Filters. A simple low-pass acoustic filter may be constructed by inserting an *enlarged* section of pipe of cross-sectional area S_1 and length L in a pipe of cross section S, as shown in Fig. 10.6. At low frequencies corresponding to $kL \ll 1$, this filter may be looked on as a side branch of acoustic compliance $C = V/(\rho_0 c^2)$, where $V = S_1 L$ is the volume of the expansion chamber, in parallel with the continuing pipe. The acoustic impedance of such a branch is a pure reactance and, therefore, $R_b = 0$ and

$$X_b = -\frac{1}{\omega C} = -\frac{\rho_0 c^2}{\omega S_1 L} \tag{10.46}$$

A substitution of these values into (10.44) leads to a transmission coefficient of

$$T_\pi = \frac{1}{1 + \left(\dfrac{1}{2}\dfrac{S_1}{S} kL\right)^2} \tag{10.47}$$

This equation predicts that at low frequencies the sound power transmission is 100 percent and then gradually decreases to zero at high frequencies. Curve (1) in Fig. 10.6 is a graph of values of the transmission coefficient calculated from (10.47) for an expansion chamber of 0.05-m length and having a cross section four times that of the original pipe. At first inspection this type of acoustic filter appears to be

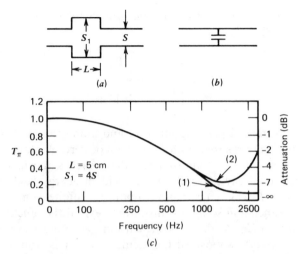

Fig. 10.6. (a) Simple low-pass acoustic filter. (b) Analogous electrical filter. (c) Power transmission curves for filter (a).

analogous to the low-pass electrical filter produced by shunting a capacitor across a transmission line, as shown in Fig. 10.6b. Actually (10.47) does not apply when $kL > 1$ and, therefore, the two filters are not truly analogous.

Let us now derive an equation for the above acoustic filter which is valid for $kL > 1$. By considering the various incident, reflected, and transmitted waves present in the three sections of pipes to be related to each other by conditions of continuity of pressure and volume velocity at the two junctions of the expanded pipe with the original pipe, it is possible to derive an equation for the transmission coefficient in a manner similar to that used in Sect. 6.3,

$$T_\pi = \frac{4}{4 \cos^2 kL + \left(\dfrac{S_1}{S} + \dfrac{S}{S_1}\right)^2 \sin^2 kL} \tag{10.48}$$

This equation is equivalent to (6.14a) when $r_1 = r_3$ and $r_2/r_1 = S_1/S$.

Curve (2) of Fig. 10.6 is a graph of values of the transmission coefficient calculated from (10.48) for the same filter section used in obtaining curve (1). At low frequencies for which $kL \ll 1$, results given by the two equations are essentially identical. The reader can show that this is to be expected by simplifying both when $kL \ll 1$. However, the more exact (10.48) indicates that the transmission coefficient reaches a minimum value of

$$T_\pi = \left(\frac{2SS_1}{S^2 + S_1^2}\right)^2 \tag{10.49}$$

for $kL = \pi/2$, when the length of the filter section is a quarter wavelength. Following this minimum, the sound power transmission gradually increases with increasing frequency until it again reaches 100 percent for $kL = \pi$. A further increase in frequency causes the transmission coefficient to run through a series of minima and maxima until finally, when ka' (where a' is the radius of the original pipe) is large in comparison with unity, it remains at 100 percent. This final attainment of 100 percent transmission is characteristic of all three of the acoustic filters in this chapter. The equations derived for power transmission are valid only when the wavelength is large compared with the radius of the original pipe, or with the dimensions of any filter section.

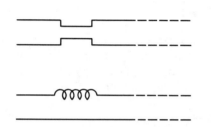

Fig. 10.7. *Construction in a pipe and its electrical analogue.*

Another type of low-pass filter is produced by a *constriction* in a pipe, as shown in Fig. 10.7. This system may be thought of as introducing an inertance in series with the pipe. However, as is the case with an enlarged section, the analogy between the electric and the acoustic cases is valid only over a limited range of frequencies. Equation (10.48) may be used for calculating the transmission coefficient for this type of filter, since its derivation is independent of whether S or S_1 is the larger.

There exist practical limitations other than those of frequency which must be taken into consideration in applying (10.48) to the design of low-pass acoustic filters. For example, it is not applicable when there is an extreme difference between the cross section of the filter and that of the original pipe. In spite of all these limitations, filters of this type are basic to the design of simple automobile mufflers, gun silencers, and sound-absorbing plenum chambers which are installed in ventilating systems.

(b) High-Pass Filters. Next consider the effect of a short length of pipe as a branch. If not only the radius a but also the length L of this pipe are small compared to a wavelength, the branch impedance of such an orifice is, from (10.2b) and (10.4b),

$$\mathbf{Z}_b = \frac{\rho_0 c k^2}{4\pi} + j\frac{\rho_0 L'\omega}{\pi a^2} \tag{10.50}$$

where $L' = L + 1.5a$. The first term results from the radiation of sound through the orifice into the external medium, and the second from the inertance of the gas in the orifice.

The ratio of the acoustic resistance of the branch to its acoustic reactance is

$$R_b/X_b = \tfrac{1}{4}ka^2/L'$$

Since the radius of the orifice has been assumed small compared to the wavelength, $ka \ll 1$, and therefore the acoustic resistance of such an orifice may be neglected compared with its acoustic reactance in calculating the transmission coefficient.

Thus, (10.44) becomes for this case

$$T_\pi = \frac{1}{1 + [\pi a^2/(2SL'k)]^2} \tag{10.51}$$

This transmission coefficient is very nearly zero at low frequencies and rises to nearly 100 percent at higher frequencies, as is indicated in Fig. 10.8. The transmission coefficient equals 50 percent when the denominator of (10.51) equals two, which yields

$$k = \pi a^3/(2SL') \tag{10.52}$$

The presence of a single orifice converts a pipe into a high-pass filter. As the radius of such an orifice is increased, the attenuation of the low frequencies is increased, as is the frequency corresponding to 50 percent power transmission. If the pipe has several orifices, located near enough to one another so that they may be

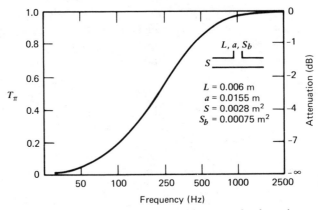

Fig. 10.8. *Attenuation produced by an orifice branch.*

considered as being at a single point (separated by a small fraction of a wavelength), the action of the group is that of their equivalent parallel impedance. If, on the other hand, the distance between orifices is an appreciable fraction of the wavelength of sound the system becomes analogous to an electrical filter network or to a transmission line across which are shunted a number of widely spaced impedances. The waves reflected from the various orifices are then out of phase with one another, and the power transmission ratio cannot be computed by (10.51) but must instead be determined by methods analogous to those of electrical filter theory. In general, the low-frequency attenuation of a number of suitably spaced orifices can be made much greater than that of a single orifice of equal total area.

The sound power transmission coefficient into a single branch is approximately

$$T_{\pi b} = \frac{2k^2S}{\pi[1 + (2SL'k/\pi a^2)^2]} \tag{10.53}$$

At the frequency corresponding to 50 percent power transmission this expression is reduced to $k^2 S/\pi$. For the orifice considered in the above example this ratio is only 1.5 percent at a frequency of 225 Hz. It is therefore quite apparent that the filtering action of an orifice does not result from the transmission of acoustic energy out of the pipe, but rather from the reflection of energy back toward the source.

The influence of an orifice may be used to explain qualitatively the action of a wind instrument, such as a flute or clarinet. When these instruments are sounded in their fundamental register the player opens all (or nearly all) the orifices lying beyond some particular distance from the mouthpiece. Since the diameters of the orifices are almost as large as the bore of the tube, this effectively shortens the length of the instrument, and the acoustic energy reflected back from the first open orifice sets up a pattern of standing waves between this orifice and the mouthpiece. In a flute, which acts essentially like an open pipe, the wavelength is approximately equal to twice the distance from the opening in the mouthpiece to the first open orifice. In a clarinet, however, the action of the vibrating reed causes the conditions at the mouthpiece to approximate those at the closed end of a tube, and hence the wavelength is nearly four times the distance from the reed to the first open orifice. In both instruments there are also a number of harmonic overtones, those of the clarinet being predominately the odd harmonics, as is to be expected from a closed pipe. When either instrument is played in a high register, the fingering is more complex; some orifices beyond the first open orifice are left closed and others opened, the purpose being to emphasize the desired standing wave pattern.

(c) **Band-Pass Filters.** If the side branch possesses both inertance and compliance, then it will act as a band-pass filter. One such side branch would be a long pipe rigidly capped on the far end. Another example is the Helmholtz resonator shown in Fig. 10.9. If we neglect viscosity losses, there is no net dissipation of energy from the pipe into the resonator: all energy absorbed by the resonator during some parts

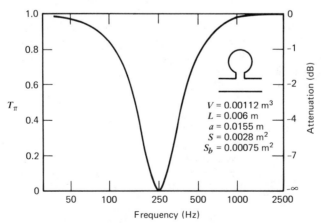

Fig. 10.9. *Attenuation produced by a Helmholtz resonator branch.*

of the acoustic cycle is returned to the pipe during other parts of the cycle so that $R_b = 0$. If the area of the opening into the resonator is $S_b = \pi a^2$, the length of its neck is L, and its volume is V, then the acoustic reactance of the branch is

$$X_b = \rho_0 \left(\frac{\omega L'}{S_b} - \frac{c^2}{\omega V} \right) \tag{10.54}$$

where $L' = L + 1.7a$.

A substitution of these values of R_b and X_b into (10.44) leads to a transmission coefficient of

$$T_\pi = \frac{1}{1 + \dfrac{c^2}{4S^2(\omega L'/S_b - c^2/\omega V)^2}} \tag{10.55}$$

This transmission coefficient becomes zero when

$$\omega = \omega_0 = c\sqrt{\frac{S_b}{L'V}}$$

the resonant frequency of the Helmholtz resonator. At this frequency large volume velocity amplitudes exist in the neck of the resonator, but all acoustic energy that is transmitted into the resonator cavity from the incident wave is returned to the main pipe with such a phase relationship as to be reflected back toward the source.

Calculated values of the transmission coefficient as a function of frequency are plotted in Fig. 10.9 for a representative resonator. Whenever the radius of the neck is rather large, as it is in this example, measured values of the power transmission ratio are found to be in excellent agreement with the values predicted by (10.55). For long narrow constrictions, however, computed and measured values are not in agreement unless (10.55) is modified to take into account the dissipative forces resulting from viscosity. One striking characteristic of the curve of Fig. 10.9 is that it indicates a material reduction in the transmission over a frequency range extending for more than an octave on either side of the resonant frequency.

10.7   FILTER NETWORKS.   By employing combinations of resonators, orifices, and enlargements or constrictions of the main pipe, it is possible to construct a wide variety of acoustic filter networks. The design of such networks is greatly facilitated by taking advantage of the analogy between acoustic and electrical filters. For example, the sharpness of cutoff of an electrical filter system can be increased by using a ladder-type network, such as that shown in Fig. 10.10 These networks are constructed by using a combination of reactances of one type of impedance Z_1 in series, with the line and reactances of another type of impedance Z_2 shunted across the line. It is demonstrated in standard treatises[2] on wave fil-

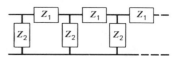

Fig. 10.10. *Ladder-type of network used as a filter.*

[2] Mason, *Electromechanical Transducers and Wave Filters*, pp. 28–31, Van Nostrand (1948).

ters that a nondissipative recurrent structure such as that of Fig. 10.10 markedly attenuates all frequencies except those for which the ratio Z_1/Z_2 satisfies the condition $0 > Z_1/Z_2 > -4$. Three simple ladder-type acoustic filters along with their electrical analogues are shown in Fig. 10.11. An application of the above condition on the ratio Z_1/Z_2 leads to cutoff frequencies of

$$f = \frac{1}{4\pi\sqrt{MC}}$$ (10.56)

for the high-pass filter of Fig. 10.11a and

$$f = \frac{1}{\pi\sqrt{MC}}$$ (10.57)

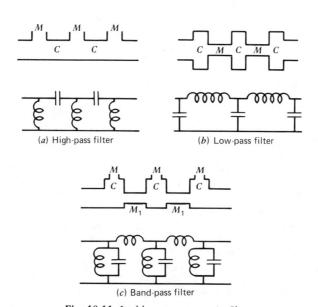

(a) High-pass filter   (b) Low-pass filter

(c) Band-pass filter

Fig. 10.11. *Ladder-type acoustic filters.*

for the low-pass filter of Fig. 10.11b. The behaviors of the acoustic filters of Fig. 10.11 will begin to deviate more and more from their analogous electrical filters as frequency increases. The primary cause is that at higher frequencies the dimensions of the system ultimately become comparable to a wavelength and the filter begins to have distributed rather than lumped-constant properties.

A knowledge of the filtering characteristics of acoustic elements is essential to the design and correction of any mechanoacoustic system such as forced air heating and ventilating systems, exhaust systems for jet engine test cells, and automobile mufflers. However, it is to be emphasized that acoustic filtering is not limited to systems using the nondissipative elements discussed above, since additional filtering action is possible by the use of frequency selective sound-absorbing materials. The capabilities and characteristics of such sound-absorbing materials will be looked into in subsequent chapters.

PROBLEMS

Unless otherwise specified, consider the fluid in the pipe to be air at $20°C$ and one atmosphere.

10.1. The sphere of a Helmholtz resonator has a diameter of 0.1 m. (a) What diameter hole should be drilled in the sphere if it is to resonate in air at 320 Hz? (b) What must be the pressure amplitude of an incident acoustic plane wave at 320 Hz if it is to produce an internal excess pressure of 20 μbar in the above resonator? (c) What will be the resonance frequency if a hole having twice the cross-sectional area of that of part (a) is drilled in the sphere? (d) What will be the resonant frequency if two independent holes having the diameter of part (a) are drilled in the sphere?

10.2. A rigid-walled back-enclosed loudspeaker cabinet has inside dimensions of 0.3 m $\times$ 0.5 m $\times$ 0.4 m. The front panel of the cabinet is 0.03-m thick and has a 0.2-m diameter hole cut in it for mounting a loudspeaker. (a) What is the fundamental frequency of the cabinet considered as a Helmholtz resonator? (b) If a direct-radiator loudspeaker having a cone of 0.2-m diameter and 0.01-kg mass, and a suspension system of 1000-N/m stiffness, is mounted in this cabinet, what is the resonance frequency of the cone? Assume that the effective mass of the system is the sum of that of the cone and of the moving air in the opening of the cabinet and that the effective stiffness is the sum of that of the cone and of the cabinet. (c) What would be the resonance frequency of the cone if it were not mounted in the cabinet and had no air loading? (d) If the cone is driven at the frequency of part (b) with an amplitude of 0.002 m, what is the acoustic power radiated? (e) Under these same conditions, what is the excess pressure amplitude inside the cabinet, and what is the amplitude of the associated force acting on one of the 0.4 $\times$ 0.5-m panels?

10.3. (a) What is the resonance frequency for a 0.01-cm radius air bubble at a depth of 10 m in the ocean? (b) Calculate Q. (c) Find the ratio of scattering to extinction cross section. Is scattering or absorption the dominant loss mechanism? (d) At resonance, compare the extinction cross section to the geometric cross section. How does this agree with the value given in Problem 7.10?

10.4. A rectangular room has internal dimensions of 2.5 m $\times$ 4.0 m $\times$ 4.0 m and walls of 0.1-m thickness. A door opening into this room has dimensions of 0.8 $\times$ 2.0 m. (a) Assuming the inertance of the door opening to be equal to that of a circular opening of equal area, calculate the resonance frequency of the room, considered as a Helmholtz resonator. (b) What is the acoustic compliance of the room and the inertance of the door opening? (c) Considering only the compliance of the room and the inertance of the door opening, what acoustic impedance is presented, at 20 Hz, by the room to a sound source within the room?

10.5. A pipe of cross section S_1 is connected to a pipe of cross section S_2. (a) Derive a general expression for the ratio of the intensity of the waves transmitted into the second pipe to that of the incident waves. (b) Under what conditions is the transmitted intensity greater than the incident intensity? Explain. (c) Derive a general expression for the standing wave ratio SWR produced in pipe S_1 in terms of the relative areas S_1 and S_2.

10.6. A plane sound wave is traveling to the right in a pipe of area S_1 containing a fluid of characteristic impedance $\rho_1 c_1$. At the end of this pipe is attached a second pipe of area S_2 containing a fluid of characteristic impedance $\rho_2 c_2$. The two fluids are separated by means of a thin rubber diaphragm. (a) Derive an expression giving the power transmission ratio from the first pipe into the second. (b) What is the condition for 100 percent power transmission?

10.7. A plane wave of pressure amplitude P is traveling to the right in a pipe of cross-sectional area S_1. At the right-hand end of this pipe is attached a smaller pipe of area S_2 and infinite length. (a) What must be the ratio of these two areas if the transmitted pressure amplitude in the second pipe is to be 50 percent greater than that of the incident wave in the first pipe? (b) If the smaller pipe is cut off at a distance of one-quarter of a wavelength from the junction of the two pipes and then covered with a rigid cap, derive an expression giving the pressure amplitude at the cap in terms of that of the incident wave in the large pipe. (c) What is the ratio of these two pressures for the area ratio as determined in (a)?

10.8. The side branch from an infinitely long main pipe of area S is another infinitely long pipe of area S_b. The main pipe is transmitting plane waves at such a frequency that their wavelength is large compared to the diameter of either pipe. (a) Derive an equation for the transmission coefficient in the main pipe. (b) Also derive an equation for the transmission coefficient into the branch pipe. (c) If the area of the main pipe is twice that of the branch pipe, calculate numerical values for the transmission coefficient into each pipe? (d) Is the sum of these two coefficients equal to unity? If not, where is the remaining power? Support your explanation by numerical computation.

10.9. A square ventilating duct has a 0.3-m length per side. A Helmholtz resonator-type of band filter is constructed around the duct by drilling a hole of 0.08-m radius in one wall of the duct leading into a surrounding closed chamber of volume V. (a) What volume V is required in order to most effectively filter sounds at a frequency of 30 Hz? (b) What will be the sound power transmission coefficient of the filter at 60 Hz?

10.10. (a) If a hole of 1-cm radius is drilled in a thin-walled pipe of 2-cm radius, what is the sound power transmission coefficient along the main pipe at a frequency of 500 Hz? (b) What will be the transmission coefficient of the pipe at this frequency if a second hole of 1-cm radius is drilled in the pipe directly across from the first hole?

10.11. Show that the expression giving the radius a of the hole that must be drilled into a thin-walled pipe of radius a_0 in order to result in a 50 percent sound power transmission coefficient at a frequency f is $a = (64/3) f a_0^2/c$.

10.12. A 300-Hz plane wave of 0.1-W power is traveling through an infinitely long pipe of 2.0-cm radius. What will be the power reflected, the power transmitted along the pipe, and the power transmitted out through a simple orifice of 0.50-cm radius?

10.13. A ventilating duct has a radius of 0.1 m and transports air at a temperature of 50°C. It is desired to reduce the intensity level of 120-Hz plane waves by 13 dB as they pass through an inserted low-pass filter section of length L and expanded radius a. (a) What length and radius of pipe used as an expansion chamber will most effectively and simply accomplish the required filtering action? (b) Compute the filtering action of the above filter at 60 Hz.

10.14. A water pipeline has a diameter of 0.04 m. It is desired to filter out plane sound waves traveling in the water by use of sections of pipe of 0.1-m diameter that are acting as expansion-chamber types of filters. (a) Calculate the minimum length of filter section that will most effectively filter out a sound of 900 Hz. (b) If the required filtering action is a reduction of 30 dB in the level of the intensity, how many sections must be used. Ignore the effects of any interactions between the individual filter sections.

10.15. Show that the attachment of the indicated branch composed of a short section of pipe of volume V with a hole of radius a drilled in its cap is equivalent to shunting the main pipe with an inertance $M = 1.7\rho_0/(\pi a)$ in parallel with a compliance $C = V/(\rho_0 c^2)$. (a) What type of filtering action will result from this filter element? (b) Given the main pipe to be of 0.005-m² cross section and the radius of the orifice on the cap to be 0.02 m, what must be the volume of the cap, if the filter element is to produce a minimum reduction in transmission of sound waves along the main pipe at 400 Hz? (c) What will be the power transmission coefficient of this filter at 200 Hz?

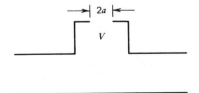

CHAPTER 11

NOISE, SIGNAL DETECTION, HEARING, AND SPEECH

11.1 INTRODUCTION. The *physical characteristics* of speech, music, and noise can be measured with considerable precision by standard acoustic instrumentation including microphones, filters, spectrum analyzers, and oscilloscopes, and the results of such measurements may be expressed in terms of precise *physical* parameters. By contrast, the *interpretative characteristics* of hearing are expressible in terms of *subjective* parameters; they are determined by experiments that lead to statistical predictions of the subjective judgments of an average listener under assumed or known conditions. For example, in attempting to measure the subjective sensation of *loudness*, one finds that different listeners will agree with surprising accuracy as to the relative loudnesses of two sounds of different frequencies; curves can be obtained relating the *subjective* parameter of loudness to the *physical* parameters of intensity and frequency. Similar investigations have determined typical values for the onset of pain, the *just noticeable differences* (jnd) for frequency and intensity, and so on. Such experiments often suffer from uncertainty as to whether or not all pertinent independent variables, including any mental bias or attitude on the part of the subject toward the experiment, are held constant. Therefore, in reading the material on hearing and in interpreting the curves that are presented, bear in mind that the data were obtained by particular experimenters using particular stimuli, presented to particular listeners under particular conditions. Other experimenters attempting to repeat a given experiment may obtain different results unless great care is taken to duplicate all factors involved in the initial experiment. For a wealth of information beyond that given here, the reader is advised to consult the references given below[1,2] and in subsequent sections.

11.2 NOISE, SPECTRUM LEVEL, AND BAND LEVEL. Until now, we have considered only monofrequency acoustic signals (tones)—a serious restriction. Whether listening to a string quartet or trying to ignore a jet aircraft, we are presented with sounds containing a distribution of frequencies. The acoustic intensity for most sounds is nonuniformly distributed over frequency, and it is convenient

---

[1] Fletcher, *Sound and Hearing in Communication*, Van Nostrand (1953).
[2] Harris (editor), *Handbook of Noise Control*, McGraw-Hill (1979).

to describe the distribution by the *spectral density*

$$\mathscr{I} = \frac{\Delta I}{\Delta f}$$

where ΔI is the intensity within the frequency interval $\Delta f = 1$ Hz. The total intensity I contained within a band with upper and lower frequencies f_2 and f_1 is

$$I = \int_{f_1}^{f_2} \mathscr{I} \, df \tag{11.1}$$

The interval $w = f_2 - f_1$ is the *bandwidth*.

For almost all noise encountered in real situations, the *instantaneous* spectral density $\mathscr{I}(t)$ is a time-varying quantity and $\mathscr{I}$ in (11.1) is a time average over some suitable interval τ, $\mathscr{I} = \langle \mathscr{I}(t) \rangle_\tau$. If $\mathscr{I}$ is constant for each frequency regardless of when the average is performed, the noise is termed *stationary*.

Many conventional acoustic filters and meters have both *fast* (1/8 s) and *slow* (1 s) integration times. The slow integration time is particularly useful in smoothing fluctuations. In general, the fluctuations in I will decrease with increasing integration time and with increasing bandwidth.

The decibel measure of $\mathscr{I}$ is the *intensity spectrum level (ISL)*,

$$ISL = 10 \log \frac{\mathscr{I} \cdot 1 \text{ Hz}}{I_{ref}} \tag{11.2}$$

where for air $I_{ref} = 10^{-12}$ W/m^2.

Obtaining the *intensity level* $IL = 10 \log I/I_{ref}$ over a desired bandwidth is straightforward if ISL is essentially constant over the interval, but becomes involved if it depends strongly on frequency. Let us examine the simpler case first.

If the ISL is constant over the bandwidth w, the total intensity calculated from (11.1) has the simple form $I = \mathscr{I} w$ and the IL for the band is

$$IL = 10 \log \frac{\mathscr{I} \cdot 1 \text{ Hz}}{I_{ref}} + 10 \log \frac{w}{1 \text{ Hz}}$$

The first term on the right is ISL

$$\boxed{IL = ISL + 10 \log w} \tag{11.3}$$

(The 1 Hz divisor of w is suppressed for convenience.)

If the intensity spectrum level ISL is a function of frequency, the problem can be handled by subdividing the total bandwidth into small intervals within each of which the ISL is relatively constant. The intensity level IL in each is calculated from (11.3) and converted to intensity I. These intensities are summed and then converted to a total intensity level,

$$IL = 10 \log \frac{\sum I_i}{I_{ref}}$$

This calculation will be simplified in the next section.

If the reference intensity and reference pressure are compatible,

$$I_{ref} = P_{ref}^2/(\rho_0 c^2)$$

then the intensity level IL and the sound pressure level SPL are equal for progressive waves and (11.3) is equivalent to

$$\boxed{SPL = PSL + 10 \log w} \tag{11.4}$$

where PSL, the *pressure spectrum level*, is equivalent to the intensity spectrum level ISL. Recall $P_{ref} = 20~\mu Pa$ for air. The physical significance of the pressure spectrum level for sounds other than tones may give the reader momentary pause. However, this is no more than the acoustic analog of the total effective voltage generated across a resistance by several series voltage sources, each independent with its own frequency. Each source V_i acting alone dissipates power V_i^2/R in the resistance. The total dissipated power is the sum $\sum V_i^2/R$ and this yields a total effective voltage $V = (\sum V_i^2)^{1/2}$, which appears across the resistance and dissipates the same power.

The sound pressure level SPL and the intensity level IL of a frequency band are each referred to as the *band level*.

Instruments used to analyze noise may be either *constant bandwidth* or *proportional bandwidth* devices.

The constant bandwidth instrument is essentially a tunable narrow-band filter with (constant) bandwidth $w = f_u - f_l$ where f_u and f_l are the upper and lower half-power frequencies. The center frequency of the filter, defined in general as

$$\boxed{f_c = \sqrt{f_u f_l}} \tag{11.5}$$

is usually variable so that the filter can be swept over the desired frequency range. Bandwidths range from a few tens of a hertz to less than a few hundredths of a hertz.

The proportional-bandwidth instrument consists of a series of relatively broadband filters with upper and lower half-power frequencies satisfying the relationship $f_u/f_l =$ constant. Each bandwidth, being *proportional* to the center frequency, increases with increasing center frequency. Most of these filters have fixed center frequencies with contiguous bands. Common instruments of this type are the *octaveband* filter with $f_u/f_l = 2$, the *1/3-octave-band* filter with $f_u/f_l = 2^{1/3}$, and the *1/10-octave-band* filter with $f_u/f_l = 2^{1/10}$. The *preferred* center frequencies and the corresponding logarithmic bandwidths for modern octave-band and 1/3-octave-band filter are shown in Table 11.1. Notice that three adjacent 1/3-octave bands span an octave.

Octave-band and 1/3-octave-band filters are used most often to analyze relatively smoothly varying spectra. If tonals are present, a 1/10-octave or a narrow-band filter should be used.

In theory, the PSL of a noise can be measured by detecting the sound with a calibrated microphone and passing the resulting voltage through a filter of 1-Hz bandwidth tunable over the desired frequency range. In practice, it is conventional

Table 11.1. Preferred octave-band and 1/3-octave-band center frequencies

Center Frequency (Hz)		10 log(Bandwidth)	
Octave	1/3-Octave	Octave	1/3-Octave
	10		3.6
	12.5		4.6
16	16	10.5	5.7
	20		6.6
	25		7.6
31.5	31.5	13.4	8.6
	40		9.7
	50		10.6
63	63	16.5	11.6
	80		12.7
	100		13.6
125	125	19.5	14.6
	160		15.7
	200		16.7
250	250	22.5	17.6
	315		18.6
	400		19.7
500	500	25.5	20.6
	630		21.6
	800		22.7
1000	1000	28.5	23.6
	1250		24.6
	1600		25.7
2000	2000	31.5	26.7
	2500		27.6
	3150		28.6
4000	4000	34.5	29.7
	5000		30.6
	6300		31.6
8000	8000	37.5	32.7

to use a filter of wider bandwidth w, determine the *SPL* of the band, and then invert (11.4) to obtain a "smoothed" spectrum level

$$\langle PSL \rangle_w = SPL - 10 \log w \qquad (11.6)$$

If w is not too great and if the spectrum does not contain strong tones, then $\langle PSL \rangle_w$ will be a relatively good estimate of *PSL*. If there is significant structure, however, $\langle PSL \rangle_w$ will vary more smoothly with frequency than *PSL*. If tones are important, then (11.6) may yield misleading results. A representative spectrum for noise containing both *broadband noise* and *tones* is given in Fig. 11.1. Although the $\langle PSL \rangle_w$ allows construction of a bar graph from which the overall band level can conveniently be calculated, the important tones are obscured by this means of presentation.

If a broadband noise has a pressure spectrum level that is independent of frequency, the noise is termed *white noise*. This noise is characterized as being rather

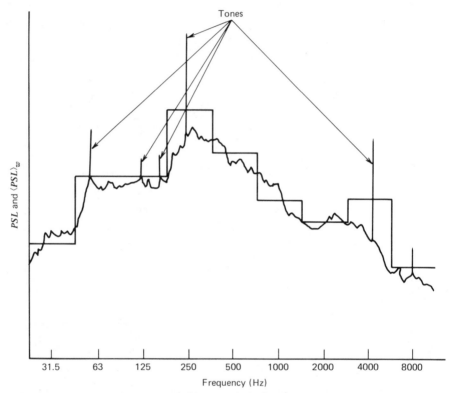

Fig. 11.1. *Representative spectrum for noise with the* $\langle PSL \rangle$ *determined over octave bands.*

sharp and "hissing." Another type of noise is *pink noise*, for which equal fractions of an octave contain the same power; the spectrum level decreases uniformly with increasing frequency with a slope of -3 dB/octave. This noise is characterized by a "hushing" sound that is less irritating than that of white noise.

11.3 COMBINING BAND LEVELS AND TONES. Calculation of the overall band level can be simplified with the help of a nomogram (Fig. 11.2) that avoids the necessity of taking antilogs and logs as discussed in the previous section. As an example of its use, Table 11.2 shows the calculation of the overall level for a noise

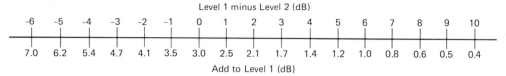

Fig. 11.2. *Nomogram for combining levels.*

Table 11.2. Calculation of an overall band level from octave-band values

Octave-Band Center Frequency	Octave-Band Level dB re 20 μPa	Calculation of Overall Band Level
31.5	70	
63	75	76.2 — 79.6
125	75	77.1
250	73	
500	76	85.1
1000	78	80.1
2000	80	83.7
4000	75	81.2
8000	65 — 65 — 65 — 65	85.1 + ≈ 85 dB re 20 μPa

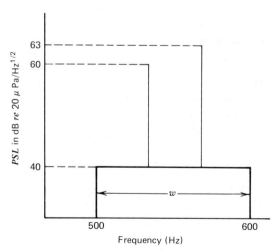

Fig. 11.3. *Example of a spectrum containing both noise and tones*

analyzed with an octave-band filter. If a more exact calculation is desired, a convenient formula for use on hand-held or desk-top calculators is

$$IL = 10 \log\left(\sum_i 10^{IL_i/10}\right)$$

where IL_i are the individual band levels for the i bands and IL is the overall band level.

Because PSL is defined over a 1-Hz interval, *the SPL of a tone is identical with its PSL.* If a spectrum contains background noise and *tones*, the band level of the continuous spectrum is combined with the PSL's of the tones. For example, consider the spectrum of Fig. 11.3 where, for calculated convenience, the background noise is constant over the bandwidth. From (11.4), the band level of the background alone is 60 dB. When this is combined with the tones of levels 60 dB and 63 dB, the overall band level is 66 dB.

11.4 DETECTING SIGNAL IN NOISE. The detection of a *signal* in the presence of *noise* ultimately reduces to a statistical problem.[3] Whether trying to focus on one voice at a noisy party or to locate a submarine underwater, a listener is attempting to isolate desired information, *signal*, from undesired information, *noise*. In humans, the complicated chain of bioacoustical, neurological, and psychoacoustic processes leading from the mechanical stimulation of the eardrum to the perception of sound in the brain constitutes a signal-processing system of stunning sophistication contained within an impressively small volume. It is designed to detect and classify sounds in the presence of noise, and in this role many of its functions and attributes can be well-described quantitatively with the aid of detection theory.

Consider a signal that is *detected, filtered* over bandwidth w, *processed* for a time τ, with an *output* presented. This output may be the time average over some selected interval of the rms voltage from a microphone, the subjective loudness of a sample of noise which may contain a tone, the instantaneous voltage developed across an electrical load by a thermosensitive bimetallic strip, or the perceived pitches of two successive samples of frequency-limited noise. Whatever the detection system, the output, from which some decision is made, will be designated by A. If the output A_i of the processor for each time $t_i = t_{i-1} + \tau$ is recorded for a long period of time, the frequency of occurrence of each value can be determined and a histogram constructed giving the probability of any particular value A being observed (Fig. 11.4). Figure 11.4a represents the *probability density function* ρ_N if *noise alone* is present, and Fig. 11.4b the probability density function $\rho_{S,N}$ for *signal with noise*. Each has its own mean A_N and $A_{S,N}$ and standard deviation σ_N and $\sigma_{S,N}$. The probability of *any* value of A_i occurring must be unity, so that

$$\int_0^\infty \rho_N \, dA = \int_0^\infty \rho_{S,N} \, dA = 1$$

For any one time interval τ, the only way to decide if a signal is present is to select a *threshold criterion* A_T and assume that if $A_i > A_T$ there is a signal present and if $A_i < A_T$ there is no signal. This yields an independent decision for each time interval, with each decision having a certain probability of being correct or incorrect. From Fig. 11.4c it is seen that the area under

[3] For a more complete introduction to detection theory, the reader is referred to Green and Swets, *Signal Detection Theory and Psychophysics*, Wiley (1966).

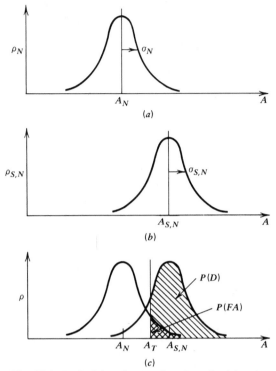

Fig. 11.4. *Probability density functions for (a) noise and (b) signal with noise. (c) The probability of detection and the probability of a false alarm for a given threshold criterion.*

the *signal-with-noise* curve to the right of A_T gives the *probability of a true detection*

$$P(D) = \int_{A_T}^{\infty} \rho_{S,N} \, dA \tag{11.7a}$$

and the area under the *noise* curve to the right of A_T gives the *probability of a false alarm*

$$P(FA) = \int_{A_T}^{\infty} \rho_N \, dA \tag{11.7b}$$

Thus, the decision as to whether a signal is present during each τ is always statistical in nature, and choosing a particular criterion A_T amounts to finding a suitable balance between the probability of a detection and the probability of a false alarm.

As A_T increases from zero to an arbitrarily large value, $P(D)$ and $P(FA)$ decrease from unity to zero (Fig. 11.5a). Which particular one of the *receiving operator characteristic (ROC)* curves for a detection system will be followed depends on the separation between the two probability functions. For example, if the value of A_T is held constant and the signal is allowed to increase, the probability density curve for signal with noise will move to the right and $P(D)$ will increase. This shifts the operating point of the system vertically upward to a

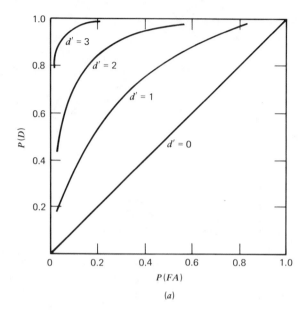

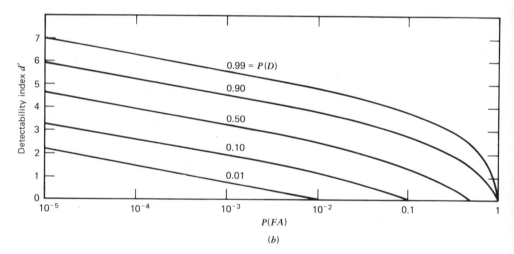

Fig. 11.5. *Receiving operator characteristics (ROC) curves for Gaussian distributions with equal standard deviations.*

higher curve in Fig. 11.5a. Each individual detection system—whether it be a hungry spider, a submarine detection system with its operator, an egg inspector, a smoke detector, or a piano tuner—has its own set of *ROC* curves. Each *ROC* curve of a set is labeled by a *detectability index* that has larger values for the higher curves.

A very important special case is that for which ρ_N and $\rho_{S,N}$ are both *Gaussian* with the same standard deviation $\sigma_{S,N} = \sigma_N = \sigma$. Under these circumstances, the *detectability index* is given by

$$d' = \frac{A_{S,N} - A_N}{\sigma}$$

and measures the number of standard deviations separating the means.

While any particular detection problem may have its own specific probability density functions, for many processes the assumption of Gaussian distributions with equal σ's is adequate or at least reasonably good. In a real situation, deviation of an experimentally determined *ROC* curve from the shapes shown in Fig. 11.5a demonstrates that this assumption is not exactly satisfied.

Determining the *ROC* curves for some detection process is an extremely laborious task. [It takes hundreds of trials to obtain good estimates of $P(D)$ and $P(FA)$ for a single point on an *ROC* curve and at least several points to determine each of the family of curves.] We summarize three methodologies that lend themselves to analysis by detection theory.

1. Yes-No Task. The subject is presented during each trial with either noise or signal with noise. Usually the signal strength is kept constant (d' fixed) so that the results will fall on a single *ROC* curve. The particular criterion is specified by the instructions to the observer; for example, "say yes only when you are sure the signal is present" sets a higher value of A_T than does "say yes when there is even a hint of the signal". By forcing the subject to change criterion between sets of trials, several points on a single *ROC* curve can be obtained. The procedure can then be repeated for different signal strengths to obtain curves for different values of d'. The decision process used as the basis of the discussion in this section, and to which the *ROC* curves of Fig. 11.5a apply, is a yes-no task.

2. Rating Task. The subject is presented with either noise or signal with noise and is asked to rate (for example, on a scale from 1 to 4) the chance that the signal is present. This amounts to applying four different values of d' to each task. If the observer remains reasonably consistent, *ROC* curves should result that are the same as those obtained from the yes-no task.

3. *n*-Alternative Forced-Choice (*n*AFC) Task. The subject is presented with n samples in each trial, only one of which has signal with noise. The observer must designate which sample is most likely to contain the signal. If the distributions of noise and signal with noise are Gaussian with $\sigma_N = \sigma_{S,N}$, then the *ROC* curves for 2AFC will be the same as for the yes-no task, but the value of d' for each curve will be $\sqrt{2}$ times that indicated in Fig. 11.5a.

Until the 1950s, there was little recognition throughout the field of psychoacoustics of the importance of *both* $P(D)$ and $P(FA)$, and the latter was rarely determined. This meant that the criterion used for establishing the threshold was not completely specified, making it difficult to compare the results of different experiments. Subsequent to the 1950s, the relevance and importance of these concepts of detection theory to an understanding of hearing have met with increasing acceptance. Detection theory, by yielding rigorously quantitative specifications of the detection process, has proven its worth.

11.5 DETECTION THRESHOLD.

Up to now we have dealt with the detection process at the *output* of the detection system. A most important task is to relate the above considerations to the *input* of the detection system. For example, the ear receives acoustic signals plus acoustic noise. There is then a succession of mechanical, neurological, and psychophysical transmutations of the combined signal and noise that results in a decision as to whether or not the signal is present.

The receiver (ear) has as its input the *signal power* S and the *noise power* N within the bandwidth of the system. We can ask, What is the minimum ratio S/N of (average) signal power to noise power that, after all the processing (mental analysis) of the received stimulus (signal plus noise) is completed, will guarantee a given $P(D)$ for a specified $P(FA)$? The decibel measure of this ratio is the *detection threshold*[4]

$$DT = 10 \log \frac{S}{N}$$
(11.8)

Relating this to the properties observed at the output (perception) is nontrivial. The calculations lie beyond the scope of this book, but we can quote two results[5] for Gaussian probability densities with equal σ's that correspond to two extremes.

1. If the signal is known exactly, then the optimum processing system is based on *cross correlation*, which searches for a facsimile of the known signal in the received signal with noise. Assume the signal $s(t)$ has duration τ and the noise is given by $n(t)$. The signal with noise is $r(t) = s(t - T') + n(t)$ where T' is some unspecified delay time. (It could be the time of flight of the signal from the source to the input of the detection system.) Cross correlation amounts to generating and studying the function

$$F(T) = \int_\tau s(t)r(t + T) \, dt$$
(11.9a)

where T is an adjustable time in the correlator. Now,

$$F(T) = \int_\tau s(t)s(t - T' + T) \, dt + \int_\tau s(t)n(t + T) \, dt$$
(11.9b)

Without going into detail, it is clear that the first integral will be maximized when $T = T'$ and the second integral (whose integrand is the product of two independently varying quantities) will, on the average, contribute very little for any value of T. Thus, $F(T)$, which peaks when $T = T'$ and is smaller for other T, plays the role of the detection system output A. If $n(t)$ and $r(t)$ are both Gaussian with equal standard deviations, the detectability index is

$$d' = \sqrt{2w\tau \ S/N} \qquad \text{(correlation)}$$
(11.10a)

It may well be that such correlation plays a role in the ability of the trained listener to "focus" on some particular instrument of an orchestra and follow its performance even during loud *tutti* passages.

2. If the salient features of the signal are unknown, *square-law* processing is best. (A detection system receives acoustic pressure, squares it to determine the acoustic energy in each observation interval, and presents that as the output A.) In the process of hearing, there is evidence that this may occur for the detection of a pure tone in noise. In the absence of the transients, associated with musical instruments as one example, or stored information on overtone prominence, the hearing process may fall back on pure energy detection. Under the restrictions $S/N \ll 1$ and $w\tau \gg 1$,

$$d' = \sqrt{w\tau \ S/N} \qquad \text{(square law)}$$
(11.10b)

---

[4] The detection threshold is often used in the form $DT_1 = 10 \log[S/(N/w)]$ where N/w is the noise power in a 1-Hz band. See, for example, Urick, *Principles of Underwater Sound*, 2nd ed., McGraw-Hill (1975).

[5] Peterson, Birdsall, and Fox, *Trans. IRE*, **PGIT4**, 171 (1954).

The detection thresholds for these two processing schemes are

$$DT = 10 \log[(d')^2/(2w\tau)] \qquad \text{(correlation)} \qquad (11.11a)$$

$$DT = 5 \log[(d')^2/(w\tau)] \qquad \text{(square law)} \qquad (11.11b)$$

For example, assume that the system is to detect an unknown signal with $P(D) = 0.5$ and $P(FA) = 0.02$. From Fig. 11.5 we see that $d' = 2$. (1) If the bandwidth of the system is 100 Hz and the processing time is $\tau = 0.5$ s, then with square-law processing $DT = -5.5$ dB and the required ratio of signal power to noise power is 0.28. If the signal were a tone, its *SPL* would be 14 dB *above* the *PSL* of the noise (assuming the noise has constant *PSL* over the bandwidth). (2) If the signal is known exactly, then with correlation processing $DT = -14$ dB and the required $S/N = 0.04$. The *SPL* of the pure tone can be 14 dB *less* than the *PSL* of the noise and still satisfy the required $P(D)$ and $P(FA)$.

In real detection systems (perhaps including some aspects of hearing), there is often a *postdetection filter* which averages the output A over a time interval T_s to reduce output fluctuations. It can be shown that the effect on the detection threshold is to increase it by $|5 \log(\tau/T_s)|$. When such a device is present, the resultant detection threshold DT' of the *combined* system of processor and postdetection filter is then increased above the DT of the processor alone,

$$DT' = DT + |5 \log(\tau/T_s)| \qquad (11.11c)$$

Note that this filter can *never decrease DT*, but if $\tau \approx T_s$ it has little deleterious effect.

11.6   THE EAR.

The properties of the human ear are phenomenal. The ear is capable of responding over a frequency range from approximately 20 Hz to 20 kHz, and, at 1 kHz, sounds that displace the eardrum only one-tenth of the diameter of the hydrogen molecule can be detected. However, it is much more than a sensitive, broadband receiver. In conjunction with the nervous system, it acts as a frequency analyzer of enviable selectivity. In this book we can give only a limited introduction to the ear and hearing. For more information, consult the references.[1,6,7]

The human ear (Fig. 11.6) is one of the most intricate and delicate mechanical structures in the human body. It consists of three main parts: the *outer, middle*, and *inner* ears.

The *pinna* of the outer ear serves as a horn collecting sound into the *auditory canal*. In a human the pinna is a relatively ineffective device, but in some animals it supplies an appreciable gain over certain frequency ranges. The auditory canal is an approximately straight tube, about 0.7 cm in diameter and 2.5 cm long, closed at its inner end by the *tympanic membrane* (eardrum). At the lowest resonance of this tube (about 3 kHz) the *SPL* at the drum is about 10 dB higher than at the entrance of the canal. Since the resonance curve is quite broad, appreciable gains are observed over a frequency range extending from about 2 to 6 kHz. If the effects of the diffraction of sound waves by the head are also taken into account, the *SPL* at the eardrum exceeds the free-field *SPL* by 15 or 20 dB at some frequencies.

The eardrum forms a flattened cone lying obliquely across the auditory canal with its apex facing inward. It is quite flexible in the center and attached around its

[6] Gelard, *The Human Senses*, 2nd ed., Wiley (1972).

[7] Jerger, *Modern Developments in Audiology*, 2nd ed., Academic Press (1973).
   Gelfand, *Hearing*, Dekker (1981).

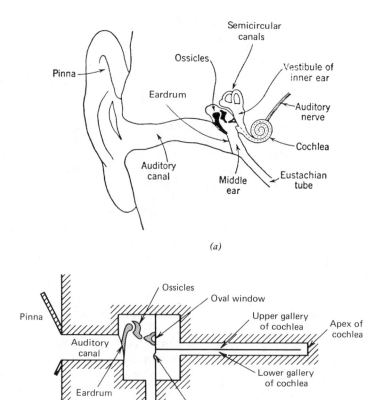

(a)

(b)

Fig. 11.6. *Sketch of the ear.*

edges to the end of the canal. This membrane is the entrance to the middle ear, an air-filled cavity of about 2 cm^3 volume which contains three *ossicles* (bones): the *malleus* (hammer), *incus* (anvil), and *stapes* (stirrup), together with their supporting muscles and ligaments. This cavity is connected to the throat through the *Eusta-chian tube* (which is normally closed, but opens during swallowing or yawning to equalize the pressure on each side of the eardrum). Vibrations of the eardrum are transmitted to the malleus, through the incus, and then to the stapes, which covers the *oval window*.

This linkage of bones, in combination with the area ratio of about 30 to 1 between the eardrum and the oval window, supplies an approximate impedance match between the air in the auditory canal and the liquid in the inner ear. This impedance match varies with the intensity of the received sound. For high inten-sities, the muscles controlling the motion of the ossicles change their tension to reduce the amplitude of motion of the stapes, thereby protecting the delicate mechanism of the inner ear from damage. This process is known as the *acoustic reflex*. Since it takes about 0.5 ms after perceiving a loud sound for the acoustic reflex to become effective, it offers no protection from sudden impulsive sounds such as gunshots, explosions, and so forth.

The inner ear (*labyrinth*) has three parts: the *vestibule* (entrance chamber), the *semicircular canals*, and the *cochlea*. The vestibule connects with the middle ear through two openings, the *oval window* and the *round window*. Both of these windows are sealed to prevent the escape of the liquid filling the inner ear; the former by the stapes and its support, and the latter by a thin membrane. With these two exceptions, the entire inner ear is surrounded by bone. (The semicircular canals play no part in the process of hearing; they provide us with a sense of balance.) The cochlea is a tube of roughly circular cross section, wound in the shape of a snail shell. This tube makes about 2.5 turns and has a total length of about 3.5 cm. Its cross-sectional area decreases in a somewhat irregular manner from its base to its apex. Its volume is about 0.05 cm^3.

The tube of the cochlea is divided by the *cochlea partition* into two longitudinal channels: the *upper gallery* (*scala vestibuli*) and the *lower gallery* (*scala tympani*). The only communication between the two galleries is through the *helicotrema*, a small opening at the apex of the cochlea. The other ends of the upper and lower galleries connect with the oval and round windows, respectively. A cross section of one of the turns of the cochlea is shown in Fig. 11.7. The *bony ledge* projects from the central portion of the shell-like structure into the liquid-filled tube and carries the *auditory nerve*. At the termination of the bony ledge the nerve fibers enter the *basilar membrane*, which continues across the tube to the farther side where it is attached to the *spiral ligament*. Lying above the basilar membrane is the *tectorial membrane*, attached along one edge to the bony ledge with its opposite edge projecting into the cochlear liquid. Running diagonally across the cochlear canal from the bony ledge to the opposite wall is *Reissner's membrane*, just two cells thick. This membrane, with the basilar membrane, isolates the majority of the cochlea (the upper and lower galleries) from the pie-shaped cross section named the *cochlea duct*. This duct is filled with *endolymphatic fluid*; the galleries are filled with *perilympatic fluid*. (The endolymph is closely related to the intercellular fluid found throughout the body, and the perilymph is similar to spinal fluid.)

Attached to the top of the basilar membrane is the *Organ of Corti* that contains four rows of *hair cells* (about 3×10^4 cells in all) spanning the entire length of the cochlea. The 3500 hair cells in the inner row are less vulnerable to damage than are the cells in the outer three rows. Several dozen small hairs extend from each hair cell to the under surface of the tectorial membrane.

When the ear is exposed to a pure tone, the motion of the eardrum is transmitted by the bones of the middle ear to the oval window, creating a fluid disturbance that travels down the upper gallery toward the apex of the cochlea, through the helicotrema, into the lower gallery, and back to the round window, which acts as a pressure-release termination. The detailed properties of this disturbance and its role in the mechanism of hearing were elucidated in a Nobel-Prize winning series of

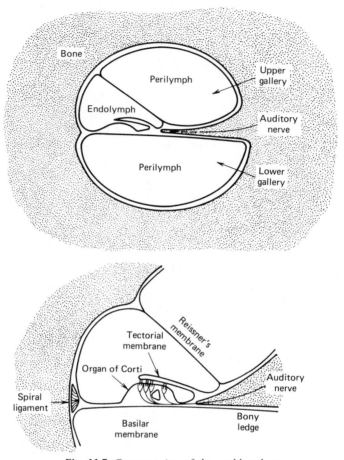

Fig. 11.7. *Cross section of the cochlea duct.*

investigations by Bekesy. His experiments demonstrated that the basilar membrane is driven into highly damped motion with a peak amplitude that increases slowly with distance away from the stapes, reaches a maximum, and then diminishes rapidly toward the apex. The peak amplitude maximizes closer to the apex for lower frequencies. A few examples are sketched in Fig. 11.8 where z is the distance in centimeters from the stapes. The relation of the position of the maximum peak amplitude to the frequency can be well-approximated by

$$f = 2.5 \times 10^{4 - 0.72z} \qquad (11.12)$$

for frequencies above about 200 Hz.[1] For lower frequencies, the distance from the apex is underestimated by (11.12). These motions of the basilar membrane occur whether the mechanical excitation is produced by airborne sounds (through the eardrum, ossicles, etc.) or by sounds conducted through the skull.

Since the Organ of Corti is attached to the basilar membrane while the tectorial membrane is attached to the bony ledge, the relative motions generated between them flex the hairs, thereby exciting the nerve endings attached to the hair cells into producing electrical impulses. These nerves do not necessarily fire at the frequency with which they are excited, but quasirandomly when they are stressed beyond certain limits, usually firing more often when highly stressed. These pulses form the information communicated from the cochlea to the brain. From each ear, information routes go to a number of interlinked processing centers within the brain.

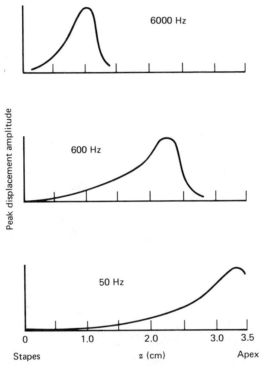

Fig. 11.8. Peak displacement amplitude of the basilar membrane for a pure tone.

There is considerable communication between processors on different sides of the brain so that the signals from both ears are processed and mixed. There are, in addition, sets of nerves that send information from the brain to the ears. (For example, the left ear transmits information to the left *superior olivary complex* which in turn has nerves going directly to the right ear.) This highly complicated, subtle, and sophisticated communication and processing network is just beginning to be understood. For further information, the reference below[8] may be a suitable start.

[8] Carteratte and Friedman, *Handbook of Perception*, Vol. 2, Academic Press (1974).

11.7 SOME FUNDAMENTAL PROPERTIES OF THE HUMAN EAR.

(a) Thresholds.   The threshold of audibility is the minimum perceptible free-field intensity level of a tone that can be detected at each frequency over the entire range of the ear. The symbol L_I is recommended by the International Standardization Organization (ISO) for IL re 10^{-12} W/m². The tone presented to both ears should have a duration of about 1 s. For tones shorter than 0.3 s the apparent loudness increases with increasing tone duration, and for tones longer than 3 s "fatigue" sets in and the apparent loudness diminishes with time.

The average threshold of audibility for the normal (undamaged) ear is shown as the lowest curve on Fig. 11.9. The frequency of maximum sensitivity is near 4 kHz. Below this, the threshold rises with decreasing frequency, the minimum power required to produce an audible sound at 30 Hz being nearly a million times as great

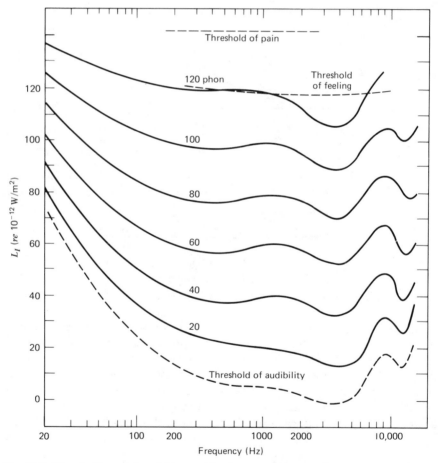

Fig. 11.9. *Thresholds and free-field, equal-loudness-level contours*[9] *for pure tones with subject facing the source.*

as at 4 kHz. For high frequencies, the threshold rises rapidly to a cutoff. It is in this frequency region that the greatest variability is observed between different listeners, particularly if they are over 30 years of age. The cutoff frequency for a young person may be as high as 20 kHz or even 25 kHz, but people who are 40 or 50 years of age can seldom hear frequencies in excess of 15 kHz, and in some the cutoff is below 10 kHz. In the range below 1 kHz, the threshold is usually independent of the age of the listener.

As the intensity of the incident acoustic wave is increased, the sound grows louder and eventually produces a tickling sensation. This level, less dependent on frequency than the threshold of audibility, has a value of approximately 120 dB and is called the *threshold of feeling*. As with the lower threshold, it varies somewhat from individual to individual, but not to so great an extent. As the intensity is still further increased, the tickling sensation gives way to one of *pain* at about 140 dB. It must be emphasized, however, that even for short exposures permanent damage to the ear can occur at levels below 100 dB.

Since the ear responds to loud sounds by reducing the sensitivity (the acoustic reflex mentioned earlier), the threshold of audibility shifts upward under exposure, the amount of shift depending on the intensity and duration of the sound. After the sound is removed, the threshold of hearing will begin to reduce and, if the ear fully recovers its original threshold, we have observed *temporary threshold shift TTS*. (The communication between right and left ears evinces itself here in that strong stimulation of one ear results in a small threshold shift in the other ear.) The amount of time required for a complete recovery increases with increasing intensity and duration of the sound. If the exposure is long enough or the intensity high enough, the recovery of the ear is not complete, and the threshold never returns to its original value; *permanent threshold shift PTS* has occurred. It is important to realize that the damage leading to *PTS* occurs in the inner ear; the hair cells are damaged with no possibility of recovery or reconstruction.

Also of importance are *differential thresholds*, one of which is the differential threshold for *intensity determination*. If two tones of almost identical frequency are sounded together, one tone much weaker than the other, the resultant signal is indistinguishable from a single frequency whose amplitude fluctuates slightly and sinesoidally (the " beat " phenomenon discussed in Sect. 1.13). The amount of fluctuation that the ear can just barely detect, when converted into the difference in intensity between the stronger and weaker portions, determines the differential threshold. As might be expected, values depend on frequency, number of beats per second, and intensity level. Generally, the greatest sensitivity to intensity changes is found for a beat frequency of about 3 beats per second. Sensitivity decreases at the frequency extremes, particularly for low frequencies, but this effect diminishes with increasing sound level. For sounds more than about 40 dB above threshold, the ear is sensitive to intensity level fluctuations of less than 2 dB at the frequency extremes, and less than about 1 dB between 100 and 1000 Hz.

Other differential thresholds involve the ability to discriminate between two sequential signals of nearly the same frequency. The frequency difference required to make the discrimination is termed the *difference limen*. Older methods of measuring the difference limen consist of exposing the ear to a frequency-modulated tone where the amount and rate of modulation can be controlled. The difference limen determined by this method depends on the intensity of the signal, its center frequency, and its rate of modulation. Greatest sensitivity seems to occur when the modulation rate is about 3 Hz. At moderate levels, the difference limen is about 1 Hz for tones with frequencies up to about 100 Hz, rising smoothly to 5 Hz for 1 kHz tones, and up to 70 Hz for 10 kHz tones. Thus, for this method, at the midrange and higher frequencies, the ear can detect changes of frequency of about 0.7 percent. More recent experiments use carefully shaped tone bursts of each frequency and yield difference limens up to about five-times smaller than those measured by the modulation method.

(b) Equal Loudness Level Contours.[9] Experiments can be performed to gauge when two tones of dissimilar frequencies, sounded alternately, appear to be equally loud. As seen in Fig. 11.9, high- and low-frequency tones require greater values of L_I to sound as loud as those in the midfrequency range. The curves resulting from such comparisons are labeled by the L_I they have at 1 kHz. Each curve is called an *equal loudness level contour* and expresses the *loudness level L_N* in *phon* which is assigned to all tones whose L_I fall on the contour. Therefore, $L_N = L_I$ for a 1-kHz tone, regardless of its level. On the other hand, a pure tone of 40 Hz with $L_I = 90$ dB (*re* 10^{-12} W/m^2) has a loudness level $L_N = 70$ phon, as does a 4-kHz tone with $L_I = 61$ dB. Notice, however, that for higher loudness levels the curves become straighter, so that L_N and L_I become more nearly equal at all frequencies.

The flattening of the equal loudness level contours for higher loudness levels explains the well-known fact that turning up the "volume" on a hi-fi system causes the frequency extremes to appear louder—there are more bass and treble. Conversely, turning the volume down removes both "body" and "brilliance," and the sound takes on a "thin" and "tinny" quality. This also reveals a problem in selecting loudspeakers of roughly equivalent quality; the one which is more efficient, and therefore louder than the other for equal signal input, may well sound more impressive merely because it is louder.

(c) Critical Bandwidth. If a subject listens to a sample of noise with a tone present, the tone cannot be detected until its L_I exceeds a value that depends on the amount of noise present. In a pivotal set of experiments, Fletcher and Munson[10] found that the masking of a tone by a broadband noise is independent of the noise bandwidth until the bandwidth became smaller than some critical value that depends on the frequency of the tone. The ear acts like a collection of *parallel filters*, each with its own bandwidth, and the detection of a tone requires that its level exceed the noise level in its particular band by some detection threshold.

In early experiments it was assumed that the signal must *equal* the noise for detection to occur ($DT = 0$). On this basis, and assuming that the sensitivity of the

[9] Robinson and Dadson, *Brit. J. of Appl. Physics*, **7**, 166 (1956) and International Standardization Organization ISO R226-1961.

[10] Fletcher and Munson, *J. Acous. Soc. Am.*, **9**, 1 (1937).

ear is constant across each bandwidth w_{cr}, the critical value of the bandwidth could be found from $w_{cr} = S/N_1$, where S is the signal power and N_1 the noise power per hertz. For reasons that will be clear in a moment, the critical values measured this way are now termed the *critical ratios* (see Fig. 11.10).

Later experiments based on the perceived loudness of noise have yielded *critical bandwidths* w_{cb} larger than the critical ratios. In some of these experiments, the loudness of a band of noise is observed as a function of bandwidth while the *overall noise level is held constant*. For noise bandwidths less than critical, the loudness will

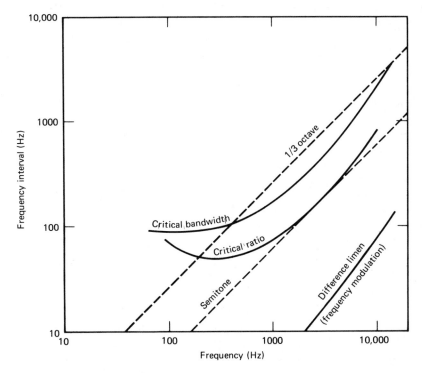

Fig. 11.10. *Critical bandwidths of the ear.*

be constant; when the bandwidth exceeds the critical bandwidth, the loudness will increase (see Sect. 11.8).

A representative curve[11] of this critical bandwidth is shown in Fig. 11.10. It is interesting to note that the critical bandwidth is nearly 1/3-octave for frequencies above about 400 Hz. Over this range, the critical bandwidth is about 2.5 times the critical ratio; one cause of the discrepancy[12] may be that the strong amplitude modulations of narrow-band noise used in the critical-ratio experiments allowed the tone to "peek through" and be detected during relatively quiet periods. It should

[11] Zwicker, Flottop, and Stevens, *J. Acous. Soc. Am.*, **29**, 548 (1957).

[12] de Boer, *J. Acous. Soc. Am.*, **34**, 985 (1962).

also be noted that the earlier experiments gave no specification of $P(D)$ and $P(FA)$ and so were incompletely documented.

A complete curve of the difference limen obtained by frequency modulation would show that the ear can perceive frequency changes of about 1/30 of the critical bandwidth. The mechanism for this frequency selectivity within intervals of a critical bandwidth cannot be explained on the basis of the simple filtering action mentioned above. One suggestion is that the signal processing occurring within the brain culminates in a "sharpening" of the nerve excitations transmitted to the brain so that there is an effective narrowing of the region of response and hence an enhancement of frequency discrimination. Another suggestion, supported by experimental evidence, is that the ear may recover frequency information from the long-term combined rate at which the nerves fire. Even though each nerve usually fires with a repetition rate far below the detected frequency, if there are many nerves firing at lower rates, the collection of all the nerve firings will yield "volleys" generated with a repetition rate equal to the frequency.

(d) Masking. Everyone is familiar with the difficulty of hearing and understanding speech in the presence of noise. The amount of *masking* is the amount the threshold of audibility of the signal is raised in the presence of noise.

It is simplest to begin by considering the masking of one pure tone by another. The subject is exposed to a single tone of fixed frequency and L_I and then asked to detect another tone of different frequency and level. Analysis yields the *threshold shift*: the increase in L_I of the masked tone above its value for the threshold of audibility before it can be detected. Figure 11.11 gives representative results for masking frequencies of 400 and 2000 Hz. The frequency range over which there is appreciable masking increases with the L_I of the masker, the increase being greater for frequencies above that of the masker. This is to be expected because the region of the basilar membrane, excited into appreciable motion at moderate values of L_I,

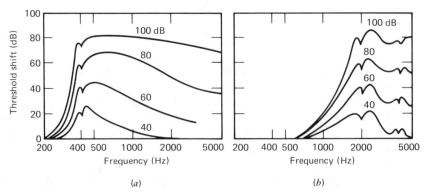

Fig. 11.11. *Masking of one pure tone by another. The absissa is the frequency of the masked tone, and the curves are labeled with the L_I of the masking tone. The frequency of the masking tone is 400 Hz in (a) and 2000 Hz in (b).*

extends from the maximum further toward the stapes than the apex. For stronger excitation of the membrane both regions grow, the region toward the stapes more significantly; it is this region that covers the frequencies higher than that of the masker: low frequencies mask high frequencies. (The notches in the curves will be explained in the discussion that follows.)

Masking of pure tones by a band of noise narrower than w_{cb} is essentially the same as that of an equally intense pure tone having the same frequency as that at the center of the band. Consequently, when the spectrum level is relatively constant, the intensity of a narrow band of noise is directly proportional to the bandwidth w and, correspondingly, the masking expressed in decibels increases as $10 \log w$. Ultimately, the bandwidth will equal the critical bandwidth, beyond which any further increase in the width of the pass band of the noise has little influence on the amount of masking of a pure tone at the center of the band.

(e) Cochlear Nonlinear Effects: Beats, Combination Tones, and Aural Harmonics. Let two tones of similar frequency f_1 and f_2 and of equal L_I be presented to one (or both) ear(s). When the two frequencies are very close together, the ear perceives a tone of single frequency $f_c = (f_1 + f_2)/2$ fluctuating in intensity at the *beat frequency* $f_B = |f_1 - f_2|$ (see Sect. 1.13). As the frequency interval between the two tones increases, the sensation of *beating* changes to *throbbing* and then to *roughness*. As the frequency interval increases further, the roughness gradually diminishes and the sound becomes smoother, finally resolving into two separate tones. For frequencies falling in the midrange of hearing, the transition from beats to throbbing occurs at about 5 to 10 beats per second and this turns into roughness at about 15 to 30 beats per second. These transitions occur for higher beat frequencies as the frequencies of the primary tones are increased. Transition to separate tones occurs when the frequency interval has increased to about the critical bandwidth. None of this occurs if each of the tones is presented to a different ear. When each ear is exposed to a separate tone, the combined sound does not exhibit intensity fluctuations as dicussed above; this kind of beating is absent. This suggests that the beats arise because the two tones generate overlapping regions of excitation on the basilar membrane, and it is not until these regions become separated by the distance corresponding to the critical bandwidth that they can be separately sensed by the ear. (When the tones are presented one in each ear, of course, each basilar membrane is separately excited, and there is no possibility of mechanical interaction.)

If the two tones (both presented together in one or both ears) are separated far enough and are of sufficient loudness, a trained observer can detect *combination tones*. These combination tones are not present in the original sound, but are *manufactured by the ear*. There is a collection of possible combination tones whose frequencies are various sums and differences of the original frequencies f_1 and f_2,

$$f\text{(combination)} = |mf_2 \pm nf_1| \qquad n, m = 1, 2, 3, \ldots \qquad (11.13)$$

Only a few of these frequencies will be sensed. One of the easiest to detect is the *difference frequency* $|f_2 - f_1|$.

A *linear* combination of two frequencies does not result in any combination tones. On the other hand, if two frequencies are supplied to a *nonlinear* system, the response will contain not only the original frequencies, but also various sum and difference frequencies. Nonlinear electrical circuits are common in radio receivers and transmitters for generating such combination tones. The cochlea is the origin of the nonlinearity creating combination tones. (The middle ear remains remarkably linear up to the threshold of pain.) One means of studying this nonlinearity is through measurement of *cochlear potentials.* When the basilar membrane is set into motion and the hair cells stressed, slight potential differences are generated that can be detected by means of an electrical probe inserted into the cochlea. The amplitudes and waveforms of the cochlear potentials appear to represent quite accurately those of the sounds heard. For two pure tones of low intensity, the cochlear potential has a waveform identical with that of the received sound. As the incident intensity is increased, however, the wave becomes distorted, indicating the generation of combination tones. In general, difference tones appear at somewhat lower intensity levels than do summation tones, but both are observed.

The fact that there are nonlinear mechanisms operating in the cochlea leads to the formation of various combination tones when an ear is exposed to two pure tones. Furthermore, there is nonlinear distortion when only one tone is received. Indeed, if m is set equal to zero in (11.13), which is equivalent to removing the tone of frequency f_2, there are still nonlinearly generated tones with frequencies nf_1. These are called *aural harmonics.* Tones having frequencies above 500 Hz and loudness levels below 40 phon do not generate aural harmonics of appreciable magnitude. For frequencies around 100 Hz, the loudness level at which distortion first appears is about 20 phon. With increasing loudness the aural harmonics appear in order; initially the second harmonic appears, then the third, and so on. In general, the loudnesses of the aural harmonics are less than that of the fundamental, and they decrease progressively with increasing order. (However, in an intense, low-pitched sound, such as a 60-Hz tone at an intensity of 100 dB, the second harmonic may be louder than the fundamental and a number of the higher harmonics may also approach the fundamental in loudness.) The strongest evidence supporting their existence is from cochlear potentials. Another method introduces a second tone whose frequency is brought close to a frequency of an aural harmonic so that beats are generated between the second tone and the harmonic. (This is less convincing because the vibratory motion of the basilar membrane is perturbed by the introduction of the second tone.) This second method, however, demonstrates the effectiveness of beats in enhancing detection. The notches in the masking curves of Fig. 11.11 show that these beats lower the threshold at the frequencies of the aural harmonics.

(f) Nonlinear Processing Effects: Consonance and the Restored Fundamental. In addition to the effects described above, there are other nonlinear effects that, while somewhat more subtle, have great importance for music.

One effect involves the perception of mistuned intervals. If two tones *almost* an octave apart are sounded simultaneously and tuned toward the octave, there will be a strong impression of beating that slows and dies away as the interval of the octave is approached. This beating is not like that for the amplitude modulation previously discussed, but is apparently a response to the slowly time-varying nature of the waveform (see Fig. 11.12). An important feature of this beating is that *it is not necessary that both tones be presented in the same ear.* If one tone is presented to one ear and the second tone to the other ear, the beating is still sensed. Thus, this interference is not the result of nonlinearities of the cochlea, but arises out of

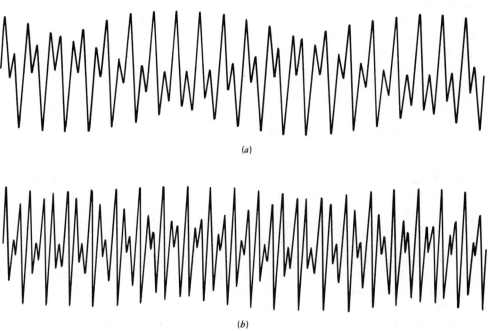

(a)

(b)

Fig. 11.12. *Waveforms for mistuned consonances. (a) Mistuned octave. (b) Mistuned fifth. (The mistuning has been greatly exaggerated for visual clarity.)*

nonlinearities in the processing of the information in the relevant centers of the brain. Besides the octave, these beats are also observed when the ratio of the two frequencies is nearly that of two integer numbers: $f_2/f_1 = 1/1,\ 3/2,\ 4/3,\ \ldots$, with the beating becoming more subtle as the integers get greater. When the frequencies are exactly aligned to these ratios, the beating vanishes and the sensation is of *consonance*. It is no accident that these ratios form the foundation of our musical scales.[13,14]

A second effect is the generation of the fundamental when two or more consonances are presented. For example, if tones of 800 and 1000 Hz are presented to the

[13] Roederer, *Introduction to the Physics and Psychophysics of Music,* Springer–Verlag (1973).

[14] Strong and Plitnik, *Music, Speech, and High Fidelity,* Brigham Young University Publications (1977).

ears, a frequency of 200 Hz can be detected. This tone is the fundamental of which the other two are adjacent harmonics. Unlike the difference tone discussed earlier which can be masked by noise, this fundamental can be perceived even when it should be masked and therefore undetectable. Furthermore, it is perceived when the signals are too weak to generate a detectable difference frequency. Finally, as in the first effect, it occurs even if the tones are fed into different ears.

The "generation of the missing fundamental" has practical application in the design of inexpensive, small radios. To eliminate the cost of expensive networks to filter out the 60-Hz line frequency and the 120-Hz harmonic generated in the rectifier, manufacturers deliberately limit the low-frequency response of such radios to eliminate frequencies below about 150 Hz. (Higher harmonics of the line frequency are filtered out quite inexpensively.) Thus, while there is no significant output below about 150 Hz, the nonlinear processing in the brain restores the fundamentals of the bass notes from the higher harmonics which are still present.

11.8   LOUDNESS LEVEL AND LOUDNESS.   Although two sounds having the *same loudness level* are judged to be *equally loud*, this does not mean that the subjective sensation called *loudness N* is proportional to *loudness level L_N*. A tone of $L_N = 60$ phon will not sound twice as loud as one of 30 phon. The unit of loudness is the *sone*, and a loudness $N = 1$ sone is equal by definition to a loudness level of 40 phon, independent of frequency. A loudness of 16 sone is twice as loud as one of 8 sone and four times as loud as one of 4 sone.

Loudness is not easy to measure, and its determination requires ingenuity. The results of a great many experiments at different frequencies have been summarized by Fletcher,[1] and a graph representing the relationship between loudness N and loudness level L_N is given in Fig. 11.13. While the lower portion of the graph is noticeably curved, the portion in excess of 1 sone is relatively straight. In the linear portion (corresponding to sounds from comfortably audible to unpleasantly loud), an increase in the loudness level of 9 phon is approximately equivalent to doubling the loudness. An empirical formula relating loudness and loudness level over the linear portion is[15]

$$N = 0.046 \times 10^{L_N/30} \tag{11.14}$$

Notice there is no dependence on frequency.

Loudness and intensity can now be related. For a frequency of 1 kHz, where by definition the loudness level L_N is equal to the intensity level L_I, substitution of this equality into (11.14) and use of $L_I = 10 \log(I/10^{-12})$ results in

$$N \ (1 \ \text{kHz}) = 460 \sqrt[3]{I}$$

where I is the intensity (MKS) of the 1-kHz tone. For any other frequency, refer to Fig. 11.10 and make a plot of L_N versus L_I. It will be seen that for any frequency the resulting curve will be very close to a straight line with a slope of $+1$. Fitting this

---

[15] For a standardized engineering approximation, ISO R131-1959, see Problem 11.20.

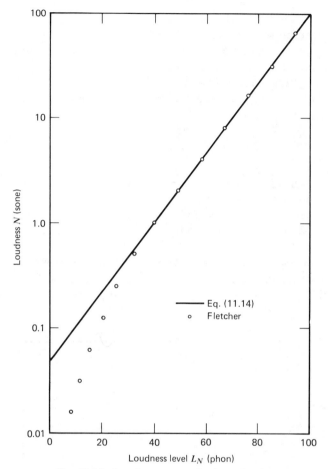

Fig. 11.13. *Loudness versus loudness level.*

curve with a straight line of this slope over the loudness levels of interest will yield

$$L_N \doteq L_I + 30 \log F(f)$$

where $F(f)$ is an empirically determined parameter depending only on frequency. Substitution into (11.14) gives

$$N \doteq 460F(f)\sqrt[3]{I} \tag{11.15}$$

This relation is one example of a general psychophysical "power law" of sensation postulated by Stevens.[16]

$$S = G(\phi - \phi_T)^{\nu} \qquad \phi \geq \phi_T$$
$$= 0 \qquad \phi < \phi_T$$

---

[16] Stevens, *Science,* **133,** 80 (1961).

where S is the sensation, ϕ the stimulus, ϕ_T the threshold value, and G and v are constants depending on the quantities represented by S and ϕ. Equation (11.15) is quite accurate for the range 500 Hz to 5 kHz and moderate loudness levels.

If two or more tones are sounded simultaneously, the total loudness depends on whether they lie within a critical band.

1. Tones lying within one critical bandwidth are sensed according to the overall power, so their intensities add and the loudness is given by

$$N \text{ (critical band)} = 460F(f)\left(\sum_i I_i\right)^{1/3}$$

2. Tones differing by more than the relevant critical bands are sensed as well-separated regions on the basilar membrane and the loudness add, $N = \sum N_i$.

3. If the tones are *considerably* different in their loudness, or widely separated in frequency, the evaluation of loudness becomes difficult, often tending to be based on the loudest of the tones.

As an example of the second case, let us determine the composite loudness of six pure tones of frequencies 125, 250, 500, 1000, 2000, and 4000 Hz and all having an *SPL* of 60 dB. This computation is summarized in Table 11.3. The result of 34 sone

Table. 11.3.   Sample calculation of loudness

Frequency f (Hz)	Intensity Level L_I (dB)	Loudness Level L_N (phon)	Loudness N (sone)
125	60	55	3.2
250	60	62	5.4
500	60	63	5.9
1000	60	60	4.7
2000	60	62	5.4
4000	60	69	9.3
Total			33.9

is equivalent to about 84 phon. This corresponds to an intensity level of 84 dB for a 1000-Hz tone. Since the sum of the intensities of the six tones corresponds to an intensity level of only 68 dB, it is apparent that the sound energy appears to be louder when it is distributed over a wide range of frequencies rather than lying within one critical bandwidth.

Calculations of the overall loudness of broadband noise (or even more complicated combinations of tones) becomes quite complex. The loudness becomes a function of the masking interactions among the sounds in the various critical bandwidths. The procedure has been systematized by Stevens through a number of permutations, and the methodology along with the associated tables necessary for the calculation is contained in reference 17.

[17] Stevens, *J. Acous. Soc. Am.,* **51**, 575 (1972).

11.9 PITCH AND FREQUENCY. Another subjective characteristic of a sound is its *pitch*. Pitch, like loudness, is a complex characteristic and is dependent on various physical quantities. While determined *primarily* by frequency, intensity and waveform are also influential. If a pure tone of about 100 Hz is sounded at a moderate and then at a high loudness level, the louder sound has a lower pitch. The most pronounced *decrease* in pitch with increasing loudness occurs for tones with frequencies below about 300 Hz. When the loudness of such a tone is increased from 40 phon to 100 phon, it may be necessary to increase the frequency by as much as 10 percent in order to maintain the pitch at a constant value.* For frequencies between 500 and 3000 Hz, the pitch of a tone is relatively independent of its loudness. For tones with frequencies above 4 kHz, the pitch *increases* with loudness, the increase being greater for higher frequencies.

For any particular loudness it is possible to assign numbers to the perceived pitches describing how "high" they sound, thereby establishing a pitch scale for pure tones. The reference frequency is usually 1 kHz, and the tone corresponding to this frequency is said to have a pitch of 1000 *mel*. A tone whose pitch is 500 mel sounds half as high, and a tone of 2000 mel sounds twice as high, as the 1000 mel tone. A curve giving the relationship between subjective pitch and frequency at a fixed loudness is shown in Fig. 11.14.

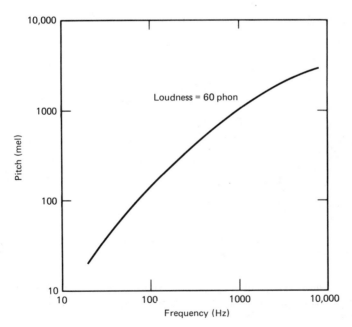

Fig. 11.14. *Pitch as a function of frequency for a 60-phon pure tone.*

---

* As noted earlier, the fundamental frequencies of adjacent semitones in the even-tempered musical scale are related by $f_1/f_2 = 2^{1/12} = 1.059$. Thus, the 10 percent change in frequency necessary to maintain the same pitch corresponds to a change of nearly a whole tone.

Musical sounds of differing loudnesses generate much smaller changes in pitch, usually not more than one-fifth as much as for similar pure tones. Furthermore, there are much smaller discrepancies between the fundamental frequency and pitch. Both these effects occur because musical sounds are usually rich in harmonics, some of which may have amplitudes exceeding that of the fundamental. Even if the fundamental lies in the frequency range where a pure tone shows a large decrease in pitch with loudness, the harmonics will have frequencies for which the pitch changes very little, so that the ear, aided by the consonances of all the harmonics, judges the sound as remaining at essentially the same pitch. The pitch of a complex sound is thus determined by the harmonics, which explains why a radio or TV, whose low-frequency response may extend down to only 150 Hz or so, still gives a reasonable approximation of the original sound (at least it is not unrecognizable).

11.10   THE VOICE.   The acoustic energy associated with speech originates in the chest muscles which, by contraction, force air from the lungs up through the various components of the vocal mechanism (Fig. 11.15). This steady stream of air may be looked on as a *carrier* of

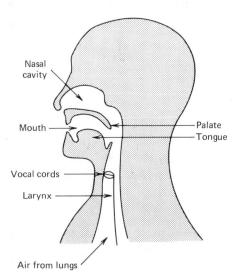

Fig. 11.15. *Sectional view of the head, showing the important elements of the voice mechanism.*

energy that must be modulated in its velocity and corresponding pressure in order to produce sounds. The requisite modulation is accomplished in two basic ways, leading respectively to *voiced* and *unvoiced* sounds.

Voiced sounds include the vowels of ordinary speech as well as tones characteristic of the singing voice. The primary modulating agent for voiced sounds is the *larynx*, across which are stretched the *vocal cords*. The vocal cords are two membranelike bands forming a diaphragm with a slitlike opening that modulates the airstream as it vibrates open and shut. The length of this opening (about 2.5 cm in men and 1.5 cm in women) and the tension to which the vocal cords are stretched determine the fundamental frequency of this modulation. The shorter and lighter vocal cords possessed by most women vibrate almost twice as rapidly as those possessed by men. This accounts for the higher pitch of most women's voices. The action of the vocal cords produces a *sawtooth* pressure waveform containing many harmonics.

The resonating cavities and orifices of the nose, mouth, and throat form an acoustic filtering network altering the relative amounts of the harmonics. Many of these are controllable at will (changing the position of the tongue or altering the configuration of the lips) and thereby a wide variety of voiced sounds may be produced.

It is also possible for the voice mechanism to produce unvoiced sounds. For instance, a steady forcible exhaling of the breath will produce a hissing sound caused by turbulences set up in the flow of air through the numerous irregularities along the vocal tract. These sounds include such unvoiced *fricative* consonants as *f* and *s* as well as the unvoiced *stop* consonants

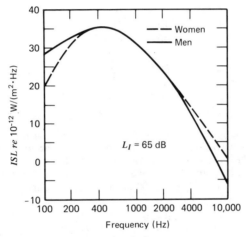

Fig. 11.16. *Spectra levels for average conversational speech measured 1 m from the mouth.*

p, *t*, and *k*. Here the sounds are produced by modulating the airstream with the lips, teeth, or tongue. As with the voiced sounds, conscious control of the tongue and lips alters the resonances of the cavities and orifices, producing a wide variety of unvoiced sounds so that recognizable speech can be generated—*whispering*. Spectrum analysis of the unvoiced sounds reveals a band of practically continuous frequency coverage largely confined to the upper portion of the audible range.

The distribution of average speech power with frequency has been measured by several investigators.[18] Typical average speech spectra are shown in Fig. 11.16. The overall L_I at 1 m is about 65 dB.

The acoustic power, averaged over 2 to 4 s, generated by a speaker at conversational level is about 10 μW. Very loud talking generates about 200 μW and shouting about 1000 μW. By contrast, whispering generates about 0.001 μW. When the power of conversational-level speech is averaged over a time interval that is short compared to the duration of a

[18] Dunn and White, *J. Acous. Soc. Am.*, **11**, 278 (1940).

syllable (about 0.2 s), sizable fluctuations in level are observed as different speech sounds are uttered. For example, the power of the vowel o as in low is about 50 μW, whereas the weak consonant v has an average power of only 0.03 μW.

PROBLEMS

11.1 An acoustic signal consists of three tones, each of different frequency and different effective pressure amplitude: $P_1 = 5 \times 10^{-2}$ Pa, $f_1 = 104$ Hz; $P_2 = 7 \times 10^{-2}$ Pa, $f_2 = 190$ Hz; $P_3 = 0.1$ Pa, $f_3 = 237$ Hz. Find the intensity in the following bands: (a) 100 to 110 Hz, (b) 100 to 150 Hz, and (c) 150 to 300 Hz.

11.2. The results of a noise analysis are as follows:

Filter #	f_l	f_u	$V = effective$ voltage output
1	100	200	7.1 mV
2	200	400	6.3
3	400	800	11.2
4	800	1600	8.9
5	1600	3200	11.2
6	3200	6400	7.9

(a) If the sensitivity of the receiver is 5×10^{-2} V/Pa, find the effective pressure of the sound within the bandwidth of each filter. (b) Find the intensity of the sound within the bandwidth of each filter. (c) Find the band level (re 20 μPa) for each filter. (d) Find the pressure spectrum level (re 20 μPa/Hz$^{1/2}$) for the bandwidth of each filter. (e) Use the intensities obtained in (b) to find the band level between 100 and 6400 Hz. (f) Use the band levels from (c) to find the total band level between 100 and 6400 Hz. Compare this to the answer for (e).

11.3. (a) A tone with a sound pressure level of 140 dB re 1 μPa is superimposed on a background noise with a constant pressure spectrum level of 150 dB re 1 μPa/Hz$^{1/2}$. Calculate the band levels obtained when the tone and the background are combined in filters of bandwidths 1, 10, or 100 Hz. Find the averaged pressure spectrum level in each of these filters. (b) Repeat for a tone with a sound pressure level of 150 dB re 1 μPa. (c) Repeat for a tone with a sound pressure level of 160 dB re 1 μPa. (d) Comment on the effect of increasing the sound pressure level of the tone.

11.4. A noise has a spectrum such that the intensity I_1 in each 1-Hz band is given by $I_1 = 10^{-6}/f$ W/m^2 where f is the center frequency of the band in hertz. (a) Compute the intensity spectrum level at 100, 500, and 1000 Hz. (b) What is the intensity level of the entire band of frequencies between 100 and 1000 Hz.

11.5. A noise is represented by an rms acoustic pressure $P_1 = 500/f$ μbar, where P_1 is the pressure in a 1-Hz band centered on the frequency f in hertz. (a) Derive a general expression for the pressure spectrum level PSL of this sound. (b) How does the pressure spectrum level change with frequency expressed in decibels per octave? (c) What is the band level of this noise in a band of 50-Hz width centered on a frequency of 2500 Hz?

11.6. A subject is presented with samples of noise and samples of noise with a signal. In each of the following cases, qualitatively explain what happens to $P(D)$, $P(FA)$, and d'. (a) The mean amplitude of the signal gets weaker and $P(FA)$ does not change. (b) The subject is instructed to say there is a signal present only "when he is sure this is true" rather than "when it may be true." (c) The mean amplitude of the noise gradually gets smaller. (d) The subject is instructed "don't give so many false alarms."

11.7. Two experiments are designed to detect the presence of a signal in noise. One is a yes-no task and the other is a 2AFC task, and both use the same samples of noise and signal with noise. In the first experiment $P(FA) = 0.002$ when $P(D) = 0.5$. For the same $P(D)$, what $P(FA)$ will be expected in the second experiment?

11.8. In a yes-no task with specific instructions to each of the subjects for making the choice, three subjects listen to samples of Gaussian noise with or without a signal of constant amplitude such that $d' = 1$. Subject 1 detected the signal 10 percent of the time when it was present, subject 2 detected the signal 70 percent of the time it was present, and subject 3 detected the signal 40 percent of the time it was present. (a) Estimate the respective $P(FA)$ for each, and (b) list the subjects in order of increasing desire to avoid a false detection.

11.9. A detection system has fixed bandwidth w and a postdetection filter of fixed integration time T_s, but the processing time τ is adjustable. For samples of noise and signal with noise with fixed d', qualitatively explain how the detection threshold changes for square-law detection.

11.10. A signal consists of a tone burst of duration $\tau = 1$ s, a frequency 200 Hz, and an effective pressure amplitude of 0.02 Pa. The noise with which it is combined has a mean pressure amplitude of 2.83×10^{-2} Pa, a bandwidth of 100 Hz, and a standard deviation of 0.41×10^{-2} Pa. (a) Calculate the sound pressure levels re 20 μPa of the signal, noise, and signal with noise. (b) Calculate d' for this case assuming a yes-no task. (c) Find $P(FA)$ for $P(D) = 0.5$. (d) If the detector is a square-law processor with a postdetection filter with $T_s = \tau$, calculate the detection threshold for the specified $P(FA)$ and $P(D)$. (e) Repeat (d) if $T_s = 500$ ms and if $T_s = 2$ s.

11.11. What must be the sound pressure level of a 200-Hz pure tone if it is to be audible in a factory noise spectrum level of 73 dB re 20 μPa/Hz$^{1/2}$ between 100 and 300 Hz? Assume that the critical ratio is the operative.

11.12. Assume that the ear can be modeled by a collection of parallel square-law detectors, each of which has a bandwidth given by w_{cr} and is followed by a postdetection filter that integrates over time T_s. In three separate experiments, using sample of 200-Hz signals of fixed duration τ in Gaussian noise, the following relationships were found to result in $P(D) = 0.6$ and $P(FA) = 0.05$:

Experiment	τ	$A_{S,N}$	A_N	σ
1	4 s	3.44×10^{-2} Pa	3×10^{-2} Pa	0.22×10^{-2} Pa
2	1	3.44×10^{-2}	3×10^{-2}	0.22×10^{-2}
3	0.1	4.51×10^{-2}	3×10^{-2}	0.22×10^{-2}

(a) Find the behavior of the detection threshold DT' as a function of τ for fixed d'. (b) Determine the detection threshold of the processor with filter for each experiment. (c) Determine the integration time T_s. (d) Assuming that T_s is independent of frequency, calculate the resultant DT' for tones of 1-s duration at 100 Hz, 200 Hz, 500 Hz, 1000 Hz, 2000 Hz, and 5000 Hz for the same $P(D)$ and $P(FA)$.

11.13. Use Figs. 11.9 and 11.11 to answer the following questions. (a) What must be the sound pressure level of a 1000-Hz tone if it is to be heard in the presence of a 2000-Hz tone of sound pressure level 80 dB re 20 μPa? (b) What must be the sound pressure level of a 5000-Hz tone in order to be heard in the presence of the 2000-Hz tone?

11.14. (a) Let the pressure exerted by acoustic waves on the eardrum be represented by $p = P \cos \omega t$. Assuming that the subjective response r may be represented by $r = a_1 p + a_2 p^2$, where all the a's are constants, prove that the response contains a constant term, and terms having angular frequencies ω and 2ω. Compute the amplitude of each term as a function of P

and the constants. (b) If the incident sound consists of two wave motions of frequency ω_1 and ω_2 which produce a pressure $p = P_1 \cos \omega_1 t + P_2 \cos \omega_2 t$, determine the frequencies and amplitudes of the response.

11.15. (a) Determine the loudness level and loudness of a pure tone having a frequency of 100 Hz and an intensity level of 60 dB *re* 1 pW/m². (b) To what intensity level must this tone be reduced in order to reduce its loudness to one-tenth of the value obtained above? (c) To what intensity level must it be increased in order to raise its loudness to ten times the value determined in part (a)?

11.16. Six pure tones have the following frequencies and intensity levels (*re* 1 pW/m²): 50 Hz at 85 dB, 100 Hz at 80 dB, 200 Hz at 75 dB, 500 Hz at 80 dB, 1000 Hz at 75 dB, and 10,000 Hz at 70 dB. (a) Determine the loudness level for each of the tones. (b) Assume the intensity level of each of the tones to be decreased by 30 dB. Determine the new values of the loudness level for each of the tones.

11.17. (a) Compute the overall intensity level for the six tones of Problem 11.16(a). (b) What is the total loudness in sone of these six tones? (c) What is the required intensity level of a single 1000-Hz tone having the loudness of part (b)?

11.18. Three pure tones have the following frequencies and sound pressure levels *re* 0.0002 μbar: 100 Hz at 60 dB, 200 Hz at 60 dB, and 500 Hz at 55 dB. (a) Which tone is the loudest? (b) What is the overall sound pressure level of these three tones when sounded simultaneously? (c) What is their total loudness level in phons?

11.19. An approximation is that a 10-dB increase in loudness level is equivalent to twice the loudness. Plot this on Fig. 11.13 normalized so that 1 sone and 40 phon are equivalent, and compare with the plot of (11.14).

11.20. For engineering applications, ISO R131-1959 recommends
$$N = 0.0625 \times 10^{0.03 L_N}.$$
Plot this on Fig. 11.13, and compare with the results of Problem 11.19.

CHAPTER 12

ENVIRONMENTAL ACOUSTICS

12.1 INTRODUCTION. The effect of noise on human emotions ranges from negligible, through annoyance and anger, to psychologically disruptive. Physiologically, noise can range from harmless to painful and physically damaging. Noise can also exert economic factors by decreasing worker efficiency, affecting turnover, altering profit margins, and so forth.

The first step in controlling noise is to compare the existing or potential noise with appropriate rating criteria. Such a comparison not only allows specification of the degree of noise suppression necessary to attain a desired noise environment, but also provides guidance as to what aspects of the noise should be attacked, and how to provide the most cost-effective solution.

The development of noise rating procedures and criteria is complicated by the variety of spectra and time histories displayed by noise and the *variability* of physiological and psychological responses not only *between* people but for the *same* person at different times.

The easiest noise environments to rate are those that are steady or slowly varying both in level and in spectral content. Examples are the noise produced by machinery (such as a ventilating system), which runs at a constant rate, and *ambient noise*, which varies slowly between day and night in a community. Rating procedures for such noises can be tailored to provide accurate predictions of the impact on an "average" individual, along with statements on the percent of the population that will be affected to varying degrees. Examples of such rating procedures include *speech interference levels* (*SIL*) and *noise rating curves* (*NC*).

Most environmental noises are not steady. Examples of unsteady noise range from *impulses* where the sound pressure level is 40 dB *re* 20 μPa or more for 0.5 s or less (a slammed door or a sonic boom), to *single events* of relatively long duration (an aircraft flyover or a motorcycle passby), to the widely *fluctuating* noise measured near a busy intersection.

Because of the number of variables involved, the rating of noise acceptable to a a community is very difficult. No single-number measure yet devised seems capable of satisfying all situations and, instead, a witches' brew of rating systems exists, each applying to a different noise or sociological condition. However, there appears to be general agreement that analysis of instantaneous spectra gives too much information and that the *A-weighted* sound level (to be discussed shortly) is an acceptable measure of noise impact. The various rating systems based on this A-weighted measurement differ only in how the time variation of the level is treated. Examples of rating procedures that use the statistical behavior of the A-weighted sound level are the *day-night averaged sound level* (L_{Adn}) the *50-percentile exceeded sound level* (L_{A50}), and the *community noise equivalent level* (*CNEL*).

One exception to the use of A-weighted sound levels is the calculation of the impact of airport noise, where the *effective perceived noise level* (L_{EPN}), calculated from the instantaneous spectra, is used to make a *noise exposure forecast* (*NEF*).

Excellent general and comprehensive treatments of environmental noise and its control are contained in the books edited by Harris[1] and Beranek.[2]

12.2 WEIGHTED SOUND LEVELS. The simplest and probably most widely used measure of environmental noise is the *A-weighted sound level* (L_A), expressed in dBA or dB(A). (The reference pressure is 20 μPa as discussed in Sect. 5.12 and will be left implicit.) A-weighting assigns to each frequency a "weight" that is related to the sensitivity of the ear at that frequency. For example, in a sound level meter the received signal is passed through a filter network with the dBA frequency characteristic, as shown in Fig. 12.1, and the level of the filtered signal is then determined and displayed. The dBA frequency characteristic was originally designed to mirror the 40-phon equal-loudness-level contour of the 1933 Fletcher-Munson data. It is also a

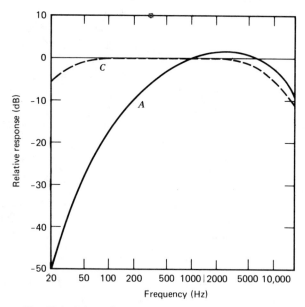

Fig. 12.1. *Filter characteristics for* A- *and* C-*weighted sound levels.*

[1] Harris (editor), *Handbook of Noise Control*, 2nd ed., McGraw-Hill (1979).
[2] Beranek (editor), *Noise and Vibration Control*, McGraw-Hill (1971).

good approximation of the 10-phon contour of the more recent Robinson-Dadson data, which is reproduced in Fig. 11.9. Accurate A-weighted levels can be obtained from octave-band levels by applying the corrections shown in Table 12.1 to each band level and then combining the corrected band levels with the nomograph of Fig. 11.2. Table 12.2 displays A-weighted sound levels for some commonly encountered noises.

Table 12.1. Correction to be added to octave-band levels to convert them to A-weighted band levels

Center Frequency (Hz)	Correction (dB)
31.5	−39.4
63	−26.2
125	−16.1
250	−8.6
500	−3.2
1000	0
2000	+1.2
4000	+1.0
8000	−1.1

Other weightings have been proposed but few have gained widespread acceptance. Most sound level meters will allow selection of either A-weighting, L_A, or C-weighting, L_C. The frequency characteristic for C-weighting (also shown in Fig. 12.1) is nearly flat, rolling off slightly at high and low frequencies. Although no single overall sound level can give information about the spectrum of a noise, measurement of both L_A and L_C will yield some information about the prominence of noise for frequencies below about 200 Hz. If there is significant energy in these lower frequencies, the L_C will be appreciably greater than the L_A.

The L_A is in widespread use mainly because it is inexpensive to obtain and easier for laypersons to appreciate than are any of the other more accurate, but more complicated, noise rating procedures. In addition, for most environmental noises L_A does correlate fairly well with the other rating procedures. While the A-weighted sound level L_A cannot completely replace more precise rating procedures, it has been shown[3] that for a wide variety of environmental noises measurements of L_A and L_C can be used to predict accurately the results of more complicated procedures. When a loudness level (for example, calculated by the method of Stevens, as mentioned in the previous chapter) is plotted against the A-weighted sound level for the same noise, the data for a given $L_C - L_A$ difference fall nearly on a line (Fig. 12.2).

---

[3] Botsford, *Sound and Vibration*, **3**, 16 (Oct. 1969).

Table 12.2. *A*-Weighted sound levels for some commonly encountered noises

A-Weighted Sound Level (dBA)	*Source of Noise*
110–120	Discotheque, rock-n-roll band
100–110	Jet flyby at 300 m (1000 ft)
90–100	Power mower,[a] cockpit of light aircraft
80–90	Heavy truck 64 km/h (40 mph) at 15 m (50 ft), food blender,[a] motorcycle at 15 m (50 ft)
70–80	Car 100 km/h (65 mph) at 7.6 m (25 ft), clothes washer,[a] TV audio
60–70	Vacuum cleaner,[a] air conditioner at 6 m (20 ft)
50–60	Light traffic at 30 m (100 ft)
40–50	Quiet residential—daytime
30–50	Quiet residential—nighttime
20–30	Wilderness area

[a] Measured at position of operator.

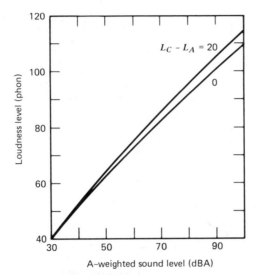

Fig. 12.2. *Relation of loudness levels (Stevens Mark VI) to A-weighted sound levels for noise with the same $L_C - L_A$. The curves fit the data with a correlation coefficient of 0.995 and a standard deviation of 1.1 phon.*

12.3 SPEECH INTERFERENCE. Noise decreases the intelligibility of speech by raising the listener's threshold of hearing while, at the same time, masking the information. This loss of information may be partially compensated for by moving closer, talking louder, or using electronic amplification.

Fortunately, speech is highly redundant. Usually, much of a sentence can be lost without seriously affecting intelligibility; meaning can still be extracted from context. To measure intelligibility, trained talkers recite, clearly and distinctly, specially selected words or sentences to trained listeners. *Intelligibility* is then rated according to the percent of correct responses. The intelligibility of *isolated* words is more strongly affected by noise, but improves markedly with increasing numbers of syllables; disyllabic words are understood about twice as easily as monosyllabic words with the same background noise. Figure 12.3 shows the intelligibility of sentences and words as a function of the relative A-weighted sound levels of speech and noise. For a sentence intelligibility of better than 95 percent, the signal level must at least equal the noise level. Since it is difficult to reconstruct proper names from contexts, a paging system should have word intelligibility over 85 percent. This requires a signal-to-noise ratio of 6 dB; a specification seldom achieved in airport lobbies let alone bus depots.

Measurements show that nearly all the information in speech is contained within the frequency interval 200 Hz to 6 kHz and that the dynamic range of speech in any bandwidth within this interval is 30 dB. If the range of levels heard within any subinterval is limited by noise, then the contribution of this subinterval to the total intelligibility is reduced.

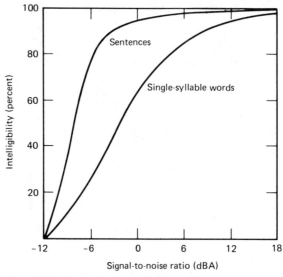

Fig. 12.3. *Percent of words and sentences correctly identified in the presence of background noise.*

A measure of intelligibility suitable for field use is the *preferred speech interference level (PSIL)*, which is the arithmetic average of the noise levels in the three octave bands centered at 500, 1000, and 2000 Hz. More precise determinations, involving the *articulation index*, require exhaustive measurements and are more suited for research purposes.[4]

If octave-band analysis is not available, the effect of noise on the intelligibility of speech can be estimated from measurements of the A-weighted and C-weighted sound levels; Fig. 12.4 gives the corrections to be added to L_A to obtain the *PSIL*.

In the absence of all other information, the A-weighted sound level can be used

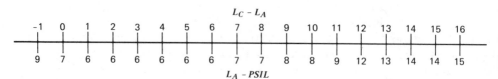

Fig. 12.4. *Nomogram for calculating preferred speech interference level (PSIL) from weighted sound levels.*

[4] American National Standards Institute (ANSI), S-3.5 (1969).

to get a *rough* estimate of the intelligibility of speech under various noise conditions. A *PSIL* estimated by

$$PSIL \sim L_A - 9 \tag{12.1}$$

will be in error by less than 4 dB for all but the most pathological noise spectra.

The relation among the quality of speech communication, the *PSIL*, and distance r between talker and listener has been studied by Webster.[5] His results for face-to-face communication can be summarized by

$$VL_A \geq \tfrac{4}{3}(PSIL + 20 \log r) - 36 \tag{12.2}$$

where VL_A is the *A-weighted voice level* measured at 1 m, which is necessary for a listener at r meters to understand virtually all sentences. Measurements made with untrained voices have established that a *normal* voice corresponds to a voice level of 57 dBA, a *raised* voice to 65 dBA, a *very loud* voice to 74 dBA, a *shout* to 82 dBA, and *maximum effort* to 88 dBA. Consequently, conditions of noise level and distance that require a voice level $VL_A < 57$ dBA can be classified as being *excellent for unassisted speech communication*. Similarly, for $57 < VL_A < 65$ dBA the *condition is satisfactory*, for $65 < VL_A < 74$ *adequate*, for $74 < VL_A < 82$ *difficult*, for $82 < VL_A < 88$ *impractical*, and for $VL_A > 88$ *impossible*.

12.4 PRIVACY. In addition to speech *intelligibility*, speech *privacy* in multi-family dwellings, offices, and open-plan classrooms is important not only to protect confidentiality, but also to prevent intrusion on the privacy of others. The degree of privacy depends not only on the sound insulation of any intervening wall or barrier, but also on the level of background noise. The insulative properties of partitions will be discussed in Sect. 12.13; here we will consider only the effect of background noise. It is seen from Fig. 12.3 that, if the A-weighted noise in the receiving room is 9 dBA above the A-weighted speech level that is transmitted through the wall, only about 10 percent of the words and 30 percent of the sentences will be understood. This should ensure adequate privacy with little distraction in an office, but an even lower speech level would be required in a dwelling unit.

Privacy can be enhanced by either decreasing the amount of sound that passes through the wall (by building a better wall), decreasing the speech level in the source room (by adding acoustic absorption), or increasing the noise level in the receiving room (by some artificial means such as a ventilator screen designed to produce aerodynamic noise). A procedure for calculating the degree of privacy that takes into account the above considerations is discussed in Beranek.[2]

12.5 NOISE RATING CURVES. A noise that is acceptable for speech interference may be annoying for other activities (listening to music or sleeping). While two continuous noises with the same frequency spectrum can be rated simply by comparing their total energies, time-varying noises with different spectra require a rating procedure that accounts not only for the increased sensitivity of the ear to the

---

[5] Webster, *J. Acous. Soc. Am.*, **37**, 692 (1965).

middle range of frequencies and the intermittent nature of the noise, but also for the subjective qualities of the noise.

The procedure generally accepted for rating steady broadband noise containing no significant tones, such as the noise produced by some kinds of machinery, ventilating systems, and distant road traffic, consists of comparing measured octave-band levels with a family of criteria curves. These criteria curves take into account the sensitivity of the ear by specifying lower acceptable levels for the middle frequencies. These curves should not exactly mirror the sensitivity of the ear, because noises with such a spectrum are usually judged to be too "rumbly" and "hissy." A noise with a more acceptable quality is obtained if the levels at both low and high frequencies are lower than those derived from the sensitivity of the ear.

Several procedures for rating steady noise are in general use, differing only in the shapes of their criteria curves. The original *noise criteria* (*NC*) of Beranek, specified in 1957, were based on the old octave bands. These were subsequently

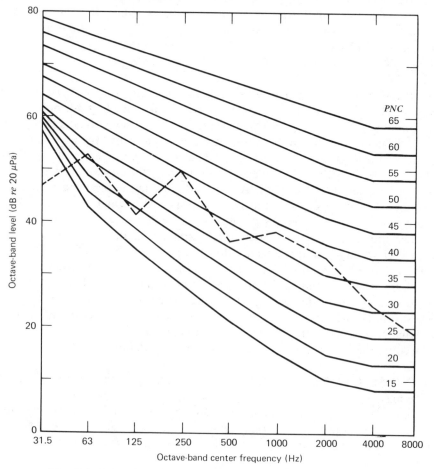

Fig. 12.5. *Octave band levels for the preferred noise criteria (PNC).*

revised to conform to the new preferred octave bands. These updated *NC* curves met with some objections and, in 1971, Beranek proposed another set of curves, the *preferred noise criteria (PNC)* shown in Fig. 12.5.

To determine the rating of a noise, its octave-band levels are plotted on the same graph as the criteria curves. The highest curve reached by any octave-band level of the noise gives the *PNC* rating of the noise. The dashed line in Fig. 12.5 is the spectrum of a noise that is rated at *PNC* = 40.

Because of the similarities of the criteria curves to each other and to A-weighted sound levels, they can be related approximately by

$$PNC \approx NC + 3.5$$

$$NC \sim 1.25(L_A - 13)$$

$$PNC \sim 1.25(L_A - 10) \tag{12.3}$$

The maximum noise rating recommended for a room depends not only on the intended use of the room, but also on such factors as the expectations of the users. For example, an urban dweller will tolerate significantly higher ambient noise levels than will a rural dweller. In fact, the urban dweller may be bothered by a low noise level because sounds otherwise masked can be clearly heard and may prove distracting. Table 12.3 shows recommended acceptable noise ratings for a variety of different room uses.

Table 12.3. Recommended acceptable noise levels in *unoccupied* rooms.

Location	Noise Criteria (NC)
Concert hall, recording studio	15–20
Music room, legitimate theater, classroom	20–25
Church, courtroom, conference room, hospital, bedroom	25-30
Library, business office, living room	30–35
Restaurant, movie theater	35-40
Retail shop, bank	40-45
Gymnasium, clerical office	45–50
Shops and garages	50–55

12.6 THE STATISTICAL DESCRIPTION OF COMMUNITY NOISE.

Figure 12.6 shows the A-weighted sound levels for two time periods in a typical suburban environment. These curves illustrate the basic characteristics of most community noises: a fairly steady *residual noise level* associated with distant, unidentifiable sources on which are superimposed discrete *noise events* of identifiable origins. The residual noise level varies slowly with time, usually displaying diurnal, weekly, and seasonal cycles but with maximum excursions rarely exceeding about 10 dBA. The

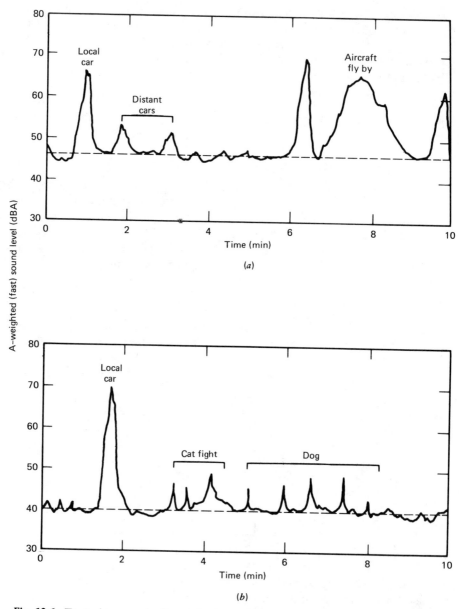

Fig. 12.6. *Typical community A-weighted sound levels in (a) daytime and (b) nighttime.*

individual noise events vary in magnitude and duration, rising as much as 40 dBA above the residual level for seconds, minutes, or even longer.

Continuous reading of the A-weighted sound level provides the basis for determining the statistics of community noise. From the recording of the A-weighted sound level over a period of time it is possible to construct a *histogram* (Fig. 12.7a), which displays the percent of the total sample time that the noise level spends within each increment of level, or a *cumulative distribution* (Fig. 12.7b), which displays the percent of the total sample time that the noise level spends *above* each value of the noise level. The statistical properties of the noise can then be determined from either of these graphs. (Continuous recording and sophisticated analysis equipment are not necessary; manual sampling of the noise level at equal time intervals gives the same results but requires more effort.)[6]

Figure 12.7 shows that the distribution of community noise is *not* Gaussian, but rises rather steeply above the residual noise level and displays a long tail, representing very noisy single events of relatively rare occurrence. The length of this tail varies markedly with location, often being more extended for measurements made in the vicinity of an airport or factory.

The measure of the environmental impact of a noise should depend on the total energy received, the rate of occurrence of noise events, and the magnitudes of noisier single events. The following are some of the A-weighted quantities used in measuring the effects of environmental noise.

(a) *Equivalent continuous sound level* (L_{Aeq}). The steady-state sound that has the same A-weighted level as that of the time-varying sound averaged in energy over the specified time interval.

(b) *Daytime average sound level* (L_{Ad12}). The L_{Aeq} calculated from 7 A.M. to 7 P.M.

(c) *Evening average sound level* (L_{Ae}). The L_{Aeq} calculated from 7 P.M. to 10 P.M.

(d) *Night average sound level* (L_{An}). The L_{Aeq} calculated from 10 P.M. to 7 A.M.

(e) *Hourly average sound level* (L_{Ah}). The L_{Aeq} calculated for any one-hour period.

(f) *Day-night averaged sound level* (L_{Adn}). The 24-hour L_{Aeq} obtained after addition of 10 dBA to the sound levels from 10 P.M. to 7 A.M.

(g) *Noise exposure level* (L_{Aex}). The level of the time integral of the squared, A-weighted sound pressure over a stated time referenced to (1 s) $\times$ (20 μPa)^2.

(h) *x-percentile-exceeded sound level* (L_{Ax}). The *fast*, A-weighted sound level equaled or exceeded x percent of the sample time. Most commonly used are L_{A10}, L_{A50}, and L_{A90} (the levels exceeded 10, 50, and 90 percent of the time, respectively).

(i) *Single event exposure level* (*SENEL*). The L_{Aex} determined for a single event.

(j) *Community noise equivalent level* (*CNEL*). The 24-hour L_{Aeq} obtained after addition of 5 dBA to the sound levels from 7 P.M. to 10 P.M. and 10 dBA to the levels from 10 P.M. to 7 A.M.

---

[6] Yerges and Bollinger, *Sound and Vibration*, 7, 23 (Dec. 1973).

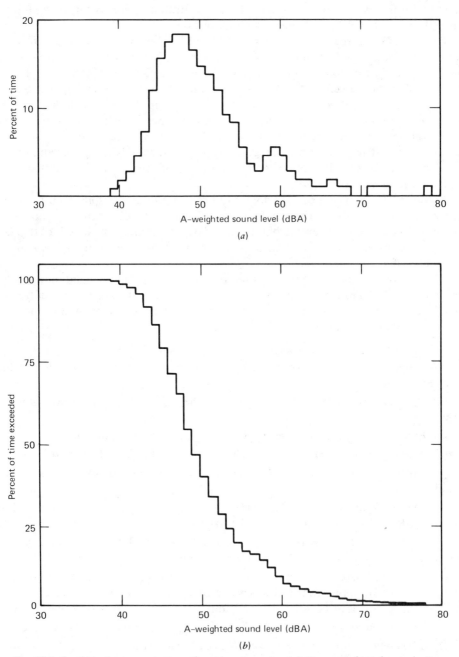

Fig. 12.7. *Statistical representation of community noise. (a) Percent of total sample time that the level is within each increment of level. (b) Percent of total sample time that level is above each value of the level.*

Sets of measurements made at the same location on different days will show different results even when allowances are made for obvious changes such as week-day versus weekend traffic or an airport approach pattern that depends on the wind direction. For example, measurements of urban noise made on four different days may show a range for L_{A50} of 9 dBA. Figure 12.8 displays, for an estimated standard deviation of any L_A, the number of runs needed to determine (with 90 percent confidence) that the average of this L_A is known within the indicated uncertainty. For example, for a sample with an estimated standard deviation of 4 dBA, 13 runs

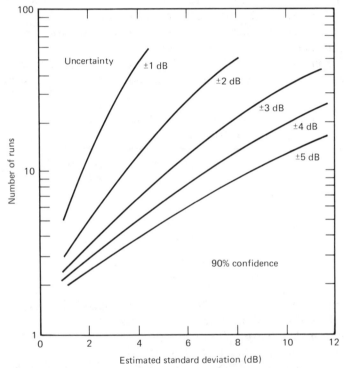

Fig. 12.8. *Uncertainty of the average level obtained from a number of runs.*

are necessary to obtain the correct average of the desired L_A, to within 2 dBA, nine times out of ten.

12.7 CRITERIA FOR COMMUNITY NOISE. Originally, complaints about community noise were taken into court where a judge would decide, on the basis of common-law precedent, if the plaintiff's use or enjoyment of his or her property was harmed. As the pervasiveness of noise sources grew and the adverse effects of noise on the general public became more obvious, local governments either included noisemakers in their general ordinances defining " disorderly conduct," or enacted new ordinances forbidding "unnecessary" or "excessive" noise. Enforcement was left to the local police officers.

As our knowledge about community noise and its effects has become more sophisticated, so have our noise ordinances. A rating procedure, suitable when simplicity of understanding and enforcement are paramount, is exemplified by the noise ordinance of Gainesville, Florida.[7] This ordinance specifies the maximum allowable noise level by district and time of day (Table 12.4), with the provision that these maximum levels are not to be exceeded for more than 3 cumulative minutes in any hour. Motor vehicles are treated separately: Measurements made 15 m (50 ft) from the center of the lane in which the vehicle is traveling must not exceed 85 dBA for trucks and buses and 79 dBA for cars and motorcycles. Exempted from these restrictions are air conditioners, lawn mowers, and construction equipment "operating within the manufacturer's specifications."

Criteria from state and federal agencies are generally more complicated and they usually rely on one of the rating procedures discussed in the previous sections.

Table 12.4. Maximum allowable noise level limits for Gainsville, Florida

	Noise Level Limits (dBA)	
Location	Day	Night
Residential	61	55
Commercial	66	60
Manufacturing	71	65

For example, in California if the one-year average of the $L_{A\mathrm{dn}}$ at the site of a proposed multifamily dwelling, hotel, or motel exceeds 60 dBA, the state requires an analysis showing that the building plans provide adequate acoustic insulation to reduce the average interior $L_{A\mathrm{dn}}$ to 45 dBA or less.

Table 12.5 summarizes the recommendations of several federal and state agencies as to the suitability of various land uses subject to various levels of external noise. For structures in use during only part of the day, the appropriate level is $L_{A\mathrm{eq}}$ measured during the period of use. For all-day use, $L_{A\mathrm{dn}}$ is the appropriate measure.

The criteria for new and rehabilitated residential construction are spelled out in more detail by the U.S. Department of Housing and Urban Development (HUD). They recommend that *interior* levels with the windows *open* (unless there is sufficient mechanical ventilation) should not exceed (a) 55 dBA for more than 60 min in any 24-h period, (b) 45 dBA for more than 30 min from 11 P.M. to 7 A.M., and (c) 45 dBA for more than 8 h in any 24-h period.

The U.S. Department of Transportation (DOT) specifies the maximum noise

[7] Schwartz, Yost, and Green, *Sound and Vibration*, **8**, 24 (Dec. 1974).

Table 12.5. Compatibility of land use with noise environment.[a]

Facility	Outdoor L_{Adn} or L_{Aeq}				
	65–70	70–75	75–80	80–85	85–90
Home	25[b]	30[b]	No	No	No
Hotel, motel, church, classroom, library	25[b]	30[b]	35[b]	No	No
Office, store, bank, restaurant	Yes	25	30	No	No
Outdoor music	No	No	No	No	No
Industry, manufacture	Yes	25[c]	30[c]	35[c]	No
Playground	Yes	Yes	No	No	No
Gymnasium	Yes	Yes	25	30	No

Source: Adapted from Air Force Environmental Planning Bulletin 125 USAF/-PREVX Env. Planning Div., 1976.

Yes = Land use compatible with noise environment. Normal construction may be used.

No = Land use not compatible with noise environmental even with special construction.

25/30/35 = Construction must provide the indicated outdoor-to-indoor noise level reduction in dBA.

[a] If facility is in use for only part of the day, use L_{Aeq} for the period of use.

[b] Use is discouraged and is acceptable only if alternative development options are not available.

[c] Noise reduction required only in areas sensitive to noise.

level of a new highway for various land-use categories. These *exterior* levels, expressed as L_{A10}, are: (a) 60 dBA for lands where quiet is of extraordinary significance such as amphitheaters and open spaces dedicated to special qualities of serenity and quiet; (b) 70 dBA for residences, motels, hotels, public meeting rooms, schools, churches, libraries, hospitals, picnic areas, recreational areas, playgrounds, active-sports areas, and parks; and (c) 75 dBA for all other developed lands not included in the above categories.

Because of their importance, highways and aircraft noise will be discussed in separate sections.

12.8   HIGHWAY NOISE.   Figure 12.9 displays the variation of the A-weighted sound level with time for low and high traffic densities. At low traffic densities, the peaks in the sound level correspond to individual vehicles passing the receiver. At high traffic densities, these peaks coalesce, increasing the average noise level so that only the peaks of the noisiest vehicles can be identified. Thus, the fluctuation of traffic noise decreases as traffic density increases. Fluctuation will also decrease for a fixed traffic density as the receiver is moved away from the roadway, since at greater distances the peaks for the individual vehicles will be broader and smoother.

Highway noise is described by the usual descriptions of community noise: L_{A10}, L_{A50}, and L_{Aeq} being the most commonly encountered.

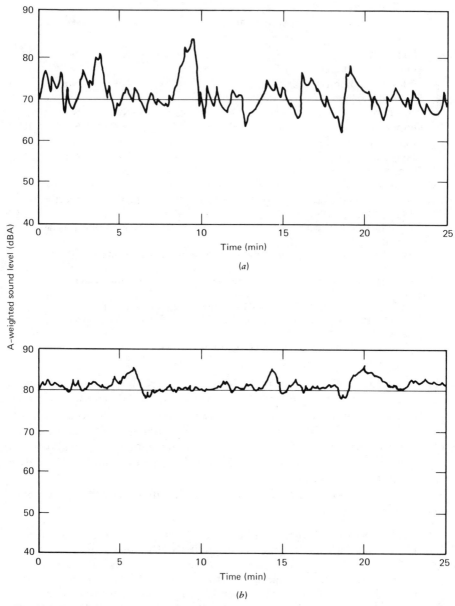

Fig. 12.9. A-weighted sound levels measured near a highway with (a) low traffic density and (b) high traffic density.

The noise radiated by a moving vehicle originates partly from the propulsion system and partly from the tires. Measurements made on a large number of automobiles and light trucks under cruising conditions give typical A-weighted sound levels at 15 m (50 ft) to be

$$L_A = 71 + 32 \log(v/88) \tag{12.4}$$

where v is the vehicle's speed in kilometers per hour (88 km/h = 55 mph). Tire noise predominates at all but the lowest speeds.

Similar measurements for motorcycles give

$$L_A = 78 + 25 \log(v/88) \tag{12.5}$$

and for trucks

$$L_A = 84 \qquad\qquad v \leqslant 48 \text{ km/h}$$
$$= 88 + 20 \log(v/88) \qquad v > 48 \text{ km/h} \tag{12.6}$$

Elaborate procedures have been developed to predict highway noise based on the expected traffic density, composition, and road geometry. As a simplified example, consider a straight, two-lane road of infinite extent, zero grade, and negligible truck traffic. The receiver is located at a distance d from the centerline of the nearest lane. Also needed for the calculation are the average speed v (in kilometers per hour) and the flow rate Q (in vehicles per hour). First, the A-weighted noise level L_{Aeq} is found for a standard receiver distance of 15 m (50 ft) from

$$L_{Aeq} = 39 + 10 \log Q + 22 \log (v/88) \tag{12.7a}$$

This level is then adjusted for cylindrical spreading (the highway is essentially a line source) and absorption over the ground to a distance d (in meters) by adding to L_{Aeq} a correction

$$\Delta L = -a \log \left[\frac{d}{15} + \left(\frac{d-15}{75} \right)^2 \right] \tag{12.7b}$$

where $a = 13.3$ over ground and 10.0 if the line of sight from the road surface to the receiver is $10°$ or more above the slope of the terrain.

From the values of L_{Aeq} determined for the times of interest, the desired sound levels (L_{A10}, L_{A50}, etc.) can be found.

As a numerical example, let us calculate L_{Aeq} for $Q = 6000/h$, $v = 88$ km/h (55 mph), $d = 61$ m (200 ft), and $a = 13.3$. From (12.7a), $L_{Aeq} = 76$ dBA and from (12.7b) $\Delta L = -8$ dBA. The equivalent continuous noise level is 68 dBA. For details on handling more realistic conditions consult Harris.[1]

12.9 AIRCRAFT NOISE RATING. Because of the large number of people severely affected by aircraft noise, and because of the great economic cost of reducing the impact of this kind of noise, aircraft noise has received more attention than any other environmental noise. The result of this attention is that the rating of aircraft noise is more complicated (and expensive) than any of the other ratings.

Since rating airport noise requires rating each class of aircraft for both takeoffs and landings, we begin by describing how single aircraft operations are rated. At each selected location in the vicinity of the airport, the 1/3-octave-band spectra for 24 bands with center frequencies from 50 to 10,000 Hz are obtained for each 0.5-s interval the aircraft noise is above the background noise. Even with modern equipment such as real-time analyzers with memory, this is a prodigious task when consideration is given to the number of locations that must be surveyed for each aircraft class and flight plan. Much reliance is therefore placed on computer simulations that model the source spectra of various aircraft classes and "fly" the aircraft along various flight paths.

Each spectrum is converted into a *tone-corrected preceived noise level* (L_{TPN}) by a procedure that takes into account details of the spectrum and adds a correction for dominant tones. The total effect of the flyby (still at one location) is then expressed as the *effective perceived noise level* (L_{EPN}) the maximum L_{TPN} plus a duration correction that accounts for the time the L_{TPN} is above a certain specified level. Readers desiring a more detailed description of the procedure should consult Harris.[1]

Once the L_{EPN} is calculated at all locations around the airport for a given aircraft class and operation, contours of equal L_{EPN} can be constructed for superposition over a map of the airport and surrounding community. Such noise *footprints* are useful for evaluating the noise impact of new aircraft or changes in flight procedures.

The L_{EPN} also serves as the input for rating the total noise environment around an airport. The *noise exposure forcast* (*NEF*) for a given·class aircraft i on a flight path j is defined as

$$NEF_{ij} = L_{EPNij} + 10 \log(N_d + 17 N_n) - 88 \tag{12.8}$$

when N_d and N_n are the numbers of daytime and nighttime events of this type. Note that a nighttime event is considered to be 17 times more significant than the same event in the daytime. The total *NEF* at a given location is the combination of all aircraft classes and flight paths

$$NEF = 10 \log \left[\sum_{i,j} \text{antilog}(NEF_{ij}/10) \right] \tag{12.9}$$

The *NEF* values at different locations are then connected to produce contours of equal *NEF*.*

On the basis of *NEF* contours for existing conditions, decisions can be made on projected land use. For example, HUD funds cannot be used in areas with $NEF > 40$, but areas with $NEF < 30$ are acceptable. For areas with *NEF* between 30 and 40 special approval is needed. Predicted *NEF* contours can be used to estimate the impact on the community of changes such as the number and timing of flight operations or changes in the mix of aircraft.

The Noise Standard for California airports is somewhat simpler. It uses the measured hourly average A-weighted sound levels L_{Ah} and calculates a *community noise equivalent level*

$$CNEL = 10 \log \frac{1}{24} \left(\sum_{7\text{A.M.}-7\text{P.M.}} 10^{L_{Ah}/10} + 3 \sum_{7\text{P.M.}-10\text{P.M.}} 10^{L_{Ah}/10} + 10 \sum_{10\text{P.M.}-7\text{A.M.}} 10^{L_{Ah}/10} \right) \tag{12.10}$$

where the sums are over the hours in the daytime, evening, and night periods. Note that evening flights are given 3 times the weight of daytime flights, and night flights 10 times the weight. According to the California Administrative Code, the *CNEL* in residential communities adjoining airports is not to exceed 70 dBA until December 1985 and 65 dBA thereafter.

12.10 COMMUNITY RESPONSE TO NOISE. The most difficult aspect of environmental acoustics (with the possible exception of getting acousticians to agree on a rating system) is predicting community response to a given noise environment. The response to noise varies greatly from person to person and all attempts to quantify this subject have had

---

* A *composite noise rating* (*CNR*), approximately $NEF + 75$, is sometimes encountered.

to rely on the subjective judgments of the investigators. A discussion of the methodology of such investigations is beyond the intent of this book, and we will have to be content with presenting a few of the more interesting conclusions.

One approach to quantifying community response to noise begins with the A-weighted sound level, adds corrections for various noise characteristics, and then compares the corrected dBA to a scale of expected reaction. Table 12.6 displays the corrections of one of the simpler versions of this technique. If the corrected level is less than 45 dBA, no community reaction is to be expected; if it is between 45 and 55 dBA, sporadic complaints are to be

Table 12.6. Corrections to be added to the A-weighted sound level to produce a measure of community reaction

Noise Characteristics	Correction in dBA
Pure tone present	+ 5
Intermittent or impulsive	+ 5
Noise only during working hours	− 5
Total duration of noise each day	
Continuous	0
Less than 30 min	− 5
Less than 10 min	− 10
Less than 5 min	− 15
Less than 1 min	− 20
Less than 15 s	− 25
Neighborhood	
Quiet suburban	+ 5
Suburban	0
Residential urban	− 5
Urban near some industry	− 10
Heavy industry	− 15

expected; between 50 and 60 dBA, widespread complaints; between 55 and 65 dBA, threats of community action; and over 65 dBA, vigorous community action is certain. A similar approach is contained in an International Standardization Organization recommendation, ISO R1996(1971).

For example, let's estimate the response to a dog kennel located in a suburban area with an ambient background level of 37 dBA. With the dogs barking, the noise level at the point of possible complaint is measured to be 72 dBA. The dogs bark at random times during the day and night for a total of 20 min per day. From Table 12.6, a correction of + 5 is applied for intermittent character, − 5 for duration less than 30 min, and 0 for surburban neighborhood; this yields a total correction of 0 and a corrected level of 72 dBA. The expected reaction would be " vigorous community action. "

Schultz[8] has made the interesting suggestion that a sufficient measure of community reaction is the percent of people "highly annoyed." By studying the results of 11 surveys of community response to transportation noise (aircraft, railroad, and street traffic) conducted in

[8] Schultz, *J. Acous. Soc. Am.*, **64**, 377 (1978).

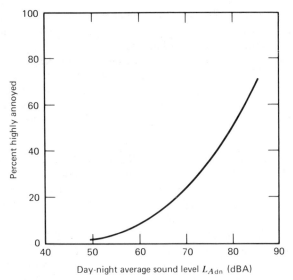

Fig. 12.10. *Estimate of the extent of public annoyance to transportation noise.*

6 countries, he found the relationship between percent highly annoyed and $L_{A\mathrm{dn}}$, which is shown in Fig. 12.10. While this relationship is based on measurements of transportation noise, Schultz speculates that it may be applicable to community noise of other kinds.

12.11 NOISE-INDUCED HEARING LOSS. *Hearing loss* is quantified by specifying the permanent threshold shift (PTS) as a function of frequency. *Hearing impairment* is a broader term specifying the loss in the ability to understand speech. A commonly used classification of hearing impairment, shown in Table 12.7, is based on the average PTS for 500, 1000, and 2000 Hz measured in the lesser impaired of the ears.

Noise-induced hearing loss occurs in two ways: (1) *Trauma*: high intensity sound, such as that from an explosion or jet engine, can rupture the eardrum,

Table 12.7. Classification of hearing impairment

Average Hearing Loss at 500, 1000, and 2000 Hz (dB)	*Classification*
Less than 25	Within normal limits
26–40	Mild or slight
41–55	Moderate
56–70	Moderately severe
71–90	Severe
More than 91	Profound

damage the ossicles, destroy the sensory hair cells, or cause collapse of a section of the Organ of Corti. Such hearing loss is sudden and always associated with a particular noise event. (2) *Chronic*: noise levels below those necessary to cause trauma, if repeated often enough or continued for a long enough time, can cause dysfunction or destruction of the hair cells. This results from metabolic stress on the maximally stimulated cells. This type of hearing loss is more insidious than that brought on by trauma because the individual may not be aware of a slowly increasing loss.

Since controlled experiments on *PTS* in humans are unacceptable, most knowledge about hearing loss is obtained from field studies of workers who have been subjected to industrial noise for periods of time, and from inferences made from laboratory studies of induced temporary threshold shifts (*TTS*).

After cessation of a noise of intensity sufficient to cause *TTS* but not *PTS*, the ear recovers its initial threshold in a characteristic pattern. The *TTS* first decreases but then increases to a maximum at about 2 min (the *bounce effect*), after which *TTS* decreases linearly with log t until it reaches 0 dB. Because of the complicated initial behavior, measurements of *TTS* are always taken 2 min after exposure ends.

For octave bands centered at 250 and 500 Hz, there is no threshold shift, no matter how long the duration of the exposure, as long as the noise is below 75 dB. For octave bands at 1, 2, and 3 kHz, there is no shift for levels below 70 dB. For band levels between 80 and 105 dB and for exposure times less than 8 h, the *TTS* increases linearly with log t, the rate of increase being proportional to the level of the noise. For durations greater than 8 h, the *TTS* approaches an asymptotic value depending on the noise level. Pre- or post-exposure to sound levels of 70 dB or lower has no effect on the amount of *TTS* or the rate of growth or decay.

For a given excitation frequency, the frequency of maximum *TTS* occurs from 1/2 to 1 octave above the source frequency. For example, a pure tone of 700 Hz will produce a maximum *TTS* at 1 kHz.

Predicting the *TTS* for fluctuating or intermittent noises is difficult. For fluctuating noise, it appears that the average pressure level is important and not the total energy. For intermittent noise with an on-time between 250 ms and 2 min, the *TTS* is proportional to the fraction of on-time. For on-times less than 250 ms, the *TTS* is generally greater than that predicted by the above rule. For on-times greater than 2 min, predictions can be made considering the usual growth and recovery properties.

Studies of industrial workers exposed to the same noise for a number of years display great individual differences. However, the median *PTS* shows regular features depending on the intensity and length of exposure. In general, hearing loss first appears at frequencies near 4 kHz—the frequency region of the greatest acuity. With continued exposure, the severity of the hearing loss increases and extends to lower and higher frequencies. The *PTS* at 4 kHz increases with the A-weighted levels of the noise and with the exposure time up to 10 years, after which it appears to reach an asymptotic value. Figure 12.11 shows the median *PTS* for a 10-year exposure as a function of the A-weighted sound level.

The effect of continuous industrial noise exposure for 8 h each working day for 10 years can be summarized as follows.

1. For the median individual, no hearing loss will occur if the sound level is below 80 dBA. While data are scarce, the Environmental Protection Agency (EPA) accepts that a level of 70 to 75 dBA will produce a small but detectable PTS at 4 kHz in 1 to 10 percent of the population.

2. At 80 dBA, shifts in the median PTS between 3 and 6 kHz begin to appear.

3. At 85 dBA, median shifts of 10 dB occur between 3 and 6 kHz, with shifts of 15 to 20 dB in the susceptible 10 percent of the population.

4. At 90 dBA, median shifts of 20 dB are expected between 3 and 6 kHz, but the thresholds at 500, 1000, and 2000 Hz are still uneffected. Thus, by definition, there is no hearing impairment.

5. Above 90 dBA, shifts at 500, 1000, and 2000 Hz begin to appear, indicating the onset of hearing impairment.

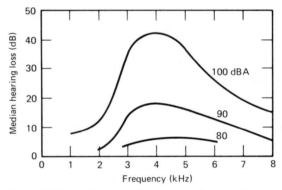

Fig. 12.11. *Median hearing loss after a 10-year exposure to industrial noise.*

From the above discussion it is seen that the amount of hearing loss depends both on the level of the noise and on its duration. In establishing criteria for noise exposure, it is necessary to determine the trade-off between level and duration. The experiments with TTS indicate that a 5-dB increase in level is equivalent to a doubling of exposure time.

On the bases of these and other findings, the Occupational Safety and Health Act (OSHA) of 1970 prescribes permissible exposures in industries doing business with the U.S. federal government. These levels are shown in Table 12.8. If the exposure consists of two or more periods at different noise levels, then the limit is that $\sum t_i/T_i$ not exceed unity, where t_i is the exposure time at a level whose total allowed exposure time is T_i.

The implicit philosophy behind these criteria should be made explicit. (1) These levels provide protection *only* for frequencies necessary for the understanding of speech, and hearing loss at 4 kHz and above must be accepted as part of the job. (2) They assume an 8-h day, 5-day week exposure time and that for the rest of the time

Table 12.8. Permissible daily noise exposure limits for industrial noise (OSHA).

Limiting Daily Exposure Time (h)	A-Weighted Sound Level Slow Response (dBA)
8	90
6	92
4	95
3	97
2	100
1.5	102
1	105
0.5	110
Less than 0.25	115

no further ear damage occurs as might accompany recreational activities with high noise levels. (3) They are designed to protect only 85 percent of the exposed population, with financial compensation provided for the hearing impairment experienced by the more susceptible 15 percent of the population. As an example of one alternative approach that adopts a different philosophy, see Reference 9.

Although the OSHA criteria provide guidelines for judging the effects of nonoccupational noise, the above limitations must be taken under consideration when estimating the possible dangers to hearing from recreational or other nonoccupa-

Table 12.9. Suggested daily noise exposure levels for nonoccupational noise.

Limiting Daily Exposure Time	A-Weighted Sound Level Slow Response (dBA)
Less than 2 min	115
Less than 4 min	110
Less than 8 min	105
15 min	100
30 min	95
1 h	90
2 h	85
4 h	80
8 h	75
16 h	70

[9] Dear and Karrh, *Sound and Vibration,* **13,** 12 (Sept. 1979).

tional noises. Table 12.9 shows recommended daily exposure times for nonoccupational noise.[10] For comparison with these levels, notice that typical levels for a rock-n-roll band are 108 to 114 dBA, for a power mower 96 dBA, inside the cockpit of a light aircraft 90 dBA, and at 7.6 m (25 ft) from a motorcycle 90 dBA.

12.12 NOISE AND ARCHITECTURAL DESIGN. An important objective of architectural design is to provide sufficient acoustic isolation to prevent externally and internally generated noise from interfering with the designated use of a space. In Sect. 12.5, noise criteria applicable to spaces intended for various uses were described. In this and the next three sections, procedures for meeting these criteria will be discussed.

The first line of defense against noise is city planning; zoning regulations should encourage the maximum possible separation between areas designated for noise-intensive use (heavy commercial, highways, airports, etc.) and noise-sensitive areas (residential, hospitals, parks, etc.).

Given an existing or potentially adverse acoustic environment, the architect can do much to alleviate economically the impact of the noise by considering acoustics *at the start.* Buildings can be located and oriented to provide noise barriers for each other; rooms requiring quiet should be placed on the side of a building away from major noise sources; within a building, noisy areas (such as kitchens, hallways, utility rooms, stairwells, and family rooms) should be separated by buffer zones from noise-sensitive areas (such as bedrooms, private offices, and study rooms).

To reduce the noise generated within the building, the architect should specify low-noise machinery (air conditioners, laundry facilities, water valves, etc.) mounted on properly designed and installed vibration isolators. Other considerations, such as resilient floor coverings in halls and stairwells, will reduce the noise produced by nonmechanical sources.

The architect's final defense against intrusive noise is to acoustically isolate the room by specifying constructions that inhibit the transmission of structure-borne and airborne sounds.

The transmission of noise into a room is aided by the multitude of paths sound can find to penetrate the architect's defenses. The most prominent paths are (1) airborne noise outside the room that sets the common wall into vibration, which in turn radiates sound into the room, and (2) noise originating in the vibration of a solid structure (machinery, footfalls, etc.) that propagates along the structure and sets surfaces in the room into vibration thereby radiating noise into the room. If the above paths are efficiently blocked by properly designed partitions and resilient mountings, then *flanking* paths can become important. Some flanking paths are obvious, such as the propagation through a false ceiling or crawl space and window-to-window transmission. Others are more insidious, such as porous cement block, poor seals between walls and ceiling or floor, gaps around wall penetrations, and back-to-back electrical outlets.

[10] Cohen, Anticaglia, and Jones, *Sound and Vibration*, **4**, 12 (Nov. 1970).

12.13   SPECIFICATION AND MEASUREMENT OF SOUND ISOLATION.

The *transmission loss* (TL) for a partition is defined as

$$TL = 10 \log (\Pi_i / \Pi_t)$$

where Π_i is the total power incident on the source side of the partition and Π_t is the total power transmitted through the partition. The transmission loss depends only on the frequency and the properties of the partition. A moment's thought will reveal that for a given partition (fixed TL), the noise reduction experienced between two rooms decreases as the area of the partition increases and increases as the absorptivity of the receiving room increases. (For the same intensity in the source room, a receiving room with little sound absorption will have a higher sound level than one with a large absorption.) A quantity of more direct architectural interest is the *noise reduction*

$$NR = 10 \log (I_1 / I_2) = L_1 - L_2 \tag{12.11}$$

where I_1 and I_2 are the intensities and L_1 and L_2 are the sound pressure levels in the source and receiver rooms, respectively.

The transmission loss and the noise reduction can be related: The power incident on a partition of area S is $\Pi_1 = I_1 S$. The power transmitted into the receiving room is equal to the rate at which energy is absorbed in the receiving room, $\Pi_2 = I_2 A$ where A is the sound absorption of the receiver room (Sect. 13.3). Combination shows that

$$TL = NR + 10 \log (S/A) \tag{12.12}$$

Rather than measure NR for each frequency, L_1 and L_2 are usually specified to be spatially averaged 1/3-octave-band levels. Spatial averaging is used to reduce the effects of standing waves in both the source and receiving rooms.

Based on studies made with noise sources typical of multifamily dwellings, the *sound transmission class* (STC) provides a fairly successful single-number specification of the acoustic isolation characteristics of a partition.

To determine the STC of a partition, its TL is measured in the 16 contiguous 1/3-octave bands between 125 and 4000 Hz, inclusive. These measured values of TL are then compared to a family of reference contours, each of which consists of three straight lines: a low-frequency segment that increases by 15 dB from 125 to 400 Hz, a middle segment that increases by 5 dB from 400 to 1250 Hz, and a horizontal segment at high frequencies (Fig. 12.12). To determine the STC of a partition, the reference contour is chosen so that the maximum deficiency (deviation of the data below the contour) at any one frequency does not exceed 8 dB and the total deficiency at all frequencies does not exceed 32 dB. The STC of the partition is then the value of the TL corresponding to the intersection of the chosen reference contour with the 500-Hz ordinate.

Wall and ceiling/floor construction as well as door and window installations can be measured in the laboratory and their TL and STC values tabulated for use by the architect. Knowing the severity of the problem, the architect can chose the construction that will provide the required isolation. Table 12.10 gives some values

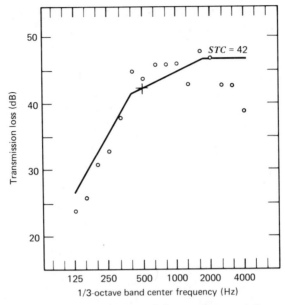

Fig. 12.12. *Determination of the sound transmission class (STC) from transmission loss (TL) measurements. The rating of this wall is STC = 42. (The TL at 4 kHz is 8 dB below the STC = 42 curve, and the total deficiency is 30 dB.*

of *STC* for representative constructions.* More extensive compilations can be found in the literature.[1,11]

The *STC* of a composite structure, for example, walls with doors and windows, can be found from the *TL*'s of the individual components. If S_i is the area of the individual component with transmission loss TL_i, its pressure transmission coefficient is

$$T_i = \text{antilog}\,(-TL_i/20) \tag{12.13}$$

If *S* is the area of the entire wall, then the effective pressure transmission coefficient T_{eff} is

$$T_{eff} = \frac{1}{S} \sum T_i\, S_i \tag{12.14}$$

and the transmission loss for the composite structure is $20 \log(1/T_{eff})$. The *STC* for the composite structure can then be calculated by the usual procedure.

Sound transmission classes measured in the field are generally less than those obtained in the laboratory. This can usually be attributed to either flanking paths or

---

* The dimensions of common constructional materials are often only *nominal*; therefore, metric equivalents must also be considered *nominal*.

[11] Doelle, *Environmental Acoustics*, McGraw-Hill (1972).

Table 12.10. Sound transmission class (*STC*) for representative wall constructions[a] (1 in. = 2.54 cm)

Construction	Mass per Unit Area (kg/m²)	STC
1. 4-in. hollow block, $\frac{1}{2}$-in. plaster on both sides	115	40
2. 4-in. brick, $\frac{1}{2}$-in. plaster on both sides	210	40
3. 9-in. brick, $\frac{1}{2}$-in. plaster on both sides	490	52
4. 24-in. stone, $\frac{1}{2}$-in. plaster on both sides	1370	56
5. $\frac{3}{8}$-in. gypsum wallboard	8	26
6. $\frac{1}{2}$-in. gypsum wallboard	10	28
7. $\frac{5}{8}$-in. gypsum wallboard	13	29
8. Two $\frac{1}{2}$-in. gypsum wallboards bonded together	22	31
9. 2 × 4 studs on 16-in. centers, $\frac{1}{2}$-in. gypsum wallboard on both sides	21	33
10. Same as 9 but with $\frac{5}{8}$-in. gypsum wallboard on both sides	26	34
11. Same as 10 but with two sheets of $\frac{5}{8}$-in. gypsum wallboard on one side and one sheet on the other side	42	36
12. Same as 10 but with $\frac{1}{2}$-in. plaster over wallboard	68	46
13. Same as 9 but with a 2-in. isolation blanket	23	36
14. Same as 10 but with a 2-in. isolation blanket	29	38
15. Same as 11 but with a 2-in. isolation blanket	44	39
16. Same as 14 but with one side resiliently mounted	29	47
17. Same as 14 but with both sides resiliently mounted	29	49
18. Double row of 2 × 4 studs on 16-in. centers, $\frac{5}{8}$-in. gypsum wallboard on both sides, and 2-in. isolation blanket	37	57
19. Double row of 2 × 4 studs on 16-in. centers, two $\frac{5}{8}$-in. gypsum wallboard on both sides, no isolation blanket	60	58
20. Same as 19 but with 2-in. isolation blanket	60	62

[a] Numbers 1 to 15 are single-leaf construction. Numbers 16-20 are double-leaf construction.

poor workmanship (improperly caulked joints, bridging between supposedly isolated elements). Even with properly constructed partitions, a 5-dB difference in *STC*'s can be expected between field and laboratory measurements.

A procedure exists for quantifying the sound insulation between two rooms without band-level analysis. With a "pink" noise generator operating in the source room, a sound level meter is used to measure the C-weighted sound level (L_C) in the source room and the A-weighted sound level (L_A) in the receiving room. The *privacy rating* (*PR*) is then

$$PR = L_C - L_A + 10 \log(2T_2) \qquad (12.15)$$

where T_2 is the *reverberation time* (see Sect. 13.3) of the receiving room in seconds. It is claimed that the *PR* will be within a few decibels of the *STC*.

Because of the different spectra associated with external noise, *STC* ratings cannot be used directly to predict isolation values of outside walls. However, a technique does exist that allows the *STC* to be used for this purpose.[1]

Besides their *STC*, floor/ceiling partitions should be considered for their *impact isolation class (IIC)*. This is a measure of their ability to isolate a room from noise produced by impact on the floor above.[1]

12.14 RECOMMENDED ISOLATION. The amount of acoustic isolation required between two rooms depends on the noise level in the source room and the level of intrusive noise acceptable in the receiving room. Both these levels depend on the designated use of the rooms and the latter depends on the ambient noise that will tend to mask the intrusive noise.

The recommendations of the Federal Housing Administration account for the difference in background noise by defining three grades of buildings.

Grade 1. Building with exterior nighttime noise levels lower than 40 dBA and recommended interior noise levels lower than 35 dBA.

Grade 2. Buildings with recommended interior noise levels of 40 dBA or lower.

Grade 3. Buildings with exterior nighttime noise levels of 55 dBA or higher and interior noise levels of 45 dBA or higher.

For walls separating different apartments, they recommend an *STC* of 55, 52, and 48 for building Grades 1, 2, and 3, respectively. For walls between rooms of the same dwelling unit, they recommend for a Grade 1 building the following: bedroom to bedroom *STC* = 48; living room to bedroom 50; bathroom to bedroom, kitchen to bedroom, and bathroom to living room 52. The recommendations for Grade 2 are 4 dB lower and those for Grade 3 are another 4 dB lower.

For walls separating apartment rooms from commonly shared service spaces (garages, laundries, party rooms), the following minimum requirements have been suggested: bedroom *STC* = 70, living room 65, and kitchen and bathroom 60.

Recommendations also exist for the impact isolation class for floors separating dwellings.

12.15 DESIGN OF PARTITIONS. Ignoring flanking and leakage, the basic mechanism of sound transmission through a wall is that sound in the source room forces the exposed surface to vibrate, this vibration is transmitted through the structure of the wall to the other surface, which in turn vibrates producing sound in the receiving room. If the two surfaces of the wall are rigidly connected so that they vibrate as a unit (a *single-leaf* partition), the transmission loss depends only on the frequency and the mass per unit area, stiffness, and intrinsic damping of the wall. If the partition consists of two unconnected walls separated by a cavity (a *double-leaf* partition), the transmission loss depends on the properties of the two walls and on the size of the cavity and its absorption.

(a) Single-Leaf Partitions. For a plane, nonporous, homogeneous, flexible wall it can be shown that

$$TL = 20 \log(f\rho_S) - 48 \qquad (12.16)$$

where ρ_S, the mass per unit area, is in kilograms per square meter and f is the frequency in hertz. This is often called the *mass law*: doubling the mass per unit area at a given frequency or doubling the frequency for a given mass per unit area increases the TL by 6 dB.

A wall that obeys the mass law (12.16) throughout the frequency range from 125 to 4000 Hz has a sound transmission class, found by the procedure described in Sect. 12.13, of

$$STC = 20 \log \rho_S + 10 \qquad (12.17)$$

As can be seen from Fig. 12.13, the STC's for single-leaf partitions are invariably below those predicted by the mass law. Part of this is caused by the porosity of the material, as seen by the improvement when concrete block is sealed, but the remainder is related to the stiffness of the panel, which is neglected in developing the mass law.

For a stiff, homogeneous panel, flexural waves propagate along its surface with a phase speed

$$c_f = \left(\frac{\pi^2}{3} \frac{Yt^2}{\rho} f^2 \right)^{1/4} \qquad (12.18)$$

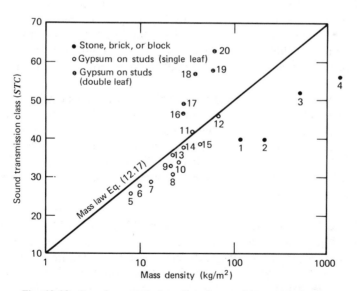

Fig. 12.13. *Sound transmission class for partitions and their comparison to the mass law. Numbers denote constructions listed in Table 12.10. Single-leaf partitions are represented by numbers 1 to 15 while numbers 16 to 20 denote double-leaf construction.*

where f is the frequency of excitation, Y the Young's modulus, t the thickness, and ρ the mass density of the panel. Since this propagation is dispersive, there exists a *critical frequency* f_c at which the wavelength of the flexural wave equals that of a wave of the same frequency in air:

$$f_c = \frac{\sqrt{3}c^2}{\pi t} \sqrt{\frac{\rho}{Y}} \qquad (12.19)$$

where c is the speed of sound in air. For all frequencies above f_c there exists an angle of incidence such that the projection of the incident wave coincides with the flexural wave

$$\lambda_f = \lambda/\sin \theta$$

where λ_f and λ are the wavelengths in the panel and air, respectively, and θ is the angle of incidence with respect to the normal. For this particular angle there is an extremely good coupling of energy from the incident wave to the flexural wave, which in turn efficiently radiates energy into the receiving room. Therefore, for frequencies above f_c averaging over all incident angles results in a transmission loss that is less than predicted by the mass law. Since λ_f is proportional to $1/f^{1/2}$ and λ to $1/f$, $\sin \theta$ is proportional to $1/f^{1/2}$ so that at high frequencies only waves incident at near-normal angles can be coincident: the TL rises to the values given by the mass law. The depth of this *coincidence dip* depends on the intrinsic damping of the panel.

To obtain an STC close to that predicted by the mass law, a wall must be designed so that the coincidence frequency occurs at frequencies either below 125 Hz, requiring a thick wall with low density and high Young's modulus, or above 4000 Hz, requiring a thin, high-density wall with low Young's modulus. Examples of the effects of the coincident dip on walls of different constructions are displayed in Fig. 12.14.

A properly designed single-leaf partition still requires a large mass per unit area to obtain a respectable STC. For example, a 15-cm (6-in.) thick concrete wall with 1.3-cm ($\frac{1}{2}$-in.) plaster on each side has a mass per unit area of 390 kg/m^2 (80 lb/ft^2) and $STC = 52$; this is 3 dB shy of that recommended for party walls between luxury apartments. (Note the mass law predicts $STC = 62$ for this same wall.) To obtain high acoustic isolation without excessive weight, it is necessary to employ double-leaf construction.

(b) Double-Leaf Partitions. As seen in Fig. 12.13, the STC for a partition of double-leaf construction is considerably higher than that of a single-leaf partition of the same mass density. This effect is dramatically illustrated in Fig. 12.15 where the transmission loss curve for two sheets of 1.3-cm ($\frac{1}{2}$-in.) gypsum board bonded together as a single-leaf is compared with that obtained for the same sheets used in a double-leaf construction. Also note the improvement when absorbing material is placed in the space between the boards.

To illustrate effects on the STC rating of various design changes, Table 12.11 outlines a sequence of partitions of similar construction. Note that without resilient

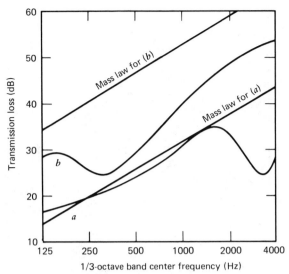

Fig. 12.14. *Effect of the properties of a wall on the coincidence frequency. (a) ½-in. gypsum wallboard (10 kg/m²). Critical frequency = 2.6 kHz. STC = 28. (b) Lightweight foamed concrete (110 kg/m²). Critical frequency = 200 Hz. STC = 35. (After Harris.)*

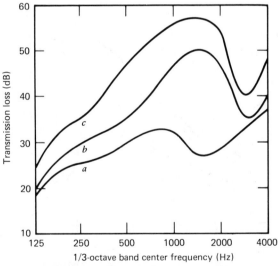

Fig. 12.15. *Transmission loss for single- and double leaf partitions constructed from similar materials. (a) Two ½-in. gypsum wallboards bonded together (22 kg/m²). STC = 31. (b) ½-in. gypsum wallboard on both sides of 4-in. steel channel studs (21 kg/m²). STC = 37. (c) Same as (b) but with a 2-in. sound isolation blanket. STC = 45. (After Harris.)*

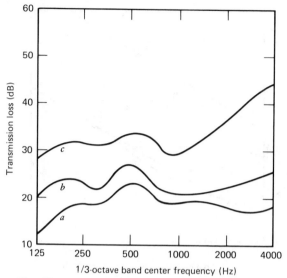

Fig. 12.16. *Transmission loss for* $1\frac{3}{4}$*-in. thick door constructions.* (*a*) *Hollow-core door, ungasketed* (7 kg/m²). *STC* = 17. (*b*) *Hollow-core door, gasketed* (7 kg/m²). *STC* = 24. (*c*) *Solid-core door, gasketed* (20 kg/m²). *STC* = 26. (*After Doelle.*)

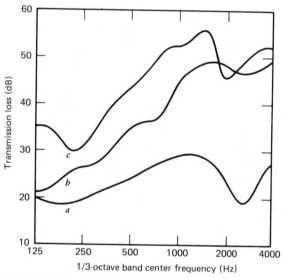

Fig. 12.17. *Transmission loss for different openable window constructions, weatherstripped with sealed edges.* (*a*) *Single-pane window, 3-mm glazing* (7.5 kg/m²). *STC* = 25. (*b*) *Double-pane window, 3-mm glazing with 10-cm air space* (15 kg/m²). *STC* = 36. (*c*) *Double-pane window, 3-mm glazing with 20-cm air space* 13 kg/m²). *STC* = 40. (*After Doelle*).

mounting, the addition of damping material adds only 4 dB to the *STC* rating but, with resilient mounting, a blanket adds 10 dB. Also note that with or without a blanket, the second resilient mounting adds only 1 or 2 dB.

Table 12.11. Sound transmission classes (*STC*) for partitions consisting of 2 × 4 studs with 16-in. centers, faced with $\frac{5}{8}$-in. gypsum wallboard which is either nailed directly to the studs or resiliently mounted on one or both sides, either with or without a 2-in. sound isolation blanket.

Isolation Blanket	Resilient Mounting	STC
No	No	34
Yes	No	38
No	One	38
No	Two	39
Yes	One	47
Yes	Two	49

To obtain even higher *STC* ratings it is necessary to use a stagered-stud construction, which provides better vibration isolation, and more layers of gypsum to add mass. A wall that can provide an *STC* = 55 consists of two sets of 2 × 4 wood studs each with 24-in. spacing on separate 2 × 4 plates, spaced 1 in. apart, with the subfloor discontinuous between plates; one 2-in. isolation blanket; and one $\frac{5}{8}$-in. gypsum sheet nailed on one side and two nailed on the other side (1 in. = 2.54 cm). If the single sheet of gypsum board is resiliently mounted instead of nailed, the *STC* increases to 60.

(c) Doors and Windows. Because of their low-surface density and the gaps around their edges, doors and windows are acoustically weak elements of a partition. Figure 12.16 illustrates the advantages of solid-core doors over hollow-core doors and the value of an acoustical seal that includes an automatic threshold closer to seal the gap at the bottom. For windows, Fig. 12.17 illustrates the advantage of using double glazing with well-sealed edges and a minimum separation of 10 to 13 cm (4 to 5 in.).

PROBLEMS

12.1. The band levels for a noise are 70 dB *re* 20 μPa at 31.5 Hz and decrease 3 dB for each octave. Find (*a*) the overall sound pressure level between 31.5 and 8000 Hz, and (*b*) the A-weighted sound pressure level for this same frequency range.

12.2. For the noise given in Problem 12.1, (*a*) Find the preferred speech interference level, and compare to the approximation given by (12.1). (*b*) Determine the voice level required for face-to-face communication at 10 m, and classify the noise conditions for unassisted speech communications.

12.3. For the noise of Problem 12.1, (*a*) Determine the preferred noise criterion from Fig. 12.5, and compare to the approximate value from (12.3). (*b*) Estimate the noise criteria (*NC*) from (12.3), and comment on the appropriateness of this noise for the rooms listed in Table 12.3.

12.4. For the community noise of Fig. 12.7, find L_{A10}, L_{A50}, and L_{A90}.

12.5. Find the *SENEL* for a single event where the A-weighted sound pressure level jumps to a constant 80 dB *re* 20 μPa for 25 s and then suddenly returns to the ambient level.

12.6. The hourly average sound level for a community noise is 60 dBA from 7 A.M. to 7 P.M., 55 dBA from 7 P.M. to 10 P.M., and 50 dBA from 10 P.M. to 7 A.M. Find (*a*)L_{Aeq} for the 24 hours, (*b*) L_{Adn}, and (*c*) *CNEL*.

12.7. Calculate the A-weighted sound pressure levels for an automobile and a motorcycle under cruising conditions measured at a distance of 15 m for speeds of 44 and 88 km/h.

12.8. If, at a given distance from a straight, two-lane road of infinite extent, zero grade, and negligible truck traffic, the maximum equivalent sound level is to remain constant, find the ratio of the maximum vehicle flow rate (*Q*) if the maximum speed is reduced from 104 km/h (65 mph) to 88 km/h (55 mph).

12.9. A worker is exposed to noise levels of 92 dBA for 4 h and 90 dBA for 4 h each working day. Do these conditions exceed the OSHA recommendations?

12.10. In measuring the sound transmission class of a 30 m^2 partition, the band level in the source room was kept constant at 90 dB, and for each 1/3-octave bands from 125 to 4000 Hz the band levels and total absorption (*A* in square meters) in the receiving room were: (band level, *A*) = (65, 15); (51, 15); (44, 16); (48, 17); (44, 19); (42, 20); (47, 25); (40, 28); (39, 30); (35, 32); (34, 36); (35, 45); (33, 53); (32, 60); (34, 60); and (37, 60). Find the *STC* of the partition.

CHAPTER 13

ARCHITECTURAL ACOUSTICS

13.1 SOUND IN ENCLOSURES. Through extensive experimental studies of the acoustic properties of a room, Sabine[1] arrived at an empirical relation among the reverberation characteristics of an enclosure, its size, and the amount of absorbing material present. His definition of the *reverberation time* as the time required for the sound pressure to drop 60 dB specifies one important acoustic parameter of a room. The Sabine equation

$$T \propto V/A \qquad (13.1)$$

relates the reverberation time T of a room to its volume V and a parameter A which specifies its total *sound absorption*. Theoretical derivations of this equation are usually based on a *ray model* wherein sound is assumed to travel outward from the source along diverging rays. At each encounter with the boundaries of the room, the rays are partially absorbed and reflected as described in earlier chapters. After a large number of reflections the sound in the room may be assumed to have become *diffuse: The average energy density $\mathscr{E}$ is the same throughout the volume of the enclosure, and all directions of propagation are equally probable.* This model oversimplifies the actual behavior of sound in a room, particularly at low frequencies, because it neglects the existence of normal modes, the distribution of absorptive materials, and the shape of the room. Nevertheless, with properly chosen values of A, (13.1) leads to valid conclusions.

When the source is turned on and then operates continuously, the acoustic intensity at any point in the room builds up to higher values than would exist if the source were operated in open air, the gain in intensity often being greater than tenfold. For any given enclosure this gain is nearly proportional to the reverberation time; hence *a long reverberation time is desirable if a weak source of sound is to be everywhere audible.* If the source is shut off, reception of sound by the direct path ceases after a short time interval $t = r/c$, where r is the distance from the source to the point of observation and c is the speed of sound in air. The reflected waves continue to be received as a succession of arrivals of decreasing intensity. The presence of this reverberant acoustic energy tends to mask the immediate recognition of any new sound, unless sufficient time has elapsed for the reverberation to have fallen some 5 to 10 dB below its initial level. Since the reverberation time is a direct measure of the persistence of such sounds, it is obvious that *a short reverber-*

---

[1] Sabine, *Collected Papers on Acoustics*, Harvard University Press (1922).

ation time is desirable to minimize masking effects. The choice of the best reverberation time for a particular enclosure must, therefore, be a compromise.

One additional acoustic factor of importance in the design of a room is its ability to screen out external sounds and thus reduce their annoyance or masking effect. Although the acoustic transmission of the walls of the enclosure is the most significant factor to be considered in this respect, it should be noted that the existence of a small reverberation time, with its attendant high total internal acoustic absorption, will tend to minimize the ambient noise level produced by those sounds that do penetrate into the enclosure. For more information than can be presented here, the references[2-5] are recommended.

13.2 A SIMPLE MODEL FOR THE GROWTH OF SOUND IN A ROOM. If a source of sound is operated continuously in an enclosure, absorption in the medium and at the surrounding surfaces prevents the acoustic pressure amplitude from becoming infinitely large. In small- and medium-sized enclosures, absorption in the medium is negligible so that both the rate at which amplitude increases and its ultimate value are controlled by surface absorption. If the total sound absorption is large, the pressure amplitude quickly reaches an ultimate value only slightly in excess of that produced by the direct wave alone. By contrast, if the absorption is small, considerable time will elapse before the ultimate, significantly higher amplitude is attained. Rooms of this latter type are known as *live* or *reverberant* rooms, and application of ray theory to such rooms yields results that are in fairly good agreement with experimental measurements.

When a source of sound is started in a live room, reflections at the walls produce a sound energy distribution that becomes more and more uniform with increasing time. Ultimately, except close to the source or to the absorbing surfaces, this energy distribution may be assumed to be completely uniform and to have essentially random local directions of flow.

Let us now derive a relationship between energy density and the energy flux across the boundaries of the room. From Fig. 13.1a, ΔS is an element of a boundary and dV an element of volume in the medium at a distance r from ΔS, where r makes an angle θ with the normal to ΔS. Let the acoustic energy density $\mathscr{E}$ be uniform throughout the region so that the acoustic energy present in dV is $\mathscr{E}\,\Delta V$. The amount of this energy that will strike ΔS by direct transmission is $\mathscr{E}\,dV$ attenuated by $4\pi r^2$ and multiplied by the projection of ΔS on the sphere of radius r centered on dV

$$\frac{\mathscr{E}\,dV}{4\pi r^2}\cos\theta\,\Delta S$$

Now let dV be part of a hemispherical shell of thickness Δr and radius r centered on

[2] Doelle, *Environmental Acoustics*, McGraw-Hill (1972).

[3] Rettinger, *Acoustic Design and Noise Control, Vol. I*, Chemical Publishing Co. (1977).

[4] Beranek, *Music, Acoustics, and Architecture*, Wiley (1962).

[5] Knudsen and Harris, *Acoustical Design in Architecture*, Wiley (1950).

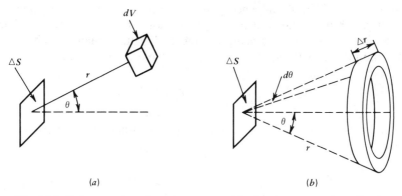

Fig. 13.1. *Volume and surface elements used in deriving expressions for the intensity of diffuse sounds.*

ΔS. The acoustic energy ΔE contributed to ΔS by this entire shell can then be obtained by assuming that energy arrives from any direction with equal probability. Integrating over the hemisphere with $dV = 2\pi r \sin\theta\, r\Delta r d\theta$ yields

$$\Delta E = \frac{\mathscr{E}\,\Delta S\,\Delta r}{2} \int_0^{\pi/2} \sin\theta \cos\theta\, d\theta = \frac{\mathscr{E}\,\Delta S\,\Delta r}{4} \qquad (13.2)$$

This energy arrives during a time interval $\Delta t = \Delta r/c$, so that (13.2) can be rewritten $\Delta E/\Delta t = \mathscr{E}c\Delta S/4$. Thus, the rate dE/dt at which energy falls on a unit area of the wall is

$$\frac{dE}{dt} = \frac{\mathscr{E}c}{4} \qquad (13.3)$$

If it is assumed that, at any point within the room, energy is arriving and departing along individual ray paths and that the rays have random phases at the point, then the energy density $\mathscr{E}$ is the sum over all rays of the energy densities $\mathscr{E}_l$ of the individual rays. Now, if the lth ray has effective pressure amplitude P_{el} we have $\mathscr{E}_l = P_{el}^2/(\rho_0 c^2)$ and thus

$$\mathscr{E} = \frac{P_r^2}{\rho_0 c^2} \qquad (13.4)$$

where $P_r = (\sum_l P_{el}^2)^{1/2}$ is the spatially averaged *effective pressure amplitude of the reverberant sound field*.

If the total *sound absorption* of the room is A, then the rate at which energy is being absorbed by all surfaces is

$$\frac{Ac}{4}\mathscr{E} \qquad (13.5)$$

from (13.3). Note that A has units of square meters; A is often given in *metric sabin* (m^2) or *English sabin* (ft^2).

This rate at which sound energy is absorbed by the surfaces, plus the rate $V \, d\mathscr{E}/dt$ at which it increases in the air throughout the interior of the room, must equal the rate Π at which it is being produced. The fundamental differential equation governing the growth of sound energy in a live room is therefore

$$V \frac{d\mathscr{E}}{dt} + \frac{Ac}{4} \mathscr{E} = \Pi \tag{13.6}$$

If the sound source started at $t = 0$, solution of this differential equation and use of (13.4) gives

$$P_r^2 = \frac{4 \, \Pi \, \rho_0 \, c}{A} (1 - e^{-t/\tau_E}) \tag{13.7}$$

where

$$\tau_E = \frac{4V}{Ac} \tag{13.8}$$

is the *time constant* governing the growth of the acoustic energy in the room.

If A is small and τ_E is large, a relatively long time will be required for the effective pressure amplitude P_r and energy density $\mathscr{E}$ to approach their ultimate values of

$$P_r^2(\infty) = \frac{4 \, \Pi \, \rho_0 \, c}{A} \tag{13.9}$$

and

$$\mathscr{E}(\infty) = \frac{4 \, \Pi}{Ac} \tag{13.10}$$

These equations indicate that a small value for A is associated with a large value for the ultimate pressure amplitude.

Since the developments of this section are based on an assumed diffuse distribution of acoustic energy, there are limitations to the resulting equations. For instance, (13.6) may not be used until a sufficient time t has elapsed to permit each initial ray to have experienced several reflections at the boundaries. For a small room this time is of the order of 50 ms, whereas for a large auditorium it may approach 1.0 s. Equation (13.9) indicates that the final pressure amplitude is independent of the volume and shape of the room, is the same at all points in the room, and depends only on the source strength and the room absorption A. This is obviously not true for rooms having well-defined, sound-focusing properties. Neither can the equations be applied to odd-shaped rooms having deep recesses, nor to rooms coupled together by an opening. Furthermore, these equations may become invalid if some large surfaces of the room are abnormally absorptive, since the energy density near such surfaces will be considerably lower than elsewhere.

13.3 REVERBERATION TIME—SABINE. The equation governing the *decay* of uniformly diffuse sound in a live room is obtained by setting $\Pi = 0$ in (13.6). If the source is turned off at $t = 0$, the pressure amplitude at any later time t is

$$P_r^2 = P_r^2(0)\, e^{-t/\tau_E} \qquad (13.11)$$

Thus, the change in sound pressure level is

$$\Delta SPL = 4.35 t/\tau_E$$

The *reverberation time* T, defined as the time required for the level of the sound to drop by 60 dB, is

$$T = 13.8\tau_E = \frac{55.2V}{Ac}$$

With $c = 343$ m/s (20°C) this becomes

$$T = \frac{0.161V}{A} \qquad (13.12a)$$

where V is in cubic meters and A in square meters (*metric sabin*). If V is expressed in cubic feet and A in square feet (English sabin), then

$$T = \frac{0.049V}{A} \qquad \text{(English)}$$

If the surface area of the room is S, the *average Sabine absorptivity* $\bar{a}$ is defined by

$$\boxed{\bar{a} = \frac{A}{S}} \qquad (13.12b)$$

and (13.12a) assumes the form

$$\boxed{T = \frac{0.161V}{S\bar{a}}} \qquad (13.12c)$$

If the reverberation time T is known, the total sound absorption A and the average Sabine absorptivity $\bar{a}$ can be calculated. However, the real goal is to be able to reverse the above logic: *given* the acoustic properties of the room, to *predict* the reverberation time. It is clear that A must depend on the areas and absorptive properties of all the materials within the room, but the functional form of this dependence is subject to a variety of simplifying assumptions. A critical discussion[6] of these matters is available.

Sabine adopted the plausible assumption that the total sound absorption is the sum of the absorptions A_i of the individual surfaces,

$$A = \sum_i A_i = \sum_i S_i a_i \qquad (13.13a)$$

[6] Young, *J. Acous. Soc. Am.*, **31**, 912 (1959).

where a_i is the *Sabine absorptivity* of the *i*th surface. With this assumption, the average Sabine absorptivity (13.12*b*) becomes

$$\bar{a} = \frac{1}{S} \sum_i S_i a_i \tag{13.13b}$$

Each a_i is to be evaluated from standardized measurements on a sample of the material in a reverberation chamber. When the chamber is empty, the measured reverberation time T will be

$$T = \frac{0.161V}{S\bar{a}} \tag{13.14a}$$

Now, if the sample of surface area S_e, is mounted on the walls of the chamber, the new reverberation time T_e will be

$$T_e = \frac{0.161V}{S\bar{a} - S_e a_0 + S_e a_e} \tag{13.14b}$$

where a_0 is the absorptivity of the covered portion of the wall and a_e is the unknown absorptivity of the sample. Combination of (13.14*a*) and (13.14*b*) yields the desired evaluation,

$$a_e = a_0 + \frac{0.161V}{S_e} \left(\frac{1}{T_e} - \frac{1}{T} \right) \tag{13.14c}$$

In practice, it is found that the measured a_e for a sample depends somewhat on its surface area and its location within the room. Nevertheless, the use of Sabine's equation and experimentally determined Sabine absorptivities is recommended as being sufficiently accurate for most engineering applications.[6,7]

One primary difficulty encountered in all reverberation measurements is the existence of local anomalies resulting from the formation of standing wave patterns. Sabine's method of overcoming this was to place near the center of the reverberation chamber a number of large reflecting surfaces that were rotated while measurements were being made. The varying standing wave patterns averaged out the local anomalies. Another approach was to make measurements at a large number of different points in the chamber. At present, the methods most commonly employed are to use either (1) a warble oscillator (which by varying the frequency of the sound emitted by a loudspeaker produces continuously changing standing wave patterns) or (2) 1/3-octave bands of noise.

The initial work of Sabine on reverberation was limited to a single frequency of 512 Hz. His later experiments included measurements in octave steps between 64 and 4096 Hz. Custom has attached so much importance to the 512-Hz frequency that, when the expression *reverberation time* is used without specification of any particular frequency, it is generally understood to refer to this frequency (or more recently 500 Hz). Because of the frequency dependence of the absorption of each

[7] Beranek (editor), *Noise and Vibration Control*, McGraw-Hill (1971).

surface, it is necessary to specify the reverberation time for representative frequencies covering the entire range that is important to speech and music. The frequencies usually chosen are 125, 250, 500, 1000, 2000, and 4000 Hz.

It is apparent from these equations that the reverberation time of a live room can be calculated if its volume and the absorptivities of its surfaces are known. Since the surfaces can readily be changed by the insertion or removal of absorptive materials at the walls, the magnitude of the reverberation time of a room is subject to precise control.

Consider a rectangular room measuring $3 \times 5 \times 9$ m whose interior boundaries have an average Sabine absorptivity $\bar{a} = 0.1$. Then $A = 17.4$ metric sabins so that $T = 1.25$ s, which corresponds to a fairly live room. Equation (13.9) gives the ultimate effective pressure amplitude developed by a 10 μW source as 0.031 Pa, which corresponds to a sound pressure level of 64 dB re 20 μPa. For comparison, the sound pressure level produced by direct transmission at a distance of 5 m from the source in free space is 45 dB re 20 μPa.

In the development of the reverberation time given by (13.12c), absorption of sound in the medium was neglected. The effect of absorption in the medium is to decrease the reverberation time. Using logic similar to that in Sect. 2.11b, it can be shown that the pressure amplitude of a standing wave experiencing only absorption in the air decreases in accordance with the equation $P = P_0 \exp(-\alpha ct)$ or $I = I_0 \exp(-mct)$ where $m = 2\alpha$ is the symbol conventionally used in architectural acoustics. Consequently, (13.11) may be rewritten

$$P_r^2 = P_r^2(0) e^{-[(A/4V) + m]ct} \tag{13.15}$$

and the expression for the reverberation time becomes

$$T = \frac{0.161 V}{S\bar{a} + 4mV} \tag{13.16a}$$

The importance of absorption in the air is determined by the relative magnitudes of the second and first terms in the denominator of (13.16). Since α increases with f whereas A remains fairly uniform above 1 kHz, air absorption may become significant at higher frequencies in rooms of large volume V (where the path lengths between successive reflections at the walls are relatively large). Also, in *extremely* reverberant rooms, the major portion of the sound absorption may occur in the air rather than at the surfaces. For relative humidities h (in percent) between about 20 and 70 percent and frequencies between about 1.5 and 10 kHz, a sufficiently accurate approximation for most architectural applications is

$$m = 5.5 \times 10^{-4}(50/h)(f/1000)^{1.7} \tag{13.16b}$$

13.4 REVERBERATION TIME—OTHER FORMULAS. One of a number of other equations proposed is that of Norris and Eyring[8] which is based on the mean

[8] Eyring, *J. Acous. Soc. Am.*, **1**, 217 (1930).

free path between reflections. It can be shown[9] that the mean distance traveled between successive reflections from the walls of a rectangular enclosure is $L = 4V/S$ so that the number of reflections per second is $N = cS/(4V)$. With each reflection, the sound is reduced in energy (on the average) by the factor $(1 - \bar{a}_E)$ where $\bar{a}_E$ is the *area-averaged random-incidence energy-absorption coefficient*. The total attenuation of the energy over a time interval of one reverberation time T must therefore be $(1 - \bar{a}_E)^{NT}$. By definition, this must correspond to a reduction in the intensity level of 60 dB, $10 \log(1 - \bar{a}_E)^{NT} = -60$, so that solution for T and use of the above expression for N yields

$$T = \frac{0.161V}{-S \ln(1 - \bar{a}_E)} \tag{13.17}$$

Since both Sabine's formula (13.12c) and that of Norris and Eyring (13.17) assume a diffuse sound field and replace the absorptive properties of the individual surfaces with a room average, they can be equated to reveal

$$\bar{a} = -\ln(1 - \bar{a}_E) \tag{13.18a}$$

For live rooms $a \ll 1$, and expansion of the natural logarithm gives

$$\bar{a} \doteq \bar{a}_E \qquad \bar{a} \ll 1 \tag{13.18b}$$

On the other hand, if all the walls of the room are perfectly absorbing, we must have $a_E = 1$ and $T = 0$. Then (13.18a) requires $a \to \infty$. [This is consistent with the Sabine formula since $T \to 0$ only if $a \to \infty$ in (13.12c).]

Determination of $\bar{a}_E$ from the individual random-incidence energy-absorption coefficient a_{Ei} of the constituents of the room depends on the assumed interaction of the rays with the absorbing surfaces. Two models are explained here.

1. If the fraction of time any ray intercepts S_i is S_i/S, then the averaged random-incidence en ergy-*reflection* coefficient $(\bar{r}_E = 1 - \bar{a}_E)$ must be the area-weighted geometric mean of the individual $r_{Ei} = 1 - a_{Ei}$,

$$1 - a_E = (1 - a_{E1})^{S_1/S}(1 - a_{E2})^{S_2/S} \dots \tag{13.19a}$$

This leads to the Millington-Sette formula[10]

$$T = \frac{0.161V}{- \sum_i S_i \ln(1 - a_{Ei})} \tag{13.19b}$$

which, in the limit of all $a_{Ei} \ll 1$, reduces to the Sabine formula with $a_i = a_{Ei}$. A most unfortunate deficiency of (13.19b) is that, if any *single* a_{Ej} is unity, then $T = 0$. This is a direct result of the statistical averaging required by (13.19a)—given enough reflections, any and all rays will eventually intercept S_j and (according to ray theory) be absorbed. The inevitability of this eventuality is built into (13.19a) by the factor $(1 - a_{Ej})^{S_j/S}$. That this is an oversimplification of the actual physics can be seen by a straightforward argument. Recall that any ray experiences, on the average,

---

[9] Bate and Pillow, *Proc. Phys. Soc.*, **59**, 535 (1947).

[10] Millington, *J. Acous. Soc. Am.*, **4**, 69 (1932) and Sette, ibid, p. 193.

$N = cS/(4V)$ reflections per second. Since the fraction of reflections the ray will experience from S_j must be about S_j/S, the time T_j it will take (on the average) for a ray to encounter S_j must be $T_j \sim (4V/cS)/(S/S_j)$. Using the Sabine relation, we have $T_j/T \sim (a/13.8)(S/S_j)$. Thus, it takes a certain amount of time for a small, highly absorptive patch to become influential. If we assert that $T_j < T$ is required for the patch to be important, it follows that the patch of area S_j is unimportant if

$$\frac{S_j}{S} < \frac{\bar{a}}{13.8}$$

Thus, if S_j is sufficiently small, a high value of a_{Ej} cannot influence the reverberation significantly.

2. If, after reflecting from the ith surface, the next reflection of the ray has a probability S_i/S of occurring at a surface with the same a_{Ei}, then the average random-incident energy-reflection coefficient is the area-weighted mean of the individual coefficients,

$$1 - \bar{a}_E = \frac{1}{S} \sum_i S_i (1 - a_{Ei}) \tag{13.20a}$$

which yields

$$T = \frac{0.161V}{-S \ln\left(1 - \dfrac{1}{S} \sum_i S_i a_{Ei}\right)} \tag{13.20b}$$

As long as the *averaged* random-incidence energy-absorption coefficient $\bar{a}_E$ is much less than unity, this also reduces to the Sabine formula with $a_i = a_{Ei}$. Thus, even though some S_i may have $a_{Ei} \sim 1$, (13.20b) is equivalent to (13.12c) with $S\bar{a} = \sum_i S_i a_{Ei}$ as long as $\bar{a} \ll 1$. Furthermore, it does not suffer the deficiency of (13.19b) if any or even several a_{Ei} are identically unity. On the other hand, the assumption leading to (13.20a) is not physically satisfying or plausible.

An approach that lies between these two possibilities would be to apply (13.19a) to each *wall* of the enclosure and to let a_{Ei} be the averaged random-incidence energy-absorption coefficient over the entire wall. For a discussion of this and other possible approaches, see the references cited in Young.[6]

13.5 SOUND ABSORPTION MATERIALS. Figure 13.2 provides representative Sabine absorptivities for various materials determined from reverberation chamber measurements and other empirical methods. For more detailed and exact information on specific materials, the reader is referred to the bulletin "Performance Data: Architectural-Acoustic Materials" published annually by the Acoustical and Insulating Materials Association and to References 2 to 5.

Sound absorbers important in acoustic design can be classified as (1) porous materials, (2) panel absorbers, (3) cavity resonators, and (4) people and furniture.

1. *Porous materials,* such as the familiar acoustic tiles, mineral wools, acoustic plasters, as well as carpets and draperies, are characterized by networks of intercon-

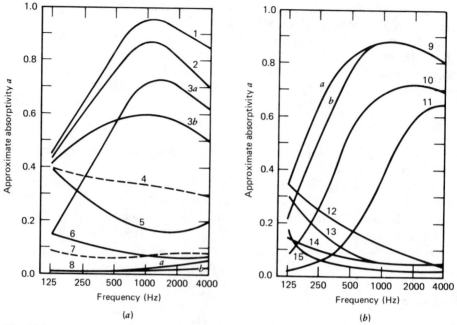

Fig. 13.2. Sabine absorptivities of common constructional materials. (1) Occupied audience, orchestra, chorus areas, including the floor beneath. (2) Well-upholstered, cloth-covered seats (perforated bottoms) without audience. (3a) Curtain (18 oz/yd²) hung to half area. (3b) Leather-covered upholstered seats, without audience, over a reflective floor. (4) Concrete-block wall, unpainted (approximate). (5) Wooden platform with air space below. (6) Wooden floor. (7) Concrete-block wall, painted (approximate). (8a) Smooth plaster on brick (but see 14). (8b) Poured concrete, unpainted. (9a) 2-in. fiberglass blanket on rigid backing. (9b) Same as 9a but with 1-in. air space between blanket and backing. (10) Heavy carpet on 40 oz (1.35 kg/m²) underpad. Unpainted acoustic tile. Unpainted acoustic plaster. (11) Heavy carpet on concrete. (12) Glass window. (13) Plaster on lath on studs. (14) 1-in. thick, damped plaster on concrete block, brick, or lath. 2-in. thick, well-fitted wooden walls. (15) Heavy plate glass window. (Adapted from Doelle,[2] Beranek,[4] Rettinger,[3] and Knudsen and Harris.[5])

nected pores. In these narrow channels and cavities viscous losses turn some of the acoustic energy into heat. The absorptivities of such materials are strong functions of frequency, being relatively small at the lower frequencies and increasing with the thickness of the material. (See Fig. 13.3a and also curves 1, 2, and 3 of Fig. 13.2a and curves 9, 10, and 11 of Fig. 13.2b.) Low-frequency absorptivity can be increased by mounting the material away from the wall. Painting acoustic plasters and tiles will invariably result in a substantial reduction in effectiveness.

    2. A *nonpous panel* mounted away from a solid backing vibrates under the influence of an incident sound, and the dissipative mechanisms in the panel convert some of the incident acoustic energy into heat. Such absorbers are very efficient at

low frequencies (Fig. 13.3b). The addition of a porous absorber in the space between the panel and the wall will further increase the efficiency of the low-frequency absorption.

3. A *cavity resonator* consists of a confined volume of air connected to the room by a narrow opening. It acts as a Helmholtz resonator, absorbing acoustic energy most efficiently in a narrow band of frequencies near its resonance (Fig. 13.3c). Such absorbers may be in the form of individual elements, such as a standard concrete block with slotted cavities; other forms consist of perforated panels and wood lattices spaced away from a solid backing with absorption blankets in between. Besides allowing for free architectural expression, these latter systems provide useful absorption over a much wider range of frequencies than is possible with individual cavity elements.

4. The sound absorption per item is given for people, upholstered seats, and wooden furniture in Fig. 13.3d. Wooden furniture includes chairs, school desks, and tables (a table providing work space for five people counts as five tables). For widely dispersed audiences with wooden desks, tables, or chairs (as are found in sparsely filled classrooms and many lecture halls), it may be more appropriate to use the absorption per body and per article of furniture given in Fig. 13.3d rather than the audience absorption of curve 1 of Fig. 13.2.

By choosing the amounts and distributions of these classes of absorbers, it is possible to tailor the behavior of the reverberation time with frequency to obtain almost any desired acoustic environment. It is even possible to design a room whose reverberation characteristics can be changed by means of sliding or rotating panels that expose surfaces of different absorption properties. Since the optimum reverberation time depends on the use of the room, it is possible in this way to design rooms for multipurpose use. However, artificial reverberation introduced electronically may be a less expensive and more flexible solution to this problem, especially in large rooms.

13.6 MEASUREMENT OF THE ACOUSTIC OUTPUT OF SOUND SOURCES IN LIVE ROOMS.

Although the most accurate methods of measuring the acoustic output of sound sources require *anechoic chambers*,* such outputs may also be measured with considerable accuracy in reverberant rooms. When the sound energy in such a room is completely diffuse, the acoustic power output Π is given by (13.9). If P_r were truly uniform throughout the room, one measurement of its magnitude would be sufficient. When it is not, either a number of measurements must be made and averaged or the microphone may be rapidly rotated on an arm by mechanical means so as to measure a mean pressure averaged over a distance of at least a quarter-wavelength. The only other unknown parameter in (13.9) is the total sound absorption A of the room. If the absorptivities of the materials constituting the walls of the room are known, A may be computed from the equations presented earlier. If not, it may be determined from (13.12a) by measuring the reverberation time T of the room. Combination of (13.9) with (13.12) to eliminate A

---

* An anechoic chamber is designed to have walls as completely absorptive as possible. This is the closest approximation to an unbounded, homogeneous medium that can be obtained in the laboratory.

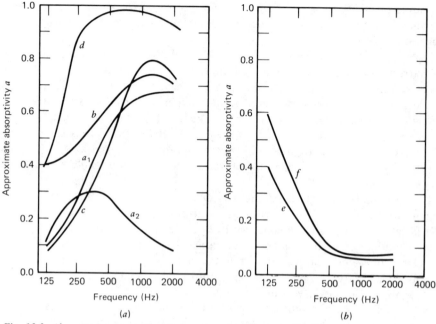

Fig. 13.3. *Absorption properties of acoustic materials.* (a_1) *Glued acoustic tile ceiling on rigid backing.* (a_2) *Material a_1 after two coats of paint (brush or roller).* (b) *Material a_1 suspended away from wall.* (c) *2.5-cm thick fiberglass (50 kg/m³) on rigid backing.* (d) *c but 10-cm thick.* (e) *6-mm plywood 75 mm from rigid backing.* (f) *e with sound isolation blanket.* (g) *Slotted two-well concrete block, single-cavity resonator.* (h) *Perforated panel resonator with isolation blanket, 10 percent open area.*

yields

$$\Pi = \frac{13.9 P_r^2 V}{\rho_0 c^2 T} = 9.7 \times 10^{-5} \frac{P_r^2 V}{T} \tag{13.21}$$

(If P_r is in μbar instead of Pa, replace 13.9 with 0.139 and the exponent -5 with -7.)

13.7 DIRECT AND REVERBERANT SOUND. Whenever a continuous source of sound is present in a room, two sound fields are produced. One is the *direct sound field*, or *direct arrival* from the source. The other, the *reverberant sound field*, is produced by the reflections from the surfaces of the enclosure. The effective pressure amplitude P_d produced by the direct sound field (assumed to be radiated uniformly in all directions) is given by

$$P_d^2 = \frac{\rho_0 c \, \Pi}{4\pi r^2} \tag{13.22}$$

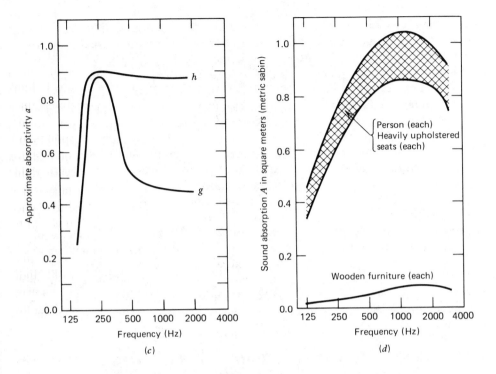

Fig. 13.3(c)(d)

where r is the radial distance from the effective center of the sound source and Π is the acoustic output of the source in watts. The effective pressure amplitude P_r of the reverberant field is obtained from (13.9) and the total mean squared pressure, $P^2 = P_d^2 + P_r^2$, is

$$P^2 = \rho_0 c\, \Pi \left(\frac{1}{4\pi r^2} + \frac{4}{A} \right) \tag{13.23}$$

The ratio of the reverberant intensity I_r to the intensity I_d of the direct arrival is

$$\frac{I_r}{I_d} = \frac{4\pi r^2}{(A/4)} = \frac{16\pi r^2}{A} \tag{13.24}$$

This equation shows that for locations very close to the source, $4\pi r^2 \ll (A/4)$, the shape or acoustic treatment of the room will have little influence on measured sound pressure levels. By contrast, at distances for which $4\pi r^2 \gg (A/4)$ the sound pressure level will be reduced by 3 dB for each doubling of the total sound absorption A. For example, a worker near a noisy machine will receive little benefit from increasing the total absorption of the room. However, the acoustic exposure of other workers at some distance from the machine will be reduced by such treatment. As another example, when two people are alone in a quiet room and relatively close together, the acoustic characteristics of the surroundings have negligible influence on their ability to carry on a conversation. However, if many talkers are present in the room, the reverberant sound pressure level will increase by 10 log N where N is the number of talkers, and will make conversation difficult unless the total absorption A is large. This explains the frequently experienced difficulty of carrying on a conversation in a large ballroom or dining hall when many people are talking. Imagine 100 talkers, each with an acoustic output of 100 μW, to be present in a room $5 \times 20 \times 40$ m with $T = 3$ s. Substitution into (13.12a) leads to $A = 215$ m². The reverberant sound pressure (13.21) is then $P_r = 0.28$ Pa, which corresponds to a sound pressure level of 83 dB re 20 μPa. This background level is much too high for a normal conversation. For instance, substitution into (13.22) will show that this same pressure is reached in the direct pressure field of an individual talker at a distance of 0.2 m. It is to be noted that, if all talkers in the room were to reduce their acoustic output to 10 μW, the background level of reverberant sound would be reduced to a more pleasant 73 dB re 20 μPa without altering the intelligibility of the conversation. Unfortunately, when a large number of talkers are present in a room, the "cocktail party effect" [11] becomes evident as each talker raises his individual acoustic output in order to be heard. On the average, this does not increase the intelligibility, but merely serves to increase the background level to an unpleasantly high value.

13.8 ACOUSTIC FACTORS IN ARCHITECTURAL DESIGN. Whether designing a music room, conference room, lecture hall, concert hall, or large auditorium, the acoustic consultant must consider several different factors whose relative importances depend on the purpose of the enclosure.

(a) The Direct Arrivals.   In any enclosure, there should be a direct and clear line of sight between the audience and the source of sound. Not only is this psychologically important, but it also guarantees that there will be a well-defined *direct arrival* of sound. Generally, in large spaces this requires that seating areas, including balconies, be raked up from front to back. (This also helps avoid the low-frequency attenuation that occurs for sounds crossing the audience at near *grazing* incidence.) Often the stage is elevated to enhance any rake of the seating area. (Too much elevation may yield difficult sight lines for the closest rows of seats.) If it is possible, raking the stage or raising its rear sections on platforms will aid in obtaining direct arrivals and reducing the grazing-incidence problem.

(b) Reverberation at 500 Hz.   As indicated in the previous section, there must be an appropriate balance between the direct arrival and the reverberant sound field. Combination of (13.24) and (13.12a) yields

$$\frac{I_r}{I_d} = 312 \frac{r^2 T}{V} \qquad\qquad (13.25)$$

[11] MacLean, *J. Acous. Soc. Am.*, **31**, 79(1959).

Since the energy of the direct arrival falls off as the square of the distance from the source, it is impossible to have a constant ratio throughout the enclosure. However, for spaces of similar geometry r^3 will be proportional to the volume V. This means that if it is desired to maintain the ratio I_r/I_d for the same relative location in enclosures of similar shape but differing volume, we must have

$$T = RV^{1/3} \qquad (13.26)$$

where R is a constant that depends on the purpose of the enclosure. Despite the approximate nature of this relationship, it is a useful empirical estimate. Table 13.1 gives the results of fitting this formula to conventionally accepted values of T and V for enclosures used for various purposes. These formulas allow the estimation of certain limits in the allowable design criteria. For example, conditions for a lecture by an accomplished speaker in a quiet hall will be very good if the reverberation

Table 13.1. Approximate values for $R = T/V^{1/3}$ for various purposes

Purpose	$R \pm 10\%$ (s/m)	Range of Volumes Conventionally Encountered $(m^3)^a$
Concert Hall	0.07	$10 \times 10^3 < V < 25 \times 10^3$
Opera house	0.06	$7 \times 10^3 < V < 20 \times 10^3$
Motion picture theater	0.05	$V < 10 \times 10^3$
Auditorium	0.06	$V < 4 \times 10^3$
Legitimate theater		
Lecture hall		
Conference room		
Recording studio	0.04	$V < 1 \times 10^3$
Broadcasting studio		

^{a} For conversion to British units, $1 \text{ m}^3 = 35.3 \text{ ft}^3$.

time does not exceed about 0.8 s, which suggests a maximum volume for a lecture hall of about $2.4 \times 10^3 \text{ m}^3$. On the other hand, for music in a concert hall the reverberation time should not exceed about 2 s which yields an estimated maximum volume of about $2.4 \times 10^4 \text{ m}^3$. It must be emphasized that these predicted values are only estimates and are untrustworthy for V lying outside the typically encountered range of values. For volumes approaching or exceeding the indicated upper limits, it is found in practice that the value of R may decrease toward its indicated lower limit. The basic trend, however, is obvious. For example, Carnegie Hall is a poor place to hold a lecture without electronic enhancement, and a symphony orchestra or rock band will be disagreeably overpowering in a lecture hall unless output is deliberately and severely limited.

The observed decay of the reverberation may correspond to different T's at different times during the decay. Evidence indicates that the *initial* apparent reverberation time is of greatest importance to the listener. This means that the first delayed arrival and those arrivals immediately following are quite important in establishing a satisfactory impression of reverberation. This can present problems in fan-shaped rooms with diverging side walls.

The problem of specifying an optimum reverberation time for an enclosure designed primarily for the playing or reproduction of music is more complex than for speech, because the optimum time varies with the size of the room, the type of music, and the desired effect. Enclosures designed as music rooms should be more reverberant than lecture halls or conference rooms of similar size. The optimum reverberation time is found to range from about 0.5 s in living rooms, through about 1.0 s in small rooms used for soloists or chamber music, up to about 2.5 s for organ music or oratorios in large cathedrals. Classical and baroque (nonorgan) music usually require reverberation times of about 1.0 to 1.4 s, whereas nineteenth-century orchestral music may require about 2.0 s of reverberation for the best effect. These values are subject to individual tastes and cultural attitudes.

In designing a recording studio for nonclassical music (especially rock, pop-rock, or country and western) or a television broadcast studio, it is advisable to hold the reverberation time to small values, particularly since these uses are usually accompanied by considerable electronic manipulation, including amplitude limiting, artificial reverberation, and frequency contouring.

One factor that *must* be considered in designing an auditorium is the effect of an audience on its reverberation time. Variations in the size of the audience may produce large changes in reverberation times at all frequencies, particularly above 250 Hz. The change is most apparent when an empty concert hall with un-upholstered seats is used for rehearsals—conductors inexperienced with this effect can suffer a rude awakening when the public performance occurs. The influence of variations in the size of an audience can be materially reduced by using seats that are well-upholstered and have perforated bottoms, since the resulting absorption is nearly the same whether the seat is empty or occupied. If economy requires that inexpensive (lightly upholstered or bare) seating be used, then a reasonable rule of thumb is to design the hall to have the desired reverberation time when it is about two-thirds occupied.

For representative* concert halls, Table 13.1 gives a ratio of direct energy to reverberant energy of about 0.07. This means that for sounds lasting long enough to allow the full reverberation of the hall to develop, the reverberant sound energy is roughly 15 times that of the direct arrival. Subjectively, the combined sound field has a loudness somewhat more than twice that for the direct sound alone. This is what generates the sense of power or grandeur associated with music of the Romantic Period. (The absence of the reverberation is what frequently makes the unamplified sound of outdoor concerts so unsatisfying.)

As a result of an extensive survey of European and New World concert halls and opera houses, an empirical relationship between reverberation time and the parameters of the enclosure has been suggested by Beranek,[4]

$$\frac{1}{T} = 0.1 + 5.4 \frac{S_T}{V} \tag{13.27}$$

* See Problem 13.15

Table 13.2. Characteristics of selected concert halls and opera houses

Hall	$V/10^3$ (m^3)	$S_T/10^3$ (m^2)	Reverberation Time (s) at Various Frequencies (Hz)						Delayed Arrival Time (ms)	Seats
			125	250	500	1000	2000	4000		
J	Jerusalem, Binyanei Ha'oomah	24.7	2.4	2.2	2.0	1.75	1.75	1.65	1.5	13–26	3100
N	New York, Carnegie Hall	24.3	2.0	1.8	1.8	1.8	1.6	1.6	1.4	16–23	2800
Bo	Boston, Symphony Hall	18.7	1.6	2.2	2.0	1.8	1.8	1.7	1.5	7–15	2600
A	Amsterdam, Concertgebouw	18.7	1.3	2.2	2.2	2.1	1.9	1.8	1.6	9–21	2200
Gl	Glascow, St. Andrew's Hall	16.1	1.4	1.8	1.8	1.9	1.9	1.8	1.5	8–20	2100
P	Philadelphia, Academy of Music	15.7	1.7	1.4	1.7	1.45	1.35	1.25	1.15	10–19	3000
V	Vienna, Grosser Musicvereinsaal	15.0	1.1	2.4	2.2	2.1	2.0	1.9	1.6	9–12	1700
Bri	Bristol, Colston Hall	13.5	1.3	1.85	1.7	1.7	1.7	1.6	1.35	6–14	2200
Bru	Brussels, Palais des Beaux Arts	12.5	1.5	1.9	1.75	1.5	1.35	1.25	1.1	4–23	2200
Go	Gothenburg, Konserthus	11.9	1.0	1.9	1.7	1.7	1.7	1.55	1.45	22–33	1400
L	Leipzig, Neues Gewandhaus	10.6	1.0	1.5	1.6	1.55	1.55	1.35	1.2	6–8	1600
Ba	Basel, Stadt-Casino	10.5	0.9	2.2	2.0	1.8	1.6	1.5	1.4	6–16	1400
C	Cambridge, Mass., Kresge Auditorium	10.0	1.0	1.65	1.55	1.5	1.45	1.35	1.25	10–15	1200
(Bu)	Buenos Aires, Teatro Colon	20.6	2.1			1.7			—	13–19	2800
(NM)	New York, Metropolitan Opera	19.5	2.6	1.8	1.5	1.3	1.1	1.0	0.9	18–22	2800
(M)	Milan, Teatro alla Scala	11.2	1.6	1.5	1.4	1.3	1.2	1.0	0.9	12–15	2500

Adapted from Beranek (Reference 4).

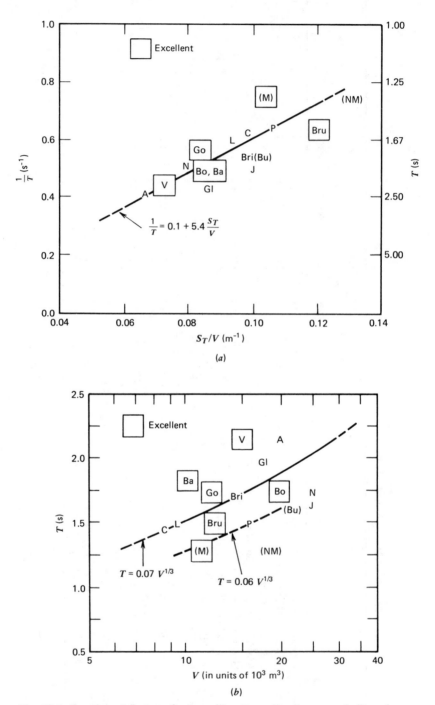

Fig. 13.4. Reverberation times for "good" to "excellent" concert halls and opera houses. For identification see Table 13.2.

This appears to describe above-average concert halls better than (13.26). The quantity S_T is the total floor area of audience, orchestra, and chorus. The absorptivity attributed to S_T is calculated from curve 1 of Fig. 13.2a. A selection of concert halls and opera houses judged "very good" to "excellent" is listed in Table 13.2, and the inverse reverberation times as functions of S_T/V are plotted in Fig. 13.4a. The close fit to the curve of (13.27) is apparent, particularly for the larger reverberation times. For comparison, the values for T and V for the same halls are plotted in Fig. 13.4b along with the appropriate curves $T = RV^{1/3}$. (The scales of the two figures have been adjusted so that they can be visually compared.) The scatter is clearly greater in Fig. 13.4b. It would therefore appear that (13.27) is an improvement over (13.26). Whether the form of (13.27) can be extended to smaller enclosures with different purposes remains to be seen.

(c) Warmth. The *warmth* of the enclosure depends on the comparison between low-frequency and mid-frequency reverberation times. The desired behaviors of the reverberation times as functions of frequency for the limiting cases of speech and music are sketched in Fig. 13.5. The cross-hatched regions indicate desired values,

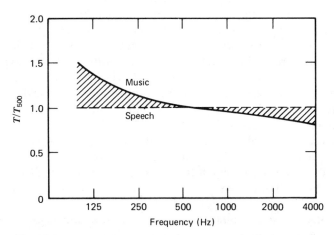

Fig. 13.5. *Relative reverberation time limits for music and speech.*

with baroque music tending toward the *speech* curve and nineteenth-century music tending toward the *music* curve. Study of Fig. 13.3b with these curves in mind will reveal that the use of thin paneling or other lightweight wall material may result in undesirably large absorption at low frequencies. For large concert halls, in which the audience is the dominant source of absorption, thick well-sealed wooden surfaces are an excellent choice.

(d) Intimacy. The *intimacy* of the perceived sound is of great importance for speech and of even greater importance for music. This quality depends on the reception of reflected arrivals immediately after the direct arrival from the source. These *early delayed arrivals* should be many and evenly distributed with time, beginning no later than about 20 ms (30 ms for opera) after the direct arrival and blending smoothly together to form the reverberation. (A 20-ms delay corresponds to a difference in path length of about 7 m.) Early arrivals from the *sides* of a hall are of greater importance than early arrivals from overhead. Obtaining the proper balance of early laterally reflected arrivals can be a problem in large or fan-shaped enclosures, particularly close to the source. Careful design of the stage, a heavy and rigid shell, judiciously positioned ceiling and wall reflectors, and suspended reflectors can enhance these early arrivals throughout the audience. Reflectors should be made of rigidly braced plywood at least 1.9-cm ($\frac{3}{4}$-in.) thick, or of some reasonable equivalent. Care must be taken that these remedies do not introduce undesired resonances, absorption, or focusing of the sound.

Scale drawings of the enclosure with its shell and reflectors must be studied to determine, at representative positions, the first few reflections of sound emitted by the source so that shadows or "hot spots" can be eliminated and early reflections can be distributed throughout the enclosure.* In particular, *echoes* must be scrupulously avoided. The back wall in any enclosure (but particularly in fan-shaped ones) and any improperly designed flat or concave surface can be troublesome. Large, flat, parallel surfaces can cause *flutter* (repetitive echoes). The use of structural irregularities on these surfaces to *scatter* the reflections, or randomly spaced absorptive elements to *suppress* reflections and provide some *diffraction*, are useful remedies. Balconies must be shallow and their undersurfaces designed to prevent focusing, allow early reflected arrivals, and allow the reverberant sound field to penetrate to the underlying audience. If possible, they should be steeply raked so that the balcony seats receive good direct paths.

For halls used primarily for unamplified speech or music, the rows of seats closest to the source should not subtend an angle measured from the source of more than about 120° unless particular attention is paid to providing early reflections from ceiling, walls, shell (if present), or specially located reflecting surfaces.

(e) Diffusion, Blend, and Ensemble.   The reverberant sound field from any source must quickly become diffuse so that there is good *blend* of the sound throughout the enclosure. This is important both for the audience and for the speaker or orchestra. There must also be a reverberant return of sound from the rest of the enclosure to the stage; otherwise, the performers will feel they are operating into a sonic void. This can cause a lecturer to speak too rapidly for the enclosure, project too strongly, or strain his voice. It can cause an orchestra and conductor to misjudge the behavior of the hall. The stage, therefore, should be designed to project the sound evenly from its boundaries back into itself as well as out into the body of the hall. This allows speakers and performers to hear themselves (and each other). This is indispensable for orchestras in obtaining good *ensemble*. In a large enclosure a stage

---

* If the enclosure is sufficiently complicated or irregular in shape, construction of a scale model and use of small, pulsed sound sources and a small probe receiver may reveal acoustic features which could otherwise be overlooked.

should be no wider than about twice its depth. If there are any problems in obtaining good diffusion, the walls and ceiling forming the stage, any shell, and relevant reflecting surfaces should be broken up with carefully chosen irregularities.

13.9 STANDING WAVES AND NORMAL MODES IN ENCLOSURES. Ray acoustics does not provide a complete theory of the behavior of sound in an enclosure. A more adequate approach must be based directly on wave theory. The wave equation has been solved (at least approximately) for simple enclosures (such as rectangular and hemispherical spaces) and new concepts have emerged from examining the transient and steady-state behaviors of sound in such enclosures. Even in complicated enclosures for which the wave equation cannot be solved, the theory has been used to supplement and extend results predicted by ray-acoustic methods.

(a) The Rectangular Enclosure.   As was seen in Sect. 9.7, the solution of the wave equation in a lossless, rigid-walled, rectangular cavity of dimensions L_x, L_y, and L_z results in the normal modes

$$\mathbf{p}_{lmn} = \mathbf{P}_{lmn}(x,\, y,\, z)e^{j\omega_{lmn}t} \tag{13.28}$$

where

$$\mathbf{P}_{lmn} = \mathbf{A}_{lmn} \cos k_{xl}x \, \cos k_{ym} y \, \cos k_{zn} z \tag{13.29}$$

The components of k are given by (9.50), the associated natural frequencies by (9.51), and the modes are labeled by the integer set (l, m, n). If *no* integer is zero, the mode is termed *oblique*. If *one* of the integers is zero, the mode is termed *tangential* because the propagation vector is parallel to one pair of surfaces. If *two* integers are zero, the mode is termed *axial* because the propagation vector is parallel to one of the axes.

(b) Damped Normal Modes.   However, if the walls of the enclosure are no longer perfectly rigid, but allow losses of acoustic energy from the system, there are two significant modifications to the above normal modes. First, since there are energy loses, the normal modes will decay with time. Each natural frequency ω_{lmn} must be replaced by a complex quantity $\boldsymbol{\omega}_{lmn}$ where

$$\boldsymbol{\omega}_{lmn} = \omega_{lmn} + j\beta_{lmn} \tag{13.30}$$

Second, the standing wave will exhibit spatial damping so that the propagation vector $\vec{k}$ (and therefore all its components) must be replaced by a complex $\mathbf{k}$ where $\mathbf{k}_{xl} = k_{xl} + j\alpha_{xl}$ and similarly for the y and z components. The dampled normal mode now has the form

$$\mathbf{p}^{D} = A \, \cos(\mathbf{k}_x x + \phi_x)\cos(\mathbf{k}_y y + \phi_y)\cos(\mathbf{k}_z z + \phi_z)e^{j\omega t} \tag{13.31}$$

and substitution into the lossless wave equation (we ignore losses in the medium) reveals that we must have

$$k_x^2 + k_y^2 + k_z^2 - (\alpha_x^2 + \alpha_y^2 + \alpha_z^2) = (\omega/c)^2 - (\beta/c)^2 \tag{13.32}$$

$$\alpha_x k_x + \alpha_y k_y + \alpha_z k_z = \omega\beta/c^2 \tag{13.33}$$

The subscripts l, m, n have been dropped for simplification in (13.31 to 13.33). The values of the k's, ϕ's, and α's must be found by application of the boundary conditions describing the lossy walls. These in turn will determine the temporal absorption coefficient β of each of the normal modes. This is no more than the three-dimensional generalization of the analysis of the fixed, resistance-loaded string in Sect. 2.11b and the freely vibrating bar in Sect. 3.6.

The normal modes and their natural frequencies depend primarily on the shape and size of the enclosure, whereas their rates of damping depend chiefly on the specific values of the normal specific acoustic impedances of the walls. This division is fortunate, for it enables us to use the simplest possible boundary condition, perfectly rigid walls with no damping, to derive the normal modes and their natural frequencies. The effect of energy absorption by the walls on the damping of the normal modes can then be considered as a *perturbation* of these simple conditions. Because of this, in dealing with the properties of the sound field, it is often possible to write the damped normal modes in the simpler form

$$\mathbf{p}_{lmn}^D \doteq \mathbf{P}_{lmn}(x, y, z)\, e^{-\beta_{lmn}t}\, e^{j\omega_{lmn}t} \tag{13.34}$$

with $\mathbf{P}_{lmn}$ given by (13.29).

(c) The Driven Standing Wave.   The previous equations have established the *homogeneous solutions* to the wave equation in an enclosure with lossy boundaries. The second step, obtaining the *particular solutions* corresponding to some source of sound within the enclosure, is not so straightforward, and the mathematical details are not of interest here. We will be content to obtain approximate forms of the standing wave resulting from a source of sound.

In a cavity with walls that are basically rigid but lossy, we may expect pressure antinodes at the boundaries, so that the standing waves must have the form (13.34) but with the driving frequency ω in place of the natural frequency ω_{lmn}. The amplitude of each standing wave must depend on the difference between the driving frequency and the resonance frequency for that standing wave, and we can borrow directly from the driven, damped harmonic oscillator of Sect. 1.7. The amplitude of the motion varies as $[R_m^2 + (\omega m - s/\omega)^2]^{-1/2}$ where R_m describes the system losses and the resonance frequency is $\sqrt{s/m}$. By analogy with (1.38), the Q of the standing wave is given by

$$Q_{lmn} = \frac{\omega_{lmn}}{2\beta_{lmn}} \tag{13.35}$$

and the resonance frequency of the standing wave is equal to the natural frequency of the corresponding damped normal mode to within a term of order β^2—they can be assumed identical. Extending this to three-dimensional standing waves, we have

$$\mathbf{p}_{lmn}^S = \mathbf{P}_{lmn}^S(x, y, z)\, e^{j\omega t} \tag{13.36a}$$

where

$$\mathbf{P}_{lmn}^S = \frac{\mathbf{B}_{lmn}\, \cos k_{xl} x\, \cos k_{ym} y\, \cos k_{zn} z}{\sqrt{(1/Q_{lmn})^2 + (\omega/\omega_{lmn} - \omega_{lmn}/\omega)^2}} \tag{13.36b}$$

Thus, the particular solution will be of the form

$$\mathbf{p}^S = e^{j\omega t} \sum_{l,\, m,\, n} \mathbf{P}_{lmn}^S(x, y, z) \tag{13.37}$$

The k's are given by (9.50), and the $\mathbf{B}$ are found from the location and configuration of the source. In practice, just those terms in the summation for which $\omega_{lmn} \sim \omega$ will be of significant amplitude.

(d) Growth and Decay of Sound from a Source. When a source of sound is turned on at $t = 0$, the complete solution must be

$$\mathbf{p} = \sum_{l,m,n} (\mathbf{p}_{lmn}^D + \mathbf{p}_{lmn}^S)$$

The constants **B** in the standing waves are determined by the form and position of the source, and the constants **A** in the normal modes by the initial condition that there be no sound field just before the source is turned on. For a reasonably "live" room with solid walls, this yields the approximate solution

$$\mathbf{p} \doteq \sum_{l,\,m,\,n} \mathbf{P}_{lmn}^S(x,\,y,\,z)(e^{j\omega t} - e^{-\beta_{lmn}t}e^{j\omega_{lmn}t}) \tag{13.38a}$$

Each standing wave is canceled by the associated, damped normal mode at $t = 0$, and as the damped normal mode dies away the combination grows to attain the final value given by the standing wave alone. The growth will be smooth if $\omega_{lmn} \approx \omega$, and more irregular the more dissimilar are the driving and natural frequencies. (For an exaggerated illustration, refer to Fig. 1.7.)

In nonrectangular rooms this same result will apply if $\mathbf{P}_{lmn}$ is reinterpreted in terms of the normal modes of the room.

Thus, a room may be treated as a resonator having numerous allowed modes of vibration, each with its natural frequency of damped free vibration. When a sound source is started in such a resonator, steady-state standing waves having the frequency of the source are set up, together with the (transient) damped, free vibrations composing the normal modes of the room. If the sound source is operating continuously, each component of the transient dies out at its own rate, eventually leaving only the steady-state standing waves.

Turning off the monofrequency signal after a long time is mathematically equivalent to turning on another monofrequency signal 180° out of phase with the original one. By superposition, the steady-state solution for the new signal annihilates the steady-state solution of the original signal, and what is left is the same collection of damped normal modes:

$$\mathbf{p} = \sum_{l,\,m,\,n} \mathbf{P}_{lmn}^S(x,\,y,\,z)e^{-\beta_{lmn}t}e^{j\omega_{lmn}t} \tag{13.38b}$$

Because each normal mode has its own natural frequency, these modes will often interfere with each other and produce beats. The collection of all these decaying normal modes is the reverberant sound field.

Unless all the excited modes of vibration decay at the same rate, there will be no unique value for the reverberation time. Instead, the rate of decay of acoustic energy will at first be rapid, corresponding to a short reverberation time, and then will decrease with longer and longer apparent reverberation times as the more weakly damped waves die out. This is readily observed at low frequencies in enclosures whose walls have distinctly different absorptivities.

(e) Frequency Distribution of Enclosure Resonances. A knowledge of the natural frequencies of a room is essential to a complete understanding of its acoustic properties. The room will respond strongly to those sounds having frequencies in the immediate vicinity of any of these natural frequencies.

It is just this characteristic that affects the output of a loudspeaker as measured in a reverberant room and causes the results to have limited significance as a true criterion of the speaker's properties. Furthermore, each standing wave has its own particular spatial pattern of nodes and antinodes. In effect, every room or enclosure superimposes its own characteristics on those of any sound source present, so that the fluctuations in sound pressure that occur as a microphone is moved from one point to another, or as the frequency of the source is varied, may completely conceal the true output characteristics of the source. It is for this reason that measurements of the response curves of loudspeakers should be carried out either in the open air or in *anechoic chambers*. If the absorption coefficient at the wall of an anechoic chamber is greater than 0.99, the reverberant field is negligible compared to the direct wave.

Each of the individual standing waves of an enclosure can only be excited to its fullest extent by a sound source located in regions where the particular standing wave pattern has a pressure antinode (recall Sect. 9.6). The pressure amplitudes of all patterns of standing waves in a rectangular enclosure are maximized in the corners of the room. Therefore, if the source is at the corner of such a room, it will be possible for it to excite every allowed mode to its fullest extent. Correspondingly, if a microphone is located in the corner of the room, it will measure the peak sound pressure for every normal mode that has been excited. By contrast, when a source is located in a region where a particular mode has a pressure node, that mode will be excited only weakly, if at all. For instance, if a loudspeaker is located in the center of a

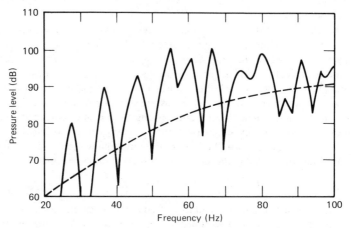

Fig. 13.6. *Normal mode response of a rectangular room at low frequencies.*

rectangular room, only those modes having even numbers simultaneously for *l*, *m*, and *n* will be excited (about 1 mode in 10) as the driving frequency is slowly varied from low to high frequencies.

An example of the transmission of sound from a loudspeaker located in one corner of a rectangular room to a microphone located in a diagonally opposite corner is given in Fig. 13.6 for the room described in Table 13.3. These curves could be obtained by slowly increasing the frequency supplied to the loudspeaker from 20 to 100 Hz and simultaneously recording the output of the microphone. The dashed line corresponds to the output of the loudspeaker measured in an anechoic chamber. The influence of the room is quite apparent. When either the loudspeaker or microphone is located in the center of the room, only those peaks corresponding to the (0, 0, 2), (0, 2, 0), and (0, 2, 2) modes would be observed below 100 Hz.

Equation (9.51) suggests that each of the natural frequencies f may be considered as a vector in *frequency space*, having components $f_x = lc/2L_x$, $f_y = mc/2L_y$, and $f_z = nc/2L_z$. A

Table 13.3. Natural frequencies below 100 Hz for a rectangular room 3.12 × 4.69 × 6.24 m with $c = 345$ m/s

l	m	n	$f\,(Hz)$	l	m	n	$f\,(Hz)$
0	0	1	27.5	1	0	2	77.5
0	1	0	36.6	0	2	1	78.5
0	1	1	45.9	0	0	3	82.5
1	0	0	55.0	1	1	2	86.0
0	0	2	55.0	0	1	3	90.2
1	0	1	61.5	0	2	2	91.5
0	1	2	66.0	1	2	0	91.5
1	1	0	66.0	1	2	1	95.5
1	1	1	71.5	1	0	3	99.0
0	2	0	73.2				

normal mode of vibration may therefore be represented by a point in this frequency space. All the normal modes of natural frequency f and below are included within the octant of frequency space between the positive $f_x, f_y,$ and f_z axes and a spherical surface of radius f. This representation is useful in computing the number of modes having natural frequencies lying within any specified frequency interval. Each lattice point occupies a rectangular block in frequency space of dimensions $c/2L_x$, $c/2L_y$, $c/2L_z$ and therefore volume $c^3/(8V)$ where $V = L_x L_y L_z$. Then, the number of points can be estimated by dividing the volume of an octant of radius f by $c^3/(8V)$. Accordingly, the number of normal modes of frequency below f is estimated by

$$N \sim \frac{4\pi V}{3c^3} f^3 \tag{13.39}$$

Differentiating (13.39) with respect to f yields the number of normal modes ΔN having frequencies in a band of width Δf centered on f,

$$\frac{\Delta N}{\Delta f} \sim \frac{4\pi V}{c^3} f^2 \tag{13.40}$$

The number of normal modes in a given frequency band of width Δf increases rapidly as the center frequency of the band (or size of the enclosure) is increased. As a result, the responses of the standing waves will overlap more and more at higher frequencies so that with increasing frequency the response of the room will become quite smooth.

Since the standing wave pattern corresponding to each natural frequency is in general associated with some particular set of direction cosines, any increase in ΔN indicates an increase in the randomness of directions of the associated waves. This is supported by the observation that reverberation equations based on a diffuse sound field, such as (13.12a), are in better agreement with experiments at high frequencies.

The response of a room is observed to become less uniform as its symmetry is increased. This results from the increase in the number of *degenerate* modes having different (l, m, n) but the same natural frequency. If it is desired to optimize the room dimensions for greatest uniformity in the distribution of normal modes, Bolt[12] has shown that acceptable length

[12] Bolt, *J. Acous. Soc. Am.*, **19**, 130 (1946).

ratios $1 : X : Y$ constitute a region in XY space. Roughly, acceptable values satisfy the joint conditions $2 < (X + Y) < 4$ and $\frac{3}{2}(X - 1) < (Y - 1) < 3(X - 1)$.

(f) Reverberation of the Normal Modes. Because we have seen from Chapter 7 that loss mechanisms are additive in the overall absorption coefficient, we will first restrict attention to an enclosure having only one lossy wall at $x = L_x$.

The question immediately arises as to what physical property of the wall should be used in applying a boundary condition to the normal mode. The simplest and most direct is to assume that the absorbing characteristics of the wall are determined by its *normal specific acoustic impedance* z_x, the ratio of pressure to the normal component of the particle velocity at the surface of the wall. (Recall the discussion of Sect. 6.6.) For ease, and not unrealistically, we assume that the reactive part of z_x can be neglected. Thus,

$$z_x = \rho_0 c v_x \tag{13.41}$$

where v_x is the dimensionless ratio of the normal specific acoustic resistance of the wall to the characteristic acoustic impedance $\rho_0 c$ of the air.

Now, define the walls at $x = 0$, $y = 0$, $z = 0$, $y = L_y$, and $z = L_z$ to be perfectly rigid and nonabsorbing, and let the wall at $x = L_x$ be quasirigid but absorbing, as described by (13.41). A normal mode given by (13.31) must then have all the values of ϕ as zero, $\alpha_y = \alpha_z = 0$ and k_y and k_z given by (9.50). The necessary conditions (13.32 and 13.33) take the forms

$$k^2 - \alpha_x^2 = \left(\frac{\omega}{c}\right)^2 - \left(\frac{\beta}{c}\right)^2 \tag{13.42a}$$

and

$$\alpha_x k_x = \frac{\omega}{c} \frac{\beta}{c} \tag{13.42b}$$

The boundary condition at $x = L_x$ becomes

$$\left. \frac{\mathbf{p}}{\mathbf{u}_x} \right|_{x = L_x} = \rho_0 \frac{j\omega - \beta}{k_x + j\alpha_x} \cot[(k_x + j\alpha_x)L_x] = \rho_0 c v_x$$

or

$$\tan[(k_x + j\alpha_x)L_x] = \frac{1}{v_x} j \frac{\omega/c + j\beta/c}{k_x + j\alpha_x} \tag{13.43}$$

For all normal modes *except* those with $l = 0$, $k_x L_x$ must be very close to $l\pi$ because the wall is assumed to be fairly rigid. Expansion of the tangent yields

$$(k_x L_x - l\pi) + j\alpha_x L_x = \frac{1}{v_x} j \frac{\omega/c + j\beta/c}{k_x + j\alpha_x} \tag{13.44}$$

and thus $k_x L_x \doteq l\pi$, $\alpha_x k_x \doteq \omega/(v_x c L_x)$ so that

$$\beta = c/(v_x L_x) \qquad l = 1, 2, 3, \ldots \tag{13.45a}$$

For the case $l = 0$ we must have $k_x L_x \ll \pi$. Now, set $l = 0$ in (13.44) and cross multiply. The result is

$$(k_x + j\alpha_x)^2 = \frac{1}{v_x} \frac{1}{L_x} j\left(\frac{\omega}{c} + j\frac{\beta}{c}\right)$$

Solution from equating imaginary parts yields $\alpha_x k_x \doteq \omega/(2v_x cL_x)$ and, therefore,

$$\beta = \tfrac{1}{2}c/(v_x L_x) \qquad l = 0 \tag{13.45b}$$

Notice that β for this "grazing" mode is *half* of that for all modes with $l \neq 0$. [Solution for the real part yields $k_x \sim \alpha_x$, and we see that for these $(0, m, n)$ modes k_x does not actually attain zero, but has a small but finite value. This is plausible on physical grounds. Because this wall is not *perfectly* rigid, any pressure on it must cause it to yield slightly; thus u_x cannot be identically zero on the wall and, therefore, there must be a finite k_x.]

The results (13.45a) and (13.45b) can be combined as

$$\beta = \varepsilon_x \frac{1}{v_x} \frac{c}{L_x} \tag{13.45c}$$

where ε_x is 0.5 for a normal mode whose propagation vector is essentially parallel to the absorbing wall at $x = L_x$ and is unity otherwise.

The energy density at any point in the standing wave is proportional to the square of the pressure and therefore decays as $\exp(-2\beta t)$ after the source of sound is shut off. Solving for the reverberation time as in Sect. 13.3 (and recognizing that $2\beta = 1/\tau_E$) yields

$$T = \frac{6.9}{c} V \frac{v_x}{\varepsilon_x S_x} \tag{13.46}$$

Before proceeding to relate v_x to the absorptivity of the wall, note that (13.46) can be immediately generalized to the more realistic situation wherein all six walls of the enclosure are individually lossy. The combined β for all is simply the sum of the β's each would contribute if it alone were absorbing, with the result that the reverberation time for one mode is

$$T = \frac{6.9}{c} \frac{V}{\sum\limits_{i=1}^{6} \varepsilon_i S_i/v_i} \tag{13.47}$$

where i labels each of the six walls, S_i is the surface area of each, and ε_i is either 0.5 or 1 depending on whether the normal mode "grazes" that surface. Notice that, by its form, this should be extensible to other than rectangular enclosures.

Now, v_x must be related to the Sabine absorptivity a of the wall. Given the assumptions that (1) the wall is described by (13.41), and (2) for a normal mode whose propagation vector makes an angle θ with respect to the normal to the wall, the plane wave transmission coefficient is given by (6.46), then this coefficient must equal the energy absorption coefficient $a_{E\theta}$ of the wall for a wave reflecting from it with this angle of incidence.

$$a_{E\theta} = \frac{4v_x \cos\theta}{(v_x \cos\theta + 1)^2} \tag{13.48}$$

Now, for this particular model of the behavior of the wall, we can relate $a_{E\theta}$ to the random incidence coefficient a_E and thereby relate v_x to a. By reasoning identical to that used in deriving (13.2), it can be shown that the fractional amount of energy absorbed by a surface element during $\Delta t = \Delta r/c$ is obtained by inserting $a_{E\theta}$ into the integrand of (13.2),

$$\frac{\mathscr{E} \, \Delta S \, \Delta r}{2} \int_0^{\pi/2} a_{E\theta} \sin\theta \cos\theta \, d\theta$$

The fractional amount of energy incident on this surface element during Δt is $\mathscr{E} \, \Delta S \, \Delta r / 4$, and hence the random incidence energy absorption coefficient is

$$a_E = 2 \int_0^{\pi/2} a_{E\theta} \sin \theta \cos \theta \, d\theta$$

If (13.48) is substituted and the integration performed, the result is

$$a_E = \frac{8}{v_x} \left[1 + \frac{1}{1 + v_x} - \frac{2}{v_x} \ln(1 + v_x) \right] \tag{13.49a}$$

When $v_x \geq 100$, this equation may be simplified to

$$v_x \approx 8/a_E \qquad a_E \leq 0.08 \tag{13.49b}$$

within an error of less than 10 percent. In the limit of these small a_E's they can be replaced by the Sabine absorptivity, $a \doteq a_E$.

Substitution of (13.49b) into (13.46) and (13.47) then yields

$$T = \frac{0.161 V}{\varepsilon_x S_x a_x}$$

for a room with only one absorbing wall, and

$$T = \frac{0.161 V}{\sum_i \varepsilon_i S_i a_i}$$

for all walls absorbing. Thus, with the exception of normal modes that graze any of the surfaces, this result is in exact agreement with the formula of Sabine.

The smaller damping of grazing modes results from the fact that the average mean square pressures produced by these modes at the surface they graze are only half those produced on the other walls. This illustrates an important general principle: an absorbing surface is most effective in damping a normal mode if it is located in a region of maximum mean square pressure. The most effective method of reducing any particular undesired mode is to place absorbing material on those parts of the walls where the pressures corresponding to this mode are the greatest. Since all normal modes have pressure maxima at the corners, absorbing material placed near the corners of a room is twice as effective, on the average, as if it were placed elsewhere.

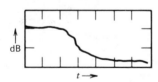

Fig. 13.7. *Influence of normal modes on a reverberation decay curve.*

Since all ε's are unity for an oblique normal mode, the reverbation time of all such modes in a live room is identical with that given by (13.12c). On the other hand, for tangential or axial modes, one or more of the ε's equals 0.5, so that the reverberation times for these waves are always somewhat greater than those for oblique modes. For example, in a room having all six walls covered with identical absorbing material, the respective reverberation times for *axial, tangential,* and *oblique* modes are related to one another as 6 : 5 : 4. In such a room the reverberation time, as measured for low driving frequencies, may vary rapidly with changes in frequency as first one and then another type of normal mode is strongly excited. At such frequencies the term *reverberation time* has a specific meaning only in connection with the damping of some particular type of normal mode. In the middle range of frequencies the measured decay in sound pressure level may be expected to approximate a broken line, such as that indicated in Fig. 13.7. Here the steeper initial part corresponds to the decay rate of the oblique modes, the middle part to the decay of the tangential modes, and the less steep final part to the decay of the axial modes. The actual length of time required for the intensity level to drop 60 dB will therefore depend on the relative amounts of acoustic energy possessed by the various types of modes.

At such high frequencies where most of the acoustic energy resides in oblique modes, the break in the curve giving the rate of decay of intensity level comes late enough so that the first 20 or 30 dB of the curve is nearly straight. The slope of this line may then be used to determine the reverberation time associated with the major portion of the acoustic energy present. Consequently, it is only in this frequency range that the concept of reverberation time is, by itself, a sufficient criterion for judging the acoustic characteristics of an enclosure.

In spite of the greater complexity of the mathematical techniques involved, the application of the methods of wave theory to architectural acoustics has markedly broadened our knowledge of the behavior of sounds in enclosures. It leads itself admirably to a consideration of such important factors as the shape of the enclosure, the effects of different distributions of absorbing surfaces, the behavior of sound in dead rooms, the effect of the position of source and receiver, and so on.

PROBLEMS

As aids in conversion, note that 10 ft = 3.05 m, 100 ft^2 = 9.3 m^2, 1000 ft^3 = 28.3 m^3, 0.0002 μbar = 20 μPa, 1 ft^2 = 1 English sabin = 0.093 m^2, and 1 m^2 = 1 metric sabin = 10.8 ft^2.

13.1. When steady-state conditions are reached in a live room, the sound pressure level is 74 dB *re* 20 μPa. (*a*) If the average absorptivity of the walls is 0.05, what is the rate at which sound energy is being absorbed per square meter of wall surface? (*b*) If the total area of absorbing surface in the room is 50 m^2, at what rate in watts is sound power being generated in the room?

13.2. The surfaces of an auditorium 200 × 50 × 30 ft have an average absorptivity $\bar{a}$ = 0.29. (*a*) What is the reverberation time? (*b*) What must be the power output of a source if it is to produce a steady-state sound pressure level of 65 dB *re* 20 μPa? (*c*) What average absorptivity would be required if a speaker having an acoustic output of 100 μW is to produce a steady-state level of 65 dB *re* 20 μPa? (*d*) Calculate the resulting reverberation time, and comment on its influence on the intelligibility of the speaker.

13.3. A room 10 × 10 × 4 m has an average absorptivity $\bar{a}$ = 0.1. (*a*) Calculate its reverberation time. (*b*) What must be the output of a source if it is to produce a steady-state sound pressure level of 60 dB *re* 20 μPa? (*c*) At what rate in watts per square meter is sound energy incident on the walls of the room?

13.4. An auditorium is observed to have a reverberation time of 2.0 s. Its dimensions are 7 × 15 × 30 m. (*a*) What acoustic power is required to produce a steady-state sound pressure level of 60 dB *re* 20 Pa? (*b*) What is the average absorptivity of the surfaces in the auditorium? (*c*) If 400 people are present, each adding 0.5 m^2 to its total absorption, what is the new reverberation time? (*d*) What will be the new sound pressure level produced by the sound source of part (*a*)?

13.5. A small reverberation chamber is constructed with concrete walls. Its internal dimensions are 6 × 7 × 8 ft. (*a*) Calculate the final steady-state sound pressure level produced by a 2000 Hz source of 7.5 μW output. Assume the absorptivity of the concrete to be 0.02. (b) How many seconds will be required from the time the source is turned on until the pressure level reaches within 3 dB of the value of part (a)? (c) When an observer enters the chamber, the steady-state level is observed to decrease by 3 dB. What is the absorbing ability of the observer in English sabin?

13.6. (a) Show that m in (13.16a) must have the units of m^{-1}. (b) Compute m at 3000 Hz in air of 38 percent relative humidity. (c) With absorption only at the walls, the reverberation time at 3000 Hz in a room of 10,000-ft^3 volume is 1.2 s. What would be the reverberation time if absorption in 38 percent relative humidity air is taken into consideration?

13.7. The reverberation time in a small reverberation chamber $9 \times 10 \times 11$ ft is 4.0 s. (a) What is the effective absorptivity of the surfaces of the chamber? (b) When 50 ft^2 of one wall is covered with an acoustic tile, the reverberation time is reduced to 1.3 s. What is the effective absorptivity of the tile? (c) What would be the reverberation time if all surfaces of the chamber were covered with this tile?

13.8. A room $3 \times 6 \times 10$ m has walls of average absorptivity 0.05. The floor is covered with a rug ($a = 0.6$), and the ceiling is wood ($a = 0.1$). (a) Calculate the average absorptivity. (b) Calculate the area-averaged random-incidence energy-absorption coefficient. (c) Calculate the reverberation time. (d) Under the assumption that $a = a_E$ for each surface, calculate the predicted reverberation times from the Millington-Sette formula and (13.20b), and compare with the answer to (c).

13.9. Given a room $12 \times 18 \times 30$ ft, (a) what is the mean free path of a ray in this room? (b) If the average decrease in intensity level per encounter with the walls of the room is 1 dB, what is the average absorptivity of the room?

13.10. Given a cubical room 10 ft on a side, (a) what is the mean free path of a sound ray in this room? (b) How many reflections per second does an average ray make with the walls? (c) If the loss in level per reflection is 1.5 dB, what is the reverberation time of the room? (d) What is the average absorptivity for the walls?

13.11. A cubical room 10 ft on a side has acoustic tile on its walls, acoustic plaster on its ceiling, and a carpeted concrete floor. (a) Using Figs. 13.2 and 13.3, calculate its reverberation time at 500 Hz. (b) Repeat the calculations at 125 Hz. (c) At 2000 Hz.

13.12. Given a noise source of 10 μW acoustic output at 125 Hz, (a) what ambient background sound pressure level will it generate in the room of Problem 13.11. (b) What will be the loudness level in phon? (c) Repeat (a) and (b) for a similar noise source of 500 Hz.

13.13. A motor produces a steady-state reverberant sound pressure level of 74 dB *re* 20 μPa in a room $10 \times 20 \times 50$ ft. The measured reverberation time of the room is 2 s. (a) What is the acoustic output of the motor? (b) How many additional sound absorbing units in English sabin must be added to this room in order to lower the sound level by 10 dB? (c) What is the new reverberation time?

13.14. A sound source has a directivity factor of 10 and an acoustic output of 100 μW in a room of 5000-ft^3 volume having a reverberation time of 0.7 s. What is the maximum distance from the sound source at which the sound pressure level in the direct field will be 10 dB above that of the reverberant field?

13.15. A concert hall has its dimensions in the ratio $1 : 1 : 2$, where the length from the stage to the rear of the audience is the long dimension. Given the rough design criterion of Table 13.1, evaluate the ratio of reverberant to direct intensity for a seat in the middle of the hall.

13.16. A concert hall has floor dimensions 20 m $\times$ 50 m and is 15-m high. The entire floor is taken up by the orchestra and audience, and there is a balcony 5-m deep across the back, 3-m deep on the sides, and extending 40 m along the sides. The sides, ceiling, and underside of the balcony are constructed of heavy, sealed wood. There is, in addition, a total of 200 m^2 of windows. Calculate the reverberation time when fully occupied as a function of frequency (assume 36 percent humidity). How does the hall compare with Beranek's criterion (13.27)?

13.17. A back-enclosed loudspeaker cabinet has internal dimensions $2 \times 3 \times 4$ ft. Its internal surfaces are lined with an absorbing material having an effective absorptivity of 0.2. (a) The influence of internal resonances on the output of the loudspeaker may be assumed to be negligible when the average number of normal modes of vibration per 1-Hz band becomes greater than one. Above what frequency does this occur? (b) The duration of any transient vibrations of the loudspeaker will be influenced by the reverberation time of the cabinet. What is the reverberation time in the cabinet? (c) Steady-state sounds radiated into the cabinet may produce very large sound pressures within it. If 0.1 W of acoustic power is radiated into the cabinet, what is the steady-state sound pressure level within it? (d) What are the frequencies of the three lowest normal modes of the cabinet?

13.18. A cubical room is 5 m on a side. (a) What is the characteristic frequency of the normal mode corresponding to the lowest-frequency axial wave? (b) The lowest-frequency tangential wave? (c) The lowest-frequency oblique wave?

13.19. Given a room of $4 \times 6 \times 10$ m, (a) what is the frequency of the (1,1,1) mode? (b) If the average absorptivity at this frequency is 0.1, what is the reverberation time for this mode? (c) What is the minimum frequency for which the average number of normal modes per 1-Hz band will exceed 5?

13.20. A cubical room is 5 m on a side. (a) Compute the characteristic frequencies between 30 and 70 Hz. (b) Locate the nodal planes for each of these frequencies.

13.21. All the walls of a rectangular room $3 \times 4 \times 7$ m have a relative normal specific acoustic impedance $v_x = 20$. (a) What is the absorptivity of the walls for randomly incident sound waves? (b) What is the reverberation time of the room for oblique waves? (c) What is the reverberation time for axial waves parallel to the longest dimension of the room?

13.22. A cubical room is 5 m on a side and has an $v_E = 32$ for the floor and ceiling and 160 for the walls. What is the reverberation time (a) for those axial waves that strike the floor and ceiling? (b) For those tangential waves that strike all four walls? (c) For the oblique waves?

13.23. The speed of sound in concrete is 3500 m/s and its density is 2700 kg/m^3. (a) What is the normal specific acoustic resistance of concrete relative to water? (b) Applying (13.49a), calculate the random-incidence energy-absorption coefficient. (c) If a concrete-walled tank 3 m on a side is filled with water, what is the indicated reverberation time for water-borne sounds? Assume the upper surface reflects all incident energy.

CHAPTER 14

TRANSDUCTION

14.1 INTRODUCTION. This chapter develops the formalism of transduction theory in reasonable detail. A more thorough treatment can be found in the book by Hunt[1] which also contains an entertaining and illuminating sketch of the historical development of this branch of science. Another source, dated in many spots but detailed and informative, is Reference 2. While the general theory developed in this chapter can be applied to any electroacoustic transducer, whether designed for use in air or water, specific examples will be confined to transducers used for audio frequencies in air. Detailed consideration of piezoelectric, ferroelectric, and magnetostrictive vibrators is too specialized for the scope of this textbook. The interested reader is referred to the article by Berlincourt, Curran, and Jaffe[3], and other sources[4-6] which do proper justice to these topics.

14.2 THE TRANSDUCER AS AN ELECTRICAL NETWORK. A transducer converting energy between electrical and mechanical forms can be viewed as a *two-port network* relating the electrical properties at one port to the mechanical properties at the other. Here are the definitions we will use throughout this chapter:

V = the voltage across the electrical inputs to the transducer

I = the current into the electrical inputs

F = the force on the radiating surface

u = the speed of the radiating surface

If *force* is considered the analog of *voltage*, and *speed* the analog of *current*, the network is as shown in Fig. 14.1a. Under certain conditions it may be more useful to consider the *mechanical dual* for which *speed* is the analog of *voltage*, and *force* the analog of *current*; the network is then that of Fig. 14.1b. Which alternative is more convenient depends on the transducer.

[1] Hunt, *Electroacoustics*, Wiley (1954).

[2] Olsen, *Elements of Acoustical Engineering*, Van Nostrand (1947).

[3] Mason (editor), *Physical Acoustics* IA, Academic Press (1964).

[4] Hueter and Bolt, *Sonics*, Wiley (1955).

[5] Bobber, *Underwater Electroacoustic Measurements*, U.S. Government Printing Office (1970).

[6] Camp, *Underwater Acoustics*, Wiley (1970).

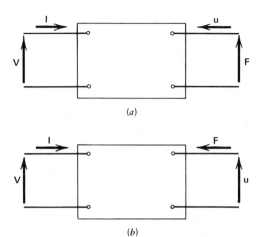

(a)

(b)

Fig. 14.1. *Two-port networks.*

Associated with the electrical and mechanical variables are a number of mechanical and electrical impedances that are measurable properties of the system:

$\mathbf{Z}_{EB} = \mathbf{V}/\mathbf{I}|_{u=0} =$ the *blocked electrical* impedance (Ω)
$\mathbf{Z}_{EF} = \mathbf{V}/\mathbf{I}|_{F=0} =$ the *free electrical* impedance (Ω)
$\mathbf{Z}_{mo} = \mathbf{F}/\mathbf{u}|_{I=0} =$ the *open-circuit mechanical* impedance (N · s/m)*
$\mathbf{Z}_{ms} = \mathbf{F}/\mathbf{u}|_{V=0} =$ the *short-circuit mechanical* impedance (N · s/m)   (14.1)

Uppercase subscripts designate electrical impedances and lowercase subscripts designate mechanical impedances; the second subscript refers to the constraint.

From the definition of $\mathbf{Z}_{EB}$ we have $\mathbf{V}(\mathbf{I}, 0) = \mathbf{Z}_{EB}\mathbf{I}$. If $\mathbf{u}$ is not zero, however, the value of $\mathbf{V}$ must be different for the same applied current. If we assume that $\mathbf{V}$ must depend *linearly on* $\mathbf{u}$, as it does on $\mathbf{I}$, then

$$\boxed{\mathbf{V} = \mathbf{Z}_{EB}\mathbf{I} + \mathbf{T}_{em}\mathbf{u}}$$ (14.2a)

where $\mathbf{T}_{em}$ is a *transduction coefficient*. Similarly, the force equation is generalized to

$$\boxed{\mathbf{F} = \mathbf{T}_{me}\mathbf{I} + \mathbf{Z}_{mo}\mathbf{u}}$$ (14.2b)

Equations (14.2a) and (14.2b) form the *canonical equations* that describe the electromechanical behavior of a transducer.

If the electrical terminals are shorted so that $\mathbf{V} = 0$, then (14.2a) can be solved to

---

* The mechanical impedance is often given units of mechanical ohms (1 mechanical ohm = 1 N · s/m).

express $\mathbf{I}$ in terms of $\mathbf{u}$ and this result can be substituted into (14.2b). The quantity $\mathbf{F}/\mathbf{u}$ is then $\mathbf{Z}_{ms}$, and further manipulation yields

$$\mathbf{Z}_{ms} = (1 - \mathbf{K}^2)\mathbf{Z}_{mo} \tag{14.3}$$

where the *coupling factor* $\mathbf{K}$ is defined by

$$\mathbf{K}^2 = \frac{\mathbf{T}_{em}\mathbf{T}_{me}}{\mathbf{Z}_{EB}\mathbf{Z}_{mo}} \tag{14.4}$$

and often has a simple form for suitable frequencies.

Similarly, imposition of $\mathbf{F} = 0$ and manipulation reveals

$$\mathbf{Z}_{EF} = (1 - \mathbf{K}^2)\mathbf{Z}_{EB} \tag{14.5}$$

Under certain conditions of symmetry the canonical equations assume simpler form, and the networks of Fig. 14.1 can be represented by reversible three-terminal networks.

(a) Reciprocal Tranducers. If $\mathbf{T}_{em} = \mathbf{T}_{me} = \mathbf{T}$, the transducer displays *electroacoustic reciprocity*. Crystal, ceramic, and electrostatic transducers are of this type. The *transformation factor* $\boldsymbol{\phi}$ is defined by

$$\boldsymbol{\phi} = \mathbf{T}/\mathbf{Z}_{EB} \tag{14.6}$$

In practice $\boldsymbol{\phi}$ is real, $\boldsymbol{\phi} = \phi$, and *constant* for most frequencies of interest. The canonical equations now become

$$\mathbf{V} = \mathbf{Z}_{EB}\mathbf{I} + \phi\mathbf{Z}_{EB}\mathbf{u}$$

$$\mathbf{F} = \phi\mathbf{Z}_{EB}\mathbf{I} + \mathbf{Z}_{mo}\mathbf{u} \tag{14.7}$$

This is a considerable simplification, since (14.7) allows use of the force-voltage and speed-current analogs suggested in Fig. 14.1a and yields the equivalent circuits of Fig. 14.2. The techniques of linear circuit analysis can now be applied. The circuit of Fig. 14.2b displays the physical significance of ϕ as the *turns ratio* of an ideal transformer linking the electrical and mechanical sides of the network. (Notice that this transformer is not dimensionless.) The coupling factor defined in (14.4) is

$$\mathbf{K}^2 = \phi^2\mathbf{Z}_{EB}/\mathbf{Z}_{mo} \tag{14.8a}$$

which when substituted into (14.3) gives

$$\mathbf{Z}_{ms} = \mathbf{Z}_{mo} - \phi^2\mathbf{Z}_{EB} \tag{14.8b}$$

(b) Antireciprocal Transducers. Another kind of coupling displays *electroacoustic antireciprocity*, $\mathbf{T}_{em} = -\mathbf{T}_{me}$. Magnetostrictive, moving-coil, and moving-armature transducers are examples of this type. With the definition

$$\boldsymbol{\phi}_M = \mathbf{T}_{em} = -\mathbf{T}_{me} \tag{14.9a}$$

(ϕ_M is usually real or complex with a very small phase angle), the canonical equations become

$$V = Z_{EB}I + \phi_M u$$

$$F = -\phi_M I + Z_{mo}u \qquad (14.9b)$$

and

$$Z_{ms} = Z_{mo} + \phi_M^2/Z_{EB} \qquad (14.10)$$

The minus sign in (14.9b) destroys the symmetry that would give a linear circuit with V and F as electrical and mechanical equivalents. However, there are two approaches that yield simple electrical circuits.

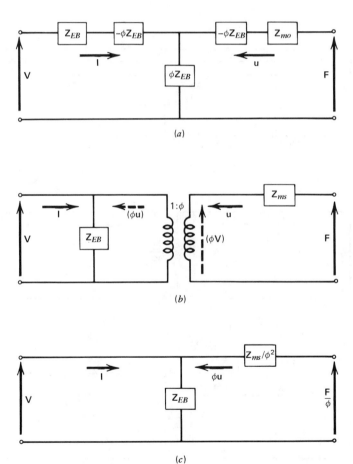

Fig. 14.2. Equivalent circuits for a reciprocal transducer.

1. *The Mechanical Dual.* Rearrangement of the canonical equations results in the symmetric pair

$$\mathbf{V} = \mathbf{Z}_{EF}\mathbf{I} + \boldsymbol{\phi}_M\,\mathbf{Y}_{mo}\,\mathbf{F}$$
$$\mathbf{u} = \boldsymbol{\phi}_M\,\mathbf{Y}_{mo}\,\mathbf{I} + \mathbf{Y}_{mo}\,\mathbf{F} \tag{14.11}$$

where

$$\mathbf{Y}_{mo} = \frac{1}{\mathbf{Z}_{mo}} \tag{14.12}$$

is the open-circuit mechanical *admittance*. Equations (14.11) now exhibit the same form as (14.7), but with **F** and **u** exchanged and admittance appearing as impedance. A few useful equivalent circuits are shown in Fig. 14.3. Note the appearance of $\boldsymbol{\phi}_M$ as an *inverse turns ratio* in the linking transformer.

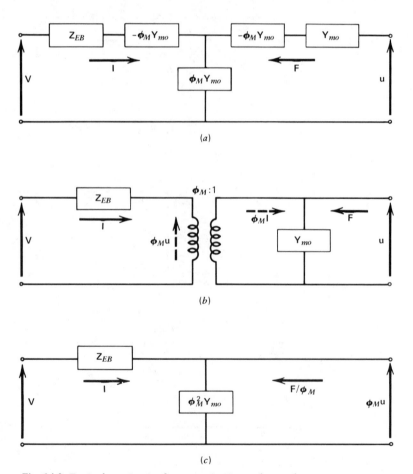

Fig. 14.3. *Equivalent circuits for an antireciprocal transducer.*

2. *Shifted Force and Speed.* The change of variables $\mathbf{F}' = \mathbf{F}/j$ and $\mathbf{u}' = \mathbf{u}/j$ allows (14.9b) to be reduced to

$$\mathbf{V} = \mathbf{Z}_{EB}\mathbf{I} + j\boldsymbol{\phi}_M\mathbf{u}'$$

$$\mathbf{F}' = j\boldsymbol{\phi}_M\mathbf{I} + \mathbf{Z}_{mo}\mathbf{u}' \tag{14.13}$$

yielding the equivalent circuit of Fig. 14.2a, but with $j\boldsymbol{\phi}_M$ in place of $\boldsymbol{\phi}\mathbf{Z}_{EB}$. This approach has the distinct advantage of preserving the $\mathbf{V}, \mathbf{F}'$ and $\mathbf{I}, \mathbf{u}'$ parallels of the reciprocal transducer, but is somewhat unphysical in replacing the observed mechanical variables with quantities shifted through a phase angle of $\pi/2$.

14.3 CANONICAL EQUATIONS FOR TWO SIMPLE TRANSDUCERS.

(a) The Electrostatic Transducer (Reciprocal). The electrostatic transducer can be modeled as a pair of capacitor plates, one of which is held stationary while the other, the *diaphragm*, moves in response to mechanical or electrical excitation, as suggested in Fig. 14.4a. The transducer is connected to an external circuit, as shown

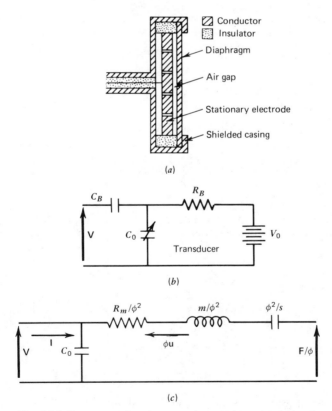

Fig. 14.4. *Representative electrostatic transducer (reciprocal).*

in Fig. 14.4b, where V_0 is a d-c polarization voltage, C_B a capacitor preventing direct current from flowing through the electrical terminals, and R_B a resistor isolating the polarization source from the alternating currents. The *electret transducer* has a diaphragm of polarized plastic so that the external voltage supply can be discarded. We will ignore C_B and R_B since, when properly chosen so that $C_B \gg C_0$ and $R_B \gg 1/(\omega C_B)$, they do not significantly alter the behavior of the transducer in the frequency range of interest. If an alternating voltage is applied across the plates, the diaphragm moves, thereby radiating an acoustic wave. If the diaphragm moves in response to an incident pressure field, an alternating current will flow in an external circuit.

The capacitance of the transducer with the plates at rest is

$$C_0 = \varepsilon S/x_0 \tag{14.14}$$

where ε is the dielectric constant for the material between the plates, S the surface area of the diaphragm, and x_0 the equilibrium spacing of the plates. The charge q_0 on the plates is found from $C_0 = q_0/V_0$. When a sinusoidal voltage V is superimposed on V_0, the instantaneous charge $q_0 + q$ and spacing $x_0 + x$ are related by

$$V + V_0 = \frac{(\mathbf{q} + q_0)(\mathbf{x} + x_0)}{\varepsilon S}$$

If $\mathbf{q} \ll q_0$, $\mathbf{x} \ll x_0$, and V varies as $\exp(j\omega t)$ so that $\mathbf{u} = j\omega\mathbf{x}$ and $\mathbf{I} = j\omega\mathbf{q}$, linearization yields

$$V = \frac{1}{j\omega C_0}\mathbf{I} + \frac{V_0}{j\omega x_0}\mathbf{u} \tag{14.15}$$

Comparison with (14.2a) yields the immediate identification

$$\mathbf{Z}_{EB} = \frac{1}{j\omega C_0} \tag{14.16}$$

and $\mathbf{T}_{em} = V_0/(j\omega x_0)$.

To obtain the second of the canonical equations, notice that the change $\mathbf{f}$ in the electrical force resulting from the change $\mathbf{q}$ of the charge on the fixed plates is $\mathbf{f} = -\mathbf{q}q_0/(\varepsilon S) = -\mathbf{q}V_0/x_0 = -\mathbf{I}V_0/(j\omega x_0)$. Application of Newton's law then gives $\mathbf{Z}_{mo}\mathbf{u} = F + \mathbf{f}$ or

$$F = \frac{V_0}{j\omega x_0}\mathbf{I} + \mathbf{Z}_{mo}\mathbf{u} \tag{14.17}$$

Comparison with (14.2b) shows that $\mathbf{T}_{em} = V_0/(j\omega x_0)$. Since the transduction coefficients are equal, $\mathbf{T} = \mathbf{T}_{em} = \mathbf{T}_{me}$, the electrostatic transducer displays electromechanical reciprocity with

$$\mathbf{T} = \frac{V_0}{j\omega x_0} \tag{14.18}$$

and the transformation factor (14.6) is a real constant

$$\phi = C_0 V_0/x_0 \tag{14.19}$$

The mechanical impedance under short-circuited conditions is

$$\mathbf{Z}_{ms} = R_m + j(\omega m - s/\omega) \tag{14.20a}$$

where m is the mass of the diaphragm, s its stiffness, and R_m the mechanical resistance. This yields the equivalent circuit given in Fig. 14.4c which follows the form of Fig. 14.2c.

Notice that use of (14.8b) and (14.16) provides an immediate expression for $\mathbf{Z}_{mo}$,

$$\mathbf{Z}_{mo} = R_m + j\left(\omega m - \frac{s + \phi^2/C_0}{\omega}\right) \tag{14.20b}$$

Thus, the only effect on the mechanical impedance of opening the shorted electrical terminals is to alter the stiffness from s to

$$s' = s + \phi^2/C_0 \tag{14.20c}$$

With the notation of (14.20), at low frequencies for which $\omega m \ll s'/\omega$ the coupling factor assumes the simple form $\mathbf{K}^2 \approx \phi^2/(C_0 s')$. If we define the electrical equivalent of s by

$$C = \phi^2/s$$

then $\mathbf{K}^2 \approx C/(C_0 + C)$. This fraction gives the ratio of stored mechanical energy to the total energy stored in the transducer when there are no losses and defines the *electromechanical coupling coefficient* k_c,

$$k_c^2 = \frac{C}{C + C_0} \tag{14.21}$$

(b) The Moving-Coil Transducer (Antireciprocal). This transducer consists of a diaphragm attached to a cylindrical coil of wire (the voice coil) that is suspended in a fixed magnetic field B, as suggested in Fig. 14.5a. If an alternating current is supplied to the coil, the interaction of the current $\mathbf{I}$ and the magnetic field will induce a force on the coil so that the diaphragm moves. Conversely, forcing the coil to move in the B field induces a voltage in the coil.

If the diaphragm is *blocked*, the electrical impedance is

$$\mathbf{Z}_{EB} = R_0 + j\omega L_0 \tag{14.22a}$$

where R_0 and L_0 are the resistance and inductance of the voice coil. Now, a current $\mathbf{I}$ in the blocked voice coil generates a force on the diaphragm of magnitude $F_e = Bl\mathbf{I}$ where l is the length of the wire in the voice coil. The external force necessary to hold the diaphragm stationary is therefore $\mathbf{F} = -Bl\mathbf{I}$. If the electrical terminals of the transducer are open, then $\mathbf{I} = 0$ and there can be no electrical effects, so that (at frequencies low enough for the diaphragm to move as a unit) it is convenient to write

$$\mathbf{Z}_{mo} = R_m + j(\omega m - s/\omega) \tag{14.22b}$$

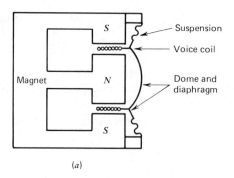

(a)

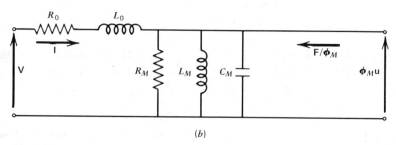

(b)

Fig. 14.5. *Representative moving-coil transducer (antireciprocal).*

and we must have $\mathbf{F} = Z_{mo}\mathbf{u}$. The linear equation expressing $\mathbf{F}$ in terms of $\mathbf{I}$ and $\mathbf{u}$ which satisfies these two special cases is

$$\mathbf{F} = -Bl\mathbf{I} + \mathbf{Z}_{mo}\mathbf{u} \qquad (14.23)$$

which is one of the canonical equations.

To obtain the second, note that motion of the coil in the magnetic field induces a voltage $Bl\mathbf{u}$ opposing the applied voltage. This leads to

$$\mathbf{V} = \mathbf{Z}_{EB}\mathbf{I} + Bl\mathbf{u} \qquad (14.24)$$

Now, (14.23) and (14.24) are the canonical equations for an *antireciprocal* transducer with a transformation factor

$$\phi_M = Bl \qquad (14.25)$$

that is real and constant. An equivalent circuit for this transducer is shown in Fig. 14.5b, where

$$R_M = \phi_M^2/R_m \qquad (14.26)$$

$$L_M = \phi_M^2/s \qquad (14.27)$$

and

$$C_M = m/\phi_M^2 \qquad (14.28)$$

For $R_0 \ll \omega L_0$ use of (14.10) yields

$$\mathbf{Z}_{ms} \rightarrow R_m + j\left(\omega m - \frac{s + \phi_M^2/L_0}{\omega}\right) \qquad (14.29)$$

which shows the change in the mechanical stiffness when the electrical terminals are shorted.

If the frequencies for which (14.29) is true lie below those of mechanical resonance and if mechanical losses are negligible, then the combination of (14.22b) and (14.29) yields

$$\frac{\mathbf{Z}_{mo}}{\mathbf{Z}_{ms}} \approx 1 - \frac{\phi_M^2}{L_0} \frac{1}{s + \phi_M^2/L_0} = 1 - \frac{L_M}{L_M + L_0}$$

We define the electromechanical coupling coefficient k_m by

$$k_m^2 = \frac{L_M}{L_M + L_0} \qquad (14.30)$$

As before, in (14.21), this gives the ratio of stored mechanical energy to total stored energy when there are no losses. Notice that k_m^2 is not related to $\mathbf{K}^2$ as simply as is k_c^2.

14.4 TRANSMITTERS. When a tranducer acts as a transmitter,* the force $\mathbf{F}$ on the diaphragm is produced by the motion of the medium

$$\mathbf{F} = -\mathbf{Z}_r \mathbf{u} \qquad (14.31)$$

where $\mathbf{Z}_r = R_r + jX_r$ is the radiation impedance.

The *sensitivity* of a transmitter can be expressed as the ratio of the acoustic pressure amplitude at 1 m from the source to the amplitude of either the driving voltage V or driving current I,

$$\mathscr{S}_V = P_{ax}(1)/V$$

$$\mathscr{S}_I = P_{ax}(1)/I \qquad (14.32a)$$

It is left as an exercise to show that

$$\mathscr{S}_V = \frac{P_{ax}(1)}{|\mathbf{u}|} \frac{T_{me}}{Z_{EB}|\mathbf{Z}_{ms} + \mathbf{Z}_r|}$$

$$\mathscr{S}_I = \frac{P_{ax}(1)}{|\mathbf{u}|} \frac{T_{me}}{|\mathbf{Z}_{mo} + \mathbf{Z}_r|} \qquad (14.32b)$$

The transmitting sensitivity level $\mathscr{SL}$ is given by

$$\mathscr{SL} = 20 \log(\mathscr{S}/\mathscr{S}_{ref}) \qquad (14.33)$$

where $\mathscr{S}_{ref}$ is the reference sensitivity, usually 1 Pa/V for $\mathscr{S}_V$ and 1 Pa/A for $\mathscr{S}_I$. (Other conventions use 1 μbar or 1 μPa instead of 1 Pa.)

---

* *Transmitter*, as used here, can refer to a loudspeaker, underwater projector, or any other device that converts electrical energy into acoustic energy.

(a) Reciprocal. For a reciprocal transmitter, use of $\mathbf{F} = -\mathbf{Z}_r \mathbf{u}$ in the equivalent circuit of Fig. 14.2c yields Fig. 14.6a. Thus, the motion of the diaphragm arises from an apparent force of magnitude $-\phi\mathbf{V}$ acting on a mechanical impedance $\mathbf{Z}_{ms} + \mathbf{Z}_r$,

$$\mathbf{u} = \frac{-\phi\mathbf{V}}{\mathbf{Z}_{ms} + \mathbf{Z}_r} \tag{14.34}$$

Over most of the frequency range of interest $\mathbf{Z}_{ms}$ is given by (14.20a). The impedance $\mathbf{Z}_{EB}$ is not just from C_0 as stated in (14.16), but is instead a parallel

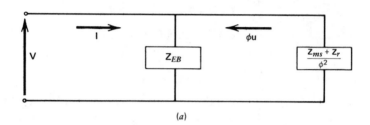

(a)

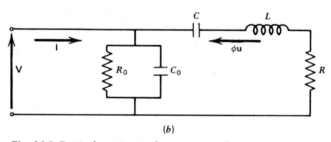

(b)

Fig. 14.6. Equivalent circuits for a reciprocal transmitter.

combination of C_0 and a large *leakage resistor* R_0. The equivalent circuit is given by Fig. 14.6b where

$$C = \phi^2/s \tag{14.35a}$$

$$L = (m + X_r/\omega)/\phi^2 \tag{14.35b}$$

$$R = (R_m + R_r)/\phi^2 \tag{14.35c}$$

Note that R_r and X_r cannot be further specified until the geometry of the diaphragm is known. In general, they are both functions of frequency. However, at high frequencies $X_r \to 0$ and $R_r \to \rho_0 cS$, where S is the area of the diaphragm and $\rho_0 c$ is the characteristic impedance of the medium. At low frequencies $R_r \to 0$ and $X_r \to \omega m_r$, where the radiation mass m_r is constant. (Recall the discussion of Sect. 8.12.)

Since the circuit of Fig. 14.6 is the parallel combination of

$$\mathbf{Z}_{EB} \quad \text{and} \quad (\mathbf{Z}_{ms} + \mathbf{Z}_r)/\phi^2$$

the *input electrical admittance* is

$$\mathbf{Y}_E = \mathbf{Y}_{EB} + \mathbf{Y}_{MOT} \tag{14.36a}$$

where the *blocked electrical admittance* $\mathbf{Y}_{EB}$ is

$$\mathbf{Y}_{EB} = \frac{1}{R_0} + j\omega C_0 \tag{14.36b}$$

and the *motional admittance* $\mathbf{Y}_{MOT}$ is found from

$$\frac{1}{\mathbf{Y}_{MOT}} = \frac{R_m + R_r}{\phi^2} + j\,\frac{\omega m - s/\omega + X_r}{\phi^2} \tag{14.36c}$$

Solving for the real and imaginary parts of $\mathbf{Y}_E = G_E + jB_E$ yields the *input electrical conductance*

$$G_E = \frac{1}{R_0} + \frac{R}{R^2 + X^2} \tag{14.37a}$$

and the *input electrical susceptance*

$$B_E = \omega C_0 - \frac{X}{R^2 + X^2} \tag{14.37b}$$

where

$$R = (R_m + R_r)/\phi^2$$

$$X = \omega L - \frac{1}{\omega C} = \omega\left(\frac{m + X_r/\omega}{\phi^2}\right) - \frac{s/\phi^2}{\omega} \tag{14.38}$$

The properties of a reciprocal transmitter can be determined by measuring the input electrical conductance G_E and susceptance B_E, shown as functions of frequency in Fig. 14.7, and then plotting them, as shown in Fig. 14.8a. At low frequencies this latter curve begins near the coordinate $(1/R_0, 0)$; as the frequency increases, the curve develops a clockwise loop and then goes to $(1/R_0, \infty)$ as $\omega \to \infty$. If the mechanical resonance is reasonably sharp, this loop is very close to being a circle. The indicated frequencies are defined as follows.

ω_0 = mechanical resonance frequency $(X = 0)$
ω_u = upper half-power frequency for the motional branch $(X = R)$ (14.39)
ω_l = lower half-power frequency for the motional branch $(X = -R)$

ω_m = frequency of maximum Y_E
ω_n = frequency of minimum Y_E (14.40)

ω_r = electrical resonance frequency, $B_E = 0$ and G_E large
ω_a = electrical antiresonance frequency, $B_E = 0$ and G_E small (14.41)

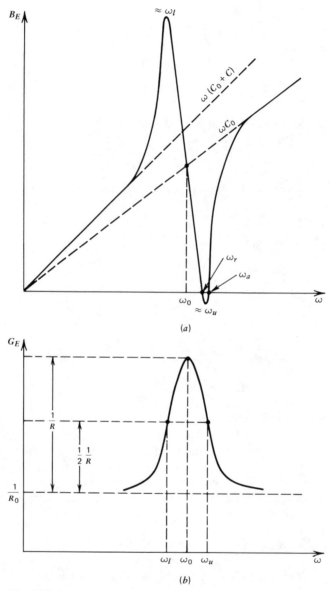

Fig. 14.7. *Reciprocal transmitter. (a) Input electrical suscep-tance. (b) Input electrical conductance.*

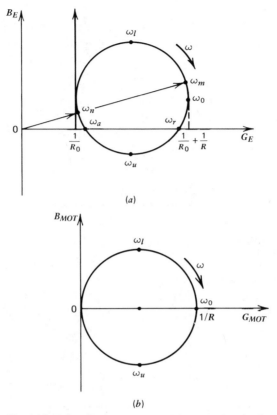

Fig. 14.8. *Reciprocal transmitter.* (a) *Input electrical admittance plot.* (b) *Motional admittance plot.*

The last two, ω_r and ω_a, will not exist if the loop does not intercept the G_E axis. Application of the definitions for ω_l, ω_0, and ω_u to the motional branch results in formulas equivalent to those for the forced, damped harmonic oscillator:

$$\omega_0 = \sqrt{s/(m + m_r)}$$

$$\omega_u \omega_l = \omega_0^2 \tag{14.42}$$

$$\omega_u - \omega_l = (R_m + R_r)/(m + m_r)$$

The *mechanical quality factor* is

$$Q_M = \frac{\omega_0}{\omega_u - \omega_l} = \frac{\omega_0 L}{R} = \frac{\omega_0(m + m_r)}{(R_m + R_r)} \tag{14.43}$$

If $G_{EB} = 1/R_0$ and $B_{EB} = j\omega C_0$ are subtracted from G_E and B_E, respectively, the results are the *motional conductance* G_{MOT} and the *motional susceptance* B_{MOT} where $Y_{MOT} = G_{MOT} + jB_{MOT}$. For R_r and $m_r = X_r/\omega$ constant across the resonance, the plot of G_{MOT} against B_{MOT} results in a circle, as shown in Fig. 14.8b.

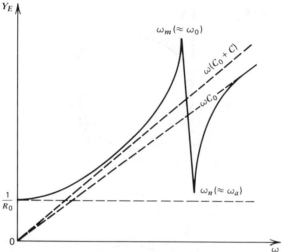

Fig. 14.9. *Input electrical admittance for a reciprocal transmitter.*

From Fig. 14.8a, the magnitude of the vector drawn from the origin to a point on the curve is Y_E. The behavior of Y_E as a function of frequency is shown in Fig. 14.9.

The *electroacoustic efficiency* η is defined as the ratio of the acoustic power radiated to the total electrical power consumed. It is

$$\eta = \frac{R_r/\phi^2}{R} \frac{R_0}{R_0 + (R^2 + X^2)/R} \tag{14.44}$$

This reduces to a simple form when $\omega = \omega_0$,

$$\eta_0 = \frac{R_r/\phi^2}{R} \frac{R_0}{R_0 + R} \tag{14.45}$$

The ratio $R_r/(R\phi^2)$ measures the conversion of mechanical energy into acoustical energy and is called the *mechanoacoustic efficiency,* η_{MA}. The second ratio measures the conversion of electrical energy into mechanical energy and is called the *electromechanical efficiency* η_{EM}.

It is possible to obtain estimates of the electromechanical coupling coefficient from the properties of the transducer near its mechanical resonance ω_0 rather than from its behavior for low frequencies. First, notice from Fig. 14.9 (or Fig. 14.7a) that the values of C_0 and $C_0 + C$ can be estimated from the slopes of the two dashed lines. Since in certain transducers, including those made with piezoelectric or ferroelectric materials, the value of C depends somewhat on frequency, the value of the electromechanical coupling coefficient obtained from the above method may differ from the value obtained at low frequencies. For this reason, the *effective* electromechanical coupling coefficient $k_c(eff)$ is used

$$k_c^2(eff) = \frac{C(\omega_0)}{C_0 + C(\omega_0)} \tag{14.46a}$$

In addition, if the transducer is of reasonably high quality and is operated in an unloaded condition so that R_r is sufficiently small, then it may be possible to treat R as very small. It can be shown analytically, and seen intuitively, that as R gets smaller the diameter of the admittance circle gets larger and the frequency interval, in which the paired values of B_E and G_E lie on the circle, gets smaller. This means that as $R \to 0$ we have $\omega_0 \to \omega_m$ and $\omega_a \to \omega_n$. Now, for small R the antiresonance frequency can be estimated from

$$\omega_a C_0 \approx \frac{1}{\omega_a L - 1/(\omega_a C)}$$

and combination with $\omega_0 = (LC)^{-1/2}$ yields

$$k_c^2(eff) \approx 1 - (\omega_0/\omega_a)^2 \approx 1 - (\omega_m/\omega_n)^2 \tag{14.46b}$$

Thus $k_c^2(eff)$ can be calculated by determining ω_0 and ω_a from Fig. 14.7a or Fig. 14.8a, or by determining ω_m and ω_n from Fig. 14.8a or Fig. 14.9.

(b) Antireciprocal. For an antireciprocal transmitter, use of $\mathbf{F} = -\mathbf{Z}_r \mathbf{u}$ in Fig. 14.3b yields the circuit of Fig. 14.10a, and the motion $\mathbf{u}$ of the diaphragm arises from an apparent force of magnitude $\boldsymbol{\phi}_M \mathbf{I}$ acting on the mechanical impedance $\mathbf{Z}_m = \mathbf{Z}_{mo} + \mathbf{Z}_r$,

$$\mathbf{u} = \frac{\boldsymbol{\phi}_M \mathbf{I}}{\mathbf{Z}_m} = \frac{\boldsymbol{\phi}_M \mathbf{I}}{\mathbf{Z}_{mo} + \mathbf{Z}_r} \tag{14.47}$$

Over most of the frequency range of interest $\mathbf{Z}_{mo}$ is given by (14.22b). The circuit representing (14.47) is shown in Fig. 14.10c.

The input electrical impedance is, from Fig. 14.10a,

$$\mathbf{Z}_E = \mathbf{Z}_{EB} + \mathbf{Z}_{MOT} \tag{14.48a}$$

where $\mathbf{Z}_{MOT}$ is *defined* by

$$\mathbf{Z}_{MOT} = \frac{\phi_M^2}{\mathbf{Z}_{mo} + \mathbf{Z}_r} \tag{14.48b}$$

(While $\mathbf{Y}_E = 1/\mathbf{Z}_E$ and $\mathbf{Y}_{EB} = 1/\mathbf{Z}_{EB}$, there is no simple relationship between $\mathbf{Y}_{MOT}$ and $\mathbf{Z}_{MOT}$ since the motional quantities are *defined* by the differences $\mathbf{Y}_E - \mathbf{Y}_{EB}$ and $\mathbf{Z}_E - \mathbf{Z}_{EB}$. Specifically, $\mathbf{Y}_{MOT}$ is *not* the reciprocal of $\mathbf{Z}_{MOT}$!)

For most antireciprocal transmitters $\mathbf{Z}_{EB}$ is given by (14.22a) and $\mathbf{Z}_{mo}$ by (14.22b). If $\boldsymbol{\phi}_M = \phi_M$, then if we define

$$R_R = \phi_M^2/R_r \tag{14.49}$$

and

$$C_R = \frac{X_r/\omega}{\phi_M^2} \tag{14.50}$$

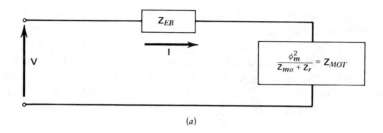

(a)

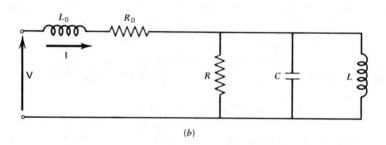

(b)

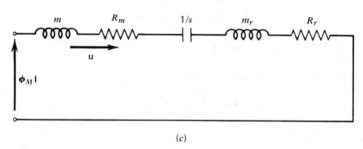

(c)

Fig. 14.10. *Equivalent circuits for an antireciprocal transmitter.*

Fig. 14.10a can be represented by Fig. 14.10b where

$$R = \frac{\phi_M^2}{R_m + R_r} = \frac{1}{1/R_M + 1/R_R}$$

$$L = L_M = \phi_M^2/s$$

$$C = (m + X_r/\omega)/\phi_M^2 = C_M + C_R \qquad (14.51a)$$

Thus, R is the parallel combination of R_M and R_R, and C is the parallel combination of C_M and C_R.

Now, $\mathbf{Z}_{MOT} = R_{MOT} + jX_{MOT}$ has the form

$$R_{MOT} = \frac{\phi_M^2(R_m + R_r)}{|\mathbf{Z}_{mo} + \mathbf{Z}_r|^2} = \frac{1/R}{\dfrac{1}{R^2} + \left(\omega C - \dfrac{1}{\omega L}\right)^2}$$

$$X_{MOT} = -\frac{\phi_M^2(\omega m - s/\omega + X_r)}{|\mathbf{Z}_{mo} + \mathbf{Z}_r|^2} = -\frac{\omega C - \dfrac{1}{\omega L}}{\dfrac{1}{R^2} + \left(\omega C - \dfrac{1}{\omega L}\right)^2} \qquad (14.51b)$$

A plot of X_{MOT} against R_{MOT} is a circle of diameter R (Fig. 14.11a) if X_r/ω and R_r are constant across the resonance. Now, $R_{EB} + R_{MOT}$ displaces the circle to the right by R_0, and $X_{EB} + X_{MOT}$ displaces each point of the circle upward by ωL_0. The resultant curve, Fig. 14.11b, has the same shape as that in Fig. 14.8a. Mechanical resonance is given by

$$\omega_0 = 1/\sqrt{CL} = \sqrt{s/(m + m_r)} \qquad (14.52a)$$

The mechanical quality factor is

$$Q_M = \omega_0 RC = \frac{\omega_0(m + m_r)}{R_m + R_r} \qquad (14.52b)$$

The acoustic power radiated by the transmitter is

$$\Pi = R_r u^2 \qquad (14.53)$$

and the total power consumed is $R_0 I^2 + (R_m + R_r)u^2$ so that the electroacoustic efficiency is, with the help of (14.47) and (14.48b),

$$\eta = \frac{R_r}{R_0 \phi_M^2/Z_{MOT}^2 + (R_m + R_r)} \qquad (14.54a)$$

At the frequency of mechanical resonance this becomes

$$\eta_{res} = \frac{R}{R_R} \frac{R}{R_0 + R} \qquad (14.54b)$$

Reasoning similar to that culminating in (14.46b) yields an effective electromechanical coupling coefficient that is well-estimated by

$$k_m^2(eff) \approx 1 - (\omega_0/\omega_r)^2 \approx 1 - (\omega_n/\omega_m)^2 \qquad (14.54c)$$

Complications arise for some real antireciprocal devices. Resistance and capacitance in the windings and hysteresis losses in magnetic material can result in a complex ϕ_M which yields a skewed impedance plot similar to Fig. 14.11c. The phase angle β of ϕ_M is called the *dip angle*. Moving armature and magnetostrictive transducers are two examples for which β is nonzero. The radius of curvature R_C is related to the losses within the magnetic material[4] and determines the asymptotic behavior of the X_E versus R_E curve for frequencies removed from mechanical resonance.

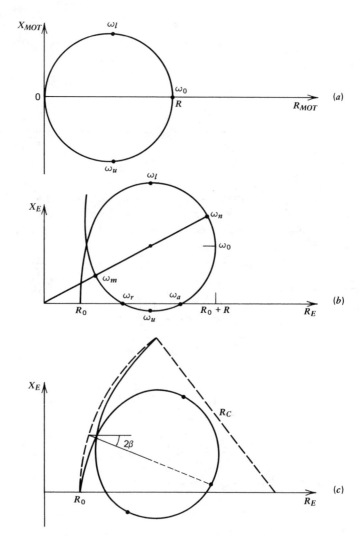

Fig. 14.11. *Impedance plots for an antireciprocal transmitter.*

The utility of these considerations for both reciprocal and antireciprocal transducers is that the mechanical and acoustic properties of a transducer can be determined from electrical measurements made at the input terminals of the transducer.

For example, consider an (antireciprocal) loudspeaker. The values of R_0 and L_0 are easily determined from measurements made either with the transducer blocked (most larger loudspeakers have their resonances at low frequencies and can be blocked by placing the hand firmly on the cone near the voice coil) or from the input electrical impedance at low and high frequencies for which Z_{MOT} becomes negligible. Several approaches are available for determining the motional properties of the transducer:

1. The motional impedance circles can be determined for the loudspeaker mounted in a large baffle and unmounted to remove the radiation loading. From these two circles, two sets of R, ω_0, and Q_M can be determined (see Fig. 14.11a), where one set is given by (14.51a), (14.52a), and (14.52b) and the other set by the same equations with $R_r = m_r = 0$. Then, if R_r or m_r is calculated for the loaded transducer [from (8.68a) for a pistonlike transducer], the combination of these two sets of values yields the desired parameters: R_m, m, s, and ϕ_M. (The same procedure can be used for an underwater antireciprocal transducer with the unloaded condition achieved by removing the transducer from the water.)

2. The second approach consists of measuring the impedance circles of the unloaded speaker, first without modification, and then with a mass M affixed to the cone near the voice coil. It is left as an exercise to show that these measurements allow determination of the parameters of the loudspeaker. This latter technique has the advantage of not requiring the assumptions necessary to calculate the radiation impedance.

The above procedures apply equally well to reciprocal transducers if admittance circles rather than impedance circles are measured.

14.5 MOVING-COIL LOUDSPEAKER.

Figure 14.12 is a simplified drawing of a typical moving-coil loudspeaker. Notice that the vibrating diaphragm (cone) is usually appreciably larger than the voice coil, to enhance the efficiency of radiation at the lower frequencies. Since a loudspeaker will become directive at high frequencies, a number of speakers with different diaphragm areas are often used in high-fidelity applications. A wide-range high-fidelity system might consist of a large, relatively massive speaker radiating the lower frequencies (*woofer*), smaller speakers for the mid-range frequencies (*squawkers*), and still smaller ones to radiate the highest frequencies (*tweeter* and *supertweeter*). These various speakers are driven through electrical filtering networks that deliver to each its appropriate range of frequencies. The squawkers, tweeters, and supertweeters may be designed more along the lines of Fig. 14.5 with no diaphragm. If the *dome* is small, then there is good dispersion of the high-frequency sounds.

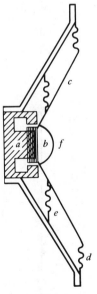

Fig. 14.12. Simple loudspeaker. (a) Magnet. (b) Voice coil. (c) Diaphragm. (d) Corrugated rim and (e) spider supplying stiffness to system. (f) Dome.

As a practical illustration of the analysis of a loudspeaker, consider a piston mounted in an infinite baffle (more realistic speaker enclosures will be discussed in a later section) and having the following physical characteristics:

Total mass of the diaphragm (including voice coil), $m = 10$ g
Radius of the diaphragm, $a = 0.1$ m (an 8-in. speaker)
Stiffness, $s = 2000$ N/m
Mechanical resistance, $R_m = 1$ N $\cdot$ s/m
Inductance of voice coil, $L_0 = 0.2$ mH
Resistance of voice coil, $R_0 = 5\ \Omega$
Length of wire in voice coil, $l = 5$ m
Magnetic field, $B = 0.9$ T (tesla)

If the diaphragm radiates as a plane piston on one side of the baffle, the radiation resistance and reactance are

$$R_r = 13R_1(0.00366f)$$

$$X_r = 13X_1(0.00366f)$$

For frequencies such that $2ka < 1$, these expressions may be approximated by (8.68b) et seq. without introducing errors in excess of 10 percent. Therefore, at

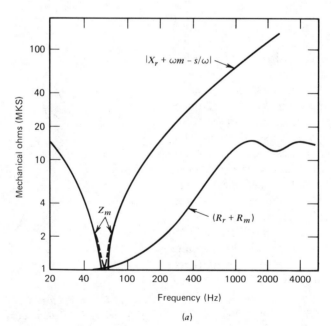

(a)

Fig. 14.13. *Properties of a representative loudspeaker.*
(a) Mechanical impedance.

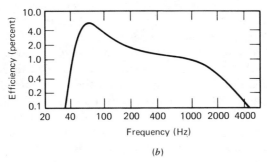

(b)

Fig. 14.13(b) *Efficiency*

frequencies below 275 Hz $R_r \approx 2.2 \times 10^{-5}f^2$ and $X_r \approx 0.02f$. For frequencies such that $2ka > 4$, $R_1(2ka)$ remains within 10 percent of its ultimate value of unity. Thus, above 1100 Hz, $R_r \approx 13$ N $\cdot$ s/m. The frequency f_0 of mechanical resonance, determined from $(\omega_0 m - s/\omega_0 + X_r) = 0$, is 62 Hz.

As sketched in Fig. 14.13a, the mechanical resistance rises slowly from 1 N $\cdot$ s/m at very low frequencies, increases rapidly between 100 and 1000 Hz as the radiation resistance begins to dominate, and then fluctuates about 14 N $\cdot$ s/m at higher frequencies. The magnitude of the reactance has large values at very low frequencies (where the stiffness dominates), reducing to zero at the resonance frequency, and then rising with higher frequencies. The magnitude Z_m of the mechanical impedance has a minimum value of 1.085 N $\cdot$ s/m at the resonance frequency, and for all other frequencies is nearly identical in magnitude to the reactance.

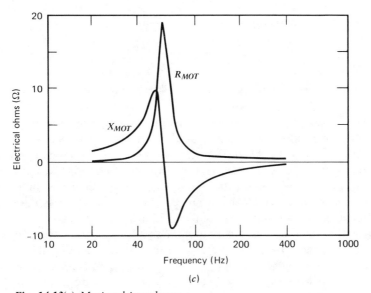

(c)

Fig. 14.13(c) *Motional impedance.*

Plotted in Fig. 14.13*b* is the electroacoustic efficiency of the speaker calculated from (14.54*a*). This efficiency rises rapidly to a maximum of 6.1 percent at mechanical resonance. Below this frequency, R_r is proportional to f^2 and Z_m to $1/f$ so that the efficiency is proportional to f^4. In the interval between 200 and 700 Hz both R_r and Z_m increase so that the efficiency remains between 1 and 2 percent. At higher frequencies the radiation resistance is nearly constant, while Z_m continues to increase with frequency; thus the efficiency rapidly decreases.

Plotted in Fig. 14.13*c* are the resistive and reactive components of the motional impedance Z_{MOT}, as computed from (14.51*b*). The motional resistance R_{MOT} rises to a maximum value of 19 Ω at mechanical resonance. Away from resonance, the motional impedance is predominately reactive, positive below resonance and negative above resonance.

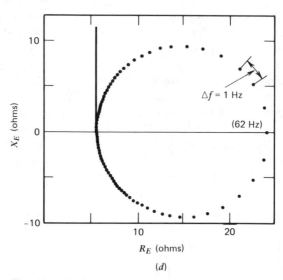

Fig. 14.13(d) *Electrical input impedance plot*

The input impedance circle is plotted in Fig. 14.13*d*. The resistive component is identical to that of Fig. 14.13*c* except that all values have been increased by 5 Ω, the resistance of the voice coil. At low frequencies the input resistance is primarily that of the motional impedance. However, at a frequency of 450 Hz the positive (inductive) reactance of the voice coil cancels the negative motional reactance and results in a second frequency at which the input reactance is zero. This is the frequency of electrical resonance, f_r. Finally, at frequencies in excess of 4000 Hz the inductive reactance of the voice coil is the predominate component of the input electrical impedance. Note that over the frequency range from 200 to 2000 Hz the input impedance is primarily resistive and nearly equals the resistance of the voice coil. Because of this constancy, many loudspeaker manufacturers do not specify the frequency at which the speaker voice coil has its rated input impedance. In general,

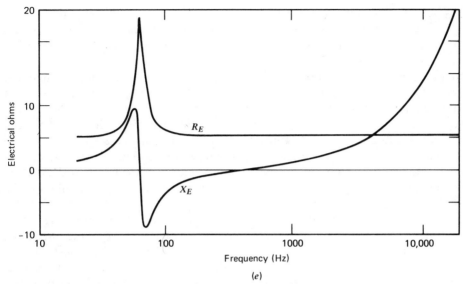

Fig. 14.13(e) Electrical input impedance.

it is measured at a frequency somewhat above the resonance frequency and is normally larger than the d-c resistance.

Plotted in Fig. 14.14 are three curves showing computed values of the acoustic output of the speaker. Curve *A* was obtained from (14.47) and (14.53) for an input current of 2 A, curve *B* was obtained from (14.54) for an input power of 20 W, and curve *C* was obtained from (14.53) and (14.47) [with $\mathbf{V} = \mathbf{IZ}_E$, where $\mathbf{Z}_E$ is found

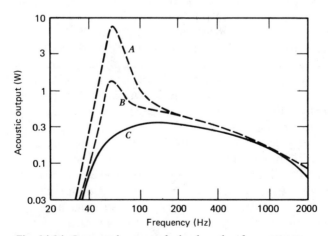

Fig. 14.14. Computed output of a loudspeaker for constant current input (curve A), constant power input (curve B), and constant voltage input (curve C).

from (14.22*a*), (14.47), and (14.51*b*)] for an input voltage of 10 V. These three particular values of current, power, and voltage were chosen to yield nearly equal acoustic outputs in the frequency range from 400 to 1000 Hz. As the frequency of
mechanical resonance at 62 Hz is approached, the three outputs differ widely. The
output at resonance for an input of 2 A is seen to be some 20 times greater than that
for an input of 10 V. A *constant-current* amplifier (which supplies the same current
at all frequencies, irrespective of the load impedance of the speaker) would produce
a peak of acoustic output at the mechanical resonance of the loudspeaker. The
smoothest acoustic output will be produced by a *constant-voltage* amplifier. Since
most high-fidelity amplifiers employ large negative voltage feedback to reduce harmonic distortion, these amplifiers have very low internal impedances and therefore
maintain almost constant voltage output; the acoustic output will then follow curve
C.

For frequencies above about 700 Hz the acoustic output shown in Fig. 14.14,
curve *C*, tends to fall off as $1/f^4$. This is partially compensated for by the directivity,
which is beginning to approach $(ka)^2$, so that the source level on the acoustic axis of
the speaker falls off only as $20 \log f$: the high-frequency output on the axis is
gradually attenuated. However, the actual behavior is more complicated than this
simple argument suggests:

1. At low frequencies the time required for a displacement of the center of the
cone to propagate to the rim is small compared to the period of vibration, so that
the cone may be assumed to vibrate as a rigid surface. (The speed of transverse
waves on a paper cone is a function of its thickness, stiffness, cone angle, and
frequency. For cone materials commonly used in commercial loudspeakers, this
speed is about 500 m/s. Consequently, it is reasonable to assume that the cone
moves as a unit for frequencies below 500 Hz.)

2. At higher frequencies, however, the cone no longer vibrates as a unit. The
amplitude of vibration at the edge of the cone is relatively small, so that the
radiation comes mostly from a central piston whose effective radius a_e and effective
mass m_e gradually decrease with increasing frequency. This decrease in effective
radius causes the radiation resistance R_r to decrease approximately as a_e^2. Since the
system is mass-controlled at the higher frequencies, m_e also decreases with a_e, and η
does not decrease so rapidly as for a rigid piston. The net result of these two effects
is to produce a significant improvement in the output of this speaker at frequencies
above 700 Hz. (Many loudspeakers have their cones corregated to take advantage
of, and even enhance, this effect.)

The problem of maintaining a uniform acoustic output from loudspeakers at
very low frequency is more difficult to solve. One method of improving the low-
frequency response is to increase the radius of the speakers. The radiation resistance
increases as the fourth power of the radius and correspondingly increases the efficiency. However, the increase in efficiency is not so great as might be expected,
because the mass of the speaker also increases with the radius.

The low-frequency response can also be enhanced by reducing the stiffness of
the suspension system, thereby lowering the frequency of mechanical resonance.
Reducing the stiffness of the suspension would be an excellent solution if it were

necessary to consider only efficiency and output power. However, if the stiffness of the mechanical system is too greatly reduced, its displacement at low frequency becomes very large, which may lead to *harmonic distortion* of the acoustic output, which results from displacement of the voice coil into nonuniform regions of the magnetic field.

Except in the immediate vicinity of mechanical resonance, (14.48) and (14.54) yield

$$\eta \approx \frac{\phi_M^2}{|Z_{mo} + Z_r|^2} \frac{R_r}{R_0} \tag{14.55}$$

This simplified equation is useful in discussing the influence on the speaker's characteristics of varying different design parameters.

Now, ϕ_M is directly proportional to the flux density in the air gap, so that increasing B will increase the efficiency of the speaker. The two most feasible methods of accomplishing this are to (1) use a more powerful field magnet, and (2) decrease the width of the air gap to as small a value as practicable. An increase in the *length* of conductor forming the voice-coil winding would be expected to improve the efficiency since $\eta \propto \phi_M^2/R_0 \propto l$. However, for any wire size there exists an optimum length beyond which the increase in mass and the decrease in B (due to the larger air gap required to accommodate the enlarged coil) are more significant than the gain resulting from the increase in l. If the *mass* of the conductor to be used in the voice coil is specified, no change in efficiency results from a change in wire size. The primary consideration involved in the choice of wire size is therefore its current-carrying capacity per unit mass, and aluminum is preferable to copper. If the *volume* occupied by the winding is the limiting factor, then copper is more efficient than aluminum.

14.6 LOUDSPEAKER CABINETS.

(a) The Enclosed Cabinet. A loudspeaker mounted in an *enclosed* cabinet can radiate acoustic energy only from the front of the speaker cone. The primary mechanical effect of such a cabinet is to contribute additional stiffness to that of the suspension system of the loudspeaker. The stiffness s_c of the cabinet at lower frequencies is that of a Helmholtz resonator (10.3b),

$$s_c = (\pi a^2)^2 \frac{\rho_0 c^2}{V} \tag{14.56}$$

where a is the radius of the speaker and V is the volume of the cabinet. This stiffness, added to the speaker stiffness, raises the mechanical resonance frequency of the system and the low-frequency roll-off begins at a higher frequency than if the speaker were mounted in an infinite baffle.

If the loudspeaker of Sect. 14.5 is mounted in a 0.05 m³ enclosed cabinet, the additional stiffness s_c is 2850 N/m and the frequency of mechanical resonance is raised from 62 to 96 Hz. Figure 14.15 shows that for frequencies below 80 Hz the output of the speaker mounted in this cabinet is much lower than that of the same speaker mounted in an infinite baffle.

One way of improving the low-frequency response of a enclosed cabinet is to make the

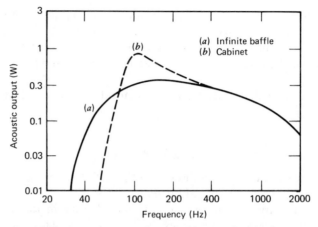

Fig. 14.15. *Acoustic output (constant input voltage) of a loudspeaker in a back-enclosed cabinet (0.05 m³) and in an infinite baffle.*

cabinet larger. Another (more practical) method is to reduce the stiffness and increase the mass of the cone. For example, the acoustic suspension loudspeaker obtains its stiffness almost completely from the air within the cabinet; its own suspension is extremely flexible. With appropriate reductions in the frequency of mechanical resonance, the acoustic output for the above speaker system can be made acceptable down to about 40 Hz. Such systems will have greatly reduced outputs above about 1 kHz and are therefore useful primarily as woofers in multispeaker systems. As a further step, the cabinet can be filled with a sound-absorbing material. The compression and rarefaction of the air contained in the cabinet then become nearly *isothermal*, because the large heat capacity of the sound-absorbing material tends to maintain the entrapped air at a constant temperature rather than allowing it to undergo the more normal adiabatic changes. Under these circumstances, the stiffness from the cabinet is further reduced by $1/\gamma \doteq 0.7$, providing additional improvement in the low-frequency output. This step has another advantage, for without this absorption the radiation from the back of the cone may set up standing waves in the cabinet at higher frequencies. Equation (14.56) no longer applies, and the varying impedance presented to the back of the cone results in an uneven acoustic output.

(b) The Open Cabinet. Another common cabinet is the open-back cabinet typical of most radio and television sets. The equivalent circuit of Fig. 14.10c as modified for an open-back

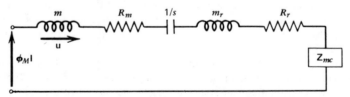

Fig. 14.16. *Equivalent mechanical circuit for a loudspeaker mounted in an open-backed cabinet.*

cabinet is shown in Fig. 14.16. The mechanical impedance Z_{mc} acting on the back of the cone can be expressed as

$$Z_{mc} = (\pi a^2)^2 Z_{Ac} \tag{14.57}$$

where Z_{Ac} is the input acoustic impedance of the cabinet seen by the back of the speaker cone. It is similar to that of the open-ended pipe discussed in Sect. 9.2.

At the fundamental resonance frequency of the cabinet, the motion of the speaker cone is enhanced. This usually occurs between 100 and 200 Hz and results in an intrinsic "boomy" quality. Open-back cabinets suffer the further disadvantage that at frequencies below their fundamental resonance the baffle size is inadequate, so that radiation from the back of the cone combines with that from the front to produce a dipole. The radiated pressure falls off at 6 dB per octave from this effect alone. If the resonance frequency of the speaker's suspension system is lower than that of the cabinet, however, the resulting increase in speaker response will somewhat compensate for this fall, shifting it to lower frequencies.

(c) Bass-Reflex Cabinet. One factor that limits the low-frequency output of all direct-radiation loudspeakers is the inefficient coupling between the cone and the air. The use of a large cone increases the radiation resistance thereby increasing this coupling, but at the expense of increased directivity. An alternative is to mount the speaker in a *phase-inverter* cabinet that enables radiation from the back of the cone to be added in phase with that from the front, thereby effectively increasing the total radiation resistance. One cabinet of this type is the *bass-reflex* cabinet of Fig. 14.17a. The equivalent circuit for this system, Fig. 14.17b, is identical with that of Fig. 14.16 except for the mechanical impedance Z_{mc} presented to the speaker by the cabinet. For the lower frequencies this loading consists of the compliance $1/s_c$ of the cavity in parallel with the series combination of the inertance m_v and resistance R_v of the vent. (Since the mass and resistance of the vent both experience the particle speed u_v of the air in the vent, these elements must be connected in series. Furthermore, if the vent is sealed, this is equivalent to m_v becoming arbitrarily large; thus u_v vanishes, and the system must

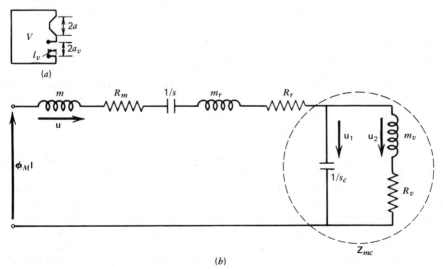

Fig. 14.17. Bass-reflex cabinet. (a) Schematic. (b) Equivalent mechanical circuit.

reduce to the enclosed cabinet with stiffness s_c.) The acoustic impedances of these elements are found from the expressions for a Helmholtz resonator whose neck is the vent, and the mechanical impedances experienced by the speaker are $(\pi a^2)^2$ times these quantities. (The assumption of a flange is plausible because of the surrounding cabinet face.) The mechanical impedance at the speaker cone is therefore given by

$$\frac{1}{Z_{mc}} = j\frac{\omega}{s_c} + \frac{1}{R_v + j\omega m_v} \tag{14.58a}$$

with

$$\frac{1}{s_c} = \frac{1}{(\pi a^2)^2}\frac{V}{\rho_0 c^2} \tag{14.58b}$$

$$R_v = (\pi a^2)^2\frac{\rho_0 c k^2}{2\pi} \tag{14.58c}$$

and

$$m_v = (\pi a^2)^2\frac{\rho_0(l_v + 1.7a_v)}{\pi a_v^2} \tag{14.58d}$$

where a_v is the radius of a circular vent. (For a rectangular vent, a_v represents the effective radius, roughly $\sqrt{S_v}/2$ where S_v is the area of the vent.)

For frequencies above the resonance frequency of Z_c (the fundamental resonance frequency of the cabinet) the impedance of #2 is larger than that of branch #1. The speaker radiates as though it were mounted in an enclosed cabinet, and it is necessary to reduce standing wave effects by lining the interior walls with absorbing material.

If radiation at the vent is to be in phase with (and therefore reinforce) that from the front of the cone, there must be a phase difference of approximately 180° between $\mathbf{u}$ and $\mathbf{u}_2$. This is true for frequencies above the resonance frequency of Z_{mc}. At frequencies below this resonance frequency, the phase of the motion of the air in the vent rapidly approaches that of the back of the cone, and the combination of the radiation from the vent with that from the front of the cone results in the poor low-frequency response characteristic of an acoustic dipole.

Many different factors must be considered in designing a bass-reflex cabinet. However, if the area of the vent is made approximately equal to that of the speaker cone, and if the resonance frequency of the cabinet is made somewhat lower than that of the speaker cone's suspension system, the low-frequency response will be better than that of the same speaker mounted in an infinite baffle. As compared to an open-back cabinet, a bass-reflex cabinet not only extends the low-frequency range but also has a smoother response. The low-frequency response can also be adjusted by filling the mouth of the vent with a *passive radiator*: a piston with properly chosen mass and stiffness.

A practical problem in the design of any type of speaker cabinet is the mechanical rigidity of its walls. It has been assumed that the walls of the enclosure are perfectly rigid, a good approximation for open-back wooden cabinets whose walls are at least 1.3-cm ($\frac{1}{2}$-in.) thick. The design requirements for enclosed and bass-reflex cabinets are, however, considerably more severe, since the acoustic pressures in these enclosures are often high enough to set the larger surfaces into mechanical vibration. The resulting spurious resonances, which usually occur at the lower audible frequencies and produce undesirable irregularities in the frequency response, can be avoided by making the cabinet rigid. The walls should therefore be thicker than those of an open-back cabinet and, if necessary, braced with stiffeners.

14.7 HORN LOUDSPEAKERS. The attachment of a properly designed horn to a pistonlike source results in a marked increase in its acoustic output at low frequencies. Such a horn is essentially an acoustic transformer matching the impedance of the air to that of the piston. In particular, the low-frequency acoustic resistance at the throat of the horn is greater than that which would act on a piston of equal area vibrating in an infinite baffle, and the acoustic output is consequently greater. At high frequencies the effect of the horn is almost negligible, for these frequencies are radiated by the piston as a narrow beam, and hence the confining effect of the walls of the horn is of limited significance.

A rigorous analysis of the propagation of waves in horns is not warranted in this book. However, the following simplified derivation of the wave equation leads to many conclusions of practical importance, even though the frequency range to which it may be applied is limited. Consider the volume element of length dx and area S_x within the horn of Fig. 14.18. By methods similar to those used in Sect. 5.3 it is possible to show that, if the change in cross-sectional area with distance is sufficiently small, then the waves are essentially planar and the equation of continuity is

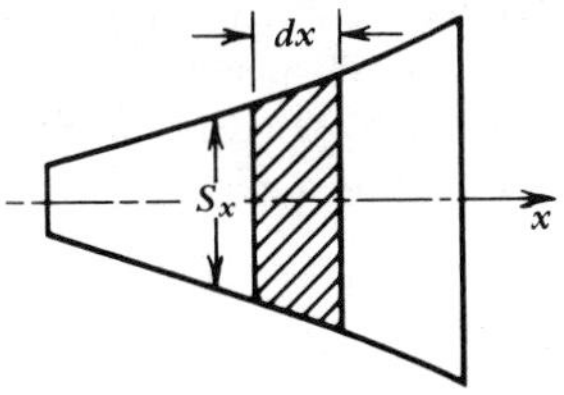

Fig. 14.18. Volume element $S_x\,dx$ of a horn.

$$s = - \frac{1}{S_x} \frac{\partial(S_x \xi)}{\partial x} \qquad (14.59a)$$

where s is the condensation and ξ the particle displacement parallel to the axis. Use of the general relationship between s and p, (5.14), yields

$$p = - \frac{\rho_0 c^2}{S_x} \frac{\partial(S_x \xi)}{\partial x} \qquad (14.59b)$$

If the net force acting on a volume element of unit cross section and length dx, $-(\partial p/\partial x)\,dx$, is equated to the product of the element's mass $\rho_0\,dx$ and its acceleration $\partial^2 \xi/\partial t^2$, a general equation for propagation of plane waves along the axis of the horn is obtained

$$\frac{\partial^2 \xi}{\partial t^2} = c^2 \frac{\partial}{\partial x}\left[\frac{1}{S_x}\frac{\partial(S_x \xi)}{\partial \xi}\right] \qquad (14.59c)$$

(Note that when S_x is constant this equation simplifies to the equation for a plane wave.)

Experience has shown that the most effective horns are those whose *rate of flare*, dS_x/dx, increases from throat to mouth. Various functions such as hyperbolas, catenaries, and exponentials have been used in constructing such horns. An *exponential horn* has a cross-sectional area given by $S_x = S_0 \exp(mx)$ where x is the distance from the throat, S_0 the throat area, and m the *flare constant*. Substitution into (14.59c) yields the wave equation for plane waves in an exponential horn

$$\frac{\partial^2 \xi}{\partial t^2} = c^2\left[\frac{\partial^2 \xi}{\partial x^2} + m \frac{\partial \xi}{\partial x}\right] \qquad (14.60)$$

If a solution of the type $\xi = A \exp[j(\omega t - \gamma x)]$ is assumed, γ must satisfy the relation $\gamma^2 + jm\gamma - k^2 = 0$ where $k = \omega/c$. The allowed values of γ are

$$\gamma = -j\frac{m}{2} \pm \sqrt{k^2 - \frac{m^2}{4}} \qquad (14.61)$$

The general harmonic solution of (14.60),

$$\xi = e^{-\alpha x}[\mathbf{A}e^{j(\omega t - \beta x)} + \mathbf{B}e^{j(\omega t + \beta x)}] \tag{14.62a}$$

represents plane waves with amplitude attenuation

$$\alpha = m/2 \tag{14.62b}$$

and phase speed

$$c_p = \frac{\omega}{\beta} = \frac{c}{\sqrt{1 - (\alpha/k)^2}} \tag{14.62c}$$

(α does not represent damping due to absorption but rather a decrease in amplitude that results as the waves spread over an increasing cross-sectional area within the horn.)

Since the phase speed c_p is a function of frequency, the air in an exponential horn is a dispersive medium. Waves will not propagate in the horn if the driving frequency is lower than the *cutoff frequency*

$$f_c = \frac{mc}{4\pi} \tag{14.63}$$

At the cutoff frequency the phase speed becomes infinite, indicating that all parts of the medium within the horn move in phase.

The usual equations may be used to obtain the acoustic pressures and particle velocities within the horn and from these the acoustic impedance

$$\mathbf{Z}_x = \frac{\rho_0 c}{S_x} \frac{1}{k} \frac{(\beta + j\alpha)\mathbf{A}e^{-j\beta x} - (\beta - j\alpha)\mathbf{B}e^{j\beta x}}{\mathbf{A}e^{-j\beta x} + \mathbf{B}e^{j\beta x}} \tag{14.64}$$

If the length of the horn x_L were infinite, there would be no reflected wave and $\mathbf{B}$ would be zero. It can be shown that the amplitude of the reflected wave is small compared to that of the incident wave whenever the radius of the mouth a_L is such that $ka_L > 3$. Such horns may be treated as infinitely long. For instance, if the mouth of the horn opens into an infinite baffle, (14.64) predicts that for $ka_L > 3$, B will be less than $A/10$. If we set $\mathbf{B} = 0$ in (14.64), the acoustic impedance at the throat becomes

$$\mathbf{Z}_0 = \frac{\rho_0 c}{S_0 k}(\beta + j\alpha) = \frac{\rho_0 c}{S_0}\left[\sqrt{1 - \left(\frac{m}{2k}\right)^2} + j\frac{m}{2k}\right] \tag{14.65}$$

To compare the throat impedance of an exponential horn with that acting on the driving pistonlike speaker when mounted in an infinite baffle, assume that the horn flares from a radius of 0.02 m at its throat to 0.4 m at its mouth and that its length is 1.6 m. The flare constant of this horn is $m = 3.7$, which corresponds to a cutoff frequency of 100 Hz. The curves of Fig. 14.19 show computed values of the throat resistance R_0 and reactance X_0 as functions of frequency. Although these curves are calculated on the assumption that the horn is of infinite length, they are applicable to the finite horn with little error for frequencies above 400 Hz. Below this frequency reflections from the mouth of the horn set up resonances that cause the horn resistance and reactance to fluctuate about the values as computed for an infinite horn. The figure also shows computed values of the acoustic resistance and reactance for the 0.02-m piston mounted in an infinite baffle. It is apparent that in the frequency interval from 100 to 3000 Hz the resistance loading at the throat of the horn is considerably greater than that on the piston mounted in a baffle. As a consequence, the attachment of the

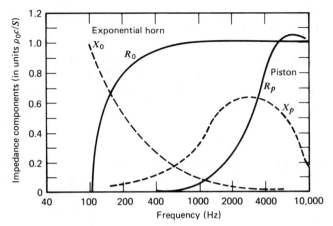

Fig. 14.19. *The acoustic resistance and reactance acting at the throat of an infinite exponential horn and on a piston mounted in an infinite baffle.*

horn to the above piston speaker will greatly enhance acoustic output at the lower frequencies.

Horns are found in high-fidelity applications; large horns for enhancing the low-frequency response of woofers and smaller horns for efficient high-level tweeters. However, the principal application of horns is in sound reinforcement systems for stadiums, large auditoriums, and public-address systems.

For more extensive treatment of the enhancement of the acoustic output of loudspeakers by properly designed cabinets and horns, the reader is referred to the excellent discussions by Beranek[7] and Olson.[2] More sophisticated modern approaches utilizing filter theory can be found in the *Journal of the Audio Engineering Society.*

14.8 RECEIVERS. The force experienced by the diaphragm of a receiver* depends on the pressure field that existed in the absence of the receiver and the radiated pressure field produced by the motion of the diaphragm in response to the excitation. We will use the following definitions in this section:

$\mathbf{p}$ = the pressure at the field point in the *absence* of the receiver

$\mathbf{p}_B$ = the pressure at the field point in the *presence* of the receiver with *blocked* diaphragm

Consideration of these definitions reveals that the total force on the moving diaphragm is

$$\mathbf{F} = \langle \mathbf{p}_B \rangle S - \mathbf{Z}_r \mathbf{u}$$

---

[7] Beranek, *Acoustics*, McGraw-Hill (1954).

* *Receiver*, as used here, can refer to a microphone, hydrophone, or any other transducer that converts acoustic energy into electrical energy. We use "receiver" and "microphone" as interchangeable generic terms.

where $\langle \mathbf{p}_B \rangle$ is the spatial average of $\mathbf{p}_B$ over the diaphragm of area S and $-\mathbf{Z}_r \mathbf{u}$ is the radiation force of the fluid on the diaphragm.

(a) Microphone Directivity. The relation between $\langle \mathbf{p}_B \rangle$ and $\mathbf{p}$ is obtained from the diffraction properties of the blocked diaphragm and its housing and their orientation in the sound field. This problem can be quite involved, but the results can be written in the form of a *diffraction factor* $\mathcal{D}$,

$$\mathcal{D} = \langle \mathbf{p}_B \rangle / \mathbf{p} \tag{14.66}$$

which is a function of frequency and the orientation of the blocked diaphragm with respect to the sound source, approaching unity when the dimensions of the microphone become much less than a wavelength. Combination yields

$$\mathbf{F} = \mathcal{D} S \mathbf{p} - \mathbf{Z}_r \mathbf{u} \tag{14.67}$$

1. Diffraction. A brief consideration of the diffraction effects of a spherical housing will show roughly what to expect from more complicated housings. Figure 14.20 shows how the acoustic pressure at a point on the surface of a sphere of radius a_h varies with frequency for various angles of incidence of the acoustic waves; P_θ is the acoustic pressure amplitude at a point on the surface of the sphere whose polar angle to the direction of incidence of the plane wave is θ; P is the pressure amplitude in the undisturbed sound wave. For normal incidence ($\theta = 0$), diffraction effects will increase the high-frequency response of a pressure-sensitive microphone by 6 dB. If a microphone is to be essentially nondiffractive throughout the audible range of frequencies, the microphone and its housing must be small, $ka_h \leq 1$. At 20 kHz, this requires $a_h \leq 0.3$ cm in air.

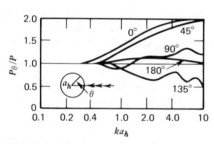

Fig. 14.20. *Influence of diffraction on the sound pressure at the surface of a sphere ensonified by plane waves*

2. Phase Interference across the Diaphragm. A second source of directivity in a pressure-sensitive microphone are the differences in phase among the forces exerted on its diaphragm by different segments of the incident wave when the direction of propagation is not perpendicular to its surface. Sensitivity to phase interference can be avoided by making the diaphragm small enough: $ka \leq \pi/4$, where a is the radius of the diaphragm. For phase interference effects to be negligible at 20 kHz in air, $a \leq 0.2$ cm.

(b) Microphone Sensitivities. The response of a microphone can be expressed in terms of the open-circuited output voltage amplitude V divided either by the amplitude of the average force per unit area F/S experienced by the diaphragm or by the pressure amplitude $P = |\mathbf{p}|$ at the point where the microphone is to be placed. Thus, the voltage sensitivities are

$$\mathcal{M}_o = \left(\frac{V}{P}\right)_{I=0}$$

$$(14.68a)$$

$$\mathcal{M}_o^D = \left(\frac{V}{F/S}\right)_{I=0}$$

It is left as an exercise to verify that these sensitivities can be expressed as

$$\mathcal{M}_o = T_{em} \frac{\mathcal{D}S}{|\mathbf{Z}_{mo} + \mathbf{Z}_r|}$$

$$\mathcal{M}_o^D = T_{em} \frac{S}{Z_{mo}} \qquad (14.68b)$$

Notice that for low frequencies, for which $\mathbf{Z}_r \to 0$ and $\mathcal{D} \to 1$, we have

$$\mathcal{M}_o \to \mathcal{M}_o^D \qquad (14.68c)$$

It is usually convenient to express microphone sensitivities in decibels. The microphone sensitivity level $\mathcal{ML}$ is given by

$$\mathcal{ML} = 20 \log(\mathcal{M}/\mathcal{M}_{ref}) \qquad (14.69)$$

where $\mathcal{M}_{ref}$ is the reference sensitivity level, 1 V/Pa. (Other conventions include 1 V/μbar in air and 1 V/μPa in water.) $\mathcal{M}$ is either $\mathcal{M}_o$ or $\mathcal{M}_o^D$ for open-circuit voltage sensitivities.

(c) Reciprocal Receiver. For a *reciprocal* receiver (14.67) reveals that the force exerted on the moving face can be obtained from the equivalent circuit given in Fig. 14.21 where the presence and orientation of the receiver is accounted for by adding an impedance $\mathbf{Z}_r$ to the mechanical side of the circuit of Fig. 14.2c and modifying the free-field pressure at the diaphragm by the diffraction factor $\mathcal{D}$. Over most of the frequency range of interest, the microphone is small compared to the wavelength so that $\mathcal{D} \approx 1$. Then

$$\mathcal{M}_o \doteq S\phi \frac{Z_{EB}}{Z_{mo}} = \frac{K^2}{\phi} S \qquad (14.70)$$

where (14.4) and (14.6) have been used.

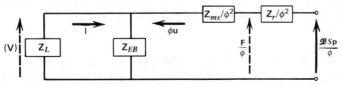

Fig. 14.21. Equivalent circuit for a reciprocal receiver.

(d) Antireciprocal Receiver. By combining (14.67) with (14.9a), and assuming that the receiver is operated under virtually open-circuit conditions so that

$$V = \phi u \tag{14.71a}$$

as well as that the wavelength is large compared to the dimensions of the receiver, we obtain

$$\mathcal{M}_o \doteq S \frac{\phi_M}{Z_{mo}} \tag{14.71b}$$

Notice that (14.70) reveals that the sensitivity of a reciprocal microphone is uniform with frequency, at least below frequencies for which $\mathscr{D}$ becomes important. On the other hand, the antireciprocal microphone sensitivity has frequency dependence through Z_{mo}, and if it is desired to obtain a sensitivity uniform with frequency, Z_{mo} must be manipulated through the introduction of additional mechanical elements to remain constant in magnitude over the frequency range of interest.

14.9 CONDENSER MICROPHONE. The diaphragm of the typical condenser microphone consists of a stretched thin membrane, usually of steel, aluminum, or metallized glass, having a radius a, and separated by a small distance x_0 from a parallel rigid plate. The rigid plate is insulated from the remainder of the microphone, and the polarizing voltage V_0 is applied between it and the diaphragm.

The canonical equations and equivalent circuit of a simple condenser microphone were developed in Sect. 14.3a, and the open-circuit microphone sensitivity can be well-approximated, except at high frequencies, by (14.70). The capacitance C_0 of the microphone in farads (abbreviated F) is given by (14.14). When substituting the *permittivity of free space* $\varepsilon_0 = 8.85 \times 10^{-12}$ F/m, this becomes

$$C_0 = 27.8 a^2 / x_0 \quad \text{pF}$$

Let us now assume that the acoustic pressure displaces the diaphragm in a manner analogous to that of the forced vibrations of a membrane discussed in Sect. 4.7. For low driving frequencies such that $ka < 1$, we can neglect $\mathbf{Z}_r$. Furthermore, in this limit the upper frequency for reasonably uniform response is given by (4.49), and (4.48) shows that the average displacement amplitude of the diaphragm is

$$\langle y \rangle = \frac{Pa^2}{8\mathscr{T}}$$

where P is the pressure amplitude in pascals and $\mathscr{T}$ is the tension in the diaphragm (in newtons per meter). Assuming that the mechanical stiffness of the diaphragm arises from the tension in the membrane, then $PA = \langle y \rangle s'$ and

$$s = 8\pi\mathscr{T} \tag{14.72}$$

For most frequencies of interest the diaphragm is operating below its resonance frequency, so that $Z_{mo} \to -j(s/\omega)$. Furthermore, by neglecting the leakage resistance, we have $Z_{EB} = 1/(j\omega C_0)$. The transformation factor ϕ is given by (14.19).

Collecting these results into (14.70), we obtain

$$\mathscr{M}_o \doteq \frac{V_0}{x_0}\frac{a^2}{8\mathscr{T}} \qquad (14.73)$$

The design of a sensitive microphone requires a large diaphragm area, high polarization voltage, small interelectrode spacing, and low stiffness. However, V_0/x_0 is limited by dielectric breakdown (arcing) and too large an a and a reduced s will degrade high-frequency response by lowering the mechanical resonance frequency.

As a numerical example, consider a microphone with an aluminum diaphragm 0.04-mm thick, having a radius $a = 1$ cm, and stretched to a tension $\mathscr{T} = 20{,}000$ N/m. If the spacing between the diaphragm and the backing plate is $x_0 = 0.04$ mm and the polarizing voltage $V_0 = 300$ V, then $\mathscr{M}_o = 4.7 \times 10^{-3}$ V/Pa and the sensitivity level is $\mathscr{ML} = -47$ dB re 1 V/Pa. Measured values of the response of such a microphone are observed to agree with the predicted response below 8 kHz. This is to be anticipated, since the limiting frequency predicted by (4.49) is 6.8 kHz. The fundamental frequency of the diaphragm, given by (4.26) is 2.4

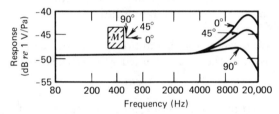

Fig. 14.22. *Free-field sensitivity of a condenser microphone.*

times this frequency, or about 16 kHz. In the vicinity of resonance the microphone may have a response of about 5 to 10 dB above the low-frequency response. The exact amount of this increase depends on the magnitude of the damping forces. Above resonance the response falls off rapidly: the motion of the diaphragm becomes mass-controlled so that the average displacement is no longer a constant but decreases inversely with frequency.

Figure 14.22 shows the open-circuit voltage sensitivity of this condenser microphone. The rise in response near resonance could be reduced by constructing the microphone so that the viscous damping associated with the motion of the air, to and from the region adjacent to the diaphragm through special slots in the fixed backing plate, becomes significant near 16 kHz.

The capacitance C_0 of the microphone in this example is 69.5 pF. Because of this small capacitance $\mathbf{Z}_{EB}$ is very high at audio frequencies. For example, its value at 100 Hz is 23 MΩ. Consequently, electrical terminals must be connected to a resistance in excess of 50 MΩ if the voltage across them is to be approximately equal to that generated in the microphone.

Because of this high internal impedance, it is necessary to provide at least one stage of preamplification that would be located in the immediate vicinity of the microphone. The preamplifier for a modern condenser microphone is usually mounted in the housing; the first input stage is an FET with a suitably high-input impedance of about 500 MΩ. If the microphone is instead connected through a long cable to its preamplifier, there is a susceptibility to electrical pickup (mostly 60-Hz line frequency and harmonics). Furthermore, the electrical capacitance of the cable is in parallel with C_0. Since the capacitance of ordinary shielded microphone cable is about 60 to 100 pF/m, the total capacitance of even relatively short runs will be greater than C_0, thus decreasing the sensitivity of the microphone. Since Z_{EB} is capacitive, the primary effect of the cable capacitance is attenuation, without appreciable frequency discrimination.

14.10 MOVING-COIL ELECTRODYNAMIC MICROPHONE. The simple moving-coil or "dynamic" microphone consists of a light diaphragm to which a small coil is rigidly attached. The action of sound waves on the diaphragm causes the coil to move in the radial field of a permanent magnet, thus generating a voltage. Basically, a moving-coil microphone is similar to a moving-coil loudspeaker; the small loudspeakers of an interoffice communication system usually serve also as the microphones of the system.

The transformation factor for this antireciprocal device is given by (14.25). As before, we will discuss only the open-circuited voltage sensitivity (14.71b)

$$\mathcal{M}_o \doteq SBl/Z_{mo} \tag{14.74}$$

where it is assumed that the wavelength is long with respect to the dimensions of the receiver.

If the voltage output for a given sound pressure amplitude on the diaphragm is to be independent of frequency, this could be done by making R_m large so that the system is resistance-controlled. On the other hand, a high-voltage sensitivity can be obtained only by making the speed amplitude as large as possible, which requires a small mechanical impedance. Therefore, it is impossible to satisfy both the requirement of high sensitivity and uniformity of response by employing a mechanical system equivalent to a simple oscillator with Z_{mo}, as given by (14.22b). Instead, additional mechanical elements must be utilized to compensate for the increase in mass reactance at frequencies above f_0 and in stiffness reactance at frequencies below f_0.

A cross-sectional view of a moving-coil microphone having the desired mechanical characteristics is shown in Fig. 14.23. The diaphragm is a dome that vibrates as a rigid piston in the useful range of frequencies. The stiffness s and resistance R are contributed by the corrugated annulus supporting the diaphragm. The stiffness s_1 is due primarily to compression of the air trapped in the chamber beneath the diaphragm. The mass m_1 and resistance R_1 arise from viscous forces opposing the flow of air through the capillaries of the silk cloth. The stiffness s_2 due to the air chamber below the silk cloth is relatively small, and we will for the moment neglect it and the influence of the tube. The equivalent mechanical system for the microphone can be

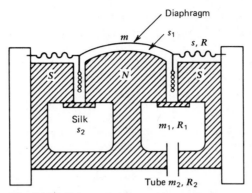

Fig. 14.23. *Schematic of a moving-coil micro-phone.*

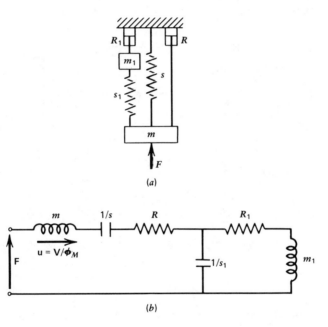

Fig. 14.24. *Moving-coil microphone. (a) Mechanical analogue. (b) Equivalent circuit.*
(c) Equivalent circuit of a more realistic moving-coil microphone.

obtained by the methods used in Chapter 1; it is shown in Fig. 14.24a. The equivalent circuit in Fig. 14.24b, although relatively complicated, is amenable to routine analysis.

The open-circuit mechanical impedance is $\mathbf{Z}_{mo} = R_{mo} + jX_{mo}$ where

$$R_{mo} = R + \frac{s_1^2 R_1}{R_1^2 \omega^2 + m_1^2(\omega_1^2 - \omega^2)^2} \tag{14.75}$$

$$\omega_1^2 = s_1/m_1 \tag{14.76}$$

and

$$X_{mo} = \omega m - \frac{s}{\omega} + \frac{s_1 \omega[m_1^2(\omega_1^2 - \omega^2) - R_1^2]}{R_1^2 \omega^2 + m_1^2(\omega_1^2 - \omega^2)^2} \tag{14.77}$$

The mechanical constants R, m, s, R_1, m_1, and s_1 can be chosen so that the absolute value of the mechanical impedance Z_{mo} is fairly uniform over a rather wide range of frequencies. Curve A of Fig. 14.25 shows Z_{mo} for the following values: $R = 1$,

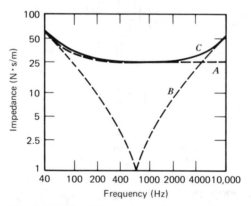

Fig. 14.25. *Possible mechanical impedances for the moving element of a moving-coil microphone.*

$R_1 = 24$ N · s/m, $s = 10^4$, $s_1 = 10^6$ N/m, $m = 0.6$, and $m_1 = 0.3$ g. For purposes of comparison, the mechanical impedance of the simple oscillator R, m, s alone has been plotted in curve B, and that of a similar oscillator for which R is 25 N · s/m in curve C. It is evident that curve A is flatter than either curve B or C and, consequently, the corresponding mechanical system will lead to a more uniform response of the microphone.

The open-circuit voltage response for a diaphragm of 5 cm² effective area attached to a coil of 10 m length and operating in a field of 1.5 T (tesla) is plotted in Fig. 14.26 for each of the three mechanical system considered above.

The low-frequency response of this microphone is further improved by the tube connecting the lower chamber to outer air, as indicated in Fig. 14.23. This modifies

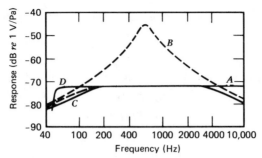

Fig. 14.26. *Sensitivity of a moving-coil micro-phone.*

the equivalent electrical circuit to that of Fig. 14.24c. Because s_2 is small, s_2/ω becomes appreciable at the lowest frequencies allowing the force $\mathbf{F}_2$ at the opening of the tube to affect the circuit. Analysis reveals that $\mathbf{F}_2$ then *increases* $\mathbf{u}$ and provides the sensitivity shown in curve D of Fig. 14.26. The open-circuit voltage sensitivity level of the above moving-coil microphone is about -72 dB *re* 1 V/Pa.

As with the condenser microphone, the dynamic microphone is usually operated into a high impedance load so that $\mathcal{M}_o$ is the appropriate sensitivity for rating response.

14.11 PRESSURE-GRADIENT MICROPHONES.

All the microphones previously considered in this chapter are classified as pressure microphones—the acoustic pressure acts primarily on one side of the moving diaphragm and the resulting driving force is basically proportional to the pressure. It is also possible to construct *pressure-gradient* microphones for which the driving force is proportional to the *difference* between the pressures acting on the two sides of a moving element.

As an introduction to the theory of this type of microphone, consider an acoustic wave $\mathbf{p}$ incident on a plane piston of area S and thickness L. Define an x-axis perpendicular to the plane of the piston. If the dimensions of the piston are small compared to a wavelength, the acoustic field is negligibly disturbed and the net force exerted perpendicular to the axis of the piston is

$$\mathbf{F} = -SL\frac{\partial \mathbf{p}}{\partial x} \tag{14.78}$$

If the piston is constrained to move perpendicular to its face and if it is mass-controlled so that $\mathbf{F} = j\omega m\mathbf{u}$ where m is the mass of the piston, then its speed is

$$\mathbf{u} = \frac{SL}{m}\rho_0 \mathbf{v}_x \tag{14.79}$$

where $\mathbf{v}_x$ is the x component of the particle velocity of the free-field acoustic signal, $\partial \mathbf{p}/\partial x = -j\omega\rho_0 \mathbf{v}_x$. If a length l of wire is attached to the piston and moves perpendicular to a magnetic field of magnitude B, the voltage generated in the conductor is

*Bl***u** and the open-circuit output voltage generated by the microphone is

$$\mathbf{V} = \frac{BlSL}{m} \, \rho_0 \, \mathbf{v}_x \tag{14.80}$$

Because the output voltage is proportional to $\mathbf{v}_x$, microphones of this design are also called *velocity* microphones. The open-circuit voltage sensitivity is

$$\mathcal{M}_o = \frac{BlSL}{m} \, \rho_0 \left| \frac{\mathbf{v}_x}{\mathbf{p}} \right| \tag{14.81}$$

If the acoustic wave is a plane wave traveling at an angle θ with respect to the x-axis, then from (6.22) it is easy to show $v_x/p = \cos\theta/(\rho_0 c)$ so that

$$\mathcal{M}_o = \frac{BlSL}{cm} \cos\theta \tag{14.82}$$

and the microphone has a directional factor $H(\theta) = |\cos\theta|$. The pressure-gradient microphone is *bidirectional* in that it favors waves incident in the vicinity of $0°$ and $180°$ and discriminates against those incident in the vicinity of $90°$ and $270°$. Substitution of $H(\theta) = |\cos\theta|$ into (8.43) with integration leads to a directivity $D = 3$ and a corresponding directivity index, (8.45), of 4.8 dB.

One example of a pressure-gradient microphone is the *ribbon* microphone of Fig. 14.27. It consists of a light corrugated ribbon suspended in a magnetic field and exposed to acoustic pressure on *both* sides. A circular baffle of radius a determines the length of the air path between the two sides of the ribbon. [In actual instruments this baffle is the magnetic-field structure. Its shape is usually quite irregular so that a theoretical calculation of the difference in pressure is extremely complex. However, if the baffle is roughly circular, the distance L of (14.82) may be considered to equal the radius of the baffle.]

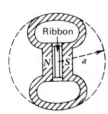

Fig. 14.27. *Simple velocity-ribbon microphone mounted in a circular baffle of radius a.*

The stiffness of the suspension system of the ribbon is made so small that its natural frequency is below the audible frequency range. As a result, the low-frequency response is quite uniform and equal to that predicted by (14.82). As the frequency increases above 1 kHz, the response falls away from the predicted value because at the higher frequencies the dimensions of the microphone are no longer small compared to the wavelength of the sound and the approximations made in deriving (14.82) are no longer valid.

A more detailed derivation of the net force on the piston, ignoring diffraction, for a plane wave traveling at the angle θ with respect to the x-axis yields $\mathbf{f} = [\mathbf{p}(x = 0) - \mathbf{p}(x = kL \cos\theta)]S$ which culminates in

$$\mathcal{M}_o = \frac{BlS}{\omega m} \, 2 \, \sin(\tfrac{1}{2}kL \cos\theta) \tag{14.83}$$

and

$$H(\theta) = \left| \frac{\sin(\frac{1}{2}kL \cos \theta)}{\sin(\frac{1}{2}kL)} \right|$$ (14.84)

These reduce to (14.82) when $kL < \pi/2$.

Curve A of Fig. 14.28, computed from (14.83) for normal incidence, shows the open-circuit voltage response of a ribbon microphone having the following constants: $m = 0.001$ g, $S = 5 \times 10^{-5}$ m^2, $l = 0.02$ m, $B = 0.5$ T (tesla), and $L = 3$ cm. Curve B shows measured values of the response of a microphone of this size. The increased response in the region from 2 to 9 kHz is caused partially by diffraction effects. The upper limit of usefulness for this particular microphone is about 9 kHz.

The velocity microphone has three principal advantages. (1) Since it discriminates by a factor of 3 against background noise and reverberation, it allows speakers to stand at $\sqrt{3}$ times greater a distance than from a nondirectional microphone, and

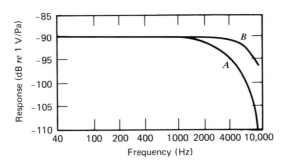

Fig. 14.28. *Normal-incidence sensitivity of a velocity-ribbon microphone. Computed (curve A). Measured in a free field (curve B).*

will still generate an electrical output having the same signal-to-noise ratio. (2) Its bidirectional characteristic permits speakers to face each other from opposite sides of the microphone and have their speech picked up with equal sensitivities. (3) The sharp nulls at 90° and 270° make it possible to orient the microphone so that the regions of zero response are pointed toward particularly annoying sources of noise, thus eliminating their direct pickup.

The velocity microphone has one peculiarity: if it is very close to a source of sound, there is an enhancement of the response at lower frequencies. This arises because the ratio of particle velocity amplitude to acoustic pressure amplitude of a spherical wave increases as the source of sound is approached. Equation (5.51) predicts that the sensitivity level should be increased by $20 \log[\sqrt{1 + (kr)^2}/(kr)]$. This peculiarity is appreciated by many signers of popular music since the strengthened bass response bestows richness to the voice.

If the output of a pressure microphone is combined in series with that of a velocity microphone, the resulting response favors the reception of sounds coming from one hemisphere. If the axial sensitivity of the velocity element is also made equal to the nondirectional response of the pressure element, then the net response of the two elements in series is

$$\mathcal{M}'_o = \mathcal{M}_o(1 + \cos\theta) \tag{14.85}$$

Figure 14.29 shows the directional responses of the individual elements and that of their combination. The combined directional characteristic is a *cardioid* of revolution, with the axis of rotation normal to the plane of the ribbon element. Microphones having a response of this type are commonly known as *unidirectional* or *cardioid* microphones. The directivity of a cardioid microphone is also $D = 3$. As a result of the large solid angle over which a cardioid microphone receives sound without appreciable discrimination, it is possible to cover widely separated speakers and musical instruments by a single microphone. Furthermore, if such a microphone

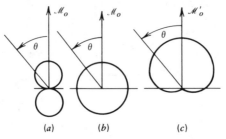

(a)   (b)   (c)

Fig. 14.29. *Directional response of various microphones. (a) Velocity microphone. (b) Pressure microphone. (c) Cardioid microphone.*

is placed near the front of a stage, its directional characteristics are exceptionally well-suited to picking up sounds from the stage and excluding those from the audience.

14.12   OTHER MICROPHONES.

(a) The carbon Microphone.   The carbon microphone is widely used for telephone and radio communication where high electrical output, low cost, and durability are of greater significance than fidelity. Its operation results from the variation in resistance of a small enclosure filled with carbon granules, the *carbon button* (Fig. 14.30a). As the diaphragm is displaced, the plunger varies the force applied to the carbon granules and hence the resistance from granule to granule, so that the total resistance across the carbon button (ordinarily about 100 Ω varies in an *approximately* linear manner with the pressure applied to the diaphragm.

In the simple equivalent circuit shown in Fig. 14.30b, let us assume that at frequencies well below the fundamental resonance of the diaphragm its motion is stiffness-controlled and that for small displacements of the diaphragm the resistance R_c of the carbon button varies linearly with the displacement x of the center of the diaphragm. Then

$$R_c = R_0 + hx = R_0 + \frac{hS}{s}p \tag{14.86}$$

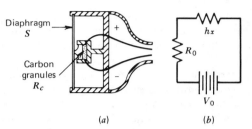

Fig. 14.30. *Carbon microphone. (a) Schematic.*
(b) Equivalent circuit.

where R_0 is the zero-displacement resistance of the button, h is its resistance constant in ohms per meter of displacement of the plunger, s is the stiffness of the diaphragm, and S is its effective area. The variation in resistance causes the current to vary as $I = V_0/R_c$ where V_0 is the voltage of the battery. If $hSp/s \ll R_0$, then $1/R_c$ may be expanded and

$$I = \frac{V_0}{R_0}\left(1 - \frac{hS}{sR_0}\,p + \cdots\right)$$

This indicates the presence of a steady direct current V_0/R_0, an alternating current $I_c = -V_0\,hSp/(sR_0^2)$, and higher harmonics of this current. The a-c component I_c generates an output voltage $V_c = I_c R_0$. If the microphone is operating under open-circuit conditions

$$\mathcal{M}_o = \frac{V_0\,hS}{R_0\,s} \tag{14.87}$$

The response increases as the battery voltage V_0 is increased or as the total resistance R_0 of the circuit is decreased. It also increases directly with the area S and inversely with the stiffness s of the diaphragm. However, there are practical limitations to such methods of increasing the sensitivity, since high values of V_0 cause excessive heating and internal noise in the carbon button, and decreasing s tends to lower the fundamental frequency of the diaphragm, thus reducing the useful frequency range of the microphone.

A measured frequency response for a carbon microphone is shown in Fig. 14.31. The peak response in this curve, which occurs near 2 kHz, is associated with the fundamental frequency of the diaphragm, and the uneven response above this frequency is due to the breaking up of the diaphragm into various modes of vibration. If the diaphragm is tightly

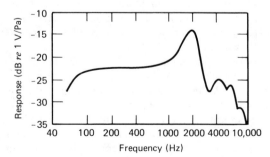

Fig. 14.31. *Sensitivity of a carbon microphone.*

stretched, its effective stiffness can be increased with a corresponding increase in the fundamental frequency. The use of a stretched diaphragm makes it possible to extend the region of relatively uniform response to about 8 kHz but only at the expense of decreased sensitivity.

(b) The Piezoelectric Microphone. Piezoelectric microphones employ crystals or ceramics which, when distorted by the action of incident sound waves, become electrically polarized and produce voltages linearly related to the mechanical strains. A mathematical analysis of the piezoelectric microphone is not warranted in this book; a nonmathematical discussion of their characteristics will suffice. Since the piezoelectric effect is reversible, all types of piezoelectric microphones will function as sources of sound when an alternating voltage is applied to their terminals. They are reciprocal transducers.

Single crystals of Rochelle salt have been widely employed in microphones. Unfortunately, such crystals are subject to deterioration in the presence of moisture and are permanently damaged if subjected to temperatures in excess of 46°C (115°F). Crystals cut from synthetic ammonium dihydrogen phosphate (ADP) are somewhat less pressure sensitive than Rochelle salt, but they can be raised to temperatures in excess of 93°C (200°F) without deterioration and they exhibit much less variation in their piezoelectric and dielectric properties with changes in temperature. Other useful substances are ceramic materials such as barium titanate. The *ferroelectric* ceramics are *polarized* (thereby behaving like piezoelectric materials) by applying a high electrostatic potential gradient of about 2000 kV/m while the temperature is above the Curie temperature of the ceramic (about 120°C) and by maintaining the external polarizing voltage during cooling. These *ceramic* microphones may be used interchangeably with crystal microphones. Their sensitivity is about 10 dB below that of Rochelle salt or that of ADP types. On the other hand, ceramic microphones have advantages: they can withstand much higher temperatures, are not as readily damaged by moisture or high humidity, and can be cast into a variety of shapes and sizes.

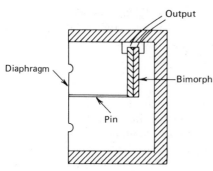

Fig. 14.32. *Diaphragm-actuated crystal microphone.*

The magnitude of the resulting potential difference produced in a piezoelectric material depends on the type of deformation and its orientation relative to the axes of the crystal (or to the axis of polarization of the ceramic). Bending, shear, and compressional deformations have been utilized. The deformation may be brought about by the sound waves acting directly on the piezoelectric material, but the principal disadvantage of a microphone of this type is its very low sensitivity. As a consequence, piezoelectric microphones are usually constructed in a manner similar to that shown in Fig. 14.32. Here, the sound waves act on a light diaphragm whose center is in turn linked to an end or corner of the piezoelectric element by means of a driving pin. Although a single element could be used, two elements are usually sandwiched together to form an assembly known as a *Bimorph*. In general, the Bimorph has a smaller mechanical impedance than does a single element producing the same voltage output. Either a series or a parallel connection of the two elements may be used—the series connection giving a larger voltage output and the parallel connection giving a lower internal impedance.

The voltage output of a Bimorph element is proportional to the amplitude of its deformation. Hence, as for condenser microphones, the moving elements of the microphone must be designed as a stiffness-controlled system. Consequently, it is necessary to arrange that the fundamental frequency of the mechanical resonance of the entire vibrating system, including the diaphragm, connecting pin, and Bimorph, be at a frequency above that of the desired range of relatively uniform response.

Piezoelectric microphones are widely used in public-address systems, sound-level meters, and hearing aids. They have a satisfactory frequency response for such applications, are relatively high in sensitivity, low in cost, and small in size. Inexpensive diaphragm-actuated types are available and can cover the frequency range from 20 Hz to 10 kHz with a maximum variation in sensitivity of less than 5 dB from their average sensitivity. A typical average sensitivity is -30 dB re 1 V/Pa. The electrical impedance of a piezoelectric microphone is that of a dielectric capacitor; a typical value is 3000 pF. This is large compared to that of a condenser microphone, so that these microphones may be connected to an audio amplifier by a cable of moderate length without an intervening preamplifier being required.

14.13 CALIBRATION OF RECEIVERS. Techniques for obtaining the sensitivity of a receiver fall into two general categories: *Absolute* techniques that require measurement only of length, force, voltage, and so on, and *relative* techniques where the sensitivity of one receiver is known *a priori*.

If a receiver of known sensitivity is on hand, any other receiver can be calibrated by *direct comparison*. In this method, (1) a transmitter produces a sound field in an anechoic environment. (The anechoic environment is not necessary at low frequencies, but becomes important at frequencies for which the wavelength of the sound is comparable to or less than the dimensions of the receiver, since for these wavelengths the orientation of the receiver with respect to the sound field must be known.) (2) Next, the receiver of known sensitivity $\mathcal{M}_{oA}$ is placed in the sound field and its output voltage V_A is recorded as a function of frequency. (3) Then receiver A is removed and the receiver of unknown sensitivity $\mathcal{M}_{oX}$ is placed in exactly the same location and its output voltage V_X is recorded for the same frequencies as in step 2. (4) At each frequency, the sensitivity of the receiver is given by

$$\mathcal{M}_{oX} = \mathcal{M}_{oA} \frac{V_X}{V_A} \tag{14.88}$$

Any limitations inherent in $\mathcal{M}_{oA}$ are also important in $\mathcal{M}_{oX}$. For example, if receiver X were much larger than A, then the calibration of X would be good for wavelengths smaller than the dimensions of X as long as the calibration of A remains valid. Of course, for wavelengths small compared to X, the orientation of the sound field with respect to X becomes important.

The most direct *absolute* technique consists of applying a known acoustic pressure to the receiver. This is easily accomplished by connecting the receiver to a chamber of accurately known volume in which a piston of known area oscillates with known displacement amplitude. If the wavelength in the chamber is much longer than the dimensions of the chamber, the pressure amplitude within the chamber can be calculated from the equation of state for the gas. While simple in concept, this technique is difficult in application because of the problem of accurately determining the volume of the chamber with the receiver in place. For certain popular makes of microphones, calibration devices, called *pistonphones*, are commercially available. However, a different *coupler* must be supplied for each type of microphone. The disadvantages of this technique are that the calibration can be performed at only one (or at most a few) frequencies and the sensitivity derived is that based on the actual pressure at the microphone face and not that which exists before the microphone disturbs the sound field. However, the latter effect will be negligible for frequencies with wavelengths long compared to the dimensions of the receiver. The convenience and simplicity of this technique make it a popular calibration procedure.

A second technique for obtaining an absolute calibration is the *far-field reciprocity calibration*. To employ this technique, it is necessary to have a transmitter T, the microphone X to be calibrated, and one reversible transducer R. Neither the transmitting nor receiving response of the reversible transducer need be known. To avoid complications which obscure the discussion, assume that both X and R are much smaller than the shortest wavelength of interest.

We will need to show that $\mathcal{M}_{oR}$ and $\mathcal{S}_{IR}$ for *any* reversible transducer R are related by the far-field reciprocity factor. Combination of (14.32*b*) and (14.68) for frequencies such that $\mathcal{D} \to 1$ leads to

$$\frac{\mathcal{M}_{oR}}{\mathcal{S}_{IR}} = \frac{uS}{P_{ax}(1)}$$

Now, uS is the source strength Q and use of (8.16*c*) yields

$$\frac{\mathcal{M}_{oR}}{\mathcal{S}_{IR}} = \frac{2\lambda}{\rho_0 c} \tag{14.89}$$

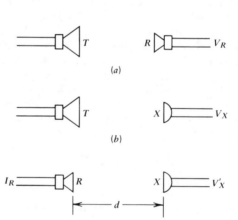

(a)

(b)

(c)

Fig. 14.33. *The three steps used in the reciprocity calibration of a microphone.*

The steps are as follows. (1) The reversible transducer R is placed at a specific position in the sound field of transmitter T (Fig. 14.33*a*). The open-circuit voltage generated by R is

$$V_R = \mathcal{M}_{oR} P_T$$

where $\mathcal{M}_{oR}$ is the sensitivity of R and P_T is the sound pressure produced at R by transmitter T. (2) The microphone X is substituted for R and its open-circuit voltage V_X is measured for the same sound pressure P_T,

$$V_X = \mathcal{M}_{oX} P_T$$

where $\mathscr{M}_{oX}$ is the desired sensitivity of X. These two equations may be combined to give

$$\mathscr{M}_{oX} = \mathscr{M}_{oR}\,\frac{V_X}{V_R} \tag{14.90}$$

(3) The reversible transducer R is substituted for the transmitter T and the open-circuit output voltage V'_X of microphone X is measured for a current I_R in the reversible transducer R. Then $V'_X = \mathscr{M}_{oX} P_R$. Combining this expression with $P_R = \mathscr{S}_{IR} I_R/r$, which gives the pressure P_R produced at r by the reversible transducer R as a function of its transmitting sensitivity $\mathscr{S}_{IR}$, we obtain

$$V'_X = \mathscr{M}_{oX}\,\mathscr{S}_{IR}\,I_R/r \tag{14.91}$$

Then from (14.90) and (14.91)

$$\mathscr{M}_{oX}^2 = \frac{\mathscr{M}_{oR}}{\mathscr{S}_{IR}}\,\frac{V_X\,V'_X}{V_R\,I_R}\,r$$

Use of (14.89) gives the sought-after open-circuit voltage sensitivity

$$\mathscr{M}_{oX} = \left(\frac{2\lambda r}{\rho_0 c}\,\frac{V_X\,V'_X}{V_R\,I_R}\right)^{1/2} \tag{14.92}$$

This method has the advantage of avoiding the necessity of attempting to produce measureable or calculable sound pressures since all the basic measurements, other than distance, are electrical. (The results obtained from (14.92) can be converted to $V/\mu\text{bar}$ by multiplying by 0.1.)

If two *identical* reversible microphones are available, the reciprocity principle makes it possible to calibrate both microphones by a single set of measurements involving only one setup. Here $V_X = V_R$, so that (14.92) reduces to

$$\mathscr{M}_{oX} = \mathscr{M}_{oR} = \left(\frac{2\lambda r}{\rho_0 c}\,\frac{V'_X}{I_R}\right)^{1/2} \tag{14.93}$$

Finally, it is possible to self-calibrate a reversible transducer at high frequencies by using short-pulse techniques. The transducer is first used as a sound source to transmit a pulse that is reflected from a perfectly reflecting surface back to the transducer. The transducer, now operating as a receiver, detects the pulse and generates an output voltage.

PROBLEMS

14.1.   Verify that the circuits of Fig. 14.2a and 14.2c yield the canonical equations (14.7) for a reciprocal transducer.

14.2.   Find the approximate relationship between $\mathbf{K}^2$ and k_m^2 for frequencies in the range $R_0/L_0 \ll \omega \ll \omega_0$ for an antireciprocal transducer.

14.3.   A transmitter generates a sound pressure level at 1 m of 100 dB *re* 1 μbar for a driving voltage of 100 V. Find the sensitivity level in dB *re* 1 μbar/V.

14.4.   A transmitter has a sensitivity level of 60 dB *re* 1 μbar/V. Find its sensitivity level *re* 1 μPa/V and *re* 20 μPa/V.

14.5.   When the voice coil of a loudspeaker is blocked, the resistive component of its input impedance is 5 Ω. When flush mounted in a large wall and driven at its frequency of mechanical resonance, the resistive component is found to be 10 Ω. When removing the speaker from the wall in order approximately to eliminate the acoustic radiation loading, the input resistance is 12 Ω at the frequency of mechanical resonance. What is the electroacoustic efficiency of the speaker at resonance when mounted in the wall?

14.6. A direct radiator loudspeaker of 0.2-m radius is mounted flush in a large flat wall. When the speaker is being driven at 1000 Hz, the sound intensity at a point 5 m from the speaker and directly out from the wall is 0.1 W/m^2. (*a*) What is the intensity level at this point? (*b*) Assuming the speaker to have the same directional pattern as a flat piston of equal radius, what is the intensity level at this same distance directly along the wall? (*c*) In a direction 30° from the wall? (*d*) In a direction 60° from the wall? (*e*) What is the total acoustic output of the speaker in watts?

14.7. For the loudspeaker of Sect. 14.5, evaluate (*a*) k_m^2, (*b*) k_m^2 (*eff*), (*c*) Q_M from the quantities given in the text.

14.8. The voice coil of a direct radiator speaker is 0.03-m in diameter and has 80 turns. Its blocked resistance is 3.2 Ω and its blocked inductance is 0.2 mH. It operates in a magnetic field of 1 T. The total mass of the cone and voice coil is 0.015 kg, the mechanical resistance R_m is 1 N $\cdot$ s/m, the radiation resistance R_r is 1 N $\cdot$ s/m, and the stiffness of the cone system is 1500 N/m. (*a*) Assuming the radiation reactance X_r to be negligible, what are the blocked electrical impedance Z_{EB}, the motional impedance Z_{MOT}, and the total electrical input impedance Z_E at a frequency of 200 Hz? (*b*) What rms driving voltage is required to produce an rms displacement amplitude of the speaker cone of 0.1 cm at this frequency? (*c*) What acoustic output in watts will be produced by the above driving voltage? (*d*) Calculate values for R, L, and C of the parallel electrical circuit having an impedance equivalent to the motional impedance of the above speaker.

14.9. (*a*) What is the frequency of mechanical resonance for the speaker of Problem 14.8? (*b*) What is the mechanical quality factor Q_M of the above speaker cone system? (*c*) What is the rms displacement amplitude of the cone for an applied voltage of 5 V at resonance? (*d*) If the driving circuit is opened at an instant of maximum displacement, what will be the rms displacement amplitude at the end of 0.02 s?

14.10. A direct radiator loudspeaker is mounted in an infinite baffle. It has a radius of 0.2 m, a mass of 0.04 kg, a voice coil having a resistance of 4 Ω, an inductance of 0.1 mH, and a transformation factor of 10 T $\cdot$ m. The suspension system of the cone has a stiffness of 2000 N/m and a mechanical resistance of 2 N $\cdot$ s/m. (*a*) If an alternating voltage of 10 V rms and 200 Hz is applied to the voice coil, what is the acoustic power output of the speaker? Assume radiation loading on just one side of the speaker cone. (*b*) Assuming the radiated beam pattern to be that of a circular piston in an infinite baffle, what axial pressure level will be produced at a distance of 10 m?

14.11. A direct radiator loudspeaker has a radius of 0.15 m. When mounted in a wall its frequency of mechanical resonance is 25 Hz. When mounted in a small back-enclosed cabinet of 0.1-m^3 volume, its frequency of mechanical resonance is raised to 50 Hz. (*a*) What is the stiffness constant of the suspension system of the speaker cone? (*b*) What is the mass of the speaker cone? In each case consider the speaker cone to be air loaded with radiation reactance on just one side.

14.12. A direct radiator loudspeaker has the following physical characteristics: mass = 0.01 kg, stiffness = 1000 N/m, mechanical resistance = 1.5 N $\cdot$ s/m, radius = 0.15 m, a voice coil of 1.5-cm radius having 150 turns of No. 34 copper wire, a flux density of the magnetic field in the air gap of 0.8 T, and 0.4-mH inductance. The speaker is mounted in an enclosed-back cabinet 0.2 $\times$ 0.5 $\times$ 1.0 m. (*a*) Considering the speaker cone to be radiation loaded on just one side, what is the frequency of mechanical resonance? (*b*) If an rms voltage of 10 V is applied to the voice coil, what is the acoustic output in watts at the resonant frequency, at 200 Hz, and at 1000 Hz?

14.13. Consider the direct radiator speaker of Problem 14.12 to be mounted in a bass-reflex cabinet. (*a*) If the vent is a circular hole of 0.15-m radius and negligible length, what must be the volume of the cabinet if its Helmholtz resonance frequency is to equal that of the speaker cone and suspension when mounted in an infinite baffle? (*b*) What is the ratio of the acoustic output at 75 Hz of the speaker when mounted in this cabinet to that when mounted in an infinite baffle? Assume the displacement of the speaker cone to be the same in each case.

14.14. A small circular piston has a radius of 0.03 m and a mass of 0.002 kg. The stiffness of its suspension system is such that its free oscillation frequency is 300 Hz. An exponential horn having a radius of 0.03 m at its throat, a length of 1.0 m, and a radius of 0.3 m at its mouth is fitted over the piston. (*a*) What is the new mechanical resonance frequency of the piston? Consider the mass loading of the piston by the horn to be that of an infinite horn having the same flare constant as the actual horn. (*b*) If the rms amplitude of the driving force acting on the piston is 5 N, what acoustic power will be radiated by this infinite horn at a frequency of 300 Hz?

14.15. One watt of acoustic power is being radiated at 250 Hz from an infinite exponential horn. The horn has a radius of 0.03 m at its throat and has a flare constant $m = 5$. (*a*) What is the cutoff frequency of the horn? (*b*) What is the required peak volume velocity at the throat in order to produce 1 W of acoustic output? (*c*) If the radius of the diaphragm in the driver unit attached to this horn is 0.05 m, what must be its peak displacement amplitude if it is to produce the above volume velocity?

14.16. The exponential horn of a tweeter has a radius of 0.01 m at its throat. The diaphragm has a radius of 0.03 m, a stiffness of 5000 N/m, and a mass of 0.001 kg. The voice coil has a resistance of 1.6 Ω, an inductance of 0.1 mH, and a transformation factor of 4 T · m. (*a*) What must be the flare constant of the horn if it is to have a cutoff frequency of 500 Hz? (*b*) What must be the peak volume-velocity amplitude at the throat of the horn, if the acoustic output is to be 0.2 W at a frequency of 1000 Hz? (*c*) What displacement amplitude of the diaphragm will produce this volume velocity at a frequency of 1000 Hz? (*d*) What is the efficiency of the speaker at this frequency? (*e*) What voltage must be applied to the voice coil in order to produce the above acoustic output?

14.17. The receiving sensitivity level of a hydrophone is -80 dB *re* 1 V/μbar. (*a*) Express this level *re* 1 V/μPa. (*b*) What will be the output voltage if the pressure field is 80 dB *re* 1 μbar?

14.18. A microphone reads 1 mV for an incident effective pressure level of 120 dB *re* 1 μbar. Find the sensitivity level of the microphone *re* 1 V/μbar.

14.19. A condenser microphone diaphragm of 0.02-m radius and 0.00002-m spacing between diaphragm and backing plate is stretched to a tension of 10,000 N/m. (*a*) If the polarizing voltage is 200 V, what is the low-frequency, open-circuit voltage response of the microphone in volts per pascal? (*b*) What is the corresponding response level in decibels relative to 1 V/Pa? (*c*) When acted on by a sound pressure of 1 Pa amplitude, what is the amplitude of the average displacement of the diaphragm? (*d*) What voltage will be generated in a load resistance of 5 MΩ by this condenser microphone at a frequency of 100 Hz when acted on by a sound wave of 10 μbar pressure amplitude?

14.20. A small condenser microphone having a diameter of 0.8 cm is to be used as a probe microphone. The steel diaphragm is 0.001-cm thick and is stretched to the maximum allowable tension of 10,000 N/m. The spacing between the diaphragm and backing plate is 0.001 cm, and the polarizing voltage is 150 V. (*a*) What is the fundamental frequency of the diaphragm? (*b*) What is the open-circuit, constant-pressure response level *re* 1 V/μbar and *re* 1 V/Pa? Find the blocked input impedance at 10 kHz. (*c*) Considering diffraction effects of the microphone to be equivalent to that of a sphere of equal diameter, what is the axial free-field response level *re* 1 V/Pa at this frequency?

14.21. A moving-coil loudspeaker is used both as a microphone and as a loudspeaker in an interoffice communication system. The constants of the speaker are as follows: $m = 0.003$ kg, $a = 0.05$ m, $R_m = 10$ N · s/m, $s = 50,000$ N/m, $B = 0.75$ T, $l = 10$ m, $R_0 = 1$ Ω, and $L_0 = 0.01$ mH. Calculate its open circuit voltage response level re 1 V/Pa at 1100 Hz.

14.22. A moving-coil microphone has a moving element of 0.0002-m^2 cross section, 0.001-kg mass, 10,000-N/m stiffness, and 20-N · s/m mechanical resistance. The coil has a resistance of 5 Ω, a length of 5 m, and moves in a magnetic field of 1.0 T. (a) What is its open-circuit, constant-pressure response level re 1μPa at 1000 Hz? (b) Repeat at 100 Hz.

14.23. A velocity-ribbon microphone is constructed by mounting a thin aluminum strip in a circular baffle of 4-cm radius. The aluminum strip is 0.001-cm thick, 0.4-cm wide, and 2.5-cm long. The magnetic field has a flux density of 0.25 T. A plane wave of 250 Hz and 2 Pa acoustic pressure is incident normally on the face of the ribbon. (a) What voltage is developed in the ribbon? (b) What is the open-circuit voltage response level at this frequency? (c) What is the amplitude of the displacement of the ribbon under the above conditions?

14.24. Given a standing plane wave $p = 2P \cos \omega t \sin kx$, derive an expression giving the net axial force acting on the cylinder of an idealized pressure-gradient microphone. Consider only the case in which the direction of propagation of the waves is along the axis of the cylinder. Show that this force is zero at the antinodes of pressure and a maximum at the antinodes of velocity.

14.25. A microphone has a directivity such that the response in any direction making an angle θ with the principal axis is proportional to $\cos^2 \theta$. (a) Compute the numerical value of the directivity. (b) What is its directivity index?

14.26. The open-circuit voltage response of a carbon microphone is -40 dB re 1 V/μbar when connected to a 12-V battery and its internal impedance is 100 Ω. Its diaphragm has an area of 0.001 m^2 and an effective stiffness of 10^6 N/m. (a) What is the value of the resistance constant h for this microphone? (b) For an incident sound wave of 100-μbar pressure amplitude, what will be the ratio of the second harmonic to fundamental voltage developed in this microphone?

14.27. A crystal microphone is rated as having an open-circuit voltage response level of -34 dB re 1 V/Pa and an internal capacitive impedance of 200,000 Ω at 400 Hz. (a) If a plane wave of this frequency with 70-dB pressure level re 20 μPa is incident on the microphone, what voltage will be generated in a 500,000-Ω load resistor connected across the output terminals of the microphone? Assume that the crystal is stiffness-controlled. (b) What power will be generated in this resistor? (c) If the area of the diaphragm is 0.0004 m^2, what is the ratio of this electrical power to the acoustic power incident on the microphone?

14.28. In a comparison calibration of a microphone, the standard microphone of sensitivity level -120 dB re 1 V/μbar gives an output voltage of 1 mV. The microphone to be calibrated gives an output of 0.2 mV when substituted for the standard. (a) What is the microphone sensitivity level of the unknown microphone? (b) What is the pressure level to which the microphones are being exposed?

14.29. A condenser microphone has a diaphragm of 2-cm radius and the separation between the diaphragm and the backing plate is 0.002 cm. The diaphragm is stretched to a tension of 5000 N/m. The polarizing voltage is 200 V. (a) What is the low-frequency, open-circuit voltage response level re 1 V/Pa of the microphone? (b) Using the electroacoustical reciprocity theorem, compute the acoustic pressure level re 1 Pa produced by the microphone, acting as a loudspeaker, at a distance of 1 m when being driven by a current of 0.01 A at 1000 Hz.

14.30. In a reciprocity calibration, the spacing between two identical reversible microphones is 2m. For 2000 Hz the measured open-circuit voltage output of one microphone is 0.0001 V for an input current of 0.01 A to the other. What is the open-circuit voltage response level of the microphones in decibels re 1 V/Pa?

14.31. A microphone is to be calibrated. From initial measurements, it is determined that its sensitivity is 5 times as great as a reversible transducer. When the reversible transducer is used as a source at a distance of 1.5 m from the microphone, the microphone is observed to have an open-circuit output of 0.001 V when a driving current of 1 A is supplied to the transducer at 500 Hz. (a) What is the open-circuit voltage response of the microphone? (b) What is the acoustic pressure acting on the microphone during the above experiment?

CHAPTER 15

UNDERWATER ACOUSTICS

15.1 INTRODUCTION. The use of sound waves in water for transmission of information is of great interest to both humans and dolphins. One of the earliest human applications of underwater sound was the installation of submerged bells on lightships. Underwater sound from these bells could be detected at considerable distance by means of a microphone mounted in the hull of a ship. If two such devices were located on opposite sides of the hull and the sounds received by each were transmitted separately to the right and the left ears of a listener, it was possible to determine the approximate bearing of the lightship. In 1912, Fessenden developed an electrodynamic underwater sound source that permitted vessels to communicate with one another by Morse code. The safety of ocean navigation was enhanced by the introduction of the *fathometer* which determined the depth of the water by measuring the time required for short pulses of sound to travel from the transmitter to the ocean bottom and return.

The greatest effort in underwater sound has been associated with detecting, tracking, and classifying submarines. It is customary to apply the name *sonar* (SOund NAvigation and Ranging) to this phase of underwater acoustics. In approaching this problem it has been necessary to develop means for the efficient conversion of electrical power into underwater sound and systems that are capable of detecting weak signals in the presence of noise. Of equal importance has been the study of fundamental phenomena that affect the transmission of sound in the ocean, such as divergence, absorption, reflection, refraction, scattering, diffraction, and so forth.

For a more comprehensive discussion of underwater acoustics, the inquisitive reader is referred to the book by Urick.[1]

15.2 SPEED OF SOUND IN SEAWATER. In Sect. 5.6 it was pointed out that the speed of sound in freshwater is a function of temperature and pressure. An additional factor in seawater is the *salinity*. Lovett[2a] has critically analyzed a large collection of laboratory measurements of the speed of sound in water of various salinities and at various temperatures and pressures, and postulated an elaborate

---

[1] Urick, *Principles of Underwater Sound*, 2nd ed., McGraw-Hill (1975).
[2a] Lovett, *J. Acous. Soc. Am.*, **63**, 1713 (1978) and **45**, 1051 (1969).

empirical equation. A reasonably accurate approximation to this equation is [2b]

$$c(\mathscr{L}, S, t) = 1449.05 + 45.7t - 5.21t^2 + 0.23t^3$$
$$+ (1.333 - 0.126t + 0.009t^2)(S - 35) + \Delta(\mathscr{L}) \qquad (15.1a)$$

where

$$\Delta(\mathscr{L}) \approx 16.3\mathscr{L} + 0.18\mathscr{L}^2 \qquad (15.1b)$$

This gives $c(\mathscr{L}, S, t)$ for a latitude of 45°. For other latitudes, the quantity $\mathscr{L}$ should be replaced with $\mathscr{L}(1 - 0.0026 \cos \phi)$ where 2ϕ is the latitude in degrees. In (15.1), S is the salinity in parts per thousand (ppt), $\mathscr{L}$ the depth in kilometers, and $t = T/10$ where T is in degrees Celsius. The combination of (15.1a) and (15.1b) with latitude correction has a standard deviation of 0.06 m/s from Lovett's equation when applied down to a depth of 4 km in oceanic waters excluding the Black Sea, Red Sea, Mediterranean Sea, and Persian Gulf. If a more accurate, but more complicated $\Delta(\mathscr{L})$ is required, the approximation

$$\Delta(\mathscr{L}) = (16.23 + 0.253t)\mathscr{L} + (0.213 - 0.1t)\mathscr{L}^2$$
$$+ [0.016 + 0.0002(S - 35)](S - 35)t\mathscr{L} \qquad (15.1c)$$

is valid over virtually all oceanic waters down to a depth of 4 km with a standard deviation of 0.02 m/s. The same correction for latitude should be applied. Note that the speed of sound in surface seawater of 35 ppt salinity is 1449 m/s at 0°C as contrasted with 1403 m/s for freshwater under the same conditions of temperature and pressure.

In many calculations it is frequently adequate to use a nominal speed of 1500 m/s which is typical of those measured in surface waters overlying the continental shelves in middle latitudes. Associated with this nominal speed is the nominal characteristic impedance $\rho_0 c = 1.54 \times 10^6$ Pa · s/m. It will be the practice in what follows to use the above values in obtaining numerical relationships among the acoustic pressure, particle velocity, and intensity in seawater. However, in calculations requiring differences in the speed of sound, (15.1) or a more accurate equation must be used.

15.3 SOUND TRANSMISSION LOSS.

The *transmission loss* is defined as

$$TL = 20 \log \frac{P(1)}{P(r)} \qquad (15.2)$$

where $P(r)$ and $P(1)$ are the acoustic pressure amplitudes measured at horizontal distances r and 1 m from the sound source.

For example, the pressure amplitude of a damped spherical wave is

$$P(r) = \frac{A}{r} e^{-\alpha(r-1)}$$

[2b] Coppens, *J. Acous. Soc. Am.*, **69**, 862 (1981).

where α is the absorption coefficient in nepers per meter. Taking the log of both sides,*

$$20 \log P(r) = 20 \log P(1) - 20 \log r - a(r - 1)$$

where $a = 8.7\alpha$ is the absorption coefficient in decibels per meter. Since for all frequencies of interest $a \ll 1$ dB/m, the transmission loss for *spherical spreading* with absorption is

$$TL = 20 \log r + ar \tag{15.3}$$

It is left as an exercise to show that, if the sound is trapped between two parallel, perfectly reflecting surfaces, the transmission loss is

$$TL = 10 \log r + ar \tag{15.4}$$

This is *cylindrical spreading* with absorption.

In general, it is convenient to separate the transmission loss into two parts

$$TL = TL(geom) + TL(losses) \tag{15.5}$$

where $TL(geom)$ represents the loss from geometrical considerations and $TL(losses)$ represents the loss due to absorption, scattering, and other nongeometrical effects. For the above examples, $TL(geom) = 20 \log r$ or $10 \log r$ and $TL(losses) = ar$.

The absorption coefficient a for sound waves in seawater has already been discussed in Sect. 7.6. For instance, at one atmosphere in seawater at 5°C, $a = 0.00006$ dB/m at 1 kHz, 0.0008 dB/m at 10 kHz, and 0.013 dB/m at 50 kHz. An approximation of (7.53) and (7.54) for seawater at 5°C and one atmosphere (zero depth) adequate for the purposes of this chapter is

$$\frac{a}{F^2} = \frac{8 \times 10^{-5}}{0.7 + F^2} + \frac{0.04}{6000 + F^2} + 4 \times 10^{-7} \tag{15.6}$$

where F is the frequency in kilohertz and a is in decibels per meter. Plotted in Fig. 15.1 are curves showing the transmission loss for spherical spreading with absorption as a function of r for each of the above frequencies. At low frequencies and short ranges the transmission loss is primarily spherical divergence. As the frequency and range increases, curves B and C show that the absorption loss assumes greater significance. It is evident that for transmission to great distances it is necessary to employ low frequencies.

When transmission loss measurements are made in the ocean, they are frequently observed to deviate from those predicted by (15.3). Factors contributing to this include: (1) geometrical effects due to divergence or convergence caused by *refraction*, or destructive and constructive *interference* associated with *multipath* propagation including reflections from the surface and bottom of the sea, and (2) enhanced attenuation from *diffraction* and *scattering* caused by inhomogeneities in the water (see Sect. 7.7). While it is possible in idealized conditions to derive equations and compute precise values for the transmission loss associated with each of

---

* Here and in what follows terms like log r really should be written log($r/1$ m). We will treat this as implicit for the sake of brevity.

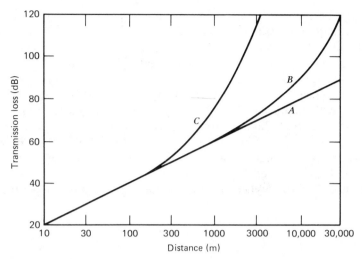

Fig. 15.1. *Dependence of transmission loss TL for spherical spreading with absorption on range and frequency. A at 1 kHz, B at 10 kHz, and C at 50 kHz.*

these factors, the oceans are so complicated that it is usually necessary either to be satisfied with simple analytical models or to rely on complex computer models for calculating transmission loss in any realistic situation.

Although many factors limit our ability to transmit sound through seawater, it should be noted that sound is immensely superior to electromagnetic waves for transmitting energy through seawater. For example, the lowest-frequency radio waves in commercial use, 30 kHz, are attenuated 1 dB in 0.3 m, and higher frequencies are attenuated even more rapidly. Similarly, the diffusion and scattering of a beam of light passing through seawater is so great that the medium is for all practical purposes opaque at distances in excess of 200 m, and the more penetrating γ-rays are reduced by 1 dB for every 1.5 cm of path. In comparison with other available means, the use of sound waves for transmitting energy through seawater is vastly superior, and it suffers only when contrasted with the far more efficient transmission of radio waves and light through air.

15.4 REFRACTION. The most important phenomenon that alters the simple spherical spreading of sound in the ocean is *refraction* resulting from spatial variations in sound speed. As has been noted, the factors influencing the speed of sound in seawater are temperature, salinity, and depth. Variations in salinity are of importance near the mouths of large rivers, where appreciable amounts of freshwater run into the sea, at the edges of large ocean currents, such as the Gulf Stream, and in water close to the surface where rain, ice melt, and evaporation have maximum effects. Variations in the speed of sound with depth are quite small; for example, the change over a depth of 100 m caused by the change in hydrostatic pressure alone is only about 0.1 percent. By contrast, variations in speed resulting from changes in temperature are quite large and are subject to wild fluctuations, especially near the surface. Differences of more than 5 C° are common in the first 100 m of the ocean, and the change in speed with temperature is about 0.2 percent per Celsius degree for temperatures in the vicinity of 15°C.

Given the dependence of the water temperature and salinity on depth, the variation of c with depth can be calculated from (15.1); alternately, the speed of sound can be measured directly as a function of position. Figure 15.2 gives a representative speed of sound profile for the deep ocean. The most pervasive feature, found at all except the highest latitudes, is a distinct minimum. In the tropics, because of the heat provided by the sun, this minimum tends to lie deep. It rises toward the surface in the higher latitudes, sometimes reaching the surface in polar

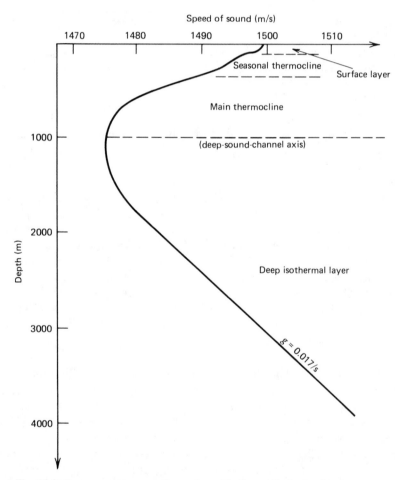

Fig. 15.2. *Representative sound-speed profile for midlatitude, deep ocean water.*

oceans. The depth at which this minimum occurs is called the *deep sound-channel axis*. Below this axis, the speed of sound increases. At great depths we find the *deep isothermal layer* where the temperature remains a constant, between -1 and $5°C$, depending on the ocean basin. In this region the sound profile becomes linear with a *positive* slope (gradient) of about $0.017 \; (m/s)/m = 0.017 \; s^{-1}$.

Above the deep sound-channel axis is the *main thermocline* (this may be absent in the Arctic). This region possesses *negative* gradients and responds slightly to seasonal changes, but is a relatively stable feature of the profile with characteristics determined primarily by latitude. Above this is the *seasonal thermocline*, also negative, which is responsive to seasonal variations. And finally, above this is the *surface layer*. This layer is quite dependent on the day-to-day, even hour-to-hour, variations in air and surface conditions. If there is sufficient surface-wave activity to mix the water near the surface, this layer becomes the *mixed layer* which is isothermal and has a positive gradient of sound speed about $0.016 \; s^{-1}$.

It is important to notice that the actual variations in c are very small compared with its magnitude. For example, the profile of Fig. 15.2 has a maximum variation of about 30 m/s, about 2 percent of the nominal value of 1500 m/s. Nevertheless, this variation has an enormous influence on the propagation of sound in the ocean.

The path of a ray through a medium in which the speed of sound varies with depth can be calculated by application of Snell's law (5.74). In the ocean, the rays of most interest are those nearly horizontal, so that it is conventional to restate Snell's law as

$$\frac{\cos \theta}{c} = \frac{1}{c_0} \tag{15.7}$$

where θ is the angle of depression made with the *horizontal* at a depth where the speed of sound is c, and c_0 is the speed at a depth (real or extrapolated) where the ray would become horizontal.

Complicated profiles such as that of Fig. 15.2 are usually simplified for analysis by separation into segments, each short enough that the gradient may be assumed constant over its length. The advantage of this is that *the path of a sound ray through a layer of water of constant sound speed gradient g is an arc of a circle whose center lies at a depth where sound speed extrapolates to zero.* To show this, consider a portion of a ray path with local radius of curvature R, as shown in Fig. 15.3. Then $\Delta z = R(\cos \theta_1 - \cos \theta_2)$. Since the gradient g for this case is

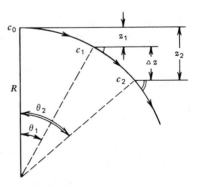

Fig. 15.3. *Diagram used in deriving relation between gradient g and the radius of curvature R of a sound ray.*

$$g = \frac{c_2 - c_1}{\Delta z} \tag{15.8}$$

combining these two equations with Snell's law (15.7) yields

$$R = -\frac{c_0}{g} = -\frac{c}{g \cos \theta} \tag{15.9}$$

The ray path is therefore a circle when g is constant because R is then constant. The center of curvature of the circle lies at the depth where $\theta = 90°$, which corresponds to $c = 0$. For the situation illustrated in Fig. 15.3, the speed gradient is negative, so that R is positive. If the speed gradient were positive, R would be negative and the path would curve upward. In everything that follows, R will stand for the *magnitude* of the radius of curvature.

Once the radius of curvature of each segment of the path is known, the actual path can be either traced graphically or computed. If the initial angle of depression of a ray is θ_1, reference to the geometry of Fig. 15.3 and use of (15.9) shows that the change in range Δr and depth Δz are

$$\Delta r = \frac{1}{g} \frac{c_1}{\cos \theta_1} (\sin \theta_0 - \sin \theta_2)$$

$$\Delta z = \frac{1}{g} \frac{c_0}{\cos \theta_0} (\cos \theta_2 - \cos \theta_0) \tag{15.10a}$$

Use of small angle approximations and elimination of θ from this pair of equations provides a convenient approximate relationship between range and depth increments along a ray for $|\theta| < 20°$

$$\Delta z = \tan \theta_1 \, \Delta r - \frac{1}{2} \frac{g}{c_1} (\Delta r)^2 \tag{15.10b}$$

15.5   MIXED LAYER.   As mentioned earlier, wave action can cause a mixing of the water in the surface layer, forming what is called a *mixed layer*. The positive sound-speed gradient in this mixed layer traps sound near the surface. Once formed, the mixed layer tends to remain until the energy input from the sun begins to heat the upper portion, decreasing the gradient. This heating effect eventually culminates in a negative gradient that leads to a downward refraction and the loss of sound from the layer. Since this usually occurs in the afternoon, the effect became known as the "afternoon effect." During the night, surface cooling and wave mixing allow the isothermal layer to be reestablished. It is rare for a positive gradient greater than about 0.016/s to occur because this requires temperature increasing with depth, a dynamically unstable condition since, for constant salinity, the density would decrease with depth.

When the mixed layer is present, the sound profile near the surface can be modeled by two linear gradients, as shown in Fig. 15.4a, where D is the depth of the layer. Figure 15.4b shows representative rays from a source in the layer at depth z_0. A ray traveling upward reflects from the water-air interface with an angle of reflection equal to the angle of incidence, whereas a ray that intercepts the lower boundary of the layer continues on a path determined by the gradient below the layer.

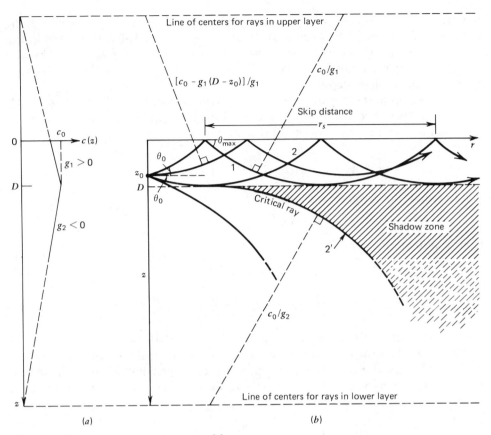

Fig. 15.4. *Sound transmission in a mixed layer.*

Notice that the path is continued smoothly with no change in θ. All rays leaving the source with angles of elevation or depression between those of the rays labeled 1 and 2 will be confined to the mixed layer. Rays 1 and 2 have the same radius of curvature and are tangent to the bottom of the layer. Ray 2' is called the *critical ray* since it demarks the inner boundary of the *shadow zone*, within which no rays are to be found. While this simple model suggests that there is no signal in the shadow zone, this is incorrect. *Scattering* from bubbles and the rough ocean surface, the presence of *internal waves* which cause D to fluctuate with horizontal distance, and the *diffraction* of sound into the shadow zone from its periphery, all contribute to provide a weak, fluctuating ensonification of the shadow zone. For frequencies in the high kilohertz region signal levels are typically at least 40 dB less than those at the edges of the zone. For low kilohertz frequencies the signal loss in the shadow zone is less severe, and for sufficiently low frequencies the shadow zone may cease to exist because of strong diffraction and the breakdown of ray theory.

All rays whose angles of elevation or depression exceed those of rays 1 and 2 penetrate to greater depths and are lost from the layer. Rays between the *limiting rays* (rays 1 and 2) which initially spread spherically will ultimately be trapped in the layer and then spread cylindrically. The range at which the change from spherical to cylindrical spreading occurs is called the *transition range* r_t. For small θ_0, this can be estimated by requiring that when $r = r_t$ the vertical extent of the beam subtended by $2\theta_0$ equals the layer depth D, $r_t \doteq D/(2\theta_0)$.

Another important parameter is the *skip distance* r_s. From Fig. 15.4b, $r_s \doteq 2R\theta_{max}$, where $R = c_0/g_1$ is the radius of the ray that grazes the bottom of the layer and $R \gg D$. For small angles, Snell's law yields

$$\frac{1}{c(D)} = \frac{1}{c(0)}(1 - \tfrac{1}{2}\theta_{max}^2)$$

and the sound speed can be approximated by

$$c(z) \doteq c(0)(1 + z/R)$$

Combination of the above equations yields

$$r_s = 2\sqrt{2RD} \tag{15.11}$$

$$r_t = \frac{1}{8}r_s\sqrt{\frac{D}{D - z_0}} \tag{15.12}$$

The nominal value of R in the mixed layer is $R = 1500/0.016 = 9.4 \times 10^4$ m.

With the help of these quantities, construction of a simple *transmission loss model* can be accomplished. For $r < r_t$ the geometric spreading is spherical, and for $r > r_t$ the spreading is cylindrical. Suitable forms that match at the transition range are

$$TL(geom) = \begin{cases} 20 \log r & r < r_t \\ 10 \log r + 10 \log r_t & r > r_t \end{cases}$$

There are various contributions to $TL(losses)$. The absorption of sound by the seawater is accounted for by ar. There are also losses from the surface duct by scattering at the rough sea surface and by leakage out of the bottom of the duct from diffraction, internal waves, and irregularities in the speed of sound profile. These contributions can be parameterized as depending on the skip distance and a "loss per bounce" b. Combination yields $TL(losses) = ar + br/r_s$. Thus the transmission loss is

$$TL = \begin{cases} 20 \log r + (a + b/r_s)r & r < r_t \qquad (15.13a) \\ 10 \log r + 10 \log r_t + (a + b/r_s)r & r > r_t \qquad (15.13b) \end{cases}$$

While all rays trapped in the duct can rise to the surface to be reflected back downward, all rays cannot reach the bottom of the channel. (For example, a ray leaving the source horizontally can never attain a depth greater than the source depth.) If a receiver is at a depth z_r greater than that of the source, it can detect only those rays reaching depths equal to or deeper than itself. This means that the transmission loss between a source and a deeper receiver must be greater than that obtained from (15.13). Recall that acoustic reciprocity requires that exchange of the source and receiver cannot alter the observed transmission loss between two points. If the source and receiver are exchanged, the source lies at the original receiver depth z_r, and the transition range is now determined from (15.12) with z_r in place of z_0. This value exceeds that calculated earlier, so the transmission loss is greater. *The larger of source and receiver depths must be used in calculating r_t.*

Since the transmission loss depends on the depth of the source (or receiver, whichever is deeper), transmission loss in a given duct is minimized by having source and receiver shallow. However, neither can lie within a few wavelengths of the surface, or interference between it and its image can become important.

Although the loss per bounce is subject to considerable variation depending on the properties of the mixed layer, very rough but representative values of b can be obtained from an equation developed empirically by Schulkin,[3]

$$b = (SS)F^{1/2} \qquad (15.14)$$

where SS is the sea state, a rating of the roughness of the sea surface (see Table 15.1), and F is the frequency in kilohertz. The range of validity of this equation is for $3 < b < 14$ dB/bounce over the frequency range 2 kHz to 25 kHz.

Because the trapping of sound in the mixed layer establishes a cylindrical wave-guide, at lower frequencies normal mode theory must be used instead of ray theory. The situation is not as simple as that studied in Sect. 9.8, but the physics is the same; there exist normal modes each with its own cutoff frequency. This means that for sufficiently low frequencies, many of the normal modes are evanescent and the ability of the mixed layer to carry energy will be reduced. As a consequence, the trapping of sound in the mixed layer does not occur for sound of sufficiently long wavelength. This cutoff frequency depends on the gradient *below* the mixed layer, but a rough approximation of the frequency below which sound will not be well-trapped is given by[1]

$$f \sim 2 \times 10^5 / D^{3/2} \qquad (15.15)$$

where the depth D is in meters and the frequency f in hertz. For frequencies of about this value or lower, the mixed layer transmission loss model developed above is increasingly suspect, and normal-mode models provide more accurate predictions.

As mentioned at the beginning of this section, rays whose angles of elevation or depression exceed θ_0 at the source are lost from the mixed layer and refract downward. These rays either *reflect* from the bottom or, because of the deep isothermal layer, *refract* back upward until they reach the surface. It often occurs that the refracted rays are *convergent* when they reach the surface, leading to enhanced signal levels at, and immediately below, the surface. This is called the *convergence zone*, and the range at which it occurs depends on the details of the sound-speed profile and varies greatly from ocean to ocean. The convergence zone usually lies between 15 and 70 km, and its width (the distance over which there is significant enhance-ment of the signal) is usually about 10 percent of the *convergence zone range*. The

---

[3] Schulkin, *J. Acous. Soc. Am.*, **44**, 1152 (1968).

reflection of rays from the bottom to the surface is also a useful propagation path in the ocean. These rays aid in spanning the interval between the outer limits of propagation in the mixed layer and any convergence zone. However, the uncertainties in the reflective properties of the bottom limit the utility of *bottom bounce* propagation paths.

15.6 DEEP SOUND CHANNEL. All rays originating near the axis of the deep sound channel and making small angles with the horizontal will return to the axis without reaching either the surface or the bottom, remaining *trapped* within the *deep sound channel* or SOFAR channel. (SOFAR is derived from "SOund Fixing And Ranging," an early application of this channel to locate by acoustic methods airmen downed at sea.) Since the absorption of low frequencies in seawater is quite small, the low-frequency components of sound from explosive charges detonated in this channel can propagate to tremendous distances; such signals have been received at distances in excess of 3000 km. The reception of these explosive signals by two or more widely separated arrays of hydrophones allowed an accurate determination of the location of the explosion by triangulation. Currently, the deep sound channel is utilized with passive sonar as an aid in monitoring activities in deep ocean areas.

The deep sound channel can be approximated by linear gradients, as indicated in Fig. 15.5. The maximum angle θ_{max} with which the most steeply inclined trapped ray crosses the channel axis found from Snell's law,

$$\frac{\cos \theta_{max}}{c_{max} - \Delta c} = \frac{1}{c_{max}}$$

where c_{max} is the greatest speed of sound found in the channel and Δc is the difference between c_{max} and the speed of sound on the axis of the channel c_{min}. For small angles

$$\theta_{max} = \sqrt{2 \, \Delta c / c_{max}} \tag{15.16}$$

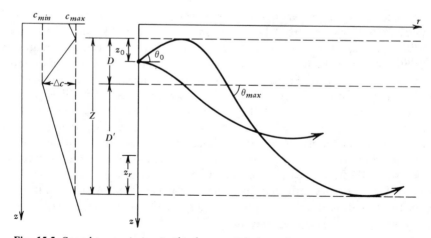

Fig. 15.5. *Sound transmission in the deep sound channel.*

For a source located above the axis and below the upper boundary, only those rays lying within some angle θ_0 will be trapped. Snell's law and the small angle approximation yield

$$\theta_0 = \theta_{max} \sqrt{z_0/D} \tag{15.17}$$

where z_0 is the depth of the source below the top of the channel. Now, analogous with the transition range in the mixed layer, the channel will be filled at some range $r'_t = Z/(2\theta_0)$, where Z is the width of the channel as indicated in Fig. 15.5. Manipulation reveals

$$r'_t = \frac{Z}{2} \sqrt{\frac{c_{max}}{2\,\Delta c}} \sqrt{\frac{D}{z_0}} \tag{15.18a}$$

As with the mixed layer, the value of r'_t must be the larger; calculated from the depth of the source z_0 and D, or the depth of the receiver z_r and D'. (If the source and receiver are both above the channel axis, then r'_t is calculated from the shallower of the two and D. If both are below the channel axis, the deeper is used with D'.)

The skip distance r'_s can be determined from Fig. 15.5 with a little geometry,

$$r'_s = 2\sqrt{2}Z \sqrt{\frac{c_{max}}{\Delta c}} \tag{15.18b}$$

For the profile of Fig. 15.2, Δc is about 30 m/s and Z about 3000 m so that the skip distance is about 60 km. Since it takes several skips for the acoustic energy to begin to be distributed over the depth of the channel, the appropriate formula for the transmission loss,

$$TL = 10 \log r + 10 \log r'_t + ar \tag{15.19}$$

will be applicable only after propagation distances of several skip distances.

The above expression is valid for very long acoustic signals. Tone bursts or explosive signals must be handled differently because of the dispersive behavior of the channel. In most oceans the speed of sound increases so rapidly with distance above and below the channel axis that those trapped rays having the greatest excursions provide the paths along which the energy travels fastest. The longest time t_{max} that can be taken for any ray to travel in the channel is thus less than or equal to r/c_{min}. Over long ranges, if the time of flight of the signal is t, the total signal of original duration τ will be *stretched* to $\tau + \Delta\tau$ where

$$\Delta\tau \sim \frac{1}{24}\left(\frac{Z}{r'_t}\right)^2 t \tag{15.20}$$

Again, the larger of the two r'_t must be used.

The energy present in the original signal will be distributed over $\tau + \Delta\tau$ at a receiver at range r, but the distribution will not be uniform. Often the arrival is relatively weak at first, with the intensity increasing with time, followed by a sudden silence. The details of these effects depend strongly on the details of the speed of sound profile. If the sound speed does not increase sufficiently rapidly above and below the axis, the sound traveling straight down the axis may arrive first.

As a rough estimate of how much time stretching can occur, for both source and receiver on the axis of the channel, the profile of Fig. 15.2 gives $\Delta\tau/t = 0.007$ or about 9 s of time stretching over a distance of 2000 km.

Those rays leaving the source with angles of elevation greater than θ_0 can ensonify the surface and constitute a *reliable acoustic path* (*RAP*) from a deep source to the surface, and vice versa. The maximum available ensonified range at the surface increases with increasing source depth. The transmission loss over this *RAP* is determined by spherical spreading with absorptive losses.

The geometry of Fig. 15.5 and the previous discussion concerning the convergence zone now reveal that the range r_{cz} to the convergence zone is r_s' if there is no mixed layer above the deep sound channel. If there is a mixed layer present, then $r_{cz} = r_s' + r_s$. (Because of the details of the ray paths from the source to the convergence zone, it turns out that r_{cz} is an estimate of the outer limit of the zone.) The transmission loss that a signal experiences in propagating from the source to the convergence zone is usually written in the form of spherical spreading, losses, and a *convergence zone gain G*,

$$TL = 20 \log r_{cz} + ar_{cz} - G \qquad \text{(convergence zone)}$$

Determination of G for a given speed of sound profile usually requires sophisticated numerical evaluation.

15.7 SURFACE REFLECTION. Whenever a nondirectional source of sound is present in the ocean, both surface- and bottom-reflected waves may arrive at a given point to combine with the direct wave. Depending on their relative phases, these waves may either reinforce each other to produce a greater pressure than that of the direct wave alone, or partially cancel each other to produce a lesser pressure. In deep water, when both source and receiver are near the surface, the bottom-reflected waves are relatively weak at short ranges. Under these circumstances the observed interference phenomena may be considered to arise from the presence of the direct and surface-reflected waves alone.

Consider a nondirectional source at a depth d. If the surface is relatively smooth compared to a wavelength, then the reflected wave acts as if it were emitted by an image located a distance d above the surface and 180° out of phase with the source. Since this is the equivalent of the acoustic doublet discussed in Sect. 8.5, the sound field *in the water* is that of an acoustic doublet of separation $2d$. For application in underwater sound it is advantageous to recast the results of Sect. 8.5, replacing the cylindrical coordinates (r, θ) with the horizontal range r between the doublet and the receiver and the depth h of the receiver. If both d and h are small compared to r,

$$P = \frac{2A}{r} \sin\left(\frac{khd}{r}\right) \qquad (15.21)$$

where A is the amplitude of the source at 1 m. It is evident that the two waves will reinforce each other when $khd/r = \pi/2, 3\pi/2, \dots$ and cancel each other when $khd/r = \pi, 2\pi, \dots$. The most important practical situation occurs when $khd/r \ll 1$. Then

$$P \doteq 2A \, khd/r^2 \qquad (15.22)$$

For large r the pressure amplitude decreases as $1/r^2$, and the transmission loss increases by 12 dB for each doubling of the distance rather than by 6 dB as for spherical divergence.

An approximate criterion for the importance of surface roughness in modifying surface interference is

$$K = \frac{4H \sin \theta_i}{\lambda} \tag{15.23}$$

where H is the average surface wave height measured from peak to trough, θ_i is the incident angle measured from the horizontal, and λ is the wavelength of the sound. If $K < 1$ the surface is smooth; if $K > 1$ the surface is rough. The surface roughness and angle of incidence determine the frequency below which there can be appreciable surface interference, and that roughness becomes unimportant as $\theta_i \to 0°$. However, for angles approaching grazing incidence, inhomogeneities near the surface and the bending of the rays because of the sound speed profile can alter or destroy the interference effects at all but short ranges.

A convenient way to characterize the effect of roughness is to assume that rays that are scattered into angles other than θ_i are removed from the image. This means that the pressure amplitude of the signal appearing to emanate from the image can be reduced by a factor $\mu < 1$, which measures the importance of the surface roughness; μ is clearly a function of angle θ_i. If in the derivation of the doublet equation the amplitude of the image is multiplied by μ, (15.22) becomes

$$P = \frac{A}{r} \sqrt{1 + \mu^2 - 2\mu \cos\left(\frac{2khd}{r}\right)} \tag{15.24}$$

It is useful to compare the transmission loss in the presence of surface roughness with that expected from spherical divergence alone. The *transmission anomaly* TA defined by

$$TL(geom) = 20 \log r - TA \tag{15.25}$$

is found to be

$$TA = -10 \log\left(1 + \mu^2 - 2\mu \cos \frac{2khd}{r}\right) \tag{15.26}$$

Sketches of TA for representative values of μ are presented in Fig. 15.6. For a more comprehensive treatment of the effects of surface roughness on acoustic waves, see Clay and Medwin.[4]

15.8 THE SONAR EQUATIONS. In all applications of underwater sound—whether detecting and localizing underwater objects, studying bathemetry, finding fish, navigating, or stalking the Loch Ness monster—the critical operation is detecting a desired acoustic signal in the presence of noise. If the level of the signal is

[4] Clay and Medwin, *Acoustical Oceanography*, Wiley (1977).

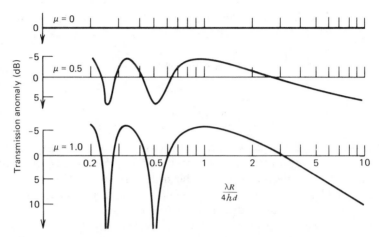

Fig. 15.6. *Transmission anomaly for reflection from a rough ocean surface.* ($\mu = 0$ *corresponds to a smooth surface.*)

the *echo level EL* and the level of the noise the *detected noise level DNL*, then the *sonar equation* is

$$EL \geq DNL + DT \tag{15.27}$$

The *detection threshold DT* (discussed in Sect. 11.4) is the value by which the echo level must exceed the detected noise level to give a 50 percent probability of detection for a specified probability of false alarm.

In signal processing for underwater sound, the *detection index d* is used in specifying detection thresholds instead of the detectability index d'. These indices are simply related,

$$d = (d')^2$$

(a) Passive Sonar. A system that listens for the noise produced by the "target" is called a *passive* sonar system. (In this case the term "echo level" is not literally appropriate, but the use is conventional.) The sound radiated by the target at a *source level SL* (Sect. 8.9b) experiences a *transmission loss TL* on its way to the receiver. The echo level is then

$$EL = SL - TL \tag{15.28}$$

With a highly directive receiver, a passive system can determine the direction from which a signal arrives. If the target can be detected simultaneously on two or more such receivers separated by a known distance, the location of the target can be determined by triangulation.

Competing with the received *signal* is *noise* from a variety of sources. The oceans are filled with noise sources such as breaking waves, snapping shrimp, surf, and shipping, which combine to produce broadband *ambient noise*. In addition, *self noise* is produced by machinery on the receiving platform and by the motion of the water around it. The combined level of these sources is the *noise level NL*. If the receiver is directional, the *detected noise level* is

$$DNL = NL - DI \tag{15.29}$$

where the *directivity index DI* (Sect. 8.9*d*) describes the ability of the receiver to discriminate against noise coming from directions other than that of the target. Combining (15.27), (15.28), and (15.29) gives the equation for passive sonar

$$\boxed{SL - TL \geq NL - DI + DT} \tag{15.30}$$

(b) Active Sonar.   For an active system, the signal is a pulse of acoustic energy that originates at the transmitter with a *source level SL*. This signal then travels to the target, accumulating a one-way transmission loss TL. At the target, a fraction of the incident signal, characterized by the *target strength TS*, is reflected toward the receiver and, suffering another transmission loss TL', arrives at the receiver. For the *monostatic* case, source and receiver are at the same location so that $TL' = TL$ and the echo level is

$$EL = SL - 2TL + TS \tag{15.31}$$

By determining the time t between the emission of a pulse and the return of the echo, the distance r to the target can be found from $r = ct/2$. If the receiver is highly directional, the bearing to the target can also be determined.

The detected noise level for an active system may be dominated by ambient or self noise. Then, combination of (15.27), (15.29), and (15.31) gives the equation for *noise-limited* (monostatic) active sonar

$$\boxed{SL - 2TL + TS \geq NL - DI + DT} \tag{15.32}$$

For active sonar there is an additional source of masking not present in a passive sonar—*reverberation*. Reverberation arises from the scattering of the emitted signal from unwanted targets, such as fish, bubbles, and the sea surface and bottom. For this case, the detected noise level is the *reverberation level RL*

$$DNL = RL \tag{15.33}$$

Combination of (15.27), (15.31), and (15.33) gives the equation for (monostatic) *reverberation-limited* active sonar

$$\boxed{SL - 2TL + TS \geq RL + DT} \tag{15.34}$$

Whether noise or reverberation dominates an active sonar system depends on the acoustic power, the range, and the speed of the target. The two possible situations are suggested in Fig. 15.7. Generally, low-power systems are *noise limited* since the maximum detection range is achieved when the echo level falls below the level at which it can be extracted from the noise (Fig. 15.7*a*). Increasing the acoustic power of a system increases both the echo level and the reverberation level at a given range, but generally reverberation decreases with increasing range less rapidly than the echo level (Fig. 15.7*b*). If, as range increases, the echo level has diminished until it is buried in the reverberation, the system is said to be *reverberation limited.*

A stratagem frequently employed to reduce the effect of reverberation is to use a *notch filter* in the receiver to eliminate the energy in a narrow frequency band, encompassing the reverberation. If the target is moving, the frequency of the echo will be different from that of the reverberation (see Sect. 15.9*c*) and the target can be more easily detected. However, the notch filter will also eliminate the echo if the target is stationary with respect to the water.

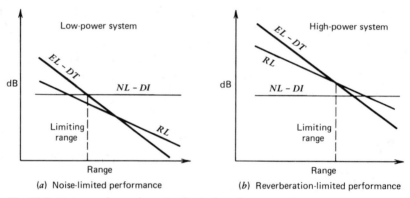

Fig. 15.7. *Noise- and reverberation-limited performance of a sonar system.*

15.9 NOISE AND BANDWIDTH CONSIDERATIONS.

From the sonar equation it is clear that sonar performance can be enhanced if the detected noise level is reduced. This can be accomplished by using knowledge of the frequency spectra of the ambient noise and of the target to select the bandwidth of the receiving system.

(a) Ambient Noise. The nominal shape of the ambient *noise spectrum level NSL* in the open ocean is sketched in Fig. 15.8. (See Sect. 11.2 for a discussion of spectrum level.) Between about 500 Hz and 20 kHz, the agitation of the local sea surface is the strongest source of ambient noise and can be characterized by specifying the local wind speed. Relations among sea state, mean wave height, and representative wind speeds are given in Table 15.1. In this range of frequencies the noise spectrum level falls at about 17 dB/decade. At lower frequencies, the major contribution to the ambient noise is from distant shipping and biological noise. The indicated limits in the figure can be exceeded considerably if the ship traffic is heavy. Below about 20 Hz, ocean turbulence and seismic noise predominate. Above 50 kHz thermal agitation of the water molecules becomes an important noise source, and the noise spectrum level increases at 6 dB/octave. In shallow water, the noise levels can be

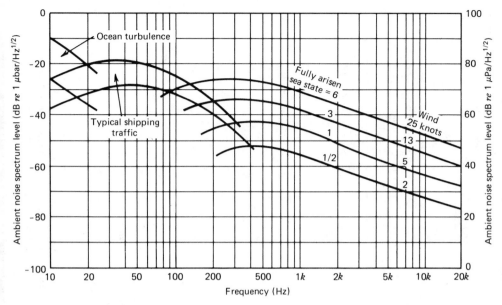

Fig. 15.8. *Deep-water ambient noise. (Adopted from Wenz, J. Acous. Soc. Am., 34, 1936 (1962) and Perrone, ibid., 46, 762 (1969).)*

considerably higher because of heavier shipping, nearby surf, higher biological noise, shore-based noises, off-shore drilling rigs, and so on.

The noise spectrum level in the figure was measured with omnidirectional receivers. If a directional receiver is used, the noise from the sea surface would be seen to arrive predominately from the vertical direction, whereas the noise from shipping arrives more horizontally. Thus, the detected noise level sensed by a directional receiver depends on its orientation.

The detected noise level for ambient noise is (15.29) and, if the bandwidth is small enough that (11.3) applies, then

$$DNL = NSL + 10 \log w - DI \tag{15.35}$$

(b) Self Noise. Self noise is noise generated by the receiving platform which interferes with the desired received signal. Self noise can reach the receiver by transmission through the mechanical structure and by transmission through the water either directly from the source or by reflection from the sea surface. Self noise usually tends to increase with the increasing speed of the platform. At low frequencies and slow speeds, machinery noise dominates, whereas at high frequencies propeller and flow noise become important. As the speed is increased, these latter sources of noise assume more importance at all frequencies. At very low speeds, self noise is usually less important than ambient noise or reverberation. At higher speeds the self noise can become the limiting factor.

Self noise is entered into the sonar equations as an *equivalent isotropic noise*

Table 15.1. Approximate relationships among the three parameters often used in characterizing the properties of the sea surface.[a]

Description	Commonly Accepted Wave Heights (ft)	Sea State	Mean Wave Heights (ft)	Average Wind Speed for Indicated Mean Wave Heights		Fully Arisen Sea	
				12-Hr Wind (knot)	Wind (knot)	Required Time (h)	Required Fetch (nautical mile)
Sea like a mirror.		0					
Ripples with the appearance of scales are formed, but without foam crests.		1/2		2	2	2	
Small wavelets, still short but more pronounced. Crests have a glassy appearance and do not break.	0–1	1	0.7	5	5	7	40
Large wavelets. Crests begin to break. Foam of glassy appearance. Perhaps scattered whitecaps.	1–2	2	2	9	9	11	100
Small waves, becoming longer; fairly frequent whitecaps.	2–4	3	3.5	14	13	14	150
Moderate waves, taking a more pronounced long form; many whitecaps are formed. (Chance of some spray.)	4–8	4	6	19	17	18	200
Large waves begin to form; the white foam crests are more extensive everywhere. (Probably some spray.)	8–13	5	9.5	24	21	23	300
Sea heaps up and white foam from breaking waves begins to be blown in streaks along the direction of the wind.	13–20	6	13.5	30	25	28	400
Moderately high waves of greater length; edges of crests begin to break into spindrift. The foam is blown in well-marked streaks along the direction of the wind.	20–30	7	18	36	29	32	500

[a] The indicated values are for deep ocean waters far from land. These values are estimated from a number of sources, primarily Wenz, *J. Acous. Soc. Am.*, **34**, 1936 (1962) and Frost, *Scientific Paper No. 25*, Great Britain Meteorological Office (1966). The sea state is estimated from a visual study of the sea surface and can therefore be accurately related to the mean wave height. The correlation between wind speed and wave height is not simple since it depends on the duration of the wind, the available fetch, and the absence of near land masses and shallow water areas. The wind speeds are measured 10 m above the sea surface. (*Fetch* is the distance necessary over deep waters that the wind must be blowing to establish the indicated mean wave height in the required time.)

spectrum level that expresses the masking level of the self noise in the bandwidth of the receiver in terms of the level of an equivalent amount of ambient noise. With this convention, the masking level for self-noise limited conditions is the same as (15.35).

(c) Doppler Shift. Anyone who has waited for a train to pass is familiar with the decrease in pitch of the train's whistle when the train rushes by. Assume a source emits a tone of frequency f when stationary. If this source and a receiver now approach each other with speeds v and u respectively (see Fig. 15.9a), the receiver will hear a frequency f' where

$$f' = f\frac{c + u}{c - v}$$

This change in frequency, the well-known *Doppler shift*, is derived in most elementary physics texts. For the case $v \ll c$ and $u \ll c$, this equation simplifies to

$$\Delta f = \frac{u + v}{c} f \tag{15.36}$$

where Δf is the change in frequency $f' - f$.

1. *Passive Sonar.* Consider two vessels traveling in different directions with different velocities, as shown in Fig. 15.9b. If vessel 1 originates a signal of frequency f_1, then by the above argument an observer stationary in the water at angle θ with respect to the motion of the source vessel will sense a signal of frequency

$$f_w = f_1\left(1 + \frac{V \cos \theta}{c}\right)$$

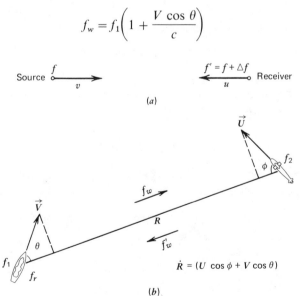

(a)

(b)

Fig. 15.9. Diagram used in deriving the relation between frequency shift and speeds, and directions of the source and receiver.

The moving vessel 2 will receive this signal and observe a frequency

$$f_2 = f_w\left(1 + \frac{U \cos \phi}{c}\right)$$

Elimination of f_w from these two equations and the assumption $U \ll c$ and $V \ll c$ yields

$$\Delta f = f_2 - f_1 = \frac{\dot{R}}{c} f_1 \tag{15.37}$$

where the *range rate*

$$\dot{R} = \frac{dR}{dt} = V \cos \theta + U \cos \phi$$

is the speed with which the two vessels are closing range.

If the source and receiver are approaching each other, $\dot{R}$ is positive and the received signal is raised in frequency ("up-Doppler"). If the source and receiver are receding from each other, the received signal is shifted down in frequency ("down-Doppler").

2. *Active Sonar.* Let the signal f_1 from vessel 1 be an active sonar pulse. Because vessel 2 is moving with respect to the water, the echo from vessel 2 will have a frequency f'_w in the water

$$f'_w = f_2\left(1 + \frac{U \cos \phi}{c}\right)$$

and vessel 1 will receive an echo having a frequency

$$f'_1 \doteq \left(1 + 2\frac{\dot{R}}{c}\right)f_1 \tag{15.38}$$

On the other hand, reverberation comes from scatterers at rest in the water, and the *reverberation frequency* f_r as observed by vessel 1 is

$$f_r \doteq f_1\left(1 + 2\frac{V \cos \theta}{c}\right) \tag{15.39}$$

Vessel 1 can compare the received frequency f'_1 of the echo either with f_1 (its sonar frequency) or with f_r (the reverberation frequency). The corresponding Doppler shifts are

$$\Delta f_1 = f'_1 - f_1 \doteq 2\frac{\dot{R}}{c} f_1 \tag{15.40}$$

$$\Delta f_r = f'_1 - f_r \doteq 2\frac{U \cos \phi}{c} f_1 \tag{15.41}$$

(d) Bandwidth Considerations. The presence of a Doppler shift Δf_1 of the echo with respect to the source places a lower limit on the bandwidth of the receiver since the bandpass filter must not ignore echos from fast-moving targets. If the maximum range rate is known, the bandwidth must be twice the associated Doppler shift (since the target may either open or close range). Thus, a receiver for an active sonar must have a total bandwidth from (15.40) of

$$w = 2.67 \times 10^{-3} \dot{R} f \quad \text{(active)} \tag{15.42a}$$

where w and f are in hertz and $\dot{R}$ is in meters per second. Comparing (15.40) with (15.37), we see that a passive system, designed to detect some specific frequency, must have a total bandwidth half of that above,

$$w = 1.33 \times 10^{-3} \dot{R} f \quad \text{(passive)} \tag{15.42b}$$

15.10 PASSIVE SONAR.

(a) The Source: Radiated Noise. The source level SL of the noise radiated by the target is obtained by extrapolating the radiated pressure from the far field to a distance 1 m from the acoustic center of the target. If the *source spectrum level* of a target is composed of a flat continuous spectrum out of which protrude tones, and if the intensity per unit bandwidth (at 1 m) for the continuous spectrum is s and the intensity of the tone (at 1 m) is I, then we must define a *source spectrum level SSL* for the continuous spectrum as the intensity level in a 1-hertz bandwidth,

$$SSL(\text{cont}) = 10 \log\left(\frac{s \cdot 1 \text{ Hz}}{I_{ref}}\right) \tag{15.43}$$

and a *spectrum level SL* for the tones (Sect. 11.3)

$$SL(\text{tone}) = 10 \log\left(\frac{I}{I_{ref}}\right) \tag{15.44}$$

If the bandwidth w of the receiver includes the tone, the total intensity received is $sw + I$, so that the overall SL is

$$SL = 10 \log \frac{sw + I}{I_{ref}} = 10 \log \frac{sw}{I_{ref}} + 10 \log\left(1 + \frac{I}{sw}\right) \tag{15.45}$$

If sw is much greater than I, the contribution from the tone is negligible. For example, if the tone were 20 dB stronger than the continuous spectrum level, the tone would contribute 3 dB to the signal level and would be completely obvious to the ear or to a filter whose bandwidth is 100 Hz or less. On the other hand, if the bandwidth of the filter were 400 Hz, the tone would add only 1 dB to the signal level received in that bandwidth.

The noise radiated by a target in some direction depends on many parameters, including the orientation of the target, its mechanical state, speed, and depth. There are some general features that seem to be present in almost all noise radiated from ships. First, the general broadband background tends to decrease at the higher frequencies at about 5 to 8 dB/octave. Thus, low-frequency signals predominate. Cavitation from the propellers adds a broadband contribution that is small at low frequencies, rises to a peak in some intermediate frequency, and falls off with increasing frequency. The region of maximum contribution shifts to lower frequencies with increasing speed or decreasing depth. Superimposed on its background are harmonic series corresponding to the machinery noise, engines, pumps, propellers, reduction gears, and other mechanical systems. Since the ship is a large acoustic source with individual noise generators at various locations, there can be considerable directionality to the radiated signal that changes with frequency and operating conditions.

The receiving system of a passive sonar can be either broadband to detect the total energy emitted by the target or narrowband so that detections are made on the tones. Thus, the passive sonar equation (15.30) can be written in two forms:

1. If the detector is broadband so that the tones do not contribute significantly to the source level, but narrow enough that (15.43) applies,

$$SSL(\text{cont}) - TL \geq NSL - DI + DT \tag{15.46}$$

2. If the bandwidth w of the receiver is small enough that $sw \ll I$, (15.44) applies,

$$SL(\text{tone}) - TL \geq NSL + 10 \log w - DI + DT \tag{15.47}$$

Detection of a tone would appear to be facilitated by having as narrow a bandwidth as possible. However, the bandwidth cannot be so narrow as to exclude a signal Doppler-shifted by the motion of the target. A narrow bandwidth and a broad frequency coverage are possible by using a set of contiguous narrowband filters and scanning the output from each. This is called *parallel processing*. If each filter has bandwidth Δw, and there are n filters, then the total bandwidth is $w = n\Delta w$.

(b) An Example.   A surfaced submarine is traveling at 4 knots. Its source spectrum level in the vicinity of 1 kHz is 120 dB re 1 μPa/Hz$^{1/2}$. The sea state is 3, and a mixed layer exists whose depth is 100 m. At what range will this submarine be detected by a 36-m deep hydrophone using broadband detection with a 100-Hz wide filter centered at 1 kHz? The directivity index is 20 dB and the detection threshold is 0 dB. *Solution*: The appropriate sonar equation is (15.46) where SSL, DI, and DT are given in the statement of the problem. The noise spectrum level at 1 kHz and sea state 3 is found from Fig. 15.8 to be $NSL = 62$ dB re 1 μPa/Hz$^{1/2}$. Substitution of these values into the sonar equation yields $TL \leq 78$ dB. Equation (15.15) shows that sound will be trapped in this 100-m mixed layer if $f > 200$ Hz. The 1 kHz signal should be well-trapped and the transmission loss is given by (15.13). For this mixed layer the skip distance and transition range (calculated with the receiver depth) are found from (15.11) and (15.12) to be 8660 m and 1350 m, respectively. From (15.14) the loss per bounce is 3. + 1 dB and from Fig. 7.5 the attenuation coefficient is 6×10^{-5} dB/m. If the target is beyond the transition range when it is detected, inserting these values into (15.13b) yields

$$10 \log r + (4.1 \times 10^{-4})r \leq 47$$

Solution by trial and error results in $r = 13.4$ km which *is* beyond the transition range 1310 m.

15.11.   ACTIVE SONAR.

(a) Target Strength.   An acoustic source sends a pulse out into the ocean which intercepts a target and illuminates it with intensity $I(r)$. The target scatters sound in all directions, some being sent in the direction of the receiver. As far as the receiver is concerned, the target has generated (by reflection) an acoustic signal that propagates from the target to the receiver. This process is suggested in Fig. 15.10.

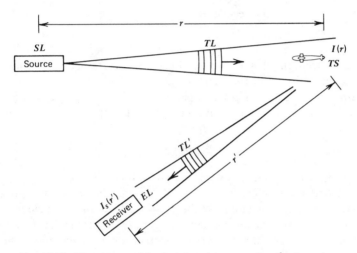

Fig. 15.10. *Diagram used in deriving the expression for target strength.*

If the reflected signal $I_s(r')$ is extrapolated from the far field to 1 m from the acoustic center of the target $(r' = 1)$, the ratio $I_s(r' = 1)/I(r)$ measures the ability of the target to reflect incident sound toward the receiver. It is conventional to write this ratio in the form

$$\frac{I_s(r' = 1)}{I(r)} = \frac{\sigma}{4\pi} \tag{15.48}$$

where σ is the *acoustic cross section* of the target.

The echo level at the receiver is

$$EL = 10 \log[I_s(r' = 1)/I_{ref}] - TL \tag{15.49}$$

or

$$EL = 10 \log\left[\frac{I(r)}{I_{ref}} \frac{\sigma}{4\pi}\right] - TL \tag{15.50}$$

Recognizing that $TL = 10 \log[I(1)/I(r)]$ from (15.2) and that $SL = 10 \log[I(1)/I_{ref}]$

from (8.39), we see that $10 \log[I(r)/I_{ref}]$ is $SL - TL$. Now, comparison with (15.31) reveals the relationship between σ and TS,

$$TS = 10 \log \frac{\sigma}{4\pi} \tag{15.51}$$

The first term on the right of (15.50) can be interpreted as an *apparent source level*

$$sL = 10 \log \frac{I(r)}{I_{ref}} + TS \tag{15.52}$$

and the echo level sent to the receiver is

$$EL = sL - TL' \tag{15.53}$$

The target strength of a reflecting object is determined primarily by its size, shape, and construction, and the frequency of the incident sound. For example, one may show that the target strength of a perfectly reflecting sphere of radius a meters, which reradiates the intercepted sound energy uniformly in all directions ($a \ll \lambda$), is given by

$$TS = 20 \log(a/2) \tag{15.54}$$

This equation indicates that a sphere of 2-m radius will have a target strength of 0 dB. Such a 0-dB target merely corresponds to one that reradiates sound with an effective source level that is equal to the pressure level of the incident sound. For larger spheres, the radiated source level is greater than the level of the incident sound. The target strength of an irregularly shaped object, such as a submarine, may be expected to depend on its orientation with respect to the incident sound. For instance, the measured target strengths of World War II fleet-type submarines are at a minimum of about 10 dB for sounds incident on the bow and stern, increasing to about 25 dB on the beam.

An incident pulse whose length in the water is short compared to the length of the target will return as a much longer echo as it reflects first from the leading edge of the target and continues to be reflected until it passes the trailing edge. This *pulse stretching* reduces the intensity of the echo because the acoustic energy in the echo is spread over a longer time interval.

Little is known about the frequency dependence of the target strength for most targets. However, high frequencies lend themselves more favorably to *classification* of a target because the short wavelengths allows some of the structure of the target to be observed in the received echo, whereas for longer wavelengths much detail in the echo is lost. Very short high-frequency pulses will show reflections from various features of the target as discrete or overlapping returns, whereas a very long pulse will establish something closer to the target strength measured under continuous wave excitation.

The wake generated by the propellers of the target can contribute to its detectability by providing a turbulent region filled with bubbles that can generate considerable scattered signal. The importance of the wake is a strong function of the speed of the target and its depth, tending to be very weak for a slow, deep submarine. A typical wake strength is between 0 and -30 dB for each meter of length illuminated.

(b) Reverberation.   When an active source ensonifies some portion of the ocean, there may be scattering from bubbles, particulate matter, fish, the sea surface or bottom, and any other inhomogeneities present. These constitute sources of unwanted signal, *reverberation,* that can compete with the echo from the target of interest.

An essential step in obtaining the *reverberation level RL* is computing the volume V (or surface A) at the range of the target from which scattered sound can

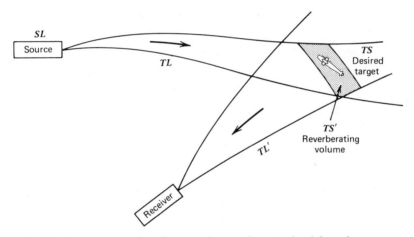

Fig. 15.11. *Diagram used in deriving the reverberation level for volume scatterers.*

arrive at the receiver *during the same time* as the echo from the desired target. Clearly, this will depend on the pulse length, the directivities of source and receiver, and the geometry. Given this volume (or surface), the reverberation level can then be calculated directly. For the moment, we will assume that V (or A) is known and obtain the desired formulas for RL.

Figure 15.11 shows the target and the surrounding reverberation volume that can scatter sound to compete with the echo at the receiver. The intensity $I(r)$ of the signal illuminating this region is related to the source level SL of the transmitter by

$$10 \log \frac{I(r)}{I_{ref}} = SL - TL \tag{15.55}$$

where TL is the transmission loss from the source to the target.

Each scatterer within the reverberating volume has an apparent source level given by

$$sL_i = 10 \log\left[\frac{I(r)}{I_{ref}} \frac{\sigma_i}{4\pi}\right]$$

where σ_i is the acoustic cross section of the ith scatterer. Assuming the individual scatterers to have random phase, the total scattered intensity is the sum of the individual intensities

$$sL = 10 \log\left[\frac{I(r)}{I_{ref}} \sum_V \frac{\sigma_i}{4\pi}\right] \tag{15.56}$$

where the summation covers all the scatterers contained in V.

The *scattering strength* S_V for a unit volume is defined by

$$S_V = 10 \log\left(\frac{1}{V} \sum_V \frac{\sigma_i}{4\pi}\right) = 10 \log \frac{s_V}{4\pi} \tag{15.57a}$$

where s_V is the *total cross section per unit volume*

$$s_V = \frac{1}{V} \sum_V \sigma_i \tag{15.57b}$$

Thus,

$$sL = 10 \log \frac{I(r)}{I_{ref}} + S_V + 10 \log V \tag{15.58}$$

Analogously, if the reverberation comes from a surface, we can define a scattering strength S_A for unit surface area. The scattered sound level from the surface is then

$$sL = 10 \log \frac{I(r)}{I_{ref}} + S_A + 10 \log A \tag{15.59}$$

Reference to (15.52) shows that the last two terms in each of (15.58) and (15.59) can be interpreted as target strengths for the reverberating region,

$$\boxed{TS_R = \begin{cases} 10 \log V + S_V \\ 10 \log A + S_A \end{cases}} \tag{15.60}$$

The reverberation level at the receiver is

$$RL = sL - TL' \tag{15.61}$$

If the source and receiver are colocated, $TL = TL'$, combination of (15.58), (15.59), (15.60), and (15.61) yields

$$\boxed{RL = SL - 2TL + TS_R} \tag{15.62}$$

and another form of (15.34) is

$$TS \geq TS_R + DT \tag{15.63}$$

This clearly reveals that reverberant interference with the desired signal is no more than the competition from undesired targets (scatterers). A striking feature of these expressions is that they are *independent* of the source strength. Thus, once the SL is large enough so that reverberation becomes more important than noise, there is no profit in increasing SL any further. Attention must be directed to reducing DI or decreasing DT to obtain any improvement in range of detection.

1. *Volume Reverberation.* Table 15.2 gives approximate values for S_V for the ocean. Notice that the *deep scattering layer*, a region of high biological activity, has a much higher S_V than ocean water in general. This layer possesses considerable structure, so its acoustic behavior is rather complicated.

Assume that source and receiver are at the same location and let the pulse duration be τ. The thickness of V must be such that all the scatterers in it contribute reverberation at the same time at the receiver. Let the source initiate the signal at time $t = 0$. Each scatterer will scatter sound for the time interval τ. Reverberation

Table 15.2. Approximate scattering strengths for deep ocean water

---

Volume Scattering

Deep Scattering Layers (1 to 20 kHz)

Scattering strength S_V ranges between -90 and -60 dB, with higher values tending to occur at higher frequencies; within the layer, there is the possibility of strong peaks at specific frequencies and depths, corresponding to distinct biological species. Reverberation from the layer shows considerable structure, including sublayers at various depths.

Water Volume (1 to 20 kHz)

S_V varies between about -100 and -70 dB, with higher values tending to occur at higher frequencies and shallower depths.

Surface Reverberation including surface layer (300 Hz to 4 kHz)

At very low grazing angles and sea states between about 1 and 4, the scattering strength S_A ranges between about -55 to -45 dB. In this range of grazing angles the bubble layer lying just beneath the sea surface assumes greatest importance, especially at the higher frequencies. For grazing angles near about $40°$, S_A increases to between -40 and -30 dB. Except near normal incidence, S_A tends to increase with frequency.

Bottom Reverberation (kHz range)

Up to about $60°$, grazing incidence scattering strength S_A is highly variable, depending on bottom and subbottom composition and roughness. S_A tends to increase with frequency, but not always. As grazing angles tend to $0°$, S_A decreases very rapidly to become negligible. For grazing angles between about 20 to $60°$, S_A lies between about -40 and -10 dB, the actual values being highly dependent on the bottom type.

Sources: Urick, *Principles of Underwater Sound*, 2nd ed., McGraw-Hill (1975). Blatzler and Vent, *J. Acous. Soc. Am.*, **41**, 154 (1967); Patterson, ibid., **46**, 756 (1969); Scrimger and Turner, ibid., **46**, 771 (1969); Urick, ibid., **48**, 392 (1970); Hall, ibid., **50**, 940 (1971); Brown and Saenger, ibid., **52**, 944 (1972).

from scatterers at range r will reach the receiver between time $t = 2r/c$ and $t + \tau$. Reverberation from scatterers at range $r + L$ will be received between times $t' = 2(r + L)/c$ and $t' + \tau$. If L is to be the thickness of V, then the received reverberation from the scatterers at range r must have just stopped when the reverberation received from those at $r + L$ just starts. This means that $t + \tau = t'$. Expressing t and t' in terms of r, L, and c and solving for L yields the result $L = c\tau/2$.

If the cross-sectional area of the reverberation volume is A_T, then its target strength is

$$TS_R = 10 \log\left(\frac{c\tau}{2} A_T\right) + S_V \tag{15.64}$$

Let the source and receiver be at the same position, have the same beam pattern, and be colinear. (A practical example would be a transmitter that also serves as the receiver.) If the directivity of the source and receiver is D, then the solid angle into which the source radiates and from which the receiver receives is approximated by $\Omega = 4\pi/D$. Since all the rays sent into Ω will pass through A_T, the geometrical transmission loss is simply the ratio of A_T to the area subtending Ω at a distance of 1 m,

$$TL(geom) = 10 \log(A_T/\Omega) \tag{15.65}$$

With this relationship, the target strength TS_R assumes the convenient form

$$TS_R = S_V + TL(geom) + 10 \log \frac{\Omega c\tau}{2} \tag{15.66}$$

A most important feature of (15.66) is that TS_R grows as the *one-way* geometrical transmission loss.

Combination of (15.62) and (15.66) yields the range dependence of the reverberation level

$$RL = SL - TL(geom) - 2TL(losses) + S_V + 10 \log \frac{\Omega c\tau}{2} \tag{15.67}$$

The reverberation level diminishes as $TL(geom) + 2TL(losses)$, whereas the echo level diminishes as $2TL(geom) + 2TL(losses)$. Thus, RL diminishes with range slower than does EL; this is the physical basis of the argument concerning Fig. 15.7.

Combination of (15.63) and (15.66) casts the sonar equation into a form showing the explicit dependence on range for reverberation-limited performance

$$\boxed{TS \geq S_V + TL(geom) + 10 \log \frac{\Omega c\tau}{2} + DT} \tag{15.68}$$

Clearly, decreasing τ and Ω allows detection of the target at larger ranges. However, recall that TS depends on pulse length: if τ is so small that the pulse length is less than the length of the target, target strength will be reduced. Decreasing Ω decreases the search rate.

2. Surface Reverberation. If reverberation arises from irregularities on or near some horizontal plane, such as the sea surface or bottom, then from (15.60) the reverberation target strength is

$$TS_R = 10 \log A + S_A \tag{15.69}$$

Since there is very little bending of the rays in the horizontal, the surface area for an acoustic beam which grazes the surface at very small angles can be seen from Fig. 15.12 to be

$$A = \frac{c\tau}{2} r\theta \tag{15.70}$$

where θ is the smaller of the source or receiver horizontal beam width. Combination yields

$$TS_R = S_A + 10 \log r + 10 \log \frac{\theta c\tau}{2} \tag{15.71}$$

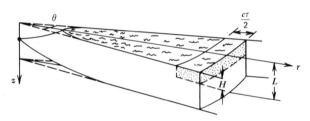

Fig. 15.12. *Diagram used in deriving the reverberation level for surface scatterers.*

The TS_R increases with range as $10 \log r$. The reverberation level for surface reverberation, obtained from combination of (15.62) and (15.71), becomes

$$RL = SL - 2TL + 10 \log r + S_A + 10 \log \frac{\theta c\tau}{2} \tag{15.72}$$

Combination of (15.63) and (15.71) yields the range form of the sonar equation for reverberation-limited performance

$$\boxed{TS \geq S_A + 10 \log r + 10 \log \frac{\theta c\tau}{2} + DT} \tag{15.73}$$

As in the case of volume reverberation, detection ability decreases with range, but for surface reverberation the geometrical transmission loss is $10 \log r$.

The *total cross section per unit area* s_A is related to the scattering strength S_A by

$$S_A = 10 \log \frac{s_A}{4\pi} \tag{15.74}$$

Reverberation from the sea surface often involves a layer beneath the surface which is heavily populated with bubbles. Rather than treating this as volume reverberation within a layer of restricted thickness, it is more convenient to lump the scattering within the layer of bubbles together with the scattering from the rough sea surface and treat the combination as occurring at the surface. The scattering strength S_A must thus include both scattering from the surface and scattering from the bubbles. Let $\sigma(a, z)$ be the acoustic cross section of a bubble of radius a and depth z, and $n(a, z)$ the number of bubbles per unit volume where a is the radius of a bubble at depth z. The total cross section per unit volume s_V integrated over z becomes the volume contribution to the surface scattering cross section,

$$s'_A = \int_0^H s_V \, dz = \int_0^H \int_0^\infty n(a, z)\sigma(a, z) \, da \, dz \qquad (15.75)$$

and

$$S_A = 10 \log \frac{s_A + s'_A}{4\pi} \qquad (15.76)$$

Scattering from the rough sea surface is important for lower frequencies and larger grazing angles, whereas at higher frequencies and very low grazing angles the layer of bubbles becomes increasingly important.

For the ocean bottom, the scattering depends on frequency, angle of incidence, roughness of the bottom, and its composition. At lower frequencies, the composition and disposition of the deeper sublayers within the bottom become increasingly important. Table 15.2 gives representative values of the scattering per unit area of ocean surface and bottom.

(c) An Example. An active sonar operates at 1 kHz with a source level of 220 dB re 1 μPa, a directivity index of 20 dB, a horizontal beam width of 10° (0.19 radian), and a pulse length of 0.1 s. Correlation detection is to be used and it is desired that the probabilities of detection and false alarm be 0.50 and 10^{-4}, respectively. Detection must be made with a single processor operating at a bandwidth wide enough not to reject a fast target. The target strength of the submarine is expected to be 30 dB and its speed may be up to 10.3 m/s (37 km/h) (20 knots). Both the source and the submarine are in a mixed layer of 100-m depth with the source at 36 m and the submarine near the surface. The sea state is 3. The scattering strength for surface scatters is -30 dB, and volume scattering is negligible. Find the maximum range of detection. *Solution:* The appropriate sonar equation is either (15.32) or (15.34), depending on whether the detection will be noise limited or reverberation limited. Both conditions must be investigated. The propagation conditions are the same as in the passive-sonar example (Sect. 15.10*b*) so, if detection is made at a range beyond the transition range,

$$TL = 10 \log r + (4.3 \times 10^{-4})r + 31.2 \qquad (15.77)$$

For a target speed of 10.3 m/s (37 km/h) the bandwidth at 1 kHz is found from (15.42a) to be 28 Hz. Figure 11.5b reveals that the detection index must be 4 to provide the required probabilities of detection and false alarm, and (11.11a) yields a detection threshold of +4 dB. Let us first assume that the detection is noise limited. The detected noise level DNL is given by (15.35), and the noise spectrum level is 62 dB re 1 $\mu Pa/Hz^{1/2}$. Thus, $DNL(noise) = 56$ dB. If the detection is reverberation limited, the detected noise level equals the reverberation level and, for surface reverberation, (15.72) gives

$$2TL = 107 + 10 \log r + (8.6 \times 10^{-4})r \qquad (15.78)$$

Combination of (15.78) and (15.77) and solution for r reveals that $r = 28.8$ km. At this range $DNL(reverberation) = 71$ dB re 1 μPa, which is much larger than $DNL(noise)$, so that the detection is reverberation limited and the above solution is the correct one.

15.12 ISOSPEED SHALLOW-WATER CHANNEL. In this and the remaining sections of the chapter, we explore a few examples using techniques that are more sophisticated than ray tracing. In many situations when the sound speed is a constant, it is possible to apply the powerful but cumbersome *method of images*. Surface interference, discussed earlier, is one example wherein the effect of the surface is replaced with an image 180° out of phase with the source. The method is valid for all frequencies; recall that *ray theory* stems from the Eikonal equation, an approximation to the wave equation valid only at higher frequencies. The solution by this method is usually expressed as an infinite or finite sum of the contributions of all the images. Profitable analysis is often accomplished only with the aid of digital computers. See, however, Tolstoy and Clay.[5]

In this section we will formulate the solutions to three propagation problems by the method images, but then make several approximations to obtain simple closed-form expressions for the transmission loss. This will reveal the underlying physics and will also provide some basis for comparison with ray-theoretic results and with normal mode analysis.

As a consequence of the multiplicity of reflections taking place both at the surface and bottom, predictions of sound transmission in shallow water are more complicated than for deep water. According to the method of images, the surface and bottom of the isospeed layer can be treated as interfaces across which sound propagates with pressure amplitude reduced by the reflection coefficient for that surface and angle of incidence. This gives a multilayered space (Fig. 15.13) wherein the various images of the source contribute to the field point along straight-line paths. The images have been labeled with the number of times the signal from the image to the field point reflects from the bottom. Simple transmission loss models can be obtained if some approximations are made.[6] The ocean surface is assumed to be a perfect reflector. If the inhomogeneities in the water and the roughness of the surface and bottom are assumed sufficient to randomize the phases of the various images, *incoherent* combination of the contributions occurs and relative phases can be ignored. In addition, each set of four

---

[5] Tolstoy and Clay, *Ocean Acoustics*, McGraw-Hill (1966).

[6] The techniques used here were developed for a more complicated situation by Macpherson and Daintith, *J. Acous. Soc. Am.*, **41**, 850 (1966).

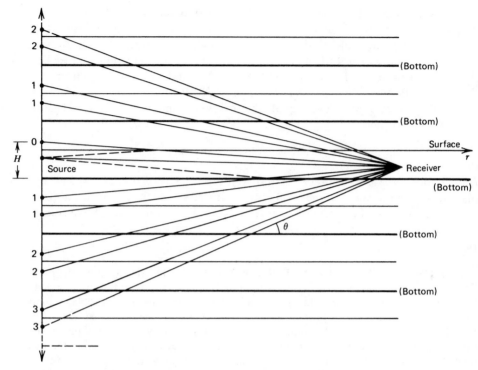

Fig. 15.13. *Source and images for an isospeed shallow-water channel. H is the layer depth.*

images whose contributions make i intersections with the bottom can be assumed to lie at an effective distance $r_i = \sqrt{r^2 + (2iH)^2}$ from the field point. This allows the images to be summed in groups of four,

$$\frac{I(r)}{I(1)} \doteq \frac{2}{r^2} + 4 \sum_{i=1}^{\infty} \frac{R^{2i}(\theta_i)}{r^2 + (2iH)^2} = \frac{2}{r^2}(1 + 2S) \tag{15.79}$$

where S is defined by

$$S = \sum_{i=1}^{\infty} \frac{R^{2i}(\theta_i)}{1 + i^2(2H/r)^2} \tag{15.80}$$

each angle of incidence θ_i is evaluated from

$$\cos \theta_i = 1/\sqrt{1 + i^2(2H/r)^2} \tag{15.81}$$

and $R(\theta_i)$ is the reflection coefficient for the bottom when the angle of incidence is θ_i.

If the range from source to receiver is much greater than the channel depth, $H/r \ll 1$, the summation can be replaced by an integral,

$$S = \int_1^{\infty} \cos^2 \theta \, R^{2u}(\theta) \, du \tag{15.82}$$

where the variable of integration u has replaced i and θ is a function of u. From (15.81) we see that $u = \frac{1}{2}(r/H)\tan\theta$, and changing the variable of integration from u to θ yields

$$S \doteq \frac{r}{2H} \int_{\tan^{-1}(2H/r)}^{\pi/2} R^{(r/H)\tan\theta}\, d\theta \qquad\qquad 15.83)$$

(a) Perfectly Rigid Bottom. If the pressure reflection coefficient at the bottom is unity for all θ, the integrand of (15.83) is unity for all θ and $r \gg H$,

$$S \doteq \frac{r}{2H} \int_{2H/r}^{\pi/2} d\theta = \frac{\pi}{4}\frac{r}{H} - 1 \qquad\qquad (15.84)$$

Substitution into (15.79) results in

$$TL(geom) \doteq 10 \log r + 10 \log \frac{H}{\pi} \qquad\qquad (15.85)$$

The divergence is cylindrical and there is a contribution to TL akin to that of r_t for the surface duct.

(b) Slow Bottom. If the bottom has a slower speed of sound than the water, the reflection coefficient is small at large grazing angles and grows to unity at grazing incidence. The major contribution to S will therefore come from the lower limit of the integration. For small incident angles the reflection coefficient can be written as $R(\theta) \sim \exp(-\gamma\theta)$ where γ is a parameter to be determined from the bottom characteristics. Since most of the value of S comes from $\theta \sim 2H/r$, replace $\tan\theta$ with θ and let the upper limit of the integration become infinitely large. Then, (15.83) becomes

$$S \doteq \frac{r}{2H} \int_{2H/r}^{\infty} e^{-(r/H)\gamma\theta^2}\, d\theta$$

A change of variable from θ to $x = \theta(\gamma r/H)^{1/2}$ yields

$$S \doteq \frac{1}{2}\sqrt{\frac{r}{\gamma H}} \int_{2\sqrt{(H\gamma/r)}}^{\infty} e^{-x^2}\, dx$$

For large r the lower limit goes to zero and $\int_0^\infty \exp(-x^2)\, dx = \sqrt{\pi}/2$. Substitution into (15.79) yields

$$\frac{I(r)}{I(1)} \doteq \frac{2}{r^2}\left(1 + \frac{1}{2}\sqrt{\frac{\pi r}{\gamma H}}\right)$$

and in the limit of large range this leads to

$$TL(geom) = 15 \log r + 5 \log \frac{\gamma H}{\pi} \qquad\qquad (15.86)$$

Thus, the effect of a slow bottom is to provide a geometrical spreading between that of spherical and cylindrical.

(c) Fast bottom. In this case the reflection coefficient is identically unity for all grazing angles less than the critical angle θ_c. All paths that reflect from the bottom with angles exceeding θ_c suffer losses and attenuate more rapidly than cylindrically. Paths more grazing than θ_c will be trapped and spread cylindrically at large ranges. Because of these considerations, (15.84) can be evaluated by integrating from the lower limit up to θ_c and setting $R(\theta) = 1$ within that range. With the restriction $r \gg 2H/\theta_c$ so that the lower limit can be replaced with zero, S becomes

$$S \doteq \frac{r}{2H} \int_0^{\theta_c} d\theta = \frac{r}{2H} \theta_c$$

and thus

$$TL(geom) = 10 \log r + 10 \log \frac{H}{2\theta_c} \qquad (15.87)$$

Notice that θ_c plays the same role that the limiting angle θ_0 plays in the mixed layer: from the ray theory approach, the angle subtended by trapped rays would be $2\theta_c$, and by the same kind of geometrical argument used in the section on the mixed layer the transition range would be $H/(2\theta_c)$.

The rather sweeping approximations applied to the acoustic fields obtained by the method of images have provided results similar to those based on ray theory. A more quantitative and exact treatment of the method of images would yield more refined results, but that is beyond our interest here.

15.13 SOLUTION OF THE POINT SOURCE IN CYLINDRICAL COORDINATES.

Normal mode theory is an *exact* method in that the normal mode solution satisfies the wave equation. The distinction between the method of images and normal mode theory is chiefly one of mathematical formalism: it can be shown that they are different, but equivalent, ways of dealing with sound propagation (see Reference 5, for example). However, the method of images is useful only if the speed of sound is a constant, whereas normal mode theory is made to order for dealing with depth-dependent profiles, particularly those leading to the trapping of sound. Normal mode theory has had its greatest utility (in underwater acoustics) in dealing with the mixed layer, SOFAR channel, arctic waters, and other ducted propagation examples. Unfortunately, normal mode solutions for profiles with nonzero gradients involve either esoteric transcendental functions or trial-and-error numerical solutions accomplished only with the aid of computers. As a result, while we will develop the theory for depth-dependent, speed-of-sound profiles, solutions will be obtained only for two simple cases: the isospeed channel with (1) a perfectly rigid bottom and (2) a fast bottom with fluidlike properties. These do not reveal the strengths of the method to best advantage, but do allow us to deal with closed-form solutions involving simple trigonometric and exponential functions of depth.

There are various approaches to normal mode theory with differing levels of rigor, complexity, and generality. The restricted and relatively simple approach presented here is applicable when the speed-of-sound profile is such that sound energy is trapped within a layer or channel and spreads cylindrically.

It is convenient to write the inhomogeneous wave equation with a point source (5.80b) in cylindrical coordinates. Therefore, express $\delta(\vec{r})$, where $\vec{r}$ is the spherical radius vector, as a product of functions in the cylindrical coordinates (r, z). Let the point source be at $r_0 = z_0 \hat{z}$, and write $\delta(r - r_0) = f_r(r)f_z(z - z_0)$. Integration over a cylindrical volume of radius a centered at $r = 0$ and of height $z_2 - z_1$ which contains z_0 gives

$$\int_{z_1}^{z_2} f_z(z - z_0)\, dz \int_0^a f_r(r)2\pi r\, dr = 1$$

Now, by inspection, appropriate choices are $f_z(z - z_0) = \delta(z - z_0)$ and $f_r(r) = \delta(r)/(2\pi r)$ so that the delta function for a point source at $\vec{r} = z_0\,\hat{z}$ is

$$\delta(\vec{r} - \vec{r}_0) = \frac{1}{2\pi r}\,\delta(r)\,\delta(z - z_0) \tag{15.88}$$

The Helmholtz equation for a point source of pressure amplitude $P(1) = 1$ is therefore

$$\left\{\frac{1}{r}\frac{\partial}{\partial r}\left(r\frac{\partial}{\partial r}\right) + \frac{\partial^2}{\partial z^2} + \left[\frac{\omega}{c(z)}\right]^2\right\}\mathbf{p} = -\frac{2}{r}\,\delta(r)\,\delta(z - z_0)e^{j\omega t} \tag{15.89}$$

The presence of the source term guarantees that

$$\mathbf{p}(r, z, t) \rightarrow \frac{\exp\{j[\omega t - k\sqrt{r^2 + (z - z_0)^2}]\}}{\sqrt{r^2 + (z - z_0)^2}} \tag{15.90}$$

in the limit $[r^2 + (z - z_0)^2]^{1/2} \rightarrow 0$. Thus a source of spherical waves at $(r, z) = (0, z_0)$ is included in the wave equation, avoiding the dynamic boundary condition that would otherwise be required.

Since it is assumed that acoustic energy is contained within a sound channel, it should be possible to represent the solution as a summation over a set of eigenfunctions,

$$\mathbf{p}(r, z, t) = e^{j\omega t} \sum_n R_n(r)Z_n(z) \tag{15.91}$$

where the $R_n(r)$ represent cylindrical spreading and the $Z_n(z)$ satisfy some Helmholtz equation, as was the case for the transverse function in waveguides of constant cross section. Let us therefore *postulate* that $Z_n(z)$ must satisfy the equation

$$\frac{d^2 Z_n}{dz^2} + \left[\frac{\omega^2}{c^2(z)} - K_n^2\right]Z_n = 0 \tag{15.92}$$

where K_n is a constant. It can be shown, subject to certain "reasonable" boundary conditions, that the $Z_n(z)$ form an *orthogonal* set of functions, *normalized* so that

$$\int Z_n(z)Z_m(z)\,dz = \delta_{nm} \tag{15.93}$$

where $\delta_{nm} = 0$ unless $n = m$ for which $\delta_{nn} = 1$. Equation (15.93) holds as long as the separation constants K_n and K_m are not equal. [There are other solutions to (15.92) that are not contained in the set $Z_n(z)$. These *continuous* eigenfunctions are characterized by eigenvalues K that are continuous rather than discrete. Since these solutions are significant only close to the source and correspond to "untrapped" energy, they will be neglected in this treatment.] Substitution of (15.91) and (15.92) into (15.89) yields

$$\sum_n \left[Z_n\frac{1}{r}\frac{d}{dr}\left(r\frac{dR_n}{dr}\right) + R_n K_n^2 Z_n\right] = -\frac{2}{r}\,\delta(r)\,\delta(z - z_0)$$

If both sides of the above equation are multiplied by $Z_m(z)$, integrated over depth, and simplified with the *orthonormality* condition (15.93), the result is

$$\frac{1}{r}\frac{d}{dr}\left(r\frac{dR_n}{dr}\right) + K_n^2 R_n = -\frac{2}{r}\,\delta(r)Z_n(z_0) \tag{15.94}$$

The solution to this inhomogeneous Helmholtz equation is $R_n(r) = -j\pi Z_n(z_0)H_0^{(2)}(K_n r)$ where $H_0^{(2)} = J_0 - jY_0$ is a Hankel function (see Appendix A4). Verification is left as a nontrivial exercise. Thus, solution to (15.89) valid for $K_n r \gg 1$ is

$$\mathbf{p}(r, z, t) \doteq -j \sum_n \sqrt{\frac{2\pi}{K_n r}} \, Z_n(z_0) Z_n(z) e^{j(\omega t - K_n r + \pi/4)} \tag{15.95}$$

where the Hankel function has been replaced by its asymptotic form. The allowed values of K_n and the form of the normalized $Z_n(z)$ are to be found from the solution of (15.92) with the desired $c(z)$ and appropriate boundary conditions. [It is important to remember that (15.95) is the solution at long ranges. It will not reduce to (15.90) at small ranges because of the existence of the untrapped modes that have been ignored.]

Each term of (15.95) corresponds to a normal mode with phase speed

$$c_{pn} = \omega/K_n \tag{15.96}$$

The values of K_n are fixed, but k is a function of space,

$$k(z) = \omega/c(z) \tag{15.97}$$

The angle of elevation or depression θ of the local direction of propagation of the waveform is found from $\cos\theta = K_n/k(z)$. Thus, each normal mode corresponds to a collection of rays traveling in the channel whose local directions of propagation at each depth z are given by the angles $\pm\theta$.

Since, from (15.90), the pressure amplitude is normalized to unit value at 1 m from the source, the transmission loss can be calculated directly as $-10 \log |p|$. The various contributions to the pressure at some field point may be summed either phase-coherently or randomly, depending on how greatly the irregularities in the medium are assumed to affect the phase of each normal mode. If coherent phasing is plausible, the geometrical transmission loss is

$$TL(geom) = -20 \log \left| \sum_n \sqrt{\frac{2\pi}{K_n r}} \, Z_n(z_0) Z_n(z) e^{-jK_n r} \right| \tag{15.98}$$

If an assumption of random phasing is more appropriate, then

$$TL(geom) = -20 \log \sqrt{\sum_n \frac{2\pi}{K_n r} \, Z_n^2(z_0) Z_n^2(z)} \tag{15.99}$$

The second of these equations is often the easier to calculate and amounts to a spatial smoothing. *Notice that TL is a function of source and receiver depth.*

These forms reveal a distinct advantage of the normal mode approach: once the set of eigenfunctions $Z_n(z)$ has been determined, the range and depth dependence of the transmission loss can be calculated directly. Ray theory, on the other hand, requires a separate ray-tracing effort for each point in the field.

Including the attenuation of sound from absorption and scattering in the normal mode theory is not an easy matter. In general, each normal mode will have its own frequency-dependent attenuation coefficient α_n. If these can be determined, the total transmission loss is found by multiplying each term in (15.98) by $\exp(-\alpha_n r)$ for coherent phasing or by multiplying each term in (15.99) by $\exp(-2\alpha_n r)$ for random phasing. Notice that the transmission loss does not separate into a geometrical term plus a loss term unless all α_n are identical, a highly unlikely event. We will ignore losses in what follows.

For all but the simplest profiles and boundary conditions (15.92) must be solved by numerical methods with the help of large computers. There are, however, some simple cases that can be solved in closed form, and the behavior of the normal modes for these cases provides considerable physical insight. The remainder of this chapter will be devoted to two such cases. For students interested in propagation, a good starting point is an article by Williams[7] and the books by Officer[8] and Tolstoy and Clay.[5]

15.14   ISOSPEED CHANNEL WITH RIGID BOTTOM.   Assume that a layer of isospeed water is bounded at $z = 0$ by a pressure release surface and is supported at $z = H$ by a perfectly rigid bottom (Fig. 15.14). The boundary conditions are that $p = 0$ on the top and $\partial p/\partial z = 0$ on the bottom. Because the variables separate, these boundary conditions can be applied to $Z_n(z)$,

$$Z_n(0) = 0 \qquad \left(\frac{dZ_n}{dz}\right)_{z=H} = 0 \quad (15.100)$$

Since the layer is isospeed, $c(z) = c_0$, and (15.92) reduced to the familiar form

$$\frac{d^2Z_n}{dz^2} + k_{zn}Z_n = 0 \qquad (15.101)$$

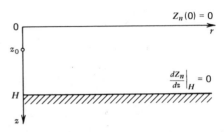

Fig. 15.14. Shallow-water channel with rigid bottom.

where $k_{zn}^2 = (\omega/c_0)^2 - K_n^2$ is not a function of space. Solution of (15.101) with the boundary conditions (15.100) yields the normalized eigenfunctions

$$Z_n(z) = \sqrt{\frac{2}{H}} \sin k_{zn} z \qquad 0 \le z \le H \qquad (15.102a)$$

with eigenvalues

$$k_{zn} = \left(n - \frac{1}{2}\right)\frac{\pi}{H} \qquad n = 1, 2, 3, \ldots \qquad (15.102b)$$

The values of K_n are obtained from the required values of k_{zn},

$$K_n = \sqrt{\left(\frac{\omega}{c_0}\right)^2 - \left[\left(n - \frac{1}{2}\right)\frac{\pi}{H}\right]^2} \qquad (15.103)$$

Since $\omega/c_0 = k$, it is seen that for values of k_{zn} exceeding ω/c_0 the associated K_n's must be imaginary. As in the case of a waveguide of constant cross section, this corresponds to modes that do not propagate, but decay rapidly with increasing range. Thus, all normal modes whose index n exceeds the greatest integer N satisfying

$$N \le \frac{H}{\pi}\frac{\omega}{c_0} + \frac{1}{2} \qquad (15.104)$$

are important only near $r = 0$; N is the greatest integer for which K_n is real. The solution can be approximated at large distances by the finite series

$$\mathbf{p}(r, z, t) \doteq -j\frac{2}{H}\sum_{n=1}^{N}\sqrt{\frac{2\pi}{K_n r}} \sin(k_{zn} z_0)\sin(k_{zn} z)e^{j(\omega t - K_n r + \pi/4)} \qquad (15.105)$$

[7] Stephen (editor), *Underwater Acoustics*, Wiley (1970).
[8] Officer, *Sound Transmission*, McGraw-Hill (1958).

The cutoff frequency ω_n for each mode occurs when $k_{zn} = k$, which is equivalent to $K_n = 0$,

$$\omega_n = \left(n - \frac{1}{2}\right)\frac{\pi}{H}c_0 \tag{15.106}$$

The phase speed ω/K_n for the nth mode is

$$c_{pn} = \frac{c_0}{\sqrt{1 - (\omega_n/\omega)^2}} \tag{15.107}$$

It is instructive to obtain a *rough* estimate of the transmission loss with the help of some rather crude assumptions. (1) Let the channel be sufficiently nonideal so that random phasing can be assumed. Then (15.99) yields

$$TL(geom) = -10\log\left(\frac{4}{H^2}\sum_{n=1}^{N}\frac{2\pi}{K_n r}\sin^2 k_{zn}z_0 \sin^2 k_{zn}z\right) \tag{15.108}$$

(2) Assume that source and receiver are well away from either the top or bottom so that the values of the $\sin^2$ terms will fluctuate between 0 and 1 pseudo-randomly as n varies from 1 to N. We can then take a statistical average and let the value of each $\sin^2$ term be $\frac{1}{2}$. This converts the above equation into the simpler form

$$TL(geom) = -10\log\left(\frac{2\pi}{H^2 r}\sum_{n=1}^{N}\frac{1}{K_n}\right) \tag{15.109}$$

If the number of allowed modes is very large, the upper limit N is given by about $(H\omega)/(\pi c_0)$, so that $K_n \doteq (\omega/c_0)\sqrt{1 - (n/N)^2}$, and the summation can be replaced by an integration,

$$\sum_{n=1}^{N}\frac{1}{\sqrt{1 - (n/N)^2}} \to N\int_0^1\frac{dx}{\sqrt{1 - x^2}} = N\int_0^{\pi/2}d\theta = \frac{H}{\pi}\frac{\omega}{c_0}\frac{\pi}{2}$$

Substitution into (15.109) gives

$$TL(geom) \doteq 10\log r + 10\log\frac{H}{\pi} \tag{15.110}$$

This compares exactly with (15.85) which was obtained on the basis of incoherent summing of the individual images. Clearly, near the top boundary the expected values of the $\sin^2$ terms will be less than $\frac{1}{2}$, so that the transmission loss will increase above (15.110). Similarly, the transmission loss will decrease if source and receiver are near the bottom where $\sin^2$ tends more toward 1.

The importance of this rough calculation is that it reveals some of the implicit assumptions and approximations underlying the approximate image result (15.85).

15.15   ISOSPEED CHANNEL WITH FLUID BOTTOM.

(a) Boundary Conditions.   It is now time to consider the requirement for "reasonable" boundary conditions mentioned earlier and to extend the formalism to include a nonrigid fluid bottom.

The functions $Z_n(z)$ are solutions to the Helmholtz equation over some interval of depth,

$$\frac{d^2 Z_n}{dz^2} + \left[\left(\frac{\omega}{c}\right)^2 - K_n^2\right]Z_n = 0 \qquad z_1 \le z \le z_2 \tag{15.111}$$

where c is a function only of z. If the Z_n are to be orthonormal (orthogonal and normalized), they must satisfy

$$\int_{z_1}^{z_2} Z_n(z)Z_m(z)\,dz = \delta_{nm} \tag{15.112}$$

Multiply (15.111) by Z_m to obtain

$$Z_m \frac{d^2 Z_n}{dz^2} + \left(\frac{\omega}{c}\right)^2 Z_m Z_n = K_n^2 Z_m Z_n$$

and then exchange the roles of n and m and take the difference between the two equations. The result is

$$\frac{d}{dz}\left(Z_m \frac{dZ_n}{dz} - Z_n \frac{dZ_m}{dz}\right) = (K_n^2 - K_m^2)Z_m Z_n$$

Integration and utilization of the *orthogonality requirement* yields an integral equation that the eigenfunctions must satisfy,

$$\int_{z_1}^{z_2} \frac{d}{dz}\left(Z_m \frac{dZ_n}{dz} - Z_n \frac{dZ_m}{dz}\right) dz = 0 \tag{15.113}$$

If each Z_n is *continuous* and has *continuous derivative*, orthogonality can be satisfied if the boundary conditions are such that either all Z's or their derivatives vanish at the boundaries.

If there is a penetrable boundary, however, then the situation becomes somewhat more involved because of the different densities of the water and the bottom. Postulate a water layer of depth H, sound speed $c_1(z)$, and uniform density ρ_1 overlying a semiinfinite bottom having sound speed $c_2(z)$ and uniform density ρ_2. Let the upper surface of the water at $z = 0$ be a smooth pressure release boundary. This is illustrated in Fig. 15.15. It is necessary to write a *pair* of Helmholtz equations,

$$\left\{\frac{d^2}{dz^2} + \left[\left(\frac{\omega}{c_1}\right)^2 - K_n^2\right]\right\}Z_{1n}(z) = 0 \qquad 0 \le z \le H$$

$$\left\{\frac{d^2}{dz^2} + \left[\left(\frac{\omega}{c_2}\right)^2 - K_n^2\right]\right\}Z_{2n}(z) = 0 \qquad H \le z \le \infty \tag{15.114}$$

Let $\mathbf{p}_1$ in the water be given by

$$\mathbf{p}_1 = -j\sum_n \sqrt{\frac{2\pi}{K_n r}}\, A_n^2 Z_{1n}(z_0)Z_{1n}(z)e^{j(\omega t - K_n r + \pi/4)} \tag{15.115}$$

and $\mathbf{p}_2$ in the bottom by

$$\mathbf{p}_2 = -j\sum_n \sqrt{\frac{2\pi}{K_n r}}\, A_n^2 Z_{1n}(z_0)Z_{2n}(z)e^{j(\omega t - K_n r + \pi/4)} \tag{15.116}$$

[Notice that $Z_{1n}(z_0)$ appears in both expressions.] The coefficients A_n have yet to be determined because Z_{1n} and Z_{2n} are *not* assumed to be normalized. Each Z_{1n} and Z_{2n} is valid *only* within the indicated intervals. The acoustic pressure and normal component of the partical velocity must be continuous across the boundary at $z = H$. The pressure must vanish at

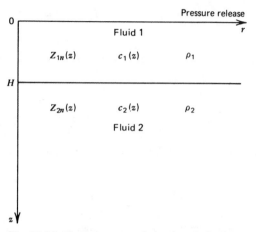

Fig. 15.15. *Shallow-water channel with fluid bottom.*

$z = 0$, be continuous across $z = H$, and vanish for $z \to \infty$. Application to (15.115) and (15.116) gives

$$Z_{1n}(0) = 0 \tag{15.117a}$$

$$Z_{1n}(H) = Z_{2n}(H) \tag{15.117b}$$

$$\frac{1}{\rho_1}\left(\frac{dZ_{1n}}{dz}\right)_H = \frac{1}{\rho_2}\left(\frac{dZ_{2n}}{dz}\right)_H \tag{15.117c}$$

$$\lim_{z \to \infty} Z_{2n}(z) = 0 \tag{15.117d}$$

From (15.114) and (15.117), Z_{1n} and Z_{2n} can be found and related to each other across the boundary at $z = H$.

Because of the discontinuity of slope (15.117c) at $z = H$, care must be taken in forming an orthonormal set of functions Z_n for the interval $0 \le z \le \infty$ from the functions Z_{1n} and Z_{2n}. It is necessary to assume that Z_n is *proportional* to Z_{1n} for $0 \le z \le H$ and to Z_{2n} for $H \le z \le \infty$, and that the proportional constants are *not necessarily the same*. Since the A_n are still undetermined, we can choose

$$Z_n(z) = \begin{cases} A_n Z_{1n}(z) & 0 \le z \le H \\ \gamma A_n Z_{2n}(z) & H \le z \le \infty \end{cases} \tag{15.118}$$

where γ is also undetermined. Now, substitution into the orthogonality condition (15.113) with $z_1 = 0$ and $z_2 = \infty$, and integration from 0 to H and H to ∞, reveals that

$$\gamma = \sqrt{\rho_1/\rho_2}$$

is required for the integral to vanish. With γ determined, $Z_n(z)$ is normalized by application of (15.112) with $n = m$. This yields the normalization constants

$$A_n^2 = \frac{1}{\displaystyle\int_0^H Z_{1n}^2(z)\, dz + (\rho_1/\rho_2)\int_H^\infty Z_{2n}^2(z)\, dz} \tag{15.119}$$

The acoustic pressure in the water layer is now found by inserting (15.119) into (15.115)

$$\mathbf{p}_1 = -j \sum_n \sqrt{\frac{2\pi}{K_n r}} \, A_n^2 Z_{1n}(z_0) Z_{1n}(z) e^{j(\omega t - K_n r + \pi/4)} \tag{15.120}$$

For the pressure in the bottom, substitute $Z_{2n}(z)$ for $Z_{1n}(z)$; recall that $Z_{1n}(z_0)$ is not replaced.

(b) Fast Bottom. We can now obtain solutions for propagation in a water layer with constant speed c_1 overlying a semiinfinite bottom with sound speed c_2, also constant. Since both speeds are constant, the solutions to the Helmholtz equations are either trigonometric or exponential functions. In order to satisfy the boundary conditions that the pressure goes to zero when $z \to \infty$, the solution Z_{2n} must decay exponentially with depth. The only way that Z_{1n} can satisfy the boundary conditions at the surface and at $z = H$ is to be a sinesoidal function of depth. Thus,

$$Z_{1n}(z) = \sin k_{zn} z \qquad 0 \le z \le H$$
$$Z_{2n}(z) = B_n e^{-\beta_n z} \qquad H \le z \le \infty \tag{15.121}$$

where

$$k_{zn}^2 = \left(\frac{\omega}{c_1}\right)^2 - K_n^2$$

$$\beta_n^2 = K_n^2 - \left(\frac{\omega}{c_2}\right)^2 \tag{15.122}$$

Both k_{zn} and β_n must be real to satisfy the requirement of waveguide propagation. This places an important restriction on K_n,

$$\omega/c_2 \le K_n \le \omega/c_1 \tag{15.123}$$

Reexpressed in terms of the phase speed $c_{pn} \doteq \omega/K_n$, this becomes

$$c_1 \le c_{pn} \le c_2 \tag{15.124}$$

(This reveals that for ducted propagation to occur, $c_1 < c_2$—a fast bottom).
Application of the boundary conditions at $z = H$ results in

$$\sin k_{zn} H = B_n e^{-\beta_n H}$$

$$\frac{1}{\rho_1} k_{zn} \cos k_{zn} H = -\frac{1}{\rho_2} \beta_n B_n e^{-\beta_n H} \tag{15.125}$$

Solution for B_n gives

$$B_n = e^{\beta_n H} \sin k_{zn} H \tag{15.126}$$

so that

$$Z_n(z) = \begin{cases} A_n \sin k_{zn} z & 0 \le z \le H \\ A_n \sqrt{\dfrac{\rho_1}{\rho_2}} (\sin k_{zn} H) e^{-\beta_n(z-H)} & H \le z \le \infty \end{cases} \tag{15.127}$$

The normalization constant A_n is obtained from substitution of (15.127) into (15.119)

$$A_n^2 = \frac{2k_{zn}}{k_{zn} H - \cos(k_{zn} H)\sin(k_{zn} H) - (\rho_1/\rho_2)^2 \sin^2(k_{zn} H)\tan(k_{zn} H)} \tag{15.128}$$

and the solution for the acoustic signal trapped in the water layer is

$$\mathbf{p}(r, z, t) = -j \sum_{n=1}^{N} \sqrt{\frac{2\pi}{K_n r}} \, A_n^2 \, \sin(k_{zn} z_0)\sin(k_{zn} z)e^{j(\omega t - K_{nr} + \pi/4)} \tag{15.129}$$

The eigenvalues k_{zn} are obtained by eliminating B_n from (15.125)

$$\tan k_{zn} H = -\frac{\rho_2}{\rho_1} \frac{k_{zn}}{\beta_n} \tag{15.130}$$

and solving this transcendental equation. With the definitions

$$y = k_{zn} H \qquad b = \frac{\rho_2}{\rho_1} \qquad a = \omega H \sqrt{\frac{1}{c_1^2} - \frac{1}{c_2^2}} \tag{15.131}$$

it assumes a simpler form,

$$\tan y = -by/\sqrt{a^2 - y^2} \tag{15.132a}$$

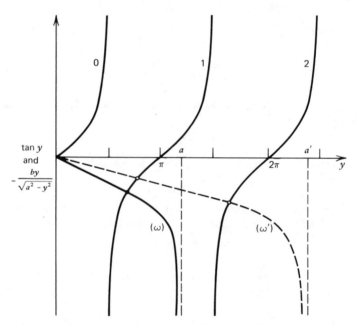

Fig. 15.16. *Graphical solution of equation (15.132a) for two values of frequency* ($\omega' > \omega$).

amenable to graphical solutions (Fig. 15.16). The values of y that solve the transcendental equation allow calculation of $k_{zn} = y/H$. With these, K_n and β_n can be found from (15.122).

Since a is a function of frequency, the tangent curves will be intersected at different points as the frequency is changed. This is suggested in Fig. 15.16 by the two curves drawn for ω and ω'. Since each curve is asymptotic to the appropriate value of $a(\omega)$, it is clear that as a increases with ω the line $y = a$ moves to the right and more normal modes can be stimulated to propagate energy. The tangent curves have been labeled 0, 1, and 2 to designate the associated normal mode.* As the input frequency increases the line $y = a$ moves to the right; the nth mode cannot support propagation until $a > (n - \frac{1}{2})\pi$. The family of cutoff frequencies is therefore found from (15.131),

$$\frac{\omega_n}{c_1} = \left(n - \frac{1}{2}\right)\frac{\pi}{H}\frac{1}{\sqrt{1 - (c_1/c_2)^2}} \tag{15.132b}$$

(Notice that in the limit $c_2/c_1 \to \infty$ this reduces to the family of cutoff frequencies predicted for the rigid bottom channel.)

Manipulation of (15.130) to extract phase speed c_p and group c_g as functions of frequency for the nth mode is nontrivial. Representative sketches of group and phase speeds as functions of frequency are given in Fig. 15.17. Some of the features in this figure and some of the properties of Z_{1n} and Z_{2n} can be understood on the basis of simple arguments.

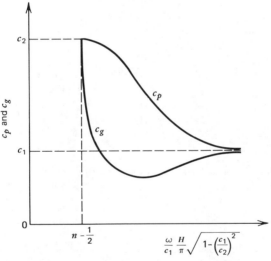

Fig. 15.17. Sketch of group and phase speeds for a normal mode propagating in an isospeed shallow-water channel with a fluid bottom.

Evaluation of K_n at the cutoff frequency ω_n yields $K_n = \omega_n/c_2$, and (14.122) shows that $\beta_n = 0$ at cutoff. Thus, the phase speed is $c_p = c_2$ and the normal mode has no depth-dependent decay. Since $a = (n - \frac{1}{2})\pi$ at cutoff, we see that $k_{zn} H = (n - \frac{1}{2})\pi$, so that there is a pressure antinode at the bottom. As frequency is increased above cutoff, K_n increases, the phase speed decreases, and β_n becomes positive real so that $Z_{2n}(z)$ decays with depth in the bottom. As the frequency becomes infinitely large, the intersection of the curves for the nth mode approaches the value $y = n\pi$ so that $k_{zn} K = n\pi$; there is a nodal plane at $z = H$ and $Z_{2n}(z) = 0$ in the bottom. In this limit we must have $K_n \to \omega/c_1$ so that the phase speed becomes asymptotic to c_1.

* The zeroth normal mode is the trivial case $\mathbf{p} = 0$ and can be ignored.

The fact that in each mode the group speed has a minimum and is high at low frequencies leads to a complicated waveform for a transient excitation (such as a small explosive charge), as suggested by Fig. 15.18. The following general features can be identified for each normal mode.

1. The *first arrival*, traveling with $c_g = c_2$, will appear at a time $t = r/c_2$ and will consist of those Fourier components of the transient having frequencies very close to the cutoff frequency of the particular mode. As time increases, signals traveling with slower group speeds can arrive, but these will still have frequencies only slightly above cutoff. This portion of the signal is called the *ground wave* and corresponds to the acoustic energy that propagates along the bottom and is reradiated back into the water layer.

2. Later, at $t = r/c_1$ the very high frequencies traveling with speed $c_g = c_1$ arrive. These will appear superimposed on the low-frequency ground wave. This portion of the signal is termed the *water wave* and corresponds to the high-frequency energy that is propagated straight down the channel.

3. For still later times the increasing frequencies in the ground wave become similar to the decreasing frequencies in the water wave and they merge into a signal traveling at the

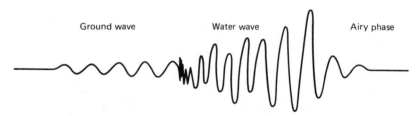

Fig. 15.18. *Signal received from a transient propagated in a shallow-water channel with a fluid bottom. (Sketch.)*

minimum group speed. This portion of the signal, known as the *Airy phase*, is complicated because of the phase interference between its two components. The signal subsequently comes to a rather abrupt termination.

By an argument exactly analogous to that at the end of Sect. 15.14, it can be shown that for *random phase* the transmission loss is well approximated by

$$TL = 10 \log r + 10 \log \frac{H}{2\sqrt{1 - (c_1/c_2)^2}} \tag{15.133}$$

Recognition of $\sin \theta_c = \sqrt{1 - (c_1/c_2)^2}$ shows this to be identical with (15.87).

PROBLEMS

Unless otherwise indicated, assume *TL(geom)* is given by spherical spreading. Useful conversions: 1 nautical mile = 6076 ft = 1852 m; 1 fathom = 1.83 m; 1 knot = 1 nm/h = 1.15 mph = 1.85 km/h = 0.514 m/s.

15.1. (*a*) From (15.1) compute the speed of sound in seawater at a depth of 4000 m and at 0° and 45° latitude if the temperature is 4°C and the salinity is 35 ppt. (*b*) Compute the sound absorption coefficient in seawater at 100 Hz, 1 kHz, and 10 kHz, for temperatures of 5 and 20°C, and depths of 0 and 2000 m from (15.6) and compare with the values obtained from (7.53).

15.2. A 30-kHz sonar transducer produces an axial sound pressure level of 140 dB *re* 1 μPa at a distance of 1000 m in seawater. (*a*) What is the axial pressure level at 1 m? (*b*) At 2000 m? (*c*) At what distance will the axial pressure level be reduced to 100 dB? (*d*) At what distance will the total transmission loss resulting from spherical divergence be equal to that caused by absorption? (*e*) At what distance is the rate of transmission loss associated with spherical divergence equal to that associated with absorption?

15.3. Show that $x = \sqrt{2c_0 d/g}$ is an approximate equation giving the horizontal distance x in which an initially horizontal ray will reach a depth d in a layer of water having a constant negative gradient of magnitude g.

15.4. The speed of sound in seawater of 35 ppt salinity decreases uniformly from a value of 1540 m/s at the surface to 1520 m/s at a depth of 50 m. (*a*) What is the gradient? (*b*) What is the average temperature gradient? (*c*) What horizontal distance is required for a horizontal ray at the surface to reach a depth of 50 m? (*d*) What will be the downward angle of such a ray on reaching this level?

15.5. An isothermal layer of seawater has a temperature of 20°C, a salinity of 35 ppt, and extends to a depth of 40 m. A sonar transducer is located at a depth of 10 m in this isothermal layer. (*a*) In what horizontal distance will a ray, leaving the transducer in a horizontal direction, reach the surface of the water? (*b*) What is the downward angle from the transducer of a ray that will become horizontal at the bottom of the isothermal layer? (*c*) In what horizontal distance will the ray of part (*b*) reach the bottom of the isothermal layer? (*d*) The speed of sound decreases below the isothermal layer at a rate of 0.2 m/s per meter. What is the depth below the surface of a ray originally starting downward at an angle of 3°, on traveling a horizontal distance of 2300 m?

15.6. A sonar transducer is at a depth of 5 m in shallow water having a flat bottom at a depth of 35 m. The speed of sound decreases regularly from a value of 1500 m/s at the surface to 1493 m/s at the bottom. (*a*) Calculate and plot the path for a ray leaving the transducer in a horizontal direction, until it strikes the bottom a second time. Assume the first reflection from the bottom to be specular. (*b*) Similarly calculate and plot the paths for rays initially 1° above and 1° below the horizontal.

15.7. A sonar transducer is at a depth of 10 m in water having a constant negative gradient of 0.2 s^{-1}. The speed of sound at the transducer depth is 1500 m/s. When the axis of the transducer is tilted down 6° from the horizontal, its beam appears to be centered on a target submarine at a horizontal distance of 1000 m. (*a*) What is the apparent depth of the submarine? (*b*) What is the true depth of the submarine?

15.8. A surface sound channel is formed by a layer of water in which the speed of sound decreases uniformly from 1500 m/s at the surface to 1498 m/s at a depth of 10 m and then increases uniformly to reach 1500 m/s again at a depth of 100 m. (*a*) What is the maximum angle with which a ray may cross the axis of the channel and still remain in the channel? (*b*) What is the horizontal distance between crossings for such a ray in the upper part of the channel? (*c*) In the lower part of the channel? (*d*) Derive a general expression giving all angles of departure θ from a sound source on the axis of the above sound channel that will result in rays recrossing the channel axis at a distance of 3000 m.

15.9. The surface layer consists of isothermal water down to 100-m depth followed by a steep negative thermocline. Calculate the minimum horizontal range at which a submarine can "escape" detection by a surface sonar, if the submarine is just below the layer.

15.10. A source 7 m below the surface is generating a 3-kHz sound in a mixed layer of

100-m depth. The receiver is at the same depth. There is a loss of 6 dB/bounce. (a) Find the transition range. (b) Find the skip distance. (c) Find the approximate range for which the transmission loss is about 80 dB. (d) Find the minimum vertical beam width of the source that takes full advantage of the propagation characteristics of the layer.

15.11. Assume a mixed layer of depth 100 m and a 3.5-kHz source at a depth of 75 m. (a) Ignoring any "leakage" from the layer, calculate the transmission loss for a receiver at a depth of 50 m and a distance of 20 km. (b) Compare this answer with the observed value of about 90 dB and estimate the leakage coefficient in decibels per kilometer. From this, calculate the loss per bounce in decibels.

15.12. The sound-speed profile consists of straight lines connecting the following points: 0 m, 1500 m/s; 800 m, 1514 m/s; 1400 m, 1470 m/s; and 5400 m (bottom), 1562 m/s. (a) What is the depth of the axis of the deep sound channel? (b) What is the maximum angle at which a ray may leave a source on the axis of the channel and still be trapped? What is the cycle distance for this ray? (c) What is the transition range? Assume the receiver is also on the axis.

15.13. A source of spherically diverging waves of 1000-Hz is located 5 m below the surface in seawater. The source produces an rms pressure amplitude of 200 μbar at 1 m from the center of the source. (a) Assuming 100 percent reflection at the surface, $\mu = 1$, what is the sound pressure level produced at a distance of 200 m at a depth of 1 m? (b) At a depth of 5 m? (c) At a depth of 10 m? (d) What would be the sound pressure level produced by the direct wave alone at a distance of 200 m? (e) Repeat the problem for $\mu = 0.5$.

15.14. (a) Assuming $\mu = 1$, derive an equation for the distance r, expressed in terms of k, h, and d, beyond which the transmission anomaly caused by the surface image effect is always greater than 10 dB. (b) What is this distance when $f = 500$ Hz, $d = 10$ m, and $h = 20$ m?

15.15. Consider a point source in water at a depth d below the surface. (a) What is the direction of the acoustic particle velocity on the surface? (b) If the pressure amplitude at 1 m from the point source alone is A, find the amplitude of the acoustic particle velocity on the surface in the limit $r \gg d$ and $kr \gg 1$.

15.16. An active sonar system has a detection threshold of -3 dB and a source level of 220 dB re 1 μPa. If the detected noise level is 70 dB re 1 μPa and the one-way transmission loss is 80 dB, what is the weakest target it can detect 50 percent of the time?

15.17. A 1-kHz sonar with a source level of 220 dB re 1 μPa receives an echo of level 110 dB re 1 μPa from a target at 1-km distance. Calculate the target strength.

15.18. A 2-kHz sonar first detects a bow aspect submarine at 5 km ($TS = 10$ dB). At what range will this same sonar first detect a bow aspect torpedo ($TS = -20$ dB)? Assume that the detection threshold is the same for both cases.

15.19. State what happens to $P(D)$, $P(FA)$, and detection index d, and explain qualitatively why it happens, in each of the following cases. (a) The target recedes and the received signal gets weaker (nothing else in the system changes). (b) With no other changes, the operator reduces the gain on the display (this is equivalent to raising the threshold).

15.20. How much must $w\tau$ change if DT is to be improved (decreased) 5 dB for (a) energy detection? (b) Correlation detection? (Assume d remains constant.)

15.21. Estimate the ambient noise level in deep water for sea state 3 and light shipping for a receiver with the following center frequencies and bandwidths: (a) 20 Hz, 0.1 Hz; (b) 200 Hz, 1 Hz; (c) 2000 Hz, 10 Hz; (d) 20 kHz, 100 Hz.

15.22. A passive receiver operates at 500 Hz with a bandwidth of 100 Hz. Its directivity index is 10 dB. Find the detected noise level in a sea state 3.

15.23. A passive detection system uses energy detection. If the bandwidth is 200 Hz and the processing time is 10 s, find the $P(FA)$ for $P(D) = 0.5$ if the detection threshold must be -10 dB.

15.24. (a) Derive an expression for determining the optimum frequency f in kilohertz to be used in reaching a given detection range r with a line hydrophone type of passive detection system. Assume the continuous-spectrum noise of the target to have the same dependence on frequency as the masking noise and that the sound absorption constant of the water is given by $a = 0.00001\, f^2$ dB/m. (b) What frequency is optimum for reaching a range of 10,000 m? (c) What range will be reached most effectively at a frequency of 1 kHz.

15.25. A submarine radiates a monofrequency signal at 250 Hz with a source level of 150 dB re 1 μPa. The signal is received by an omnidirectional receiver in the presence of a sea state 3. (a) If the submarine is capable of 14.7 m/s (28.6 knots), estimate the requisite bandwidth of the receiver. (b) Evaluate the ambient detected noise level at the receiver. (c) It is desired to have detection at 10 km. What is the maximum allowed detection threshold. (d) If the false alarm rate of 0.2 percent at 50 percent probability of detection is acceptable, find the required observation time assuming energy processing. (e) Repeat (b), (c), and (d) if the receiver consists of five parallel processors each of 1-Hz bandwidth. Assume the same overall false alarm rate.

15.26. A passive sonar system operates at 500 Hz. It must be able to detect a target generating a 500-Hz tone up to a range rate of 30 knots (15.4 m/s). (a) What must be the minimum bandwidth? (b) If the receiver contains 10 parallel filters of equal bandwidths, what is the bandwidth of each? (c) A detection is made in the highest frequency filter. What are the possible values for the range rate of the detected target?

15.27. (a) A conventional submarine at periscope depth is traveling at 4 knots (2.06 m/s). The sea state is 3 and the layer depth is 100 m. At what range will the submarine be detected by a 36-m deep hydrophone listening at 1 kHz if $DI = 20$ dB, $DT = 0$ dB, and the SSL of the submarine is 120 dB re 1 μPa/Hz$^{1/2}$ in this frequency range? (b) Repeat with $DI = 0$. (c) Repeat (a) with $f = 100$ Hz. Assume heavy shipping and $SSL = 133$ dB re 1 μPa/Hz$^{1/2}$.

15.28. A receiving platform has a self-noise spectrum level of 120 dB re 1 μPa between 200 Hz and 2 kHz with a tone with sound pressure level 140 dB re 1 μPa at 300 Hz and another of 160 dB re 1 μPa at 600 Hz. Find the band levels between (a) 200 and 299 Hz, (b) 250 and 350 Hz, (c) 1 and 2 kHz, (d) 200 Hz and 2 kHz. (e) What is the dominating contributor to the self noise in each of the above bandwidths?

15.29. A passive sonar has 90 beams. Each beam is sent through the receiver, which has 20 parallel processors, each of 0.1-Hz bandwidth, and the integration time in each processor is 10 s (the processors for each beam operate simultaneously). (a) How long does it take to get one full bearing search if the beams are formed sequentially? (b) How many frequency-bearing cells are there? (c) There must be no more than one false alarm per hour. What is $P(FA)$? (d) Assume the processing is energy detection. Use Fig. 11.5 to find the detection threshold.

15.30. A sonar is capable of echo-ranging at 20 kHz out to a maximum detection range of 3000 m on a given submarine target. (a) If the source level of the transducer is increased by 20 dB, what will be the new maximum detection range of the sonar for the same target? (b) If the transmitted frequency is lowered to 10 kHz without changing the physical dimensions of the transducer or its acoustic output from that leading to a 3000-m detection range, what will be the new maximum detection range? Assume the transducer to produce a searchlight type of beam and that the detection is masked by ambient noise.

15.31. The bearing-range recorder of an active sonar divides a circle of 15-km radius into cells 3.6° wide and 100 m long. (a) What is the maximum repetition rate of the transmitted pulse (i.e., how many pulses per second) if an echo must be able to return from 15 km before the next ping is initiated? (b) If $P(D)$ is 0.5, what is $P(FA)$ if no more than one false alarm is experienced every ping? (c) Repeat (b) for the case of no more than one false alarm each minute. (d) What are the required values of d for (b) and (c)? Assume that Fig. 11.5 applies.

15.32. (a) Derive a general equation for the echo level returned from the ocean bottom at a depth of r meters for a fathometer having a transducer of source level SL and directivity index DI. Assume the bottom to have a sound power reflection coefficient R_π and to reflect the incident sound uniformly over a hemisphere. (b) If $SL = 100$ dB re 1 μPa, $DI = 20$ dB, $f = 30$ kHz, and $R_\pi = 0.1$, what echo level will be returned from the bottom at a depth of 1000 m?

15.33. One second after a transmitter with source level 220 dB re 1 μPa emits a pulse of 3 kHz, the reverberation level is 90 dB re 1 μPa. If the volume of water contributing to the reverberation at this instant is 10^7 m^3, calculate the scattering strength.

15.34. The detected noise level for ambient noise is 80 dB re 1 μPa. The detected noise level for reverberation is given by $113 - 10 \log r - 10^{-4}r$ dB re 1 μPa. (a) Find the detected noise level for reverberation and the total detected noise level for the combination of ambient noise and reverberation for ranges of 500 m, 1 km, 2 km, 5 km, 10 km, and 20 km. (b) Within what range is reverberation more important than ambient noise?

15.35. An active sonar operates at 1 kHz with a source level of 220 dB re 1 μPa, a directivity index of 20 dB, a horizontal beam width of 10°, and a pulse length of 0.1 s. Correlation detection is used and it is desired to have $P(D) = 0.50$ with $P(FA) = 10^{-4}$. The target strength of a submarine is expected to be 30 dB and its speed may be up to 10.6 m/s. Both the sonar and the submarine are in a mixed layer of depth 100 m with the submarine shallow and the source at 36 m. The sea state is 3, the scattering strength for surface scatterers is -20 dB, and volume scattering is negligible. For simplicity, assume that the detection threshold is the same for both noise-limited and reverberation-limited performance. Find the maximum range of detection.

15.36. Assume that the near-shore ocean can be considered to be modeled by two nonparallel planes: a horizontal pressure-release surface and a sloping, rigid bottom. (a) Sketch the position and indicate the phase of the images representing the paths from a source within the layer that reflect once off the top; once off the bottom; first off the top then off the bottom; and first off the bottom and then off the top. (b) Show that all images will lie on a circle passing through the source and centered at the shore.

15.37. A 50-Hz source of spherical waves is located at a depth of 30 m in water having a hard bottom at a depth of 60 m. The acoustic pressure produced at a distance of 1 m from the center of the source is 1000 Pa. For simplicity, assume the pressure reflection coefficient at the surface is $\mu = 0.8$ and that at the bottom is $\mu = 0.5$. (a) Compute the pressures produced at a horizontal distance of 100 m from the source at a depth of 30 m, for the direct path and for the reflected paths coming from the first four images. (b) Considering the speed of sound to be 1500 m/s, compute the phase differences between the above rays and the resultant pressure amplitude of their combination.

15.38. Find the range beyond which (15.87) becomes valid assuming a 90-m channel whose bottom is like coarse silt.

15.39. For the isospeed shallow-water channel with perfectly reflecting surfaces, show that conservation of energy yields $TL(geom) = 10 \log r + 10 \log (H/2)$. Compare with (15.110) and comment.

15.40. For the isospeed channel over a fast bottom, the cutoff frequency is given by (15.132b). For the lowest mode, derive this equation from ray acoustic concepts utilizing the critical angle θ_c and the fact that $k_z = \pi/(2H)$ near cutoff of the lowest mode.

15.41. (a) Assuming the simple normal mode treatment of Sect. 15.15, evaluate the transmission loss at 20 km for an isospeed channel of depth 100 m carrying a 3.5-kHz signal. Assume the bottom is fluid with $\rho_2 = 1.25\rho_1$, $c_2 = 1.6 \times 10^3$ m/s, $c_1 = 1.5 \times 10^3$ m/s, and that only the lowest mode is important at the range of interest. Assume source and receiver are at a depth of 50 m, and comment on the results for other depths. (b) Find the distance over which the lowest two normal modes will go through a 180° phase shift with respect to each other.

APPENDIX

CONVERSION FACTORS BETWEEN MKS AND CGS

Quantity	Multiply SI	by	to obtain CGS
Length	meter (m)	10^2	centimeter (cm)
Mass	kilogram (kg)	10^3	gram (g)
Time	second (s)	1	second (s)
Force	newton (N)	10^5	dyne
Energy	joule (J)	10^7	erg
Power	watt (W)	10^7	erg/s
Volume density	kg/m^3	10^{-3}	g/cm^3
Pressure	pascal (Pa)	10	$dyne/cm^2$
Speed	m/s	10^2	cm/s
Energy density	J/m^3	10	erg/cm^3
Elastic modulus	Pa	10	$dyne/cm^2$
Coefficient of viscosity	Pa · s	10	$dyne \cdot s/cm^2$
Volume velocity	m^3/s	10^6	cm^3/s
Acoustic intensity	W/m^2	10^3	$erg/(s \cdot cm^2)$
Mechanical impedance	N · s/m	10^3	dyne · s/cm
Specific acoustic impedance	Pa · s/m	10^{-1}	$dyne \cdot s/cm^3$
Acoustic impedance	$Pa \cdot s/m^3$	10^{-5}	$dyne \cdot s/cm^5$
Mechanical stiffness	N/m	10^3	dyne/cm
Magnetic flux density	tesla (T)	10^4	gauss

OTHER USEFUL CONVERSIONS

1 in. = 2.5400 cm

1 ft = 0.3048 m

1 yd = 0.9144 m

1 fathom = 1.829 m

1 mi (statute) = 1.609 km

1 mi (nautical international) = 1 nm = 6076 ft = 1.852 km

1 mph = 1.6093 km/h = 0.4470 m/s

1 knot = 1 nm/h = 1.1508 mph = 1.852 km/h = 0.5144 m/s

1 atm = 14.70 psi = 29.92 in Hg (32°C) = 33.90 ft H_2O = 1.033 × 10^4 kgf/m^2 = 1.0133 × 10^6 bar

1 μbar = 1 × 10^6 dyne/cm = 14.50 psi = 1 × 10^5 Pa

A2

COMPLEX NUMBERS

$$\sin x = x - \frac{x^3}{3!} + \frac{x^5}{5!} - \cdots$$

$$\cos x = 1 - \frac{x^2}{2!} + \frac{x^4}{4!} - \cdots$$

$$e^x = 1 + x + \frac{x^2}{2!} + \frac{x^3}{3!} + \cdots$$

Therefore

$$\boxed{\cos \theta + j \sin \theta = e^{j\theta}} \qquad \text{Euler's equation}$$

$$\sin \theta = \frac{e^{j\theta} - e^{-j\theta}}{2j} \qquad \cos \theta = \frac{e^{j\theta} + e^{-j\theta}}{2}$$

If

$$x + jy = \sqrt{x^2 + y^2}\, e^{j \tan^{-1} (y/x)} = A e^{j\theta}$$

then

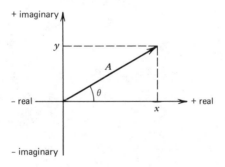

If

$$\mathbf{z} = x + jy = A e^{j\theta}$$

then

$$\text{Re}\{\mathbf{z}\} = x = A \cos \theta$$

$$\text{Im}\{\mathbf{z}\} = y = A \sin \theta$$

$$|\mathbf{z}| = \sqrt{x^2 + y^2} = A$$

$$|\mathbf{z}_1 \mathbf{z}_2| = |\mathbf{z}_1||\mathbf{z}_2|$$

A3

FUNCTIONS OF COMPLEX ARGUMENT

Let $z = x + jy$.

(a) Relations to Hyperbolic Functions.

$$\sin(jy) = j \sinh y \qquad \sinh(jy) = j \sin y$$
$$\cos(jy) = \cosh y \qquad \cosh(jy) = \cos y$$

(b) Circular Functions in Terms of Real and Imaginary Parts.

$$\sin z = \sin x \cosh y + j \cos x \sinh y$$
$$\cos z = \cos x \cosh y - j \sin x \sinh y$$
$$\tan z = \frac{\sin(2x) + j \sinh(2y)}{\cos(2x) + \cosh(2y)}$$

(c) Useful Relationships.

$$\sin^2 \theta + \cos^2 \theta = 1$$
$$\sin(\alpha + \beta) = \sin \alpha \cos \beta + \cos \alpha \sin \beta$$
$$\cos(\alpha + \beta) = \cos \alpha \cos \beta - \sin \alpha \sin \beta$$
$$\sin \alpha \cos \beta = \tfrac{1}{2}[\sin(\alpha + \beta) + \sin(\alpha - \beta)]$$
$$\sum_{n=0}^{N-1} \cos(n\alpha) = \frac{\sin\left(\dfrac{N}{2}\alpha\right) \cos\left(\dfrac{N-1}{2}\alpha\right)}{\sin \dfrac{\alpha}{2}}$$
$$\sum_{n=1}^{N-1} \sin(n\alpha) = \frac{\sin\left(\dfrac{N}{2}\alpha\right) \sin\left(\dfrac{N-1}{2}\alpha\right)}{\sin \dfrac{\alpha}{2}}$$

A4

BESSEL FUNCTIONS OF INTEGER ORDER

Solutions to the differential equation

$$\left[x^2 \frac{d^2}{dx^2} + x \frac{d}{dx} + (x^2 - n^2) \right] f(x) = 0$$

are (1) the Bessel functions of the first kind $J_n(x)$ for all x, and (2) the Bessel functions of the second kind $Y_n(x)$ (often called Neuman functions) and the Bessel functions of the third kind $H_n^{(1)}(x)$ and $H_n^{(2)}(x)$ for all x greater than zero.

(a) Relationships Between Solutions.

$$H_n^{(1)} = J_n + jY_n$$

$$H_n^{(2)} = J_n - jY_n$$

$$J_{-n} = (-1)^n J_n$$

$$Y_{-n} = (-1)^n Y_n$$

$$J_{n+1} Y_n - J_n Y_{n+1} = \frac{2}{\pi x}$$

(b) Series Expansions for J_0 and J_1.

$$J_0 = 1 - \frac{x^2}{2^2} + \frac{x^4}{2^2 \cdot 4^2} - \frac{x^6}{2^2 \cdot 4^2 \cdot 6^2} + \cdots$$

$$J_1 = \frac{x}{2} - \frac{2x^3}{2 \cdot 4^2} + \frac{3x^5}{2 \cdot 4^2 \cdot 6^2} - \cdots$$

(c) Approximations for Small Argument, $x < 1$.

$$J_0 \rightarrow 1 - \frac{x^2}{4}$$

$$J_1 \rightarrow \frac{x}{2} - \frac{x^3}{16}$$

$$Y_0 \rightarrow \frac{2}{\pi} \ln x$$

$$Y_1 \rightarrow -\frac{2}{\pi} \frac{1}{x}$$

(d) Approximations for Large Argument, $x > 2\pi$.

$$J_n \rightarrow \sqrt{\frac{2}{\pi x}} \cos\left(x - \frac{n\pi}{2} - \frac{\pi}{4}\right)$$

$$Y_n \rightarrow \sqrt{\frac{2}{\pi x}} \sin\left(x - \frac{n\pi}{2} - \frac{\pi}{4}\right)$$

$$H_n^{(1)} \rightarrow \sqrt{\frac{2}{\pi x}} \exp\left[j\left(x - \frac{n\pi}{2} - \frac{\pi}{4}\right)\right]$$

$$H_n^{(2)} \rightarrow \sqrt{\frac{2}{\pi x}} \exp\left[-j\left(x - \frac{n\pi}{2} - \frac{\pi}{4}\right)\right]$$

(e) Recursion Relations.

These relationships hold for C being any of the Bessel functions of first, second, or third kind (or linear combinations of these functions).

$$C_{n+1} + C_{n-1} = \frac{2n}{x} C_n$$

$$\frac{dC_0}{dx} = -C_1$$

$$\frac{dC_n}{dx} = \tfrac{1}{2}(C_{n-1} - C_{n+1})$$

$$\frac{d}{dx}(x^n C_n) = x^n C_{n-1}$$

$$\frac{d}{dx}\left(\frac{1}{x^n} C_n\right) = -\frac{1}{x^n} C_{n+1}$$

(f) Modified Bessel Functions.

$$I_n(x) = j^{-n} J_n(jx)$$

$$I_0(x) = J_0(jx) = 1 + \frac{x^2}{2^2} + \frac{x^4}{2^2 \cdot 4^2} + \frac{x^6}{2^2 \cdot 4^2 \cdot 6^2} + \cdots$$

$$\int x I_0(x)\, dx = x I_1(x)$$

$$\int I_1(x)\, dx = I_0(x)$$

$$I_0(x) - I_2(x) = \frac{2}{x} I_1(x)$$

A5

TABLES OF BESSEL FUNCTIONS, ZEROS, AND INFLECTION POINTS

(a) Bessel Functions of Order 0, 1, and 2.

x	$J_0(x)$	$J_1(x)$	$J_2(x)$	$I_0(x)$	$I_1(x)$	$I_2(x)$
0.0	1.0000	0.0000	0.0000	1.0000	0.0000	0.0000
0.2	0.9900	0.0995	0.0050	1.0100	0.1005	0.0050
0.4	0.9604	0.1960	0.0197	1.0404	0.2040	0.0203
0.6	0.9120	0.2867	0.0437	1.0921	0.3137	0.0464
0.8	0.8463	0.3688	0.0758	1.1665	0.4329	0.0843
1.0	0.7652	0.4401	0.1149	1.2661	0.5652	0.1358
1.2	0.6711	0.4983	0.1593	1.3937	0.7147	0.2026
1.4	0.5669	0.5419	0.2074	1.5534	0.8861	0.2876
1.6	0.4554	0.5699	0.2570	1.7500	1.0848	0.3940
1.8	0.3400	0.5815	0.3061	1.9895	1.3172	0.5260
2.0	0.2239	0.5767	0.3528	2.2796	1.5906	0.6890
2.2	0.1104	0.5560	0.3951	2.6292	1.9141	0.8891
2.4	$+0.0025$	0.5202	0.4310	3.0492	2.2981	1.1111
2.6	-0.0968	0.4708	0.4590	3.5532	2.7554	1.4338
2.8	-0.1850	0.4097	0.4777	4.1575	3.3011	1.7994
3.0	-0.2601	0.3391	0.4861	4.8808	3.9534	2.2452
3.2	-0.3202	0.2613	0.4835	5.7472	4.7343	2.7884
3.4	-0.3643	0.1792	0.4697	6.7848	5.6701	3.4495
3.6	-0.3918	0.0955	0.4448	8.0278	6.7926	4.2538
3.8	-0.4026	$+0.0128$	0.4093	9.5169	8.1405	5.2323
4.0	-0.3971	-0.0660	0.3641	11.302	9.7594	6.4224
4.2	-0.3766	-0.1386	0.3105	13.443	11.705	7.8683
4.4	-0.3423	-0.2028	0.2501	16.010	14.046	9.6259
4.6	-0.2961	-0.2566	0.1846	19.097	16.863	11.761
4.8	-0.2404	-0.2985	0.1161	22.794	20.253	14.355
5.0	-0.1776	-0.3276	$+0.0466$	27.240	24.335	17.505
5.2	-0.1103	-0.3432	-0.0217	32.584	29.254	21.332
5.4	-0.0412	-0.3453	-0.0867	39.010	35.181	25.980
5.6	$+0.0270$	-0.3343	-0.1464	46.738	42.327	31.621
5.8	0.0917	-0.3110	-0.1989	56.039	50.945	38.472

Table continued on p. 452.

(a) Bessel Functions of Order 0, 1, and 2. (continued)

x	$J_0(x)$	$J_1(x)$	$J_2(x)$	$I_0(x)$	$I_1(x)$	$I_2(x)$
6.0	0.1507	−0.2767	−0.2429	67.235	61.341	46.788
6.2	0.2017	−0.2329	−0.2769	80.717	73.888	56.882
6.4	0.2433	−0.1816	−0.3001	96.963	89.025	69.143
6.6	0.2740	−0.1250	−0.3119	116.54	107.31	84.021
6.8	0.2931	−0.0652	−0.3123	140.14	129.38	102.08
7.0	0.3001	−0.0047	−0.3014	168.59	156.04	124.01
7.2	0.2951	+0.0543	−0.2800	202.92	188.25	150.63
7.4	0.2786	0.1096	−0.2487	244.34	227.17	182.94
7.6	0.2516	0.1592	−0.2097	294.33	274.22	222.17
7.8	0.2154	0.2014	−0.1638	354.68	331.10	269.79
8.0	0.1716	0.2346	−0.1130	427.57	399.87	327.60

(b) Zeros of $J_m : J_m(j_{mn}) = 0$.

$$j_{mn}$$

m \ n	0	1	2	3	4	5
0	—	2.40	5.52	8.65	11.79	14.93
1	0	3.83	7.02	10.17	13.32	16.47
2	0	5.14	8.42	11.62	14.80	17.96
3	0	6.38	9.76	13.02	16.22	19.41
4	0	7.59	11.06	14.37	17.62	20.83
5	0	8.77	12.34	15.70	18.98	22.22

(c) Inflection points of $J_m : (dJ_m/dx)_{j'_{mn}} = 0$.

$$j'_{mn}$$

m \ n	1	2	3	4	5
0	0	3.83	7.02	10.17	13.32
1	1.84	5.33	8.54	11.71	14.86
2	3.05	6.71	9.97	13.17	16.35
3	4.20	8.02	11.35	14.59	17.79
4	5.32	9.28	12.68	15.96	19.20
5	6.41	10.52	13.99	17.31	20.58

A6

TABLES OF DIRECTIVITIES AND IMPEDANCE FUNCTIONS FOR A PISTON

x	Directivity Functions $(x = ka \sin \theta)$		Impedance Functions $(x = 2ka)$	
	Pressure	Intensity	Resistance	Reactance
	$\dfrac{2J_1(x)}{x}$	$\left[\dfrac{2J_1(x)}{x}\right]^2$	$R_1(x)$	$X_1(x)$
0.0	1.0000	1.0000	0.0000	0.0000
0.2	0.9950	0.9900	0.0050	0.0847
0.4	0.9802	0.9608	0.0198	0.1680
0.6	0.9557	0.9134	0.0443	0.2486
0.8	0.9221	0.8503	0.0779	0.3253
1.0	0.8801	0.7746	0.1199	0.3969
1.2	0.8305	0.6897	0.1695	0.4624
1.4	0.7743	0.5995	0.2257	0.5207
1.6	0.7124	0.5075	0.2876	0.5713
1.8	0.6461	0.4174	0.3539	0.6134
2.0	0.5767	0.3326	0.4233	0.6468
2.2	0.5054	0.2554	0.4946	0.6711
2.4	0.4335	0.1879	0.5665	0.6862
2.6	0.3622	0.1326	0.6378	0.6925
2.8	0.2927	0.0857	0.7073	0.6903
3.0	0.2260	0.0511	0.7740	0.6800
3.2	0.1633	0.0267	0.8367	0.6623
3.4	0.1054	0.0111	0.8946	0.6381
3.6	0.0530	0.0028	0.9470	0.6081
3.8	+0.0068	0.00005	0.9932	0.5733
4.0	−0.0330	0.0011	1.0330	0.5349
4.5	−0.1027	0.0104	1.1027	0.4293
5.0	−0.1310	0.0172	1.1310	0.3232
5.5	−0.1242	0.0154	1.1242	0.2299

Table continued on p. 454.

| x | Directivity Functions $(x = ka \sin \theta)$ | | Impedance Functions $(x = 2ka)$ | |
	Pressure	Intensity	Resistance	Reactance
	$\dfrac{2J_1(x)}{x}$	$\left[\dfrac{2J_1(x)}{x}\right]^2$	$R_1(x)$	$X_1(x)$
6.0	-0.0922	0.0085	1.0922	0.1594
6.5	-0.0473	0.0022	1.0473	0.1159
7.0	-0.0013	0.00000	1.0013	0.0989
7.5	$+0.0361$	0.0013	0.9639	0.1036
8.0	0.0587	0.0034	0.9413	0.1219
8.5	0.0643	0.0041	0.9357	0.1457
9.0	0.0545	0.0030	0.9455	0.1663
9.5	0.0339	0.0011	0.9661	0.1782
10.0	$+0.0087$	0.00008	0.9913	0.1784
10.5	-0.0150	0.0002	1.0150	0.1668
11.0	-0.0321	0.0010	1.0321	0.1464
11.5	-0.0397	0.0016	1.0397	0.1216
12.0	-0.0372	0.0014	1.0372	0.0973
12.5	-0.0265	0.0007	1.0265	0.0779
13.0	-0.0108	0.0001	1.0108	0.0662
13.5	$+0.0056$	0.00003	0.9944	0.0631
14.0	0.0191	0.0004	0.9809	0.0676
14.5	0.0267	0.0007	0.9733	0.0770
15.0	0.0273	0.0007	0.9727	0.0880
15.5	0.0216	0.0005	0.9784	0.0973
16.0	0.0113	0.0001	0.9887	0.1021

A7

VECTOR OPERATORS

(a) Cartesian Coordinates.

$$\nabla f = \hat{x}\,\frac{\partial f}{\partial x} + \hat{y}\,\frac{\partial f}{\partial y} + \hat{z}\,\frac{\partial f}{\partial z}$$

$$\nabla \cdot \vec{A} = \frac{\partial A_x}{\partial x} + \frac{\partial A_y}{\partial y} + \frac{\partial A_z}{\partial z}$$

$$\nabla^2 f = \frac{\partial^2 f}{\partial x^2} + \frac{\partial^2 f}{\partial y^2} + \frac{\partial^2 f}{\partial z^2}$$

$$\nabla \times \vec{A} = \hat{x}\left(\frac{\partial A_y}{\partial z} - \frac{\partial A_z}{\partial y}\right) + \hat{y}\left(\frac{\partial A_z}{\partial x} - \frac{\partial A_x}{\partial z}\right)$$

$$+\, \hat{z}\left(\frac{\partial A_x}{\partial y} - \frac{\partial A_y}{\partial x}\right)$$

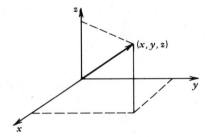

(b) Cylindrical Coordinates.

$$\nabla f = \hat{r}\,\frac{\partial f}{\partial r} + \hat{\theta}\,\frac{1}{r}\,\frac{\partial f}{\partial \theta} + \hat{z}\,\frac{\partial f}{\partial z}$$

$$\nabla \cdot \vec{A} = \frac{1}{r}\,\frac{\partial}{\partial r}\,(rA_r) + \frac{1}{r}\,\frac{\partial}{\partial \theta}\,A_\theta + \frac{\partial}{\partial z}\,A_z$$

$$\nabla^2 f = \frac{1}{r}\,\frac{\partial}{\partial r}\left(r\,\frac{\partial f}{\partial r}\right) + \frac{1}{r^2}\,\frac{\partial^2 f}{\partial \theta^2} + \frac{\partial^2 f}{\partial z^2}$$

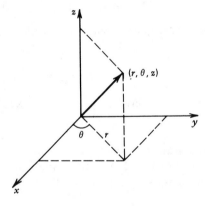

(c) Spherical Coordinates.

$$\nabla f = \hat{r}\,\frac{\partial f}{\partial r} + \hat{\theta}\,\frac{1}{r}\,\frac{\partial f}{\partial \theta} + \hat{\phi}\,\frac{1}{r\sin\theta}\,\frac{\partial f}{\partial \phi}$$

$$\nabla \cdot \vec{A} = \frac{1}{r^2}\,\frac{\partial}{\partial r}\,(r^2 A_r) + \frac{1}{r\sin\theta}\,\frac{\partial}{\partial \theta}\,(A_\theta \sin\theta)$$

$$+\, \frac{1}{r\sin\theta}\,\frac{\partial A_\phi}{\partial \phi}$$

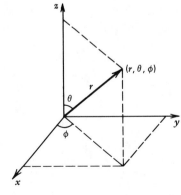

$$\nabla^2 f = \frac{1}{r^2}\,\frac{\partial}{\partial r}\left(r^2\,\frac{\partial f}{\partial r}\right) + \frac{1}{r^2 \sin\theta}\,\frac{\partial}{\partial \theta}\left(\sin\theta\,\frac{\partial f}{\partial \theta}\right) + \frac{1}{r^2 \sin^2\theta}\,\frac{\partial^2 f}{\partial \phi^2}$$

A8

GAUSS' LAW AND GREEN'S THEOREM

Gauss' law, a special case of the transport theorem, is

$$\int_V \nabla \cdot \vec{F} \, dV = \int_S \vec{F} \cdot \hat{n} \, dS$$

where $\hat{n}$ is the outward normal to the surface S of the volume V. (See any standard text on vector analysis.)

Greens' theorem

$$\int_V (U \, \nabla^2 V - V \, \nabla^2 U) \, dV = \int_S (U \, \nabla V - V \, \nabla U) \cdot \hat{n} \, dS$$

is a consequence of the vector identities

$$\nabla \cdot (U \, \nabla V) = \nabla U \cdot \nabla V + U \, \nabla^2 V$$

$$\nabla \cdot (V \, \nabla U) = \nabla V \cdot \nabla U + V \, \nabla^2 U$$

and Gauss' law. To prove this, take the difference of the above identities;

$$U \, \nabla^2 V - V \, \nabla^2 U = \nabla \cdot (U \, \nabla V - V \, \nabla U)$$

Integrate this over the volume V within the surface S and use Gauss' law with $\vec{F} = U \, \nabla V - V \, \nabla U$ to change the volume integral of the gradient into a surface integral to obtain Green's theorem.

A9

A LITTLE THERMODYNAMICS AND THE PERFECT GAS

The change in energy ΔE of a thermodynamic system can be expressed as the sum of the heat Δq added to the system and the work $-\mathscr{P}\,dV$ done on the system,

$$\boxed{\Delta E = \Delta q - \mathscr{P}\,dV} \tag{A9.1}$$

(The minus sign arises since if V decreases, so that ΔV is negative, then the system has gained energy from the work done on it.)

Now, we can define certain quantities useful when the energy of the system is changed under certain constraints. First, assume that the thermodynamic system has a mass M where M is the *molecular weight in grams*. This amount of material is defined as one *mole*. If an amount of heat Δq is added to a one-mole system whose volume is held constant (for example, a mole of gas in a sealed box of constant dimensions), then the temperature T of the system* will change by

$$\Delta q = C_V\,\Delta T \qquad (\Delta V = 0)$$

where C_V is defined as the (molal) *heat capacity at constant volume*. From (A9.1),

$$\Delta E = C_V\,\Delta T$$

or

$$C_V = \left(\frac{\partial E}{\partial T}\right)_V \tag{A9.2}$$

Analogously, introduce Δq into a one-mole system under the constraint that the pressure remain fixed. The associated temperature change is then related to Δq by

$$\Delta q = C_{\mathscr{P}}\,\Delta T \qquad (\Delta\mathscr{P} = 0)$$

where $C_{\mathscr{P}}$ is the (molal) *heat capacity at constant pressure*. Use of (A9.1) then reveals

$$\Delta E = C_{\mathscr{P}}\,\Delta T - \mathscr{P}\,\Delta V \qquad (\Delta\mathscr{P} = 0)$$

---

* For notational simplicity, we write T instead of T_K for absolute temperature in kelvin in this part of the appendix.

or

$$C_{\mathscr{P}} = \left(\frac{\partial E}{\partial T}\right)_{\mathscr{P}} + \mathscr{P}\left(\frac{\partial V}{\partial T}\right)_{\mathscr{P}} \tag{A9.3}$$

From the above, the energy can be considered a function of T and V,

$$\Delta E = \left(\frac{\partial E}{\partial T}\right)_{V} \Delta T + \left(\frac{\partial E}{\partial V}\right)_{T} \Delta V \tag{A9.4}$$

where the partial derivatives are constants to be evaluated for the equilibrium temperature and volume of the system. If some process now changes the temperature of a system, but holds the pressure constant, we can write

$$\left(\frac{\partial E}{\partial T}\right)_{\mathscr{P}} = \left(\frac{\partial E}{\partial T}\right)_{V} + \left(\frac{\partial E}{\partial V}\right)_{T}\left(\frac{\partial V}{\partial T}\right)_{\mathscr{P}} \tag{A9.5}$$

This allows us to relate C_V and $C_{\mathscr{P}}$ by combining (A9.2), (A9.3), and (A9.5),

$$C_{\mathscr{P}} - C_V = \mathscr{P}\left(\frac{\partial V}{\partial T}\right)_{\mathscr{P}} + \left(\frac{\partial E}{\partial V}\right)_{T}\left(\frac{\partial V}{\partial T}\right)_{\mathscr{P}} \tag{A9.6}$$

This rather unsatisfying equation will yield an extremely simple result when applied to a perfect gas.

A perfect gas can be considered as a collection of infinitesimally small, perfectly-rigid particles that exert forces on each other only when they collide (for example, a collection of perfectly elastic, rapidly moving billiard balls). In such a system, the energy is only kinetic. The absence of interparticle forces means that there can be no potential energy. The energy of the system is then just the sum of the energies of all the particles, and application of the *kinetic theory of gases* reveals two important results.

1. The energy of the gas is a function *only* of temperature, which has the immediate consequence

$$\left(\frac{\partial E}{\partial V}\right)_{T} = 0 \tag{A9.7a}$$

so that

$$C_{\mathscr{P}} - C_V = \mathscr{P}\left(\frac{\partial V}{\partial T}\right)_{\mathscr{P}} \tag{A9.7b}$$

for a perfect gas.

2. The pressure, volume, and temperature of a mole of perfect gas are related by

$$\boxed{\mathscr{P}V = \mathscr{R}T} \tag{A9.8}$$

where $\mathscr{R}$ is the *universal gas constant*

$$\mathscr{R} = 8.315 \text{ J/(mol} \cdot \text{K)} \tag{A9.9}$$

(If there are n moles, then $\mathscr{P}V = n\mathscr{R}T$ since for the same $\mathscr{P}$ and T the volume must vary as n.) In terms of the density ρ, we have $\rho V = M$ so that

$$\mathscr{P} = \rho r T \qquad r = \mathscr{R}/M \tag{A9.10}$$

where r is the gas constant for the particular gas in question. (If r is to be in MKS units, then M must be expressed in kilograms.)

Two important conclusions about the thermodynamic behavior of perfect gases can now be drawn.

1. Since E is a function only of T, $C_{\mathscr{P}}$ and C_V are related by (A9.7b). Use of the equation of state (A9.8) reveals

$$\left(\frac{\partial V}{\partial T}\right)_{\mathscr{P}} = \left(\frac{\partial}{\partial T}\frac{\mathscr{R}T}{\mathscr{P}}\right)_{\mathscr{P}} = \frac{\mathscr{R}}{\mathscr{P}}$$

so that

$$\boxed{C_{\mathscr{P}} - C_V = \mathscr{R}} \tag{A9.11}$$

2. For an *adiabatic process* there is no gain or loss of heat so that $\Delta q = 0$. We then have

$$\Delta E = -\mathscr{P}\,\Delta V$$

Now, (A9.4), (A9.7a), and (A9.2) yield $\Delta E = (\partial E/\partial T)_V\,\Delta T = C_V\,\Delta T$ so that $-\mathscr{P}\,\Delta V = C_V\,\Delta T$ for a perfect gas. Use of (A9.8) gives

$$-\frac{\mathscr{P}\,\Delta V}{\mathscr{P}V} = \frac{C_V\,\Delta T}{\mathscr{R}T}$$

or

$$-\mathscr{R}\frac{\Delta V}{V} = C_V\frac{\Delta T}{T}$$

Integration of both sides gives $-\mathscr{R}\ln(V/V_0) = C_V\ln(T/T_0)$ or

$$\left(\frac{V_0}{V}\right)^{\mathscr{R}} = \left(\frac{T}{T_0}\right)^{C_V}$$

Again using the perfect gas law (A9.8), this yields the adiabatic equation of state

$$\frac{\mathscr{P}}{\mathscr{P}_0} = \left(\frac{V_0}{V}\right)^{(\mathscr{R}+C_V)/C_V} = \left(\frac{\rho}{\rho_0}\right)^{1+\mathscr{R}/C_V}$$

The conventional definition

$$\boxed{\gamma = C_{\mathscr{P}}/C_V} \tag{A9.12}$$

and (A9.11) gives the familiar form

$$\boxed{\frac{\mathscr{P}}{\mathscr{P}_0} = \left(\frac{\rho}{\rho_0}\right)^{\gamma}} \tag{A9.13}$$

The ratio γ is commonly referred to as the ratio of *specific heats*.

A10

TABLES OF PHYSICAL PROPERTIES OF MATTER

(a) Solids

Solid	Density (kg/m³) ρ_0	Young's Modulus (Pa) Y ×10^10	Shear Modulus (Pa) $\mathscr{G}$ ×10^10	Adiabatic Bulk Modulus (Pa) $\mathscr{B}$ ×10^10	Poisson's Ratio σ	Speed (m/s) c Bar	Speed (m/s) c Bulk	Characteristic Impedance (Pa·s/m) $\rho_0 c$ Bar ×10^6	Characteristic Impedance (Pa·s/m) $\rho_0 c$ Bulk ×10^6
Aluminum	2700	7.1	2.4	7.5	0.33	5150	6300	13.9	17.0
Brass	8500	10.4	3.8	13.6	0.37	3500	4700	29.8	40.0
Copper	8900	12.2	4.4	16.0	0.35	3700	5000	33.0	44.5
Iron (cast)	7700	10.5	4.4	8.6	0.28	3700	4350	28.5	33.5
Lead	11300	1.65	0.55	4.2	0.44	1200	2050	13.6	23.2
Nickel	8800	21.0	8.0	19.0	0.31	4900	5850	43.0	51.5
Silver	10500	7.8	2.8	10.5	0.37	2700	3700	28.4	39.0
Steel	7700	19.5	8.3	17.0	0.28	5050	6100	39.0	47.0
Glass (Pyrex)	2300	6.2	2.5	3.9	0.24	5200	5600	12.0	12.9
Quartz (X-cut)	2650	7.9	3.9	3.3	0.33	5450	5750	14.5	15.3
Lucite	1200	0.4	0.14	0.65	0.4	1800	2650	2.15	3.2
Concrete	2600	—	—	—	—	—	3100	—	8.0
Ice	920	—	—	—	—	—	3200	—	2.95
Cork	240	—	—	—	—	—	500	—	0.12
Oak	720	—	—	—	—	—	4000	—	2.9
Pine	450	—	—	—	—	—	3500	—	1.57
Rubber (hard)	1100	0.23	0.1	0.5	0.4	1450	2400	1.6	2.64
Rubber (soft)	950	0.0005	—	0.1	0.5	70	1050	0.065	1.0
Rubber (rho-c)	1000	—	—	0.24	—	—	1550	—	1.55

(b) Liquids

Liquid	Temperature (°C) T	Density (kg/m³) ρ_0	Bulk Modulus (Pa) $\mathscr{B}$ $\times 10^9$	Ratio of Specific Heats γ	Speed (m/s) c	Characteristic Impedance (Pa · s/m) $\rho_0 c$ $\times 10^6$	Coefficient of Shear Viscosity (Pa · s) η
Water (fresh)	20	998	2.18	1.004	1481	1.48	0.001
Water (sea)	13	1026	2.28	1.01	1500	1.54	0.001
Alcohol (ethyl)	20	790	—	—	1150	0.91	0.0012
Caster (oil)	20	950	—	—	1540	1.45	0.96
Mercury	20	13600	25.3	1.13	1450	19.7	0.0016
Turpentine	20	870	1.07	1.27	1250	1.11	0.0015
Glycerin	20	1260	—	—	1980	2.5	1.2
Fluid-like sea bottoms							
Red clay		1340	—	—	1460	1.96	—
Calcareous ooze		1570	—	—	1470	2.31	—
Coarse silt		1790	—	—	1540	2.76	—
Quartz sand		2070	—	—	1730	3.58	—

(c) Gases

Gas[a]	Temperature (°C) T	Density (kg/m³) ρ_0	Ratio of Specific Heats γ	Speed (m/s) c	Characteristic Impedance (Pa · s/m) $\rho_0 c$	Coefficient of Shear Viscosity (Pa · s) η
Air	0	1.293	1.402	331.6	428	0.000017
Air	20	1.21	1.402	343	415	0.0000181
Oxygen	0	1.43	1.40	317.2	453	0.00002
CO_2 (low freq.)	0	1.98	1.304	258	512	0.0000145
CO_2 (high freq.)	0	1.98	1.40	268.6	532	0.0000145
Hydrogen	0	0.09	1.41	1269.5	114	0.0000088
Steam	100	0.6	1.324	404.8	242	0.00013

[a] At a pressure of 1.013×10^5 Pa.

GLOSSARY OF SYMBOLS

The following list will help the reader identify the symbols that are not necessarily identified every time they appear in the text. After the symbol is identified, its definition may be found by reference to the Index.

A	sound absorption
a	acceleration; Sabine absorptivity; absorption coefficient in decibels per meter
a_E	energy absorption coefficient
B	magnetic field intensity; susceptance
b	loss per bounce
$b(\theta, \phi)$	beam pattern
$\mathscr{B}$	adiabatic bulk modulus
$\mathscr{B}_T$	isothermal bulk modulus
C	electrical capacitance; acoustic compliance
C_P	heat capacity at constant pressure
C_V	heat capacity at constant volume
c_P	phase speed
c_g	group speed
c	thermodynamic speed of sound
$CNEL$	community noise equivalent level
CNR	community noise rating
D	directivity
$\mathscr{D}$	diffraction factor
d'	detectability index
DI	directivity index
DNL	detected noise level
DT	detection threshold
E	total energy
E_p	potential energy
E_k	kinetic energy
$\mathscr{E}_i$	instantaneous energy density

464

$\mathscr{E}$	time-averaged energy density
EL	echo level
F	peak force amplitude
F_e	effective force amplitude
f	instantaneous force; frequency (Hz)
G	conductance
g	spectral density of a transient function; speed of sound gradient
$\mathscr{G}$	adiabatic shear modulus
$H(\theta, \phi)$	directional factor
I_i	instantaneous acoustic intensity
I	time-averaged acoustic intensity; current in Chapter 14
I_{ref}	reference intensity
$\mathscr{I}$	time-averaged spectral density of intensity
IIC	impact isolation class
IL	intensity level
ISL	intensity spectrum level
ITC	impact transmission class
k	wave number
$\vec{k}$	propagation vector
L	inductance
L_A	A-weighted sound level
L_{Ax}	statistical A-weighted sound level
L_C	C-weighted sound level
L_{EPN}	effective perceived noise level
L_I	intensity level re 10^{-12} W/m^2
L_N	loudness level
L_{TPN}	tone-corrected perceived noise level
M	acoustic inertance
$\mathscr{ML}$	microphone sensitivity level
m	mass; flare constant
m_r	radiation mass
N	loudness
NC	noise criteria
NEF	noise exposure forecast
NL	noise level

NR	noise reduction
NSL	noise spectrum level
P	peak pressure amplitude
P_e	effective pressure amplitude
P_{ref}	reference pressure
p	acoustic pressure
$\mathscr{P}$	hydrostatic pressure
PNC	preferred noise criteria
PR	privacy rating
$PSIL$	preferred speech interference level
PSL	pressure spectrum level
PTS	permanent threshold shift
Q	quality factor; source strength
R	electrical resistance; reflection coefficient; radius of curvature
R_m	mechanical resistance
R_r	radiation resistance
$\dot{R}$	range rate
R_I	intensity reflection coefficient
R_π	power reflection coefficient
$\mathscr{R}$	universal gas constant
r	gas constant; characteristic acoustic impedance
r_t	transition range
r_s	skip distance
RL	reverberation level
ROC	receiver operating characteristic
S	surface area; salinity
S_V	scattering strength (volume)
s_V	scattering cross section per unit volume
S_A	scattering strength (surface)
s_A	scattering cross section per unit area
$\mathscr{SL}$	transmitter sensitivity level
s	spring constant; condensation
$SENEL$	single event noise exposure level
SIL	speech interference level

SL	source level
SPL	sound pressure level
SSL	source spectrum level
STC	sound transmission class
SWR	standing wave ratio
T	period of motion; temperature in degrees Celsius; tension; transmission coefficient; reverberation time
T_I	intensity transmission coefficient
T_π	power transmission coefficient
T_K	temperature in kelvin
T_{em}, T_{me}	transduction coefficients
$\mathscr{T}$	membrane tension
TL	transmission loss
TS	target strength
TS_R	target strength for reverberation
TTS	temporary threshold shift
U	peak speed amplitude; volume velocity
U_e	effective speed amplitude
$\tilde{u}$	particle velocity
u	particle speed
V	volume
VL	voltage level; voice level
X	electrical reactance
X_m	mechanical reactance
X_r	radiation reactance
Y	admittance; Young's modulus
Z	acoustic impedance
Z_m	mechanical impedance
Z_r	radiation impedance
z	specific acoustic impedance

Greek Letter Symbols

α	spatial absorption coefficient
β	temperal absorption coefficient
γ	ratio of heat capacities
η	coefficient of shear viscosity; efficiency
η_B	coefficient of bulk viscosity

θ	angle of incidence; grazing angle; horizontal beam width
κ	thermal conductivity
λ	wavelength
ξ	particle displacement
Π	time average power
Π_i	instantaneous power
ρ	acoustic density; probability density
ρ_0	volume density (kg/m^3)
ρ_L	linear density (kg/m)
ρ_S	surface density (kg/m^2)
σ	extinction cross section; standard deviation
σ_s	scattering cross section
σ_a	absorption cross section
τ	relaxation time; pulse duration; processing time
τ_E	time constant
Φ	velocity potential
ϕ	transformation factor
ω	angular frequency
ω_0	natural angular frequency
ω_d	damped angular frequency
Ω	solid angle

ANSWERS TO ODD-NUMBERED PROBLEMS

1.1. (a) $(1/2\pi)\sqrt{2s/m}$; (b) $(1/2\pi)\sqrt{s/(2m)}$; (c) $(1/2\pi)\sqrt{s/(2m)}$; (d) $(1/2\pi)\sqrt{2s/m}$.

1.3. Proof.

1.5. (a) $AB\cos(2\omega t + \theta + \phi)$; (b) $(A/B)\cos(\theta - \phi)$; (c) $AB\cos(\omega t + \theta)\cos(\omega t + \phi)$; (d) $2\omega t + \theta + \phi$; (e) $\theta - \phi$.

1.7. $R_m = 1.0$ kg/s, $\omega_d = 9.85$ rad/s, $A = 0.0402$ m, $\phi = -5.8°$.

1.9. Proof.

1.11. (a) $x = F[(1/s) - 1/(m\omega^2)]\sin \omega t$; (b) proof; (c) plot.

1.13. Proof.   **1.15.** Proof.

1.17. (a) $-(\omega F/Z_m)\sin(\omega t - \phi)$; (b) $\omega_0/\sqrt{1 - [R_m^2/(2\omega_0^2 m^2)]}$

1.19. Proof.

1.21. (a) $g(w) = F\delta(w - \omega)$; (b) $\mathbf{u} = (F/Z)\exp(j\omega t)$.

1.23. $\omega^2 mL/Z$ where $Z^2 = R_m^2 + [\omega(M + m) - s/\omega]^2$.

2.1. (a) $\partial^2 y/\partial t^2 = [T/\rho_L(x)]\partial^2 y/\partial x^2$; (b) $\partial^2 y/\partial t^2 = g(\partial/\partial x)(x\partial y/\partial x)$.

2.3. Sketch.

2.5. $\rho_L c - j\rho_L c \cot(kL)$.

2.7. (a) $-j2\rho_L c \cot(kL/2)$; (b) Proof; (c) $(F/2kT)\sin(kL/4)/\cos(kL/2)$.

2.9. (a) $\frac{1}{2}nc/L$; (b) $\frac{1}{2}nc/L$; (c) $\frac{1}{2}(n - \frac{1}{2})c/L$; (d) no.

2.11. $0.79h, 0.198h, 0, 0.049h$.   **2.13.** Proof.

2.15. $k_n L = 2.03, 4.91, 7.98, \ldots$.

3.1. (a) Proof; (b) 2525 Hz; (c) proof; (d) $A_1 = 2.1 \times 10^{-4}$ m, $A_3 = -2.3 \times 10^{-5}$ m, $A_5 = 8.3 \times 10^{-6}$ m.

3.3. (a) 6.8 kHz; (b) 0.185 m; (c) 1.91; (d) 15.9 kHz.

3.5. $0.35M$.

3.7. (a) $A = [F/(YSk \sin kL)]\cos k(L - x)$; (b) $\mathbf{Z} = j\rho_0 cS \tan kL$; (c) $\mathbf{Z} = \rho_0 cS$; (d) plot.

3.9. (a) 179 Hz; (b) 0.033 m.

4.1. (a) $0.406A$; (b) $0.5 = \sin(\pi x/a)\sin(\pi z/a)$; (c) no.

4.3. (a) $\sqrt{\mathcal{T}/\rho_s}\sqrt{5}(4L)$; (b) $f_{nm} = \sqrt{\mathcal{T}/\rho_s}\sqrt{(2n + 1)^2 + 4m^2}/(4L)$, $n = 0, 1, 2, \ldots$, $m = 1, 2, 3, \ldots$, $y_{nm} = A_{nm}\sin[\pi(n + \frac{1}{2})x/L]\sin(m\pi z/L)\exp(j\omega t)$; (c) sketch.

4.5. (a) 11.1 kHz; (b) 5.55×10^{-3} m, 1.28×10^{-2} m; (c) 6.24×10^{-6} m.

4.7. (a) 5.42 kHz; (b) 1.36×10^{-2} cm^3.

4.9. (a) 10.4 kHz; (b) 13.8 kHz.   **4.11.** (a) 153 Hz; (b) 194 Hz.

4.13. Plot.   **4.15.** 68.5 percent.

4.17. (a) 1230 Hz; (b) double the frequency; (c) quarter the frequency.

4.19. (a) -0.0025; (b) $y_2 = A_2 \cos(\omega_2 t + \phi_2)J_0(6.3r/a) - 0.0025 I_0(6.3r/a)$; (c) plot; (d) 0.38.

5.1. (a) $\mathscr{B} = \mathscr{P}_0 \gamma$; (b) $\mathscr{B} \propto T_K$.

5.3. (a) Proof; (b) $\rho[\partial \hat{u}/\partial t + (\hat{u} \cdot \nabla)\hat{u}] = -\nabla \mathscr{P}$; (c) $c = \infty$.

5.5. (a) no, yes; (b) $c = \sqrt{r T_K}$; (c) 344 m/s, 291 m/s.

5.7. (a) $P/(\rho_0 c^2)$; (b) $|s|$.

5.9. (a) $(P/c^2)\exp[j(\omega t - kx)]$; (b) $(P/\rho_0 c)\exp[j(\omega t - kx)]$; (c) $j(P/\rho_0 \omega)\exp[j(\omega t - kx)]$; (d) $(P^2/\rho_0 c^2)\cos^2(\omega t - kx)$; (e) $P^2/(2\rho_0 c)$.

5.11. (a) $-j(\rho_0 \omega/r)\cos kr \exp(j\omega t)$; (b) $-(k/r)[\sin kr + (1/kr)\cos kr]\exp(j\omega t)$; (c) $j\rho_0 c \cos kr/[\sin kr + (1/kr)\cos kr]$; (d) $-\rho_0 c(k/r)^2[\sin kr + (1/kr) \cos kr] \cos kr \sin \omega t, \cos \omega t$; (e) 0.

5.13. (a) 4.8×10^{-3} W/m^2, 96.8 dB re 10^{-12} W/m^2; (b) 7.7×10^{-6} m; (c) 4.82×10^{-3} m/s; (d) 1.41 Pa; (e) 97 dB re 20 μPa.

5.15. (a) Proof; (b) 6.75 kW/m^2; (c) 59.7.

5.17. (a) Proof; (b) 430 Pa · s/m, 378 Pa · s/m; (c) $+13.8$ percent; (d) $+0.56$ dB, 0 dB.

5.19. (a) Proof; (b) c_0/g, yes; (c) 1.48 km.

5.21. Proof.

6.1. (a) 0.0281 Pa; (b) 1.69 mW/m^2, 1.9 μW/m^2; (c) 29.5 dB; (d) 66.5 Pa, 1.5 mW/m^2, 0.5 dB; (e) 0.109.

6.3. (a) 3.62×10^3, 1.10×10^{-3}; (b) 2, 1.10×10^{-3}; (c) -65 dB, -30 dB, $+6$ dB, -30 dB.

6.5. (a) 1 dB; (b) 0.2; (c) 0.2 dB, 0.05.

6.7. Sketch.   **6.9.** (a) 13.5°; (b) 0.84.

6.11. (a) 46 Pa; (b) 146 Pa; (c) 0.212; (d) 47.7°.

6.13. (a) 73°; (b) 0.30; (c) 0.53.

7.1. (a) 3.2×10^{-10} s; (b) 500 MHz; (c) Plot.

7.3. 1.6×10^{-10} s.

7.5. (a) 3.18×10^{-5} s.
(b) and (c)

Frequency (kHz)	α_M (Np/m)	α (Np/m)
1	0.00123	0.00124
2	0.00441	0.00446
5	0.0160	0.0163
7	0.0212	0.0219
10	0.0256	0.0270

(d) Insignificant; within computational uncertainties.

7.7. $C_{\mathscr{P}} = 35.1$ J/mol°C, $C_V = 26.8$ J/mol°C, $C_e = 20.3$ J/mol°C, $C_i = 6.5$ J/mol°C.

7.9. (a) 4.0×10^7 m; (b) 1.0×10^5 m; (c) 1.7×10^5 m, 4.5×10^3 m; (d) 4.1×10^3 m, 130 m; (e) 2900 m, 204 m.

8.1. (a) 0.628 W; (b) 5 W/m^2, 64.5 Pa, 0.863 m/s; (c) 0.2 W/m^2, 12.9 Pa, 0.046 m/s.

8.3. (a) $\rho_0 c(1 + j)/2$, $1/\sqrt{2}$; (b) $\propto \omega^{-1}$, constant.

8.5. (a) $\frac{1}{2}$; (b) $\frac{1}{4}$.

8.7. (a) $\sin \theta_1 = 3.83/(ka)$; (b) $(r/a) = (ka/4\pi)[1 - 4\pi^2/(ka)^2]$; (c) impossible.

8.9. (a) $14.8°$; (b) 114 dB *re* 1 μbar.

8.11. Proof.   **8.13.** (a) 200 m; (b) 11 dB.

8.15. (a) 0.0328 m/s; (b) 0.0556 kg; (c) $18.6°$; (d) 25 dB.

8.17. (a) $(1/2\pi)\sqrt{s/m}$; (b) sketch, mass-controlled when $\omega m \gg s/\omega$ and $\gg R_m + S\rho_0 c$, compliance-controlled when $s/\omega \gg \omega m$ and $\gg R_m + S\rho_0 c$.

8.19. 13.2.   **8.21.** Proof.

8.23. (a) 2.42 m; (b) straight line at $\theta = 0°$; (c) $42°$; (d) 19 dB.

9.1. 0.218 m.

9.3. Plot.

9.5. (a) $0.029 + j13.2$ N $\cdot$ s/m; (b) 62 N; (c) 0.32 W.

9.7. (a) 0.25; (b) 4.44×10^6 Pa $\cdot$ s/m.

9.9. (a) 12.3; (b) 0.38 m, 1.24 m.

9.11. 0.81 dB/m, 0.0025 dB/m; 2.6 dB/m, 0.025 dB/m; 8.1 dB/m, 1.4 dB/m.

9.13. (a) $\alpha = [(SWR)_1 - (SWR)_2]/[(SWR)_1(SWR)_2 \times (x_1 - x_2)]$; (b) 2.77×10^{-2} Np/m; (c) 2.14×10^{-2} Np/m; (d) either near 44 percent or 3 percent.

9.15. Proof.

9.17. 27.6, 36.8, 46.0, 55.3, 55.3, 61.8, 66.4, 66.4, 71.9, 73.6 Hz.

9.19. 37.5 Hz.

10.1. (a) 1.94 cm; (b) 0.34 μbar; (c) 380 Hz; (d) 452 Hz.

10.3. (a) 46 kHz; (b) 9.5; (c) 0.18, absorption; (d) 2000.

10.5. (a) $4S_1^2/(S_1 + S_2)^2$; (b) $S_1 > S_2$; (c) S_1/S_2 for $S_1 > S_2$, S_2/S_1 for $S_1 < S_2$.

10.7. (a) 0.33; (b) $2S_1 P/S_2$; (c) 6.

10.9. (a) 0.49 m^3; (b) 0.5.

10.11. Proof.

10.13. (a) 0.75 m, 0.29 m; (b) -10.2 dB.

10.15. (a) Band pass; (b) 6.90×10^{-4} m^3; (c) 0.64.

11.1. (a) 6×10^{-6} W/m^2; (b) 6×10^{-6} W/m^2; (c) 3.6×10^{-5} W/m^2.

11.3. (a) 150.4 dB *re* 1 μPa, 150.4 dB *re* 1 μPa/Hz$^{1/2}$

160	150
170	150
(b) 153	153
160.4	150.4
170	150

(c) 160.4   160.4
    163   153
    170.4   150.4

(d) comments.

11.5. (a) $128 - 20 \log f$ dB re 20 μPa/Hz$^{1/2}$; (b) -6 dB/octave;
(c) 77 dB re 20 μPa

11.7. 0.03.

11.9. If $\tau > T_S$, DT' is constant; if $\tau < T_S$, DT' increases with decreasing τ.

11.11. 90 dB re 20 μPa.

11.13. (a) 22 dB re 20 μPa; (b) 53 dB re 20 μPa.

11.15. (a) 2.1 sone; (b) 40 dB re 10^{-12} W/m^2; (c) 85 dB re 10^{-12} W/m^2.

11.17. (a) 85 dB re 10^{-12} W/m^2; (b) 78 sone; (c) 97 dB re 10^{-12} W/m^2.

11.19. Plot.

12.1. (a) 73.0 dB re 20 μPa; (b) 60.8 dBA re 20 μPa.

12.3. (a) 57; (b) 60, inappropriate.

12.5. 94 dB re 1 s $\cdot$ (20 μPa)^2.

12.7. 61, 71, 67, 78 dBA re 20 μPa.

12.9. Yes.

13.1. (a) 3.01×10^{-7} W/m^2; (b) 1.51×10^{-5} W.

13.3. (a) 1.79 s; (b) 8.7×10^{-6} W; (c) 2.4×10^{-7} W/m^2.

13.5. (a) 77.6 dB re 20 μPa; (b) 0.142 s; (c) 5.83 ft^2.

13.7. (a) 0.020; (b) 0.52; (c) 0.15 s.

13.9. (a) 11.6 ft; (b) 0.23.

13.11. (a) 0.17 s; (b) 1.14 s; (c) 0.12 s.

13.13. (a) 1.4×10^{-4} W; (b) 2205 ft^2; (c) 0.2 s.

13.15. 13.8

13.17. (a) 2.2 kHz; (b) 0.113 s; (c) 116 dB re 20 μPa; (d) 141, 187, 234 Hz.

13.19. (a) 54.3 Hz; (b) 1.56 s; (c) 259 Hz.

13.21. (a) 0.35; (b) 0.28 s; (c) 0.46 s.

13.23. (a) 6.4; (b) 0.64; (c) 0.018 s.

14.1. Proof.

14.3. 60 dB re 1 μbar/V.

14.5. 14.3 percent.

14.7. (a) 0.98; (b) 0.98; (c) 4.7.

14.9. (a) 50.3 Hz; (b) 2.37; (c) 1.9×10^{-3} m; (d) 5×10^{-4} m.

14.11. (a) 2.37×10^3 N/m; (b) 0.085 kg.

14.13. (a) 0.69 m^3; (b) 1.08.

14.15. (a) 136 Hz; (b) 4.0×10^{-3} m^3/s; (c) 3.2×10^{-4} m.

14.17. (a) -180 dB re 1 V/μPa; (b) 1.0 V.

14.19. (a) 0.05 V/Pa; (b) -26 dB re 1 V/Pa; (c) 5×10^{-9} m; (d) 0.028 V.

14.21. -50 dB *re* 1 V/Pa.

14.23. (*a*) 5.4×10^{-5} V; (*b*) -91 dB *re* 1 V/Pa; (*c*) 5.5×10^{-6} m.

14.25. (*a*) 5; (*b*) 7 dB.

14.27. (*a*) 1.17×10^{-3} V; (*b*) 2.7×10^{-12} W; (*c*) 6.7×10^{-4}.

14.29. (*a*) -20 dB *re* 1 V/Pa; (*b*) $+4$ dB *re* 1 Pa.

14.31. (*a*) 5.0×10^{-3} V/Pa; (*b*) 0.2 Pa.

15.1. (*a*) 1534.42 m/s, 1534.60 m/s.

(*b*)

		100 Hz	1 kHz	10 kHz
0 m	5°C	1.27×10^{-6} dB/m	5.57×10^{-5} dB/m	8.12×10^{-4} dB/m
		1.20	5.41	7.75
	20°C	1.00	6.03	6.00
		1.20	5.41	7.75
2000 m	5°C	1.26	5.45	6.88
		1.20	5.41	7.75
	20°C	0.99	5.94	5.20
		1.20	5.41	7.75

15.3. Proof.

15.5. (*a*) 1.36 km; (*b*) 1.5°; (*c*) 2.39 km; (*d*) 154 m.

15.7. (*a*) 115 m; (*b*) 184 m.   **15.9.** 4.34 km.

15.11. (*a*) 80 dB; (*b*) 4.5 dB/bounce.

15.13. (*a*) -13.6 dB; (*b*) 0 dB; (*c*) 4.8 dB; (*d*) 0 dB; (*e*) -5.6 dB, -1.2 dB, 2.4 dB, 0 dB.

15.15. (*a*) perpendicular to surface; (*b*) $2dA/(\rho_0 cr^2)$.

15.17. 10 dB.

15.19. (*a*) $P(D)$ decreases, $P(FA)$ remains the same, d decreases; (*b*) $P(D)$ decreases, $P(FA)$ decreases, d remains the same.

15.21. (*a*) 58 dB *re* 1 μPa; (*b*) 65 dB *re* 1 μPa; (*c*) 67 dB *re* 1 μPa; (*d*) 60 dB *re* 1 μPa.

15.23. 10^{-5}.

15.25. (*a*) 4.9 Hz; (*b*) 75 dB *re* 1 μPa; (*c*) -5 dB; (*d*) 20 s; (*e*) 68 dB *re* 1 μPa, 2 dB, 5.2 s.

15.27. (*a*) 13.5 km; (*b*) 0.80 km; (*c*) 8.0 km.

15.29. (*a*) 15 min; (*b*) 1800; (*c*) 1.39×10^{-4}; (*d*) 5.8 dB.

15.31. (*a*) 20 s; (*b*) 6.67×10^{-5}; (*c*) 2.22×10^{-5}; (*d*) 16, 18.

15.33. -88 dB.   **15.35.** 3.0 km.

15.37. (*a*) 10, 6.8, 4.3, 2.6, and 2.6 Pa; (*b*) $-18°$, $-198°$, $-134°$, $-134°$, 10 Pa.

15.39. Proof.   **15.41.** (*a*) 80 dB; (*b*) 31 km.

INDEX